SURVEY LAW IN CANADA

A Collection of Essays on the Laws
Governing the Surveying of Land in Canada

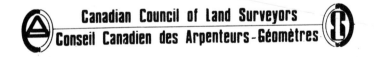

1989
CARSWELL
Toronto ■ Calgary ■ Vancouver

Canadian Cataloguing in Publication Data

Main entry under title:

Survey law in Canada

ISBN 0-459-33191-4

1. Surveying — Law and legislation — Canada.
I. Canadian Institute of Surveying.

KE742.S87 1989   346.7104′32   C89-093635-8
KF683.S87 1989

All rights reserved. No part of this publication may be reproduced or transmitted, in any form or by any means, electronic, mechanical, photocopying, recording or otherwise, or stored in any retrieval system of any nature, without the prior written permission of the copyright holder and the publisher, application for which shall be made to the publisher, The Carswell Co. Ltd., 2330 Midland Avenue, Agincourt, Ontario, Canada, M1S 1P7.

© The Carswell Co. Ltd. 1989

Foreword

The Canadian Institute of Surveying and Mapping is proud to have been entrusted with the publication of *Survey Law in Canada* and grateful to the Honorable Gerald S. Merrithew, Minister of State for Forestry and Mines, for officially presenting the final text from the Government of Canada to CISM.

It is indeed appropriate for CISM to sponsor final publication, not only because the Institute represents surveyors of every specialization from all across Canada, but also because it was the Institute that started taking positive action toward production of this reference volume more than a decade ago. It was in 1975 that we formed the committee that clearly established the feasibility of producing such a textbook and even more emphatically — the urgent need to produce it. It is extremely gratifying to us now to see this work brought to fruition and to be able to play a part in bringing it to the many people across Canada who have waited so long for it.

As Minister Merrithew so aptly observed in his remarks at the presentation to CISM, a long standing need of both the professional land surveyors of Canada and the Canadian Bar Association will be met by publication of this book.

On behalf of CISM I wish to express sincere appreciation and gratitude to all who participated in this venture with particular thanks to Charlie Weir, Chairman of the Survey Law Text Steering Committee whose unrelenting yet always encouraging persistence provided the necessary impetus, and to Bill Blackie, Editor-in-Chief, who patiently and tactfully provided the editorial leadership and encouragement to the several authors.

iv Survey Law in Canada

At this time I also wish to acknowledge the contribution of the Canadian Council of Land Surveyors in promoting the distribution of this work to the land surveyors of Canada.

Michel Brunet
President, CISM
February, 1989

Preface

For many years practicing surveyors, as well as those engaged in teaching the legal principles of boundaries, bemoaned the lack of any single organized body of reference material which reflected Canadian law and practice. The need for a comprehensive and authoritative book on survey law in Canada was first raised formally at a 1974 meeting of the Canadian Council on Surveying and Mapping.

As a result, the Council requested the Canadian Institute of Surveying and Mapping to undertake the task of producing a legal reference text for Canadian land surveyors. The Institute accordingly launched a feasibility study which was chaired by A.W. McLaughlin, then Assistant Director of Surveys for the Province of New Brunswick. The study was supported by the Canadian Council of Land Surveyors, by the Canadian Bar Association, and by the Canadian Council on Surveying and Mappling itself. The McLaughlin Committee concluded that it was feasible to produce a single-volume textbook that would be be useful in all parts of Canada, but the implications of cost were such that the Canadian Institute of Surveying and Mapping was unable to undertake the project at that time.

The matter was revived again in 1979 at the Conference on Research and Development Requirements in Surveys and Mapping, which was sponsored by the National Advisory Committee on Control Surveys and Mapping. This body, which met annually, was composed of senior members of the Surveys and Mapping Branch of the federal government together with the principal survey officers of the provinces. At the conference the need for a Canadian text on survey law was raised in several of the presented briefs and by other participants. As a result, the conference recognized this as something greatly needed by all surveying jurisdictions in Canada, and recommended to the Director General, Surveys and Mapping Branch, that funds be found for the publication of a reference book on Canadian survey law.

As a direct result of this recommendation the Director General proposed that the Department of Energy, Mines and Resources sponsor the research necessary for such a book. This proposal was endorsed by both the Canadian Council of Land Surveyors and the Canadian Institute of Surveying and Mapping. Accordingly, a Survey Law Text Steering Committee was established which was composed of leading personalities from the Canadian survey community. This group identified the specific chapters and outlines for the content of each, marshalled a group of contributing authors, and established an Editorial Review Board. Authors' meetings were held, editorial guidelines established, and appropriate reference material distributed to all authors. It is to these authors that the final credit is due for the production of *Survey Law in Canada*.

The need for a comprehensive and authoritative book on Canadian survey law and practice is conspicuously evident. No single work currently exists in Canada; practitioners and students have had to depend on American and British works, on case reports distributed or discovered intermittently, and on occasional commentaries about individual Canadian cases. This volume will accordingly be of significant value to both the legal and surveying communities in Canada. Its production represents a unique undertaking for it embraces the law as it affects surveying and boundaries within 11 separate legislative jurisdictions as well as the common law and civil law regimes. The ten provinces of Canada are autonomous in matters of the law of real property; the federal government has equivalent responsibilities in the Yukon, the Northwest Territories, certain federal lands within the provinces, and in the offshore; the province of Quebec, of course, operates within the terms of the *Civil Code* while the other provinces follow the common law.

Survey Law in Canada does not purport to be a definitive work; rather, its intent is to offer guidance on the general principles of those laws which affect the surveyor's daily practice. It is for this reason that the text is replete with references that can be used for more detailed analysis of particular legal questions as they arise. It is a fervent wish that this volume will also serve as a catalyst and lead to more searching publications, perhaps on more particular themes, that will further close the gap in available reference works on Canadian survey law.

That this work has finally been brought to fruition is a distinct credit to the authors who contributed their time and talent for very modest honoraria,

Preface vii

to the members of the Editorial Review Board for their unstinting toil and guidance, and to members of the Survey Law Text Steering Committee for their constant encouragement, advice and support throughout the life of the project. The names of all participants are listed below but special mention must be accorded to J.F. Doig for undertaking the arduous task of detailed editing and compilation of the index. Finally, acknowledgement must be made of the contribution of R.E. Moore, Director General, Surveys and Mapping Branch and Chairman of the National Advisory Committee on Control Surveys and Mapping. It was he who broke the financial barrier which for so many years stifled the production of this book. The federal grant of funds through the Department of Energy, Mines and Resources in support of this undertaking is due directly to Mr. Moore's initiative, determination and persuasiveness. Without his continual support and encouragement this book could not have been written.

G.H.W. and W.V.B.
July 1988.

PARTICIPANTS

Steering Committee

Charles H. Weir — Chairman	Edmonton, Alberta
Robert G. Code — Vice-Chairman	Toronto, Ontario
John D. Barber	Toronto, Ontario
William V. Blackie	Ottawa, Ontario
Grégoire Girard	St-Hyacinthe, Quebec
Timothy E. Koepke	Whitehorse, Yukon
John D. McLaughlin	Fredericton, New Brunswick
Lorraine Petzold	Toronto, Ontario

Editorial Review Board

William V. Blackie	Ottawa, Ontario
Isaak deRijcke	Guelph, Ontario
James F. Doig	Wolfville, Nova Scotia
Grégoire Girard	St-Hyacinthe, Quebec
David W. Lambden	Mississauga, Ontario
John D. McLaughlin	Fredericton, New Brunswick
William L. MacLellan	Ottawa, Ontario
Louis M. Sebert	Ottawa, Ontario

Authors

G. Kenneth Allred	Edmonton, Alberta
Brian M. Campbell	Toronto, Ontario

Izaak DeRijcke   Guelph, Ontario
James S. Dobbin   Saint John, New Brunswick
James F. Doig   Wolfville, Nova Scotia
Grégoire Girard   St-Hyacinthe, Quebec
André Laferrière   Montreal, Quebec
David W. Lambden   Mississauga, Ontario
Donald McLaurin Lamont   Canmore, Alberta
John D. McLaughlin   Fredericton, New Brunswick
Susan E. Nichols   Fredericton, New Brunswick
Gérard Raymond   Ottawa, Ontario

The Authors

Brian Campbell

Mr. Campbell is a partner with the firm of Bogart, Russell, Campbell, Robertson and has practised exclusively in the area of administrative law and civil litigation since his call to the Bar in 1975.

Mr. Campbell has lectured at Canadian Bar Association seminars, continuing education and insight seminars on several occasions and currently teaches public law at the Bar Admission course in Toronto.

In addition Mr. Campbell has instructed at the intensive trial and advocacy program at York University and is an instructor in trial advocacy with the Advocacy Institute.

Mr. Campbell obtained his B.A. in political science and economics in 1967 from the University of Toronto, followed by an M.A. in political science in 1968 and an M.Phil. (Masters of Philosophy and International Politics) in 1970. He obtained his LL.B. from York University (Osgoode) in 1973.

James S. Dobbin

Mr. Dobbin is a graduate of the University of New Brunswick with a Bachelor's Degree in Survey Engineering (1977). He became a member of the Association of the New Brunswick Land Surveyors in 1979 and has served in various committees of the Association including the Discipline Committee. He graduated from the University of New Brunswick Law School in 1980 and was admitted to the Bar in 1981. He currently practises law with the firm of Barry & O'Neil in Saint John, New Brunswick with a practise restricted almost entirely to Real Property Law.

Donald McLaurin Lamont Q.C.

Employed by the Province of Manitoba as a lawyer on the staff of the

Land Title's Office in Winnipeg from 1949 to 1981. Deputy Registrar General of Manitoba from 1952 to 1969. Registrar General of Manitoba from 1969 to 1981. Inspector of Land Title's Offices for Alberta from 1981 to 1984. Formerly a part-time lecturer on Real Property Law at the Manitoba Law School, and on Land Titles Practice for the Bar Admission Course provided by the Law Society of Manitoba.

David W. Lambden

David W. Lambden, a native of Ontario, graduated B.Sc.F. from the University of New Brunswick in 1950, and received a graduate Diploma in Town and Country Planning from the University of Sydney in 1963. He qualified: O.L.S., D.L.S., N.S.L.S., Registered Surveyor, Vic., N.S.W., Qld. and N.Z. He is a Fellow of the Royal Institution of Chartered Surveyors and of the Institution of Surveyors, Australia, and was a co-editor of the Australian Surveyor for several years. He has practiced privately and with governments, and in Canada, Australia, New Zealand and the United States, taught at the University of California and the University of New South Wales. In 1974, he was appointed professor at the University of Toronto, Erindale Campus, for survey law and related topics.

Izaak de Rijcke

Born in the Netherlands in 1952, Izaak de Rijcke immigrated as a child to Canada and was raised in southern Ontario. He attended the Survey Science Program at the Erindale Campus of the University of Toronto and graduated with a B.Sc. degree in 1976. He articled to M.P. Van Harten and received his commission as an O.L.S. in 1978. He then attended Faculty of Law at the University of Windsor and received the LL.B. degree in 1981. After articling to Jane Pepino in Toronto, he was called to the Bar in Ontario in 1983. Izaak has been in private practice in Guelph since that time. His professional contributions have included continuing education seminars for land surveyors, the co-authorship of *Boundaries*, 1985, and acting as adjunct professor of survey science at the University of Toronto since 1982.

Susan E. Nichols

Susan Nichols is a lecturer in cadastral studies at the University of New Brunswick where she is currently completing a Ph.D. degree in Surveying Engineering. She holds a B.Sc.(Hon.) in Mathematics from Acadia University and is a graduate of the Nova Scotia Land Survey Institute (now College of Geographic Sciences). In 1983, she received a M.Eng. degree in Surveying Engineering from U.N.B. for her research on tidal boundary delimitation. She is a professional engineer and is the Canadian representative to Commission VII of Fédération Internationale des Géomètre (FIG).

James F. Doig

James F. Doig is a Nova Scotia Land Surveyor and a Canada Lands Surveyor. Principal of the Nova Scotia College of Geographic Sciences from 1968 to 1985, he has written extensively on survey law. As well, he has compiled handbooks of case reports related to surveys and boundary disputes within their respective jurisdictions, for the Association of Nova Scotia Land Surveyors and the Association of New Brunswick Land Surveyors. For several years Mr. Doig has been examiner in survey law for the Atlantic Provinces Examining Board for Land Surveyors (APEBLS) and for the Board of Examiners for Nova Scotia Land Surveyors. He is Associate Editor, Cadastral Surveys for *CISM Journal ACSGC*, the quarterly publication of the Canadian Institute of Surveying and Mapping; he is also Legal Editor for the *Nova Scotian Surveyor*. Mr Doig is a past president (1972) of the Association of Nova Scotia Land Surveyors, a former chairman of the APEBLS, and has recently been appointed to the Board of Examiners for Nova Scotia Land Surveyors. He has been a Director of the Land Registration and Information Service, Council of Maritime Premiers since 1980. A graduate of Acadia University with bachelor degrees in Science and in Education, he lives in Wolfville, Nova Scotia.

Grégoire Girard, Q.L.S.

Grégoire Girard was born in Sainte-Rosalie, Quebec. In 1945, he graduated with a Bachelor of Arts degree from the University of Montreal. In 1948, he obtained a Bachelor degree in Land Surveying and in 1949, a Bachelor degree in Forestry sciences, both from the University of Laval.

In 1951, he was commissioned as a Quebec Land Surveyor and opened a private office in St-Hyacinthe, Quebec.

He was elected mayor of the City of St-Hyacinthe in 1971 for a five year term. In 1982, he was appointed Director of La Survivance Life Insurance Company, with its head office in St-Hyacinthe, Quebec.

He has served as Director and as President of the "Ordre des arpenteurs-géomètres du Québec" (1965-1971), as Treasurer of the "Fédération Internationale des Géomètres" (FIG) 1984-1985 and as President of "La Société de Cartographie du Québec" (1978-1987).

For the last three years, he has served as invited teacher in the Geomatic Program of the Faculty of Forestry and Geodesy at the University of Laval.

J. André Laferrière

J. André Laferrière was born in Montreal where he received his primary and secondary school education. In 1949, he was commissioned as a Quebec Land Surveyor and has since been in private practice. He is presently a member of the "Ordre des arpenteurs-géomètres du Québec" and a member of the Order of Quebec Town Planners.

He has served on the Board of Directors of the "Ordre des arpenteurs-géomètres du Québec" (1960-67) and as its President (1964-66).

Aside from his professional practise he is interested in survey education and for the past several years has been teaching in the Geomatic Program of the Faculty of Forestry and Geodesy at the University of Laval.

Gérard Raymond

Gérard Raymond was born in Terrebonne, Quebec, and received his primary and secondary school education in Montreal. He graduated in 1965 from the University of Laval with a Bachelor of Science (Survey) degree. In 1966, he obtained a commission as a Quebec Land Surveyor, and joined Legal Surveys Division, Surveys and Mapping Branch, Department of Energy, Mines and Resources, Ottawa.

In 1970 he obtained a license in Civil Law from the University of Ottawa. He is presently a member of the Quebec Bar and the "Ordre des arpenteurs géomètres du Québec".

He has served on the Board of Directors and as President of the "Ordre des arpenteur-géomètres du Québec" (1976-83), the Canadian Council of Land Surveyors (1979-86), and the Institute for Land Information (1980-86).

In 1974 he was commissioned as a Canada Lands Surveyor and in 1978 was made the Assistant Surveyor General of Canada Lands. In 1985 he was appointed Surveyor General of Canada Lands. For several years, he has also been teaching law in the Geomatics Program of the Faculty of Forestry and Geodesy at the University of Laval.

G.K. Allred

G.K. (Ken) Allred is an Alberta and Canada Lands Surveyor. He is

Executive Director of the Alberta Land Surveyors' Association, has served as Secretary Treasurer and President of the Canadian Council of Land Surveyors and currently represents Canada on Commission I of Fédération Internationale des Géomètre (FIG).

Mr. Allred has been employed by the Alberta Land Titles Office, as a city surveyor, with a private practice firm, and in resource exploration. He has served as an Alderman with the City of St. Albert and has been active on regional planning commissions and development appeal boards. He also holds the position of adjunct professor in Geography at the University of Alberta, where he lectures in Cadastral Studies.

John McLaughlin

John McLaughlin received his B.Sc. and M.Sc. degrees from the University of New Brunswick and a Ph.D. from the University of Wisconsin. He is a commissioned land surveyor and professional engineer in New Brunswick, and has served as president of the Association of New Brunswick Land Surveyors, the Canadian Council of Land Surveyors and of the Canadian Institute of Surveying and Mapping. Professor McLaughlin joined the Department of Surveying Engineering at the University of New Brunswick in 1972 and is presently the chairman of the department.

Table of Contents

Foreword.. iii
Preface... v
The Authors.. ix
Table of Cases... xxxv
Table of Statutes.. xlv

Chapter 1:	**The Canadian Legal System**	
	— *Brian Campbell*...............................	1
	Introduction — The Nature of Law, §**1.02**...........	1
	Precedent and Stare Decisis, §**1.09**................	3
	Sources of Law, §**1.15**.............................	5
	Rules and Principles of Statutory Interpretation, §**1.19**..	6
	The English Legal Legacy, §**1.30**...................	9
	Divisions of Law, §**1.35**...........................	11
	Substantive Law, §**1.35**......................	11
	Administrative Law, §**1.41**....................	12
	Constitutional Basis of Authority, §**1.54**..........	16
	Structure of the Canadian Constitutional System, §**1.54**................................	16
	The British North America Act, §**1.57**...........	17
	The Unwritten Component of the Constitution, §**1.60**................................	18
	Federal and Provincial Powers, §**1.62**..........	18
	Powers of the Federal Government, §**1.63**....	18
	Powers of the Provincial Government: Section 92, §**1.64**...................	19
	The Constitution Act, 1982, §**1.66**..............	20
	Amending Formula, §**1.67**.....................	20
	The Charter of Rights, §**1.68**..................	20
	Constitutional Basis of Judicial Authority, §**1.74**......	23
	System of Courts, §**1.76**......................	23
	Conclusion, §**1.77**.................................	25

Chapter 2:	**Real Property Law in Common Law Provinces** — *James S. Dobbin*...............................	27
	Introduction, **§2.01**...............................	27
	History and Evolution of Real Property Law, **§2.04**..	28
	Some Background, **§2.04**.....................	28
	Norman Conquest, **§2.07**.....................	28
	Early English Developments, **§2.11**.............	29
	Transfer to Canada, **§2.15**....................	31
	Modern Developments, **§2.19**..................	32
	Aboriginal Title, **§2.21**.......................	32
	Theories of Ownership, **§2.25**.....................	33
	Introduction, **§2.25**..........................	33
	Tenure, **§2.26**...............................	33
	Estates, **§2.30**..............................	34
	Seisin, **§2.38**...............................	35
	Legal and Beneficial Ownership, **§2.39**..........	36
	Other Rights, **§2.42**..............................	36
	Introduction, **§2.42**..........................	36
	Easements, **§2.43**...........................	36
	Prescription, **§2.54**..........................	39
	Highways, **§2.57**............................	40
	Licences, **§2.59**............................	40
	Profits à Prendre, **§2.62**......................	41
	Options and Rights of First Refusal, **§2.65**.......	42
	Covenants, **§2.68**...........................	42
	Conveyancing, **§2.74**.............................	44
	Introduction, **§2.74**..........................	44
	Preliminaries, **§2.75**..........................	44
	Representation by Solicitors, **§2.79**.............	45
	Mortgages, **§2.84**...........................	46
	Legal Principles Affecting Ownership, **§2.87**..........	47
	Introduction, **§2.87**..........................	47
	Limitations of Actions, **§2.88**..................	47
	Colour of Title, **§2.93**........................	49
	Prescription, **§2.94**..........................	49
	Expropriation, **§2.99**.........................	50
	Concurrent Ownership, **§2.105**.................	51
	Mechanics Liens, **§2.110**.....................	53
	Trusts, **§2.113**..............................	54
	Matrimonial Legislation, **§2.116**................	55
	Planning and Land Use Control, **§2.120**.........	56
	Condominiums and Air Space Parcels, **§2.126**....	58
	Wills and Estates, **§2.133**.....................	59

	Escheat and Forfeiture, §**2.136**...............	60
	Future Developments, §**2.138**...................	60
Chapter 3:	**Land Registration Systems**	
	— *Donald McLaurin Lamont*.....................	63
	Introduction, §**3.01**.............................	63
	Deed Registry System, §**3.03**....................	64
	Basic Principles, §**3.03**......................	64
	Overriding Interests, §**3.06**...................	64
	Defective or Missing Deeds, §**3.07**.............	65
	Indexing Methods, §**3.08**.....................	65
	Priority, §**3.12**..............................	66
	Functions of the Registrar, §**3.17**..............	68
	Instruments which may be Registered, §**3.18**.....	69
	Advantages and Disadvantages, §**3.21**..........	70
	Land Surveys, §**3.28**.........................	73
	Title Registration Systems, §**3.30**.................	73
	Torrens Land Titles System, §**3.31**.................	74
	Historical Background, §**3.31**.................	74
	The Certificate of Title, §**3.32**................	74
	Powers and Duties of the Registrar, §**3.33**........	74
	Correction of the Register, §**3.35**..............	75
	Rejection of Instruments, §**3.36**................	76
	Appeals from the Decisions of the Registrar,	
	§**3.37**...................................	77
	First Registration, §**3.38**......................	77
	Fundamentals of the System, §**3.39**.............	78
	Indefeasibility, §**3.39**.....................	78
	Nature of Fraud which would Invalidate a	
	Certificate of Title, §**3.43**.............	80
	Two Certificates of Title for Same Land,	
	§**3.45**...........................	82
	Misdescription, §**3.46**....................	82
	Historical Searches, §**3.47**................	83
	Completeness of the Register — Overriding	
	Interests, §**3.48**....................	83
	Assurance Fund, §**3.50**...................	84
	Registration of Instruments (except plans),	
	§**3.52**...................................	84
	Transfers and Leases, §**3.52**...............	84
	Transmission of Titles, §**3.54**..............	85
	Mortgages and Encumbrances, §**3.56**.......	86
	Restrictive Covenants, §**3.57**..............	86
	Easements, §**3.58**.......................	87

Table of Contents

	Statutory Rights of Way for Utilities, §**3.59**...	87
	Removal of Interests from the Register, §**3.60**.............................	88
	Registration of Plans, §**3.61**....................	88
	Plans of Subdivision, §**3.62**................	88
	Expropriation Plans, §**3.69**.................	92
	Condominium Plans, §**3.70**...............	92
	Search Facilities, §**3.71**......................	93
	Caveats, §**3.72**...............................	93
	Registration of Mineral Interests, §**3.81**..........	96
	Land Registration Systems in the Province of Ontario, §**3.91**......................................	100
	Land Registration System in Newfoundland, §**3.101**....	103
	Land Registration in the Maritime Provinces, §**3.103**...	104
	Conclusion, §**3.105**............................	106
Chapter 4:	**Boundaries**	
	— *David W. Lambden* and *Izaak de Rijcke*...........	107
	Boundary Defined — Nature and Function, §**4.01**.....	107
	History and Geography — Canadian Land Resources, §**4.07**.....................................	109
	Original Surveys of the Crown, §**4.23**...............	114
	The Origin of Boundaries, §**4.30**...................	116
	Surveys, Plans and Evidence, §**4.38**................	120
	Survey Systems and Statutes, §**4.47**................	122
	True and Unalterable Boundaries: Priorities of Evidence, §**4.62**............................	128
	Property Rights and Boundaries, §**4.65**..............	129
	Easements, §**4.66**............................	129
	Prescriptive Rights — Lost Modern Grant, §**4.74**.................................	133
	Adverse Possession, §**4.78**....................	135
	Standards of Measurements, §**4.89**................	139
	Coordination and Integration of Surveys, §**4.110**......	147
	Lost Boundaries, §**4.113**.........................	148
	Conventional Lines, §**4.118**...................	150
	Estoppel, §**4.122**............................	151
	Re-definition of Boundaries: Resurvey, Retracement, Restoration, §**4.137**........................	156
	Modern Surveys, Titles and Land Dealings, §**4.147**....	159
	Survey Law: The Future, §**4.158**...................	162
	Appendix A......................................	164

Chapter 5:	**Water Boundaries — Coastal**	
	— *Susan Nichols*.............................	167
	Introduction, §**5.01**.............................	167
	Water Boundaries, §**5.03**...................	168
	Natural and Artificial Boundaries, §**5.05**.....	169
	Property and Jurisdictional Boundaries, §**5.10**.............................	170
	Tidal Waters, §**5.13**.............................	171
	Classification of Waters, §**5.14**...............	172
	Tidal and Nontidal waters, §**5.15**...........	172
	Navigable and Nonnavigable Waters, §**5.18**...	173
	Tidal and Tidal Reference Surfaces, §**5.22**.......	174
	Daily Tides, §**5.23**......................	175
	Spring, Neap and Long Period Tides, §**5.25**...	175
	Tidal Reference Surfaces, §**5.28**............	178
	Spatial and Temporal Variations, §**5.35**......	181
	Ordinary Tides, §**5.38**.......................	182
	Coastal Land Law, §**5.41**.........................	183
	Early Development, §**5.43**...................	184
	Private and Public Rights, §**5.46**...............	186
	Title to the Shore, §**5.47**..................	186
	Riparian Rights, §**5.50**...................	187
	Public Rights, §**5.55**.....................	189
	Effect of Legislation, §**5.58**...............	190
	Federal and Provincial Jurisdiction, §**5.60**........	191
	Public Harbours, §**5.61**...................	191
	The Seabed, §**5.63**......................	192
	Tidal Boundary Delimitation, §**5.66**.................	193
	Boundary Definitions, §**5.67**..................	194
	Ordinary High Water Mark and Related Definitions, §**5.68**...................	194
	Mean High Water Line and Related Definitions, §**5.71**...................	196
	Boundary Surveys, §**5.73**.....................	197
	The Water Mark as a Physical Feature, §**5.74**............................	197
	Establishing the Water Line Using Tidal Data, §**5.78**.......................	200
	Ambulatory Boundaries, §**5.83**.................	203
	Effect of Measurements, §**5.84**.............	203
	Establishing Former Boundaries, §**5.86**......	204
	Apportionment of Accretion, §**5.88**.........	205
	Maritime Boundary Delimitation, §**5.90**..............	206
	Development of the Law of the Sea, §**5.92**.......	207

Emergence of a New Accord, §**5.93**........ 207
Maritime Zones, §**5.96**.................. 209
Baselines, §**5.97**............................. 212
Normal Baselines, §**5.98**.................. 213
Closing Lines, §**5.99**..................... 213
Straight Baselines, §**5.100**................. 214
International Boundaries, §**5.102**................ 215
Equadistance Principle, §**5.103**............ 215
Special Circumstances, §**5.105**............. 216
Canada's Boundaries, §**5.107**.............. 217
The Surveyor's Role, §**5.110**................... 220

Chapter 6: **Water Boundaries — Inland**
 — David W. Lambden........................... 221
Riparian Rights and Their Development, §**6.01**....... 221
Constitutional Basis, §**6.05**...................... 222
Summary of Basic Constitutional Provisions,
§**6.05**................................. 222
Jurisdiction and Reception of Laws in Ontario,
§**6.06**................................. 224
The Ontario Developments, §**6.10**................. 225
Navigability, Water Power and the Bed of
Navigability Waters Act (1911), §**6.10**....... 225
A Further Ontario Provision Excluding Land
Under Certain Waters, §**6.13**.............. 226
Current Ontario Law Respecting Riparian
Entitlement, §**6.15**...................... 227
The Legal Situation in the Prairie Provinces, §**6.17**....: 228
Statutory Change of Common Law Rules,
§**6.17**................................. 228
Boundaries: Provisions for the Survey of
Dominion Lands, §**6.25**.................. 231
The Ad Medium Filum Rule, §**6.29**................. 232
The Specific Problem of Inland Water Boundaries,
§**6.33**....................................... 234
Terminology, §**6.33**........................... 234
Water Levels in Rivers and Lakes, §**6.38**........ 235
The Interpretation of Water Boundary Terms,
§**6.44**...................................... 237
Summary: the Natural Boundary of Lakes and
Rivers, §**6.65**........................... 243
Survey Systems and the Survey of Water Boundaries,
§**6.73**....................................... 245

	The Concept of True and Unalterable Boundaries, §**6.73**...............................	245
	Lot and Parcel Lines Run to the Water Boundary, §**6.76**...............................	246
	Shore Road Allowances and Shore Reservations, §**6.78**...............................	247
	Significance of the Date of Grant, §**6.82**.........	248
	Synonymity of Terms and Accuracy of Surveys, §**6.83**.......................................	249
	The Doctrine of Accretion, §**6.87**.................	251
	Principles, §**6.87**............................	251
	The Mechanics of Accretion, §**6.88**.............	252
	Riparian Rights and Title Records, §**6.89**............	254
	The Survey of Upland Riparian Parcels, §**6.95**........	255
Chapter 7:	**Evidence**	
	— *Izaak de Rijcke*...............................	259
	Introduction, §**7.01**..............................	259
	Admissibility, §**7.11**.............................	262
	Exclusionary Rules, §**7.18**........................	266
	Origins of the Hearsay Rule, §**7.20**................	266
	Exceptions to the Hearsay Rule, §**7.23**..............	267
	Declarations Against a Party's Interest, §**7.23**.....	267
	Declarations Made in the Course of a Business Duty, §**7.24**...........................	268
	Declarations as to Reputation, §**7.30**............	270
	Statements in Ancient Documents, §**7.33**........	271
	Statements in Public Documents, §**7.34**.........	271
	Admissions of a Party, §**7.35**..................	271
	Privilege, §**7.38**.................................	272
	Parol Evidence, §**7.43**............................	274
	Best Evidence Rule as yet Another Exclusionary Principle, §**7.47**.............................	276
	Opinion Evidence as an Exclusionary Rule, §**7.51**.....	277
	Expert Opinion, §**7.54**............................	278
	Presumptions, §**7.59**.............................	279
	Burden of Proof, §**7.61**...........................	280
	Documents, §**7.63**................................	281
	Standards of Proof, §**7.67**........................	282
	Real Evidence, §**7.69**.............................	282
Chapter 8:	**Settlement of Boundary Uncertainties**	
	— *James F. Doig*.................................	285
	Introduction, §**8.01**..............................	285
	Aim, §**8.01**.................................	285

Settlement, §**8.02**............................ 286
Transfers in Writing, §**8.04**..................... 287
Exceptions, §**8.05**............................ 288
Causes of Uncertainty, §**8.06**................... 289
After Retracement, §**8.07**...................... 289
Sources and Nature of Uncertainties, §**8.08**.......... 289
    Title, §**8.09**................................ 290
        Overriding Interests, §**8.11**................. 291
    Boundaries, §**8.14**............................ 293
    Responsibilities of Surveyors, §**8.15**.............. 293
        Impartiality, §**8.16**........................ 294
        Best Evidence, §**8.17**...................... 294
        Consultation, §**8.17**....................... 294
        Client's Affairs, §**8.19**..................... 295
Resolution of Uncertainties, §**8.20**................. 295
    Courts, §**8.20**............................... 295
        Surveyor's Authority, §**8.21**................ 295
        Quality and Extent of Title, §**8.22**........... 295
        Land Title Systems, §**8.23**................. 296
        Registry Systems, §**8.24**................... 296
    Conventional Lines, §**8.30**..................... 298
        Doctrine, §**8.30**.......................... 298
        Application, §**8.31**........................ 299
        Future, §**8.32**............................ 299
        No Alienation, §**8.33**...................... 300
        Crown Cannot be Party, §**8.34**.............. 300
        Owners Only can Agree, §**8.35**.............. 300
        Boundary Discoverable, §**8.36**.............. 301
        Essentials, §**8.37**......................... 301
        Evidence of Agreement, §**8.38**.............. 302
        Establishing Line, §**8.39**................... 303
        Agreement Binds Successors, §**8.40**.......... 303
        Line Run by Surveyor, §**8.41**................ 304
        Survey Line Wrong, §**8.42**................. 304
        Survey Line Correct or Not, §**8.43**........... 305
        Lines at Variance with Descriptions, §**8.44**... 305
        Lines Never Run, §**8.45**................... 305
    Estoppel, §**8.46**............................. 306
        General, §**8.46**........................... 306
        Defence, §**8.47**........................... 306
        Concept, §**8.48**........................... 306
        Elements, §**8.49**.......................... 307
        Improvements, §**8.50**..................... 307
        No Improvements, §**8.51**................... 307

    Occupation Only, §**8.52**.................. 308
    Admission, §**8.53**...................... 308
Apportionment of Accretion, §**8.54**............ 309
    General, §**8.54**........................ 309
    Methods of Division, §**8.55**............... 310
    Representative Baseline, §**8.56**............ 310
    Proportional Shorelines, §**8.57**............ 310
Improvements or Encroachments, §**8.58**......... 311
Fences, §**8.61**.............................. 312
    General, §**8.61**........................ 312
    Purpose, §**8.62**........................ 313
    Limits, §**8.64**.......................... 314
    Convenience, §**8.65**.................... 314
    Long Acceptance, §**8.66**................. 314
    Conventional Line, §**8.67**................ 314
    Protraction, §**8.68**...................... 315
    Parallelism, §**8.69**...................... 315
    Conflict with Monuments, §**8.70**........... 316
    Confirmation by Plan, §**8.71**.............. 316
    Onus of Proof, §**8.72**.................... 316
Statutes, §**8.73**............................. 317
    Special Surveys, §**8.74**.................. 317
    Quieting Titles, §**8.77**................... 319
    Resurveys, §**8.80**...................... 322
    The Boundaries Act, §**8.81**............... 323
    Replotting, §**8.83**...................... 325
    The Boundary Lines and Line Fences Act of
        Manitoba, §**8.84**.................... 325
Miscellaneous, §**8.85**....................... 326
    Quit Claim Deeds, §**8.85**................. 326
    Statutory Declarations, §**8.86**............. 326
    Tax Deeds, §**8.87**...................... 327
Bornage, §**8.88**............................. 327
    General, §**8.88**........................ 327
    Development, §**8.89**.................... 327
    Definition, §**8.91**....................... 328
    Necessary Conditions, §**8.92**.............. 329
    Circumstances Appropriate, §**8.93**.......... 329
    Two Stages, §**8.94**...................... 329
    Delimitation, §**8.95**..................... 330
    Report of Survey, §**8.96**.................. 330
    Demarcation, §**8.97**..................... 330
Future, §**8.98**.............................. 331

xxiv Table of Contents

Chapter 9: **Liability in Negligence and Contracts**
— *Izaak de Rijcke* 333
Introduction, §**9.01** 333
Liability in Negligence, §**9.14** 336
Liability in Contract, §**9.55** 347
Liability Insurance, §**9.59** 348

Chapter 10: **The Law in Québec**
— *Grégoire Girard, J. André Laferrière and Gérard Raymond* 353
Introduction, §**10.01** 353
The Civil Code, §**10.04** 354
    Introduction, §**10.04** 354
    The Historical Roots of Québec Private Law, §**10.05** 354
        The French Regime, §**10.05** 354
        The English Regime, §**10.06** 355
        The Québec Act of 1774, §**10.07** 355
        The Civil Code of Lower Canada, §**10.08** 355
        The Revision of the Civil Code, §**10.09** 356
    The Sources of Law, §**10.10** 356
        Objective Law, §**10.10** 356
        Subjective Law, §**10.11** 357
    Obligations, §**10.12** 357
        Introduction, §**10.12** 357
        Obligations Defined, §**10.13** 357
        Sources of Obligations, §**10.14** 358
        Distinction Between Results and Means, §**10.15** 358
        Kinds of Obligations, §**10.16** 358
        Conditional Obligations, §**10.17** 358
        Obligations with a Term, §**10.18** 359
        Joint and Several Obligations, §**10.19** 359
        Obligations with a Penal Clause, §**10.20** 359
        Notice of Default, §**10.21** 359
    Contracts, §**10.22** 360
        Introduction, §**10.22** 360
        The Role of Will in Contracts, §**10.23** 360
        Conditions as to the Form of Contracts, §**10.24** 360
        Requirements for the Validity of Contracts, §**10.25** 360
        The Legal Capacity of the Parties, §**10.26**.... 361
        Consent, §**10.27** 361

Irregularities of Consent, §**10.28**........... 361
Error, §**10.29**.......................... 362
Fraud, §**10.30**......................... 362
Violence and Fear, §**10.31**................ 362
Lesion, §**10.32**......................... 362
Object and Cause of Contracts, §**10.33**...... 363
The Interpretation of Contracts, §**10.34**..... 363
Civil Liability, §**10.35**....................... 363
Introduction, §**10.35**..................... 363
General Principles, §**10.36**............... 364
Contractual and Delictual Responsibility,
§**10.37**............................ 364
Essential Elements of Civil Liability,
§**10.38**............................ 364
Fault, §**10.39**.......................... 364
Damage, §**10.40**....................... 365
The Causal Link, §**10.41**................. 365
Matrimonial Regimes, §**10.42**................. 365
Introduction, §**10.42**..................... 365
Definition, §**10.43**....................... 365
The Legal Regime, §**10.44**................ 366
Partnership of Acquests, §**10.45**............ 366
Separation as to Property, §**10.46**.......... 366
Community of Property, §**10.47**............ 367
Introduction to Property Law, §**10.48**........... 368
Distinction Between Things and Property,
§**10.48**............................ 368
Distinction Between Movables and
Immovables, §**10.49**................. 368
Kinds of Immovables, §**10.50**.............. 368
Kinds of Movables, §**10.51**................ 369
Implications of the Distinction Between
Movables and Immovables, §**10.52**...... 369
Property Law, §**10.53**....................... 370
Introduction, §**10.53**..................... 370
Attributes of Ownership, §**10.54**............ 370
Traditional Limitations of Ownership Rights,
§**10.55**............................ 370
Modern Limitations of Ownership Rights,
§**10.56**............................ 370
Abuse of Rights and Difficulties Between
Neighbours, §**10.57**.................. 371
Co-ownership, §**10.58**....................... 371
Kinds of Co-ownership, §**10.58**............. 371

Ordinary Co-ownership, §**10.59**	371
Forced Co-ownership, §**10.60**	372
Co-ownership of Immovables by Declaration, §**10.61**	372
Superficiary Rights, §**10.62**	372
Corollary Property Rights, §**10.63**	373
Introduction, §**10.63**	373
The Concept of Servitudes, §**10.64**	373
Usufruct, Use, and Habitation, §**10.65**	373
Definition of Usufruct, §**10.65**	373
Characteristics of Usufruct, §**10.66**	374
Rights and Obligations of the Parties, §**10.67**	374
Distinction Between the Rights of a Usufructuary and Those of a Tenant, §**10.68**	374
Use of Habitation, §**10.69**	374
Emphyteusis, §**10.70**	374
Essential Elements of Emphyteusis, §**10.70**	374
Rights and Obligations of the Parties, §**10.71**	375
Introduction to the Study of Real Servitudes, §**10.72**	375
Kinds of Servitudes, §**10.73**	375
Natural Servitudes, §**10.74**	376
Legal Servitudes, §**10.75**	376
Common Areas or Structures, §**10.76**	377
Introduction, §**10.76**	377
Acquisition of Common Ownership, §**10.77**	377
Presumption of Common Ownership, §**10.78**	377
Spacing Requirements for Plantations and Structures, §**10.79**	378
The View Onto the Property of a Neighbour, §**10.80**	378
Right of Way, §**10.83**	379
Conventional Servitudes, §**10.85**	380
Introduction, §**10.85**	380
Kinds of Conventional Servitudes, §**10.86**	380
Continuous and Discontinuous Servitudes, §**10.87**	380
Apparent and Unapparent Servitudes, §**10.88**	381

Modes of Establishing Servitudes, §**10.89**.... 381
Use and Extent of Servitudes, §**10.90**........ 382
Rights and Obligations of the Proprietor of
    the Servient Land, §**10.91**............. 383
Privileges and Hypothecs, §**10.92**.............. 383
    Introduction, §**10.92**..................... 383
    Preliminary Provisions, §**10.93**............. 383
    Privilege Defined, §**10.94**................. 384
    Privileges upon Movables, §**10.95**.......... 384
    Privileges upon Immovables, §**10.96**........ 384
    Definition and Characteristics of Hypothecs,
        §**10.97**............................. 385
    Kinds of Hypothecs, §**10.98**............... 385
    Legal Hypothecs, §**10.99**................. 385
    Judicial Hypothecs, §**10.100**............... 385
    Conventional Hypothecs, §**10.101**.......... 386
Possession, §**10.102**......................... 386
    Introduction, §**10.102**.................... 386
    Definition and Characteristics of Possession,
        §**10.103**............................ 386
    Protection of Possession, §**10.104**........... 386
    Area of Application of Possession, §**10.105**... 387
    Elements of Possession, §**10.106**........... 387
    Precariousness, §**10.107**.................. 387
    Interversion of Title, §**10.108**.............. 387
    Qualities of Possession, §**10.109**............ 388
    Possession must be Continuous, §**10.110**..... 388
    Possession must be Uninterrupted, §**10.111**... 388
    Possession must be Peaceable, §**10.112**...... 388
    Possession must be Public, §**10.113**......... 388
    Possession must be Unequivocal, §**10.114**.... 389
    Possession must be Proprietorial, §**10.115**.... 389
    Judicial Effects of Possession, §**10.116**....... 389
    Possessory Actions, §**10.117**............... 389
    Petitory Action, §**10.118**.................. 390
Prescription, §**10.119**........................ 390
    Introduction, §**10.119**.................... 390
    Areas of Application, §**10.120**............. 391
    Imprescriptible Things, §**10.121**............ 391
    Acquisition of Immovables by Prescription,
        §**10.122**............................ 392
    Thirty-year Prescription, §**10.123**........... 392
    Ten-year Prescription, §**10.124**............. 392
    The Condition of Good Faith, §**10.125**...... 392

xxviii   Table of Contents

    Translatory Title of Ownership, §**10.126**..... 393
    The Joining of Possession, §**10.127**.......... 393
    Arrests in the Course of Prescription,
     §**10.128**............................. 393
    Interruption of Prescription, §**10.129**........ 394
    Suspension of Prescription, §**10.130**......... 394
    Judicial Recognition of Ownership Acquired
     by Prescription, §**10.131**.............. 395
  The Cadastre, §**10.132**........................... 396
    Introduction, §**10.132**....................... 396
    Division and Concession of Land under the
     French Regime, §**10.133**................. 397
      The Seigniories, §**10.133**................. 397
      Plans and Land Registers, §**10.134**.......... 397
    Division and Concession of Land under the
     English Regime, §**10.135**................. 397
      Townships, §**10.135**...................... 397
      Methods of Subdividing Townships,
       §**10.136**............................. 398
     Concession of Lots, §**10.137**............... 398
    Origins of the Cadastre, §**10.138**............... 398
     Introduction, §**10.138**.................... 398
     The Ordinance of 1841 to Prescribe and
      Regulate the Registering of Titles,
      §**10.139**............................. 399
     The Seigniorial Act of 1854, §**10.140**........ 399
     The Act of 1860 and the Establishment of
      Official Cadastres, §**10.141**............ 400
     Instructions for Establishing Cadastres,
      §**10.142**............................. 401
     The Civil Code of Lower Canada, §**10.143**... 402
     Cadastral Units, §**10.144**.................. 402
    The Cadastre Today, §**10.145**.................. 403
     Distinction Between an Original Cadastre
      and a Subdivision, §**10.145**............ 403
     Parts of the Cadastre, §**10.146**.............. 403
     Deposit of an Original Cadastre and Its
      Effects, §**10.147**...................... 404
    Types of Cadastres, §**10.148**................... 404
     Original Cadastres, §**10.148**................ 404
     Cadastral Revision, §**10.149**................ 404
     Cadastral Renewal, §**10.150**................ 405
     Cadastral Reform, §**10.151**................. 405
    Cadastral Operations, §**10.152**................. 406

Introduction, §**10.152**	406
Division, §**10.153**	406
Subdivision, §**10.154**	407
Redivision, §**10.155**	407
Replacement, §**10.156**	408
Cancellation, §**10.157**	408
Addition of an Original Lot, §**10.158**	409
Correction, §**10.159**	409
Vertical Cadastres and Co-ownership, §**10.160**	410
Effects of the Cadastre, §**10.161**	410
Registration, §**10.164**	413
Introduction, §**10.164**	413
Definition, §**10.164**	413
Origins and Evolution, §**10.165**	413
Laws Governing Registration, §**10.166**	413
Documents Subject to Registration, §**10.167**	413
Rights that Require Registration, §**10.168**	414
The Form of Acts Subject to Registration, §**10.169**	415
Custody of Documents, §**10.170**	415
Principle Registers, §**10.170**	415
Entry-Book, §**10.171**	415
Index of Names, §**10.172**	415
Index of Immovables, §**10.173**	416
Other Registers, §**10.174**	416
Modes of Registration, §**10.175**	416
Introduction, §**10.175**	416
Registration by Deposit, §**10.176**	416
Registration by Memorial, §**10.177**	417
Place of Registration, §**10.178**	417
Duties of a Registrar, §**10.179**	417
Cancellation, §**10.180**	418
Certificate of Search, §**10.181**	418
Effects of Registration, §**10.182**	418
Introduction, §**10.182**	418
Order of Priority of Principal Real Rights, §**10.183**	419
Order of Priority of Accessory Real Rights, §**10.184**	419
Boundary Determination, §**10.185**	420
Introduction, §**10.185**	420
Nature of Boundary Determination, §**10.187**	420
Definition and Nature of Boundary	

Determination, §**10.188**................. 421
Conditions for Boundary Determination, §**10.189**.. 421
Categories of Boundary Determination, §**10.190**... 422
Capacity of the Parties to Determine a Common
    Boundary, §**10.191**...................... 422
Formal Request, §**10.192**..................... 423
Grounds for Recusation, §**10.195**............... 424
Oath of Office, §**10.196**...................... 424
Procedure for Determining Boundaries, §**10.197**... 424
Surveying Operations, §**10.201**................. 425
Evidence, §**10.202**........................... 425
    Introduction, §**10.202**..................... 425
    Titles, §**10.203**.......................... 426
    Witnesses, §**10.204**...................... 428
    Admission, §**10.205**...................... 428
    Possession, §**10.206**...................... 429
    Secondary Documents, §**10.207**............ 429
    Presumptions, §**10.208**.................... 430
The Report, §**10.209**......................... 430
    Introduction, §**10.209**..................... 430
    Content of the Report, §**10.210**............. 430
    Documents Attached to the Report, §**10.211**.. 431
Acceptance of the Conclusions of the Report by
    the Parties, §**10.212**...................... 431
Demarcation, §**10.213**........................ 432
    Notice to the Parties, §**10.213**.............. 432
    Presence of the Parties, §**10.214**............ 432
    Placing of Monuments, §**10.215**............ 432
Minutes of a Boundary Determination, §**10.216**... 433
    Contents of the Minutes, §**10.216**........... 433
    Legal Force of Minutes of a Boundary
        Determination, §**10.217**............... 434
Costs for Boundary Determination, §**10.218**...... 434
The Profession of Land Surveying, §**10.219**.......... 435
    Introduction, §**10.219**....................... 435
    Role of the Land Surveyor, §**10.219**......... 435
    Scope of the Land Surveyor's Activities,
        §**10.220**........................... 435
    Outline of Topics, §**10.221**................ 435
    Surveying Operations on Crown Property,
        §**10.222**..................................... 436
        Division of Land in Québec, §**10.222**........ 436
        Grant of Concessions in the Form of
            Seigniories, §**10.223**................. 436

Survey of Seigniorial Boundaires, §**10.224**.... 436
Survey of Lateral Boundaries of Seigniorial
   Lots, §**10.225**....................... 436
Survey of Township Boundaries, §**10.226**..... 437
Survey of Lateral Boundaries of Township
   Lots, §**10.227**....................... 437
Delimitation of Private Properties, §**10.228**...... 438
   Procedure for Determining Boundaries,
      §**10.228**.......................... 438
   Monumentation, §**10.229**................. 438
Determining Bundaries Between Private and
   Crown Property, §**10.230**................. 438
Analysis of the State of Immovables: the
   Certificate of Location, §**10.231**........... 439
Nature of the Certificate of Location, §**10.232**.... 439
Importance of the Certificate of Location,
   §**10.233**.............................. 440
   Importance of the Certificate of Location for
      the Examiner of Titles, §**10.233**........ 440
   Importance of the Certificate of Location for
      Investors, §**10.234**................... 441
   Importance of the Certificate of Location for
      Owners, §**10.235**.................... 441
   Importance of the Certificate of Location for
      the Courts, §**10.236**.................. 442
   Importance of the Certificate of Location for
      Municipalities, §**10.237**............... 442
   Other Uses for the Certificate of Location,
      §**10.238**.......................... 442
Contents of the Certificate of Location, §**10.239**.. 442
Surveying and Locating Operations, §**10.240**..... 443
The Report, §**10.241**........................ 443
   Mandatory Contents, §**10.241**............. 443
   Non-Apparent Encumbrances, §**10.242**...... 444
   Non-Apparent Encumbrances Not Mentioned
      in Titles, §**10.243**.................... 444
   The Reserve of Three Chains, §**10.244**...... 444
   Views Onto Neighbouring Properties,
      §**10.245**.......................... 444
   Encroachments Suffered or Exercised,
      §**10.246**.......................... 445
   Encroachments On or Above Public Roads,
      §**10.247**.......................... 445
   Party Walls, §**10.248**.................... 446

Research in the Registry Office, §**10.249**..... 446
Regulations Concerning the Laying Out of
  Lots, §**10.250**....................... 446
Zoning Regulations, §**10.251**.............. 446
Enforcement of Municipal Regulations,
  §**10.252**............................ 447
MRC Regulations, §**10.253**................ 447
New Laws, §**10.254**...................... 447
The Act to Preserve Agricultural Lands,
  §**10.255**............................ 447
The Act Respecting the Régie du Logement,
  §**10.256**............................ 448
Place and Date of the Closing of the Minute,
  §**10.257**............................ 448
Use of the Certificate of Location, §**10.258**... 448
The Plan, §**10.259**........................... 449
Optional Contents of the Certificate of Location,
  §**10.260**................................ 449
Ownership Title, §**10.261**.................. 449
History of the Cadastre, §**10.262**........... 449
Location Measurements (on the Plan),
  §**10.263**............................ 449
Co-ownership of Immovables, §**10.264**.......... 450
Copies of the Certificate of Location, §**10.265**.... 451
Staking an Immovable, §**10.266**................ 451
Updating the Certificate of Location, §**10.267**.... 451
The Certificate of Location: Concluding
  Remarks, §**10.268**...................... 452
Custody of the Land Surveyor's Records,
  §**10.269**................................ 453
The Records of a Land Surveyor, §**10.269**.... 453
Confidentiality of the Records, §**10.270**...... 453
Copies of Documents, §**10.271**............. 453
Regulation Respecting the Custody of
  Records, §**10.272**.................... 453
Minuted Documents, §**10.273**.............. 454
Documents in Original Form, §**10.274**....... 454
Provisional Documents, §**10.275**........... 454
Custody of Records, §**10.276**.............. 454
Professional Status of the Land Surveyor,
  §**10.277**................................. 454
The Land Surveyor as Public Officer,
  §**10.277**............................ 454
The Land Surveyor as Survey Expert,
  §**10.278**............................ 455

The Law Respecting Bodies of Water in the Province
  of Québec, §**10.279**........................... 455
  Jurisdiction of the Province Over Bodies of Water,
    §**10.279**................................. 455
  Jurisdiction of the Federal Government Over
    Bodies of Water, §**10.283**.................. 456
  The Law in Québec, §**10.285**................... 458
    The French Regime, §**10.285**............... 458
    The English Regime, §**10.286**.............. 458
  The Concept of Navigability and Floatability,
    §**10.289**................................. 459
    Introduction, §**10.289**.................... 459
    The Criterion of Navigability and
      Floatability, §**10.290**................. 460
    Artificial Navigability, §**10.291**........... 461
    The Criterion of the High Water Mark,
      §**10.292**............................. 461
    The Beach, §**10.293**...................... 462
    The Criterion of the Middle Thread,
      §**10.294**............................. 462
  Individual Rights over Bodies of Water,
    §**10.295**................................ 463
    Introduction, §**10.295**................... 463
    Rain Water and Run-off, §**10.296**.......... 463
    Underground Water, §**10.297**.............. 463
    Spring Water, §**10.298**................... 463
    Flowing Water, §**10.299**.................. 464
  Rights of Accession, §**10.300**................. 464
    Introduction, §**10.300**................... 464
    Alluvion, §**10.301**....................... 464
    Accretions, §**10.302**..................... 464
    Avulsion, §**10.303**....................... 465
    Islands, Islets and Deposits, §**10.304**....... 465
    Change in the Course of a River, §**10.305**.... 465
  Riparian Rights, §**10.306**.................... 466
  Public Ownership of Bodies of Water,
    §**10.307**................................ 466
  The Reserve of Three Chains, §**10.308**.......... 467
    Introduction, §**10.308**................... 467
    Origin of the Reserve, §**10.309**............ 467
    Legal Provisions, §**10.310**................ 468
  Characteristics of Property Rights of the
    Province, §**10.311**....................... 469
    Introduction, §**10.311**................... 469
    Crown Property is Imprescriptible, §**10.312**... 469

xxxiv   Table of Contents

                Fiscal Immunity of Crown Property,
                  §10.313 469
                Alienability of Crown Property, §10.314 469

Chapter 11: **The Surveying Profession**
      — G.K. Allred 471
      Concepts of Professionalism, §11.01 471
      Legislation Governing Professions in Canada, §11.11 .. 475
          Jurisdiction, §11.11 475
          Policies Relating to Professional Organization,
            §11.17 478
          The Professional Association, §11.21 479
      Features of Canadian Survey Legislation, §11.26 481
          Exclusive Field of Practice, §11.26 481
          Qualifications, §11.29 482
          Codes of Ethics, §11.33 486
          Standards of Practice, §11.39 488
          Discipline, §11.42 489
      The Professional Survey Practioner, §11.48 493
          Powers and Responsibilities, §11.52 494
          Responsibilities, §11.59 497
          Quasi-Judicial Function, §11.62 497
          Professional and Technical Affiliations, §11.64 499
          Fédération International des Géomètres, §11.65 ... 499
          Canadian Institute of Surveying and Mapping,
            §11.66 500
          Canadian Council of Land Surveyors, §11.67 500
          Association of Canada Lands Surveyors, §11.68 ... 501
          Provincial Professional Associations, §11.69 501
          Other Organizations, §11.71 501
      Conclusion, §11.72 502
      Appendix 'A' 503

Chapter 12: **The Future**
      — John McLaughlin 507
      Introduction, §12.01 507
      The Management of Land Resources, §12.02 508
      The Role of Information, §12.06 509
      The Surveyor as an Information Specialist, §12.13 512
      Emerging Issues, §12.19 514
      Cadastral Systems Issues, §12.20 514
      Records Management Issues, §12.23 515
      The Privacy Issue, §12.28 517
      Other Issues, §12.31 518

Index 521

Table of Cases

References are to paragraph numbers in the text.

Abell v. Woodbridge and York (1917), 39 O.L.R. 382, 37 D.L.R. 352, reversed
on other grounds, 45 O.L.R. 79, 46 D.L.R. 513, but restored 61 S.C.R.
345, 57 D.L.R. 81 .... 4.74
Abley v. Dale (1851), 11 C.B. 378 .... 1.21
Adamson v. Bell Telephone Co. (1920), 55 D.L.R. 157, 48 O.L.R. 24 (C.A.) 2.46
Afton Band of Indians v. A.G. of Nova Scotia (1978), 29 N.S.R. (2d) 226, 45
A.P.R. 226, 3 R.P.R. 298, 85 D.L.R. (3d) 454 (T.D.) 8.78
Alexander v. McKillop & Benjafield, [1912] 1 W.W.R. 871, 45 S.C.R. 551,
20 W.L.R. 850, 1 D.L.R. 586 .... 3.72
Arrow River v. Pigeon River Co., [1932] S.C.R. 495, 39 C.R.C. 161, [1932]
2 D.L.R. 250 .... 10.282
Assets Co. v. Mere Roihi, [1905] A.C. 176 (P.C.) 3.43
Attersley v. Blakely, [1970] 3 O.R. 303, 13 D.L.R. (3d) 39 (Co. Ct.), affirmed
[1970] 3 O.R. 313, 13 D.L.R. (3d) 49 (C.A.) 6.50
A.G. v. Chambers (1854), 4 De G.M. & G. 206, 43 E.R. 486, [1843-60] All
E.R. Rep. 941 .... 5.10, 5.39, 5.80
A.G. v. McCarthy, [1911] 2 I.R. 260 .... 6.88
A.G. for B.C. v. A.G. for Can., [1914] A.C. 153, 5 W.W.R. 878, 15 D.L.R. 308
(P.C.) 5.18
A.G. for B.C. v. Miller, [1975] 1 S.C.R. 556, [1974] 4 W.W.R. 545, 45 D.L.R.
(3d) 376, reversing [1973] 2 W.W.R. 201 (B.C. C.A.) 5.85
A.G. of B.C. v. Neilson, [1956] S.C.R. 819, 5 D.L.R. (2d) 449, reversing 16
W.W.R. 625, [1955] 5 D.L.R. 56, which affirmed 13 W.W.R. 241 .... 5.10, 5.52,
5.54, 8.54
A.G. of Can. v. Acadia Forest Products Ltd. (1985), 37 R.P.R. 184 (Fed. T.D.) 5.49
A.G. Can. v. A.G. B.C., [1984] 1 S.C.R. 388, [1984] 4 W.W.R. 289, 54 B.C.L.R.
97, 8 D.L.R. (4th) 161, 52 N.R. 335 .... 5.65, 5.96
A.G. of Can. v. A.-G. of Ont.; Provincial Fisheries, Re, [1898] A.C. 700 (P.C.) 5.62
A.G. of Can. v. Higbie, [1945] S.C.R. 385, [1945] 3 D.L.R. 1 .... 5.01, 5.62
A.G. for N.S.W. v. Dickson, [1904] A.C. 273 (P.C.) 6.81
A.G. for Ont. v. Booth (1923), 53 O.L.R. 374 (C.A.) 8.34
A.G. of Ont. v. Hamilton Street Railway, [1903] A.C. 524, 2 O.W.R. 672, 7
C.C.C. 326 (P.C.) 1.65
A.G. for Ont. v. Walker, [1975] 1 S.C.R. 78, 42 D.L.R. (3d) 629, 1 N.R. 283,

Table of Cases

affirming [1972] 2 O.R. 558, 26 D.L.R. (3d) 162 (C.A.), which affirmed
[1971] 1 O.R. 151, 14 D.L.R. (3d) 643 (H.C.).............6.35, 6.55, 6.62, 6.68, 6.75
A.G. for Quebec v. Healy, [1983] Que. C.A. 573, affirmed [1987] 1 S.C.R. 158......10.310
A.G. for Quebec v. MacLaren (1911), 21 Que. Q.B. 42, affirmed (1912), 46
S.C.R. 656, 8 D.L.R. 800, reversed [1914] A.C. 258 (P.C.)..................10.310
A.G. for Southern Nigeria v. Holt & Co., [1915] A.C. 599, [1914-15] All E.R.
Rep. 444 (P.C.)...6.87, 6.88
A.G. and Southland County Council v. Miller (1906), 26 N.Z.L.R. 348, 9 G.L.R.
145..6.81
Auty v. Thompson (1903), 5 Gaz. L.R. 541 (N.Z. S.C.)..........................6.91
Badgely v. Bender (1834), 3 O.S. (C.A.) 221..6.85
Badgely v. Dickson (1886), 13 O.A.R. 494..9.14
Barnes v. Belyea (1880), 19 N.B.R. 541 (S.C.)......................................8.80
Bartlett v. Delaney (1913), 29 O.L.R. 426, 17 D.L.R. 500 (C.A.)..............6.31
Bateman v. Potruff, [1955] O.W.N. 329 (C.A.)...................................11.63
Bea v. Robinson (1977), 18 O.R. (2d) 12, 3 R.P.R. 154, 81 D.L.R. (3d) 423
(H.C.)..4.119, 8.31, 8.32, 8.36
Beebe v. Robb (1977), 81 D.L.R. (3d) 349 (B.C. S.C.).............................9.42
Belhumeur v. Discipline Ctee. of Que. Bar Assoc. (1983), 34 C.R. (3d) 279
(C.S. Qué.)..11.44
Bell v. Howard (1957), 6 U.C.C.P. 292 (C.A.)......................................4.55
Belyea v. Belyea (1857), 8 N.B.R. 588 (C.A.).......................................8.68
Benoit v. Benoit (1976), 15 N.B.R. (2d) 233 (S.C.), affirming 15 N.B.R. (2d)
59 (C.A.)..8.14, 8.50
Bernard v. Gibson (1874), 21 Gr. 195 (C.A.)......................................4.120
Biron v. Caron (1895), S.C.R. 451..10.203
Black Brook Salmon Club Inc. v. Seder (1981), 34 N.B.R. (2d) 474, 85 A.P.R.
474 (C.A.)..5.20
Boehner v. Hirtle (1912), 46 N.S.R. 231, 6 D.L.R. 548, reversing 9 E.L.R. 258
(C.A.), reversed without written reasons 50 S.C.R. 264, 18 D.L.R. 794.......8.02, 8.10
Bonnasserre v. The National Harbour Board, [1972] Que. S.C. 713...............10.306
Borax Consolidated Ltd. v. Los Angeles (City of) (1935), 296 U.S. 10..............5.71
Borys v. C.P.R., 7 W.W.R. 546, [1953] A.C. 217, [1953] 1 All E.R. 451, [1953]
2 D.L.R. 65..3.87
Bouillon v. R. (1913-17), 16 Exch. C.R. 443...................................10.290
Boyd v. Fudge (1965), 46 D.L.R. (2d) 679, 50 M.P.R. 384 (N.B. C.A.)...........5.20, 5.84
Boyd v. Luscombe (1986), 57 Nfld. & P.E.I.R. 242, 170 A.P.R. 242 (Nfld. Dist.
Ct.)...8.14, 8.41
Bragg v. Rogers (1875), 25 U.C.C.P. 156 (C.A.)..................................8.72
Brean v. Thorne (1982), 52 N.S.R. (2d) 241, 106 A.P.R. 241 (T.D.)................8.59
Brew Island, Re, [1977] 3 W.W.R. 80 (B.C. S.C.)................................8.57
Brighton & Hove General Gas Co. v. Hove Bungalows Ltd., [1924] 1 Ch. 372,
[1923] All E.R. Rep. 369...6.88
Brookman v. Conway (1903), 35 N.S.R. 462, affirmed 35 S.C.R. 185.............8.64
Brown v. Norbury, [1931] 2 W.W.R. 863, 255 Alta. L.R. 591, [1931] 4 D.L.R.
507 (C.A.)...8.48
Brown v. Phillips, [1964] 1 O.R. 292, 42 D.L.R. (2d) 38 (C.A.).....................4.81
Bruce v. Johnson, [1953] O.W.N. 724, [1954] 1 D.L.R. 571 (Co. Ct.).......5.85, 6.30, 6.88
Buchanan v. Saulnier (1984, N.S. S.C.)..8.36
Bulman, Re (1966), 56 W.W.R. 225, 57 D.L.R. (2d) 658 (B.C.)...................6.88
Buram v. Violette (1893), 32 N.B.R. 68 (C.A.)......................................8.52

Table of Cases   xxxvii

Burchell v. Bigelow (1904), 40 N.S.R. 493 (C.A.)................................3.17
Cairncross v. Lorimer (1860), 3 Macq. 827 (H.L.)...............................8.48
Capital Trust Corp. v. Gordon, [1954] O.R. 277 (S.C.)..........................8.48
Cambell v. Hall (1774), 1 Cowp. 204, 98 E.R. 1045...........................10.307
C.P.R. v. District Registrar of Dauphin Land Titles Office (1956), 4 D.L.R.
   (2d) 518, 18 W.W.R. 241, 64 Man. R. 76 (Q.B.)..............................3.77
Canadian Western Natural Gas Co. v. Pathfinder Surveys Ltd. (1980), 21 A.R.
   459...9.44
Candler v. Crane, Christmas & Co., [1951] 2 K.B. 164, [1951] 1 All E.R. 426
   (C.A.)..9.49 9.51
Carbonneau v. Godbout (1920), 31 Que. K.B. 69.............................10.163
Carnegy v. Godin (1982), 52 N.S.R. (2d) 697, 106 A.P.R. 697 (T.D.)...............8.87
Carrigan v. Lawrie (1909), 7 E.L.R. 108 (N.S.)..................................8.41
Carroll v. Empire Limestone Co. (1919), 45 O.L.R. 121, 48 D.L.R. 44 (C.A.),
   affirmed (1919), 17 O.W.N. 295 (S.C.C.)....................................6.68
Carson v. Musialo, [1940] O.W.N. 398, [1940] 4 D.L.R. 651 (C.A.)................8.64
Crabb v. Arun District Council, [1976] Ch. 179, [1975] 3 W.L.R. 847, [1975]
   3 All E.R. 865 (C.A.)...2.48
Central Trust Co. v. Thistle (1986), 58 Nfld. & P.E.I.R. 1, 174 A.P.R. 1 (Nfld.
   Dist. Ct.)..8.47
Charbonneau v. McCusker (1910), 22 O.L.R. 46 (C.A.)..........................8.68
Chaulk v. South River (Town of) (1971), 2 Nfld. & P.E.I.R. (Nfld. S.C.)...........5.47
Chuckry v. R., [1973] S.C.R. 694, [1973] 5 W.W.R. 339, 35 D.L.R. (3d) 607,
   4 L.C.R. 61, reversing [1972] 3 W.W.R. 561, 27 D.L.R. (3d) 164, 2 L.C.R.
   249 (Man. C.A.)...6.30, 6.87 6.90, 8.54
Clarke v. Edmonton (City of), [1930] S.C.R. 137, [1929] 4 D.L.R. 1010, reversing
   [1928] 1 W.W.R. 553, [1928] 2 D.L.R. 154, 23 Alta. L.R. 233........5.19, 5.52, 5.53,
                                                       5.84, 5.85, 8.54, 11.41
Coleman and A.-G. for Ont., Re (1983), 27 R.P.R. 107, 12 C.E.L.R. 104, 143
   D.L.R. (3d) 608 (Ont. H.C.)..6.98
Crossland v. Dorey (1977), 27 N.S.R. (2d) 139, 41 A.P.R. 139, 83 D.L.R. (3d)
   54, affirmed 28 N.S.R. (2d) 91, 43 A.P.R. 91, 90 D.L.R. (3d) 40 (C.A.)......8.32, 8.38
Crowley v. Feeney (1932), 5 M.P.R. 248 (N.B. C.A.).............................8.41
Davies, Re. Se Re Ellenborough Park; Re Davies; Powell v. Madison.
Davison v. Benjamin (1874), 9 N.S.R. 474 (C.A.)................................8.02
Davison v. Kinsman (1853), 2 N.S.R. 1 and 69 (C.A.).......................8.33, 8.39
Dawe v. Avalon Coal & Salt Ltd. (1950), 26 M.P.R. 112 (Nfld. S.C.)...............5.49
Deerfield v. Arms (1835), 34 Mass. (17 Pick.) 41, 28 Am. Dec. 276 (Supreme
   Court of Massachusetts)..8.57
Delamatter v. Brown (1908), 13 O.W.R. 58, varied on other grounds (1909),
   13 O.W.R. 862 (Div. Ct.)...9.61
Delap v. Hayden, 57 N.S.R. 346, [1924] 3 D.L.R. 11, reversing [1923] 4 D.L.R.
   1102 (C.A.)..5.47
Delima v. Paton (1971), 1 Nfld. & P.E.I.R. 317, 19 D.L.R. (3d) 351 (P.E.I. S.C.).....8.04
Dell v. Howe (1857), 6 U.C.C.P. 292 (C.A.).....................................8.30
Dennison v. Chew (1836), 5 O.S. 161 (Ont. C.A.)................................4.57
DeSantis and Alta. Land Surveyors Act, Re (1964), 48 W.W.R. 50, 44 D.L.R.
   (2d) 749 (Alta. S.C.)..11.43
Devereaux v. Saunders (1977), 26 N.S.R. (2d) 301, 4 R.P.R. 267, 4 A.P.R. 301
   (T.D.)...8.87
Diehl v. Zanger, 39 Mich. 60...8.63

xxxviii Table of Cases

Dill v. Wilkins (1853), 2 N.S.R. 113 (C.A.)...................................8.35, 8.50
Dir. of Investigation & Research and Can. Safeway Ltd., Re, 26 D.L.R. (3d)
  745, [1972] 3 W.W.R. 547 (sub nom. Dir. of Investigation & Research
  v. Can. Safeway Ltd.), 6 C.P.R. (2d) 41 (B.C. S.C.)...........................7.39
Dixon v. Crowhurst (1976), 14 N.B.R. (2d) 401 (C.A.)...........................8.27
Dison v. Snetsinger (1873), 23 U.C.C.P. 235 (Ont. C.A.)...................6.09, 6.10
Doe d. Fry v. Hill (1853), 7 N.B.R. 587 (C.A.).....................................5.47
Doe & Giblert v. Ross (1840), 7 M. & W. 102, 151 E.R. 696.....................7.50
Doherty v. McDevitt (1892), 31 N.B.R. 526 (C.A.)................................8.61
Dolphin Land Assoc. Ltd. v. Southampton (Town of) (1975), 37 N.Y. (2d) 292........5.71
Donnelly v. Vroom (1909), 42 N.S.R. 327, affirming (1907), 40 N.S.R. 585
  (C.A.)..5.54
Drysdale v. Drysdale (1977), 18 N.B.R. (2d) 429 (Q.B.).........................8.14
Dunphy v. Phillips (1929), 1 M.P.R. 227 (N.B. S.C.)............................8.38
Dunphy v. Williams (1874), 15 N.B.R. 350 (C.A.)................................6.88
Dunstan v. Hell's Gate Enterprises Ltd., 22 D.L.R. (4th) 568, [1986] 3 C.N.L.R.
  47 (B.C. S.C.), reversed (1987), 20 B.C.L.R. (2d) 29, 45 D.L.R. (4th) 677
  (C.A.)..6.64, 6.66
Eastwood v. Ashton, [1915] A.C. 900 (H.L.).....................................7.46
Ellenborough Park Re; Re Davies; Powell v. Madison, [1956] Ch. 131, [1955]
  3 All E.R. 667, [1955] 3 W.L.R. 892 (C.A.)..................................4.67
Elliot v. University of Alta. Governors, [1973] 4 W.W.R. 195, 37 O.L.R. (3d)
  197 (Alta. S.C.)..11.44
Ellis v. Abell (1884), 10 O.A.R. 226...7.44
Ernst v. Dartmouth (City of) (1970), 3 N.S.R. (2d) 254 (T.D.)..................5.03
Esson v. Mayberry (1841), 1 N.S.R. 186 (C.A.)............................5.06, 5.47
Ewing v. Publicover (1975), 13 N.S.R. (2d) 346, affirmed 14 N.S.R. (2d) 159
  (C.A.)..8.29
Fang v. College of Physicians and Surgeons of Alta., 66 A.R. 352, [1986] 2
  W.W.R. 380, 42 Alta. L.R. (2d) 89, 25 D.L.R. (4th) 632 (C.A.).........11.01, 11.44
Finley v. Sutherland (1969), 2 N.S.R. 1965-69 197, 4 D.L.R. (3d) 586 (C.A.)..........8.14
Fitzpatrick v. McSorley (1920), 48 N.B.R. 162 (C.A.)...........................8.72
Flewelling v. Johnson, 16 Alta. L.R. 409, [1921] 2 W.W.R. 374, 59 D.L.R. 419
  (C.A.)..6.19, 6.41
Ford v. Hopkins (1701), 1 Salk. 283, 91 E.R. 250...............................7.47
Forrest v. Turnbull (1909), 14 O.W.R. 478, affirmed 14 O.W.R. 930, 1 O.W.N.
  150 (Div. Ct.)..8.41
Fort George Lumber Co. v. Grand Trunk Pacific Ry. (1915), 24 D.L.R. 527,
  9 W.W.R. 17 (B.C. S.C.)...5.18
Fraser v. Cameron (1854), 2 N.S.R. 189 (C.A.)............................5.06, 5.84
Fraser v. Kirk (1858), 3 N.S.R. 290 (C.A.).....................................8.35
Frazer v. Walker, [1967] 1 A.C. 569, [1967] 1 All E.R. 649, [1967] 2 W.L.R.
  411 (P.C.)..3.41
Garton v. Hunter, [1969] All E.R. 451..7.49
George v. Bates (1858), 7 U.C.C.P. 116 (C.A.)..................................5.19
Georgian Cottagers' Assoc. v. Flos (Twp.), [1962] O.R. 429, 32 D.L.R. (2d)
  547 (H.C.)..6.56
Gibbs v. Messer, [1891] A.C. 248 (P.C.)..3.40
Gifford v. Lord Yaborough (1828), 5 Bing. 163, 130 E.R. 1023 (H.L.).............6.88
Gorgichuk v. American Home Assurance Co., 5 C.P.C. (2d) 166, 14 C.C.L.I.
  32, [1985] I.L.R. 1-1984 (Ont. H.C.)..7.58

Goss v. Nugent (1833), 5 B. & Ad. 58, 110 E.R. 713............................7.44
Gov't. of State of Penang v. Beng Hong On, [1972] A.C. 425, [1971] 3 All E.R.
   1163 (P.C.)..6.88
Grand Hotel Co. v. Cross (1879), 44 U.C.R. 153...............................2.64
Grasett v. Carter (1884), 10 S.C.R. 105.........................4.35, 8.33, 8.40, 8.50
Grey v. Pearson (1857), 10 E.R. 1216, 6 H.L. Cas. 61 (H.L.)......................1.24
Griffin v. Catfish Creek Conservation Aurthority, [1968] 1 O.R. 574 (Co. Ct.)........8.61
Guinn v. Leathem, [1901] A.C. 495..8.01
Hackworth v. Baker, [1936] 1 W.W.R. 321 (Sask. C.A.).........................3.39
Haggerty v. Latreille (1913), 29 O.L.R. 300, 14 D.L.R. 532 (C.A.)..................6.12
Haigg v. Bamford, [1977] 1 S.C.R. 466, [1976] 3 W.W.R. 331, 9 N.R. 43, 27
   C.P.R. (2d) 149, 72 D.L.R. (3d) 68.....................................9.52, 9.53
Hanley and Fall v. Bank of N.S. (1976), 13 Nfld. & P.E.I.R. 261, 29 A.P.R. 261
   (P.E.I. S.C.)..8.47
Harries Hall & Kruse v. South Sarnia Properties Ltd., 63 O.L.R. 597, [1929]
   2 D.L.R. 821 (S.C.)..9.18
Hebb v. Hebb, 17 M.P.R. 276, [1944] 2 D.L.R. 255 (N.S. C.A.)....................8.87
Hedley, Byrne v. Heller, [1964] A.C. 465 (H.L.)...................9.50, 9.51, 9.52, 9.53
Henrietta Edwards v. Att. Gen. of Can., [1930] A.C. 124.........................1.23
Hermanson v. Martin, [1982] 6 W.W.R. 312, 25 R.P.R. 245, 18 Sask. R. 430,
   140 D.L.R. (3d) 512 (Q.B.)...3.41
Herriman v. Pulling & Co. (1906), 8 O.W.R. 149 (T.D.).........................6.81
Heydon's Case (1584), 3 Co. Rep. 7a, 76 E.R. 637..............................1.27
Hindson v. Ashby, [1896] 1 Ch. 78, reversed on the facts [1896] 2 Ch. 1
   (C.A.)..6.27, 6.33, 6.36, 6.40, 6.45, 6.66
Holden v. Strickland (1976), 13 Nfld. & P.E.I.R. 439, 29 A.P.R. 439 (Nfld. Dist.
   Ct.)...8.14
Holman v. Green (1881), 6 S.C.R. 707..5.62
Home Bank v. Might Directories Ltd. (1914), 31 O.L.R. 340, 20 D.L.R. 977
   (C.A.)...11.63
Houson v. Austin (1923), 23 O.W.N. 603(H.C.)................................11.63
Howard v. Ingersoll (1851), 13 Howard 381, 54 U.S. 381 (U.S. Sup. Ct.)............6.46
Hudson's Bay Co. v. Kenora. See Keewatin Power Co. v. Kenora.
Huebner v. Wiebe, [1984] 1 W.W.R. 272, 25 Man. R. (2d) 70
   (Q.B.)...11.04, 11.8, 11.22, 11.63
Hugg v. Low, [1929] 2 W.W.R. 55, 23 Sask. L.R. 592, [1929] 3 D.L.R. 725
   (C.A.)...4.128
Hughes v. Washington (1966), 67 Wash Dec. (2d) 787, 410 P. (2d) 20...............5.82
Hull & Selby Railway, Re (1839), 5 M. & W. 327, 151 E.R. 139....................6.87
Hunter, Re (1978), 23 N.B.R. (2d) 130, 44 A.P.R. 130, reversing 19 N.B.R. (2d)
   710, 30 A.P.R. 710 (C.A.)..8.51
Hunter v. Ronne (1870), 8 N.S.R. 113 (C.A.)..................................8.64
Hurdman v. Thompson (1895), 4 Q.B. 409............................10.290, 10.291
Hutchinson v. Gillespie (1838), 2 R.J.R.Q. 313................................10.289
Inch v. Flewelling (1890), 30 N.B.R. 19 (C.A.)................................8.40
Irving Refining Ltd. v. Eastern Trust Co. (1967), 51 A.P.R. 155 (N.B. S.C.)..5.06, 5.47, 5.49,
   5.67, 5.72, 5.79, 5.87
Iverson and Greater Winnipeg Water District, Re, 31 Man. R. 98, [1921] 1 W.W.R.
   621, 57 D.L.R. 184 (C.A.)..6.19
James v. Stevenson, [1893] A.C. 162 (P.C.)...................................4.59
Janes, Re (1977), 17 N.B.R. (2d) 600 (Q.B.)..................................8.14

xl Table of Cases

Jollymore v. Acker (1915), 49 N.S.R. 148, 24 D.L.R. 503 (C.A.)....................8.35
Jones v. Jones (1843), 4 N.B.R. 265 (C.A.).......................................4.74
Jones v. Morgan (1882), 22 N.B.R. 338 (C.A.).....................................8.40
Joyce v. Smith (1984), 66 N.S.R. (2d) 406, 152 A.P.R. 406 (T.D.)............8.40, 8.48
K. & W. Enterprises Ltd. v. Smith (1974), 7 N.S.R. (2d) 411 (S.C.).................8.69
Kaneen v. Mellish (1922), 70 D.L.R. 327 (P.E.I. C.A.).............................8.40
Keewatin Power Co. v. Kenora (1906), 13 O.L.R. 237 varied on appeal (1908),
    16 O.L.R. 184, 11 O.W.R. 266 (C.A.)..............................5.19, 6.10 6.20
Kingdon v. Hutt River Bd. (1905), 25 N.Z.L.R. 145, 7 G.L.R. 634 (S.C.).........6.61, 6.66
Kingston v. Highland (1919), 47 N.B.R. 324 (K.B.).........................8.44, 11.63
Kirton and Frolak, Re, [1973] 2 O.R. 185, 33 D.L.R. (3d) 281 (H.C.)...............8.87
Labrador Boundary, Re, 43 T.L.R. 289, 2 D.L.R. 401 (P.C.)...................4.31, 8.30
Lake Erie Excursion Co. v. Bertie (1912), 4 O.W.N. 111, 6 D.L.R. 853, affirming
    3 O.W.N. 1191, 4 D.L.R. 585 (C.A.)...8.72
Lambert v. St-Lauveur, 20 R.L. 46..10.206
Larson v. Sask. Land Surveyors' Assoc. (1987), 30 Admin. L.R. 33, 63 Sask.
    R. 119, 45 D.L.R. (4th) 583 (Q.B.)...11.45
Lavoie v. Michaud, [1981] R.C.S. 445, 19 R.P.R. 71, 38 N.R. 171..................10.124
Lawrence v. McDowall (1838), 2 N.B.R. 442 (C.A.)....................8.39, 8.40, 8.45
LeBlanc v. DeWitt (1984), 34 R.P.R. 196 (N.B.Q.B.)...............................4.42
Lee v. Arthurs (1918), 46 N.B.R. 185 (S.C.), affirmed (1919), 46 N.B.R. 482,
    48 D.L.R. 78 (C.A.)...5.40
Legge v. Scott Paper Co. (1970), 3 N.S.R. (2d) 206 (T.D.).........................8.77
Les Soeurs de Misericorde v. Tellier, [1932] 2 W.W.R. 357, 40 Man. R. 351,
    [1932] 3 D.L.R. 715 (C.A.)..8.42, 8.65
Locher v. Howlett (1894), 13 N.Z.L.R. 584......................................3.44
Lord v. The Commissioners of Sydney (1859), 12 Moo. P.C. 473, 14 E.R. 991
    (P.C.)..6.44
Lundrigans Ltd. and Prosper, Re (1982), 38 Nfld. & P.E.I.R. 10, 132 D.L.R.
    (3d) 727 (Nfld. Dist. Ct.)...8.78
Luness, Re (1919), 51 D.L.R. 114, 46 O.L.R. 320 (C.A.)...........................8.46
Lutz v. Kawa (1980), 15 R.P.R. 40, 112 D.L.R. (3d) 271, 23 A.R. 9 (Alta. C.A.).....8.58
Louis Lavoie v. Arthur Chartier (18 January 1967), District of Hull 2976, (Mag.
    Ct.)...10.266
MacDougall v. Layes (1969), 2 N.S.R. 1965-69 96 (S.C.)..........................8.14
MacLaren-Elgin Corp. v. Gooch, [1972] 1 O.R. 474, 23 D.L.R. (3d) 394 (H.C.).......9.27
MacMain v. Hurontario Mgmt. Services Ltd. (1980), 14 R.P.R. 158 (Ont. Co.
    Ct.)...8.33
MacMillan v. Campbell, 28 M.P.R. 112, [1951] 4 D.L.R. 265 (N.B. C.A.)............8.37
Mahon v. McCully (1868), 7 N.S.R. 323 (C.A.)...................................5.53
Marr v. Davidson (1867), 26 U.C.Q.B. 641 (Ont. C.A.)............................8.46
Martin v. Weld (1860), 19 U.C.Q.B. 631 (C.A.)...................................8.43
McCormick v. Pelée (Township) (1890), 20 O.R. 288 (Ch. D.).....................6.81
McDonald v. Linton, 53 N.B.R. 107, [1926] 3 D.L.R. 779 (C.A.)...................5.20
McDonald v. Mahoney (1895), 31 N.S.R. 523 (C.A.)..............................8.38
McDonald v. McDonald (1867), 7 N.S.R. 42 (C.A.)...............................8.42
McGrath v. Williams (1912), 12 S.R. (N.S.W.) 477................................6.81
McGregor v. Webber (1917), 51 N.S.R. 226 (C.A.)................................8.38
McGugan v. Turner, [1948] O.R. 216, [1948] 2 D.L.R. 338 (H.C.).............8.48, 8.52
McIntyre v. White (1911), 10 E.L.R. 248, affirming 40 N.B.R. 591 (C.A.)...........8.53
McIsaac v. McKay (1916), 49 N.S.R. 476, 27 D.L.R. 184 (C.A.)..............8.63, 8.69

Table of Cases xli

McKie v. K.V.P. Co., [1948] O.R. 398, affirmed [1948] O.W.N. 812, [1949]
　1 D.L.R. 39, affirmed [1949] S.C.R. 698, [1949] 4 D.L.R. 497 .... 4.69
McKinney v. Bd. of Governors of the University of Guelph (1986), 57 O.R.
　(2d) 1, 14 C.C.E.L. 1, 22 Admin. L.R. 29, 9 C.H.R.R. D/4685, 87 C.L.L.C.
　17009, C.E.B. & P.G.R. 8038, 32 D.L.R. (4th) 65 .... 1.72
McLean v. Jacobs (1834), 1 N.S.R. 9 (C.A.) 8.39
McLennan v. Nova Scotia Steel & Coal Co. (1908), 18 Que. K.B. 317 .... 10.163
McNeil v. Jones (1894), 26 N.S.R. 299 (C.A.) 5.20
McNichol, Re (1976), 20 N.B.R. (2d) 240, 34 A.P.R. 240 (Q.B.) 5.40
Mercer v. Denne, [1905] 2 Ch. 538, [1904-07] All E.R. Rep. 71, affirming [1904]
　2 Ch. 534 .... 6.91
Merriman v. New Brunswick (1974), 7 N.B.R. (2d) 612, 45 D.L.R. (3d) 464
　(C.A.) 6.75, 6.83, 8.29
Merritt v. Toronto (1913), 48 S.C.R. 1, 12 D.L.R. 734, affirming 27 O.L.R. 1,
　6 D.L.R. 152, which affirmed 23 O.L.R. 365 .... 8.41
Mersereau v. Swim (1914), 42 N.B.R. 497 (C.A.) 8.34
Mineral and Other Natural Resources of the Continental Shelf, Re (1983), 145
　D.L.R. (3d) 9, 41 Nfld. & P.E.I.R. 271 (Nfld. C.A.) 5.64
Monashee Enterprises Ltd. v. Min. of Recreation & Conservation (B.C.) (1981),
　28 B.C.L.R. 260, 21 R.P.R. 184, 23 L.C.R. 19, 124 D.L.R. (3d) 372, reversing
　(1978), 7 B.C.L.R. 388, 7 R.P.R. 197, 16 L.C.R. 212, 90 D.L.R. (3d) 521
　(C.A.) 5.57, 6.81
Moneypenny v. Hartland (1824), 171 E.R. 1227, 1 Car. & P. 351 .... 9.17
Mooney v. McIntosh (1887), Cam. S.C. 171, 14 S.C.R. 740, affirming 19 N.S.R.
　419 .... 8.35, 8.36, 8.40
M'taggart v. M'Douall (1867), 5 M. 534 .... 8.56
Muller v. Mamchur, 15 W.W.R. 468, [1955] 4 D.L.R. 184 (Sask. C.A.) 11.63
Murray v. McNairn, 30 M.P.R. 200, [1953] 1 D.L.R. 128 (N.B. C.A.) 8.51
Myers v. Brennan (1973), 37 D.L.R. (3d) 79, affirmed 10 N.S.R. (2d) 391, 53
　D.L.R. (3d) 388 (C.A.) 8.47
Nash v. Newton (1891), 30 N.B.R. 610 (C.A.) 5.16, 5.20
National Bank of New Zealand v. National Mortgage and Agency Co. (1885),
　N.Z.L.R. 3 S.C. 257 .... 3.44
Naugle v. Naugle (1970), 2 N.S.R. (2d) 309 (C.A.) 8.63
Nelson v. Pac. Great Eastern Ry., [1918] 1 W.W.R. 597, 25 B.C.R. 259 (S.C.) 5.67, 5.75
Nelson v. Varner (1977), 20 N.S.R. (2d) 181 (S.C.) 8.66
New Hamburg v. Waterloo (1892), 20 O.A.R. 1 (C.A.), reversed on other grounds
　(1893), 22 S.C.R. 296 .... 6.66
New Hampshire v. 6.0 Acres of Land (1958), 139 A. (2d) 75 (Supreme Court
　of New Hampshire) 8.57
Niles v. Burke (1873), 14 N.B.R. 237 (C.A.) 5.20
Niagara Falls Park v. Howard (1892), 23 O.R. 1, affirmed (1896), 23 O.A.R.
　355 .... 4.11
Nourse v. Clark, [1936] O.W.N. 563 (C.A.) 8.66
Offshore Mineral Rights (B.C.), Re, [1967] S.C.R. 792, 62 W.W.R. 21, 65 D.L.R.
　(2d) 353 .... 5.65
O'Neil v. Steele (1979), 33 N.S.R. (2d) 514, 47 A.P.R. 514 (T.D.) 8.38
O'Melia v. Himmelman (1985), 69 N.S.R. (2d) 271, 163 A.P.R. 271 (T.D.) 8.32, 8.69
O'Toole v. Walters (1979), 34 N.S.R. (2d) 246, 7 R.P.R. 213 (T.D.) 3.16
Ouimet v. Desmarais, 21 R. de Jur. 96 .... 10.217
Palmer v. Thornbeck (1877), 27 U.C.C.P. 291 (Ont. C.A.) 4.44, 4.52, 8.28, 8.72

Table of Cases

Parker v. Elliott (1852), 1 U.C.C.P. 470 (Ont. C.A.).....................6.34, 6.49, 6.67
Parrot v. Thompson, [1984] 1 S.C.R. 57, 51 N.R. 161, 28 C.C.L.T. 76..........9.33, 10.41
Patenaude v. Edward & Co. (1915), 21 R.L. (N.S.) 523.........................10.310
Paul v. Bates (1934), 48 B.C.R. 473 (Co. Ct.)................................5.88, 8.56
Penney v. Gosse (1974), 6 Nfld. & P.E.I.R. 344 (Nfld. S.C.)........................8.41
Perry v. Patterson (1874), 15 N.B.R. 367 (C.A.).....................................8.67
Phillips v. Montgomery (1915), 43 N.B.R. 229, 25 D.L.R. 499 (C.A.)................8.40
Piers v. Whiting, 50 N.B.R. 363, [1923] 3 D.L.R. 879 (C.A.).......................8.38
Pipi Te Ngahuru v. The Mercer Road Bd. (1887), 6 N.Z.L.R. 19....................6.81
Pitre v. Robinson (1978), 15 Nfld. & P.E.I.R. 63, 38 A.P.R. 63, reversing (1977),
　19 Nfld. & P.E.I.R. 475, 50 A.P.R. 475 (P.E.I. S.C.)...........................8.14
Placements Miracle Inc. v. Gérard Larose (16 February 1978), District of Montréal
　500-05-015927-73 (Que. S.C.)..................................10.232, 10.278
Plumb v. McGannon (1871), 32 U.C.Q.B. 8 (Ont. C.A.)...........................6.66
Powell v. Madison. See Re Ellenborough Park; Re Davies; Powell v. Madison.
Price v. La Cie de pulpe de Chicoutimi (1906), 30 Que. S.C. 293, 16 Que. K.B.
　142..10.293
Provincial Fisheries, Re (1895), 26 S.C.R. 444...................................6.10
Quartermain v. Stevens (1972), 4 N.B.R. (2d) 266 (S.C.).........................8.48
Ratte v. Booth (1887), 14 O.A.R. 419, affirming 11 O.R. 491, which reversed
　10 O.R. 351...5.19
Ravid v. Jasmin (1939), 66 Que. K.B. 279....................................10.266
Reddy v. Strople (1910), 44 N.S.R. 332, reversed on other grounds 44 S.C.R.
　246..8.43
R. v. Carleton (1882), 1 O.R. 277 (C.A.)...6.31
R. v. Drybones, [1970] S.C.R. 282, 10 C.R.N.S. 334, 71 W.W.R. 161, 9 D.L.R.
　(3d) 473, [1970] 3 C.C.C. 355...1.68
R. v. Fares, [1932] S.C.R. 78, [1932] 1 D.L.R. 421, reversing [1929] Ex. C.R.
　144...6.18, 6.21
R. v. London (City of), [1892] 1 Q.B. 273 (C.A.)................................1.24
R. v. Lord (1864), 1 P.E.I. 245...5.51
R. v. Miller, [1941] 1 W.W.R. 483, 75 C.C.C. 398 (Alta. Dist. Ct.)................6.19
R. v. Mojelski (1968), 65 W.W.R. 565, reversing [1969] 1 C.C.C. 69, 1 C.R.N.S.
　392, 60 W.W.R. 355, [1968] 1 C.C.C. 222 (Sask. C.A.)........................1.25
R. v. Price Bros. & Co., [1926] S.C.R. 28, [1925] 3 D.L.R. 595....................7.66
R. v. Robertson (1882), 6 S.C.R. 52, affirming 1 Ex. C.R. 374..............517, 5.19, 5.20
R. v. Ross (1985), 72 N.S.R. (2d) 381, 173 A.P.R. 381 (C.A.)....................11.57
R. v. Weremy, [1943] Ex. C.R. 44, [1943] 1 D.L.R. 9 (Ex. Ct.)....................8.71
Richards v. Gaklis (1984), 63 N.S.R. (2d) 230, 141 A.P.R. 230 (T.D.).............8.69
Riddiford v. Feist (1902), 22 N.Z.L.R. 105, 5 Q.L.R. 43..........................8.57
Risser's Beach, Re (1975), 20 N.S.R. (2d) 479 (T.D.)............................8.14
Roach v. Ware (1886), 19 N.S.R. 330 (C.A.).....................................8.42
Robertson v. R. (1882) 6 S.C.R. 52...10.309
Robertson v. Steadman (1876), 16 N.B.R. 612 (C.A.)..............................5.20
Robertson v. Watson (1874), 27 U.C.C.P. 579 (C.A.)..............................6.52
Robertson and Saunders, Re (1977), 75 D.L.R. (3d) 507 (Man. Q.B.)...........8.58, 8.60
Robinson v. Murray (1851), 3 Nfld. L.R. 184....................................5.20
Rochon v. Gaudreau, [1968], Que. Q.B. 889....................................10.206
Rogers v. Hosegood, [1900] 2 Ch. 388, [1900-3] All E.R. Rep. 915 (C.A.)............2.71
Rollings v. Smith (1977), 2 R.P.R. 10, was affirmed 15 Nfld. & P.E.I.R. 128,
　38 A.P.R. 128 (P.E.I. C.A.)..................................4.135, 8.27, 8.49

Ross v. Hunter (1882), 7 S.C.R. 289..3.16
Ross v. McKenzie (1872), 9 N.S.R. 69 (S.C.)..................................8.40
Rotter v. Canadian Exploration Ltd., [1961] S.C.R. 15, 33 W.W.R. 337, 26
    D.L.R. (2d) 133, reversing (1959), 30 W.W.R. 446, 23 D.L.R. (2d) 136
    (B.C. C.A.)..5.19, 6.21
Ruptash v. Zawick, [1956] S.C.R. 357, 2 D.L.R. (2d) 145........................3.76
Russell v. McKercher, [1905] 1 W.L.R. 138 (Man. C.A.).........................9.14
St. Catherine's Milling & Lbr. Co. v. R. (1883), 13 S.C.R. 577, affirmed 14 App.
    Cas. 46, 4 Cart. B.N.A. 107 (P.C.)...4.16
St. Francis Hydro-Elec. Co. v. R., [1937] 2 All E.R. 541, 66 Que. K.B. 374,
    [1937] 2 D.L.R. 353...10.290
Saueracker v. Snow (1974), 14 N.S.R. (2d) 607, 47 D.L.R. (3d) 577 (T.D.).......5.06, 5.84
Schuler A.G. v. Wickman Machine Tool Sales Ltd., [1974] A.C. 235, [1973]
    2 All E.R. 39, [1973] 2 W.L.R. 683 (H.L.)..................................4.35
Scott v. Crerar (1887), 14 O.A.R. 152, reversing (1886), 11 O.R. 541.................7.11
Scott v. Smith (1979), 36 N.S.R. (2d) 541, 13 R.P.R. 73, 64 A.P.R. 541 (C.A.),
    affirming 7 R.P.R. 10 (T.D.)..8.10, 8.87
Seabed and Subsoil of the Continental Shelf Offshore Nfld., Re, [1984] 1 S.C.R.
    86, 51 N.R. 362, 5 D.L.R. (4th) 385...5.64
Secretary of State for India v. Foucar & Co. (1933), 50 T.L.R. 241 (P.C.).............6.88
Sen v. Sask. College of Physicians and Surgeons (1969), 69 W.W.R. 201, 6
    D.L.R. (3d) 520 (Sask. C.A.)...1.48
Servos v. Stewart (1907), 15 O.L.R. 216 (H.C.)..........................6.62, 6.68
Shaw v. R., [1980] 2 F.C. 608 (T.D.)....................5.32, 5.35, 5.54, 5.76, 5.77, 5.87
Sherren v. Pearson (1887), 14 S.C.R. 581..2.92
Shey v. McHeffey (1868), 7 N.S.R. 350 (S.C.)..................................8.54
Sim E. Bak v. Ang Yong Huat, [1923] A.C. 429 (P.C.)............................5.21
Simpson Sand Co. v. Black Douglas Contractors Ltd., [1964] S.C.R. 333, 45
    D.L.R. (2d) 19..5.18
Smith v. Anderson, 16 M.P.R. 287, [1942] 2 D.L.R. 242 (N.S. C.A.)..............8.35
Smith v. Renwick (1882), 3 L.R. (N.S.W.) 398 (Sup. Ct. F.C.)......................6.81
Smith v. Smith, [1952] 2 S.C.R. 312, [1952] 3 D.L.R. 449........................7.68
Snowball v. Ritchie (1888), 14 S.C.R. 741, reversing 26 N.B.R. 258................8.41
South Australia State v. Victoria State, [1914] A.C. 283, 83 L.J.P.C. 137, 110
    L.T. 720, 30 T.L.R. 262, 18 C.L.R. 115, 20 A.L.R. 74 (P.C.), affirming
    (1911), 12 C.L.R. 667, 17 A.L.R. 206 (H.C.).....................4.32, 4.44, 8.30
Spearwater v. Seaboyer (1984), 65 N.S.R. (2d) 280, 147 A.P.R. 280 (T.D.)........8.6, 8.53
Stafford v. Bell (1881), 6 O.A.R. 273 (C.A.).....................................9.14
Steeper v. Harding (1884), 24 N.B.R. 143 (C.A.)....................8.40, 8.41, 8.67
Steeves v. Downey (1972), 4 N.B.R. (2d) 646 (S.C.)..............................8.14
Stevens v. Stevens, 2 M.P.R. 209, [1930] 3 D.L.R. 762 (N.S. S.C.).................8.67
Stollmeyer v. Trinidad Lake Petroleum Co., [1918] A.C. 485, 87 L.J.P.C. 77
    (P.C.)..6.61
Stover v. Lavoia (1906), 8 O.W.R. 398 (H.C.), affirmed 9 O.W.R. 117 (C.A.).....6.51, 6.68
Stuart v. Bank of Montreal (1909), 41 S.C.R. 516...............................1.11
Sullivan v. Lawlor (1981), 45 N.S.R. (2d) 325, 86 A.P.R. 325 (C.A.)............8.32, 8.38
Sutherland v. Campbell (1923), 25 O.W.N. 409 (C.A.)...........................8.51
Swadron v. North York (1985), 8 O.A.C. 204 (Div. Ct.).........................7.70
Taylor v. Pac Petroleums Ltd. (1977), 6 A.R. 200 (Dist. Ct.).....................11.53
Tellier v. R., [1967] 209 Que. S.C..10.284
Thomas v. Thomas, [1938] 4 D.L.R. 566, 13 M.P.R. 280 (N.B. S.C.)..............3.15

Thompson & Purcell Surveying Ltd. v. Burke (1977), 39 N.S.R. (2d) 181, 71
A.P.R. 181 (Co. Ct.)..8.14
Throop v. Cobourg & Peterboro·Ry. (1856), 5 U.C.C.P. 509, affirmed (1859),
2 O.A.R. 212n (C.A.)...6.69
Tickle v. Brown (1836), 111 E.R. 826..2.95
Tourigny v. Park, [1959] R.P. 385...10.232, 10.278
Tuckwell v. Guay, 34 D.L.R. 106, [1917] 1 W.W.R. 1229, affirmed 37 D.L.R.
805, 27 Man. R. 529 (C.A.)...8.58
Turnbull v. Saunders (1921), 60 D.L.R. 666 (N.B. C.A.)...5.74
Turta v. C.P.R. Co. and Imperial Oil Ltd. (1952), 5 W.W.R. (N.S.) 529, affirmed
[1953] 4 D.L.R. 87, 8 W.W.R. (N.S.) 609 (Alta. C.A.), which was affirmed
[1954] S.C.R. 427, [1954] 3 D.L.R. 1, 12 W.W.R. (N.S.) 97...............3.35, 3.46
Ukranian Greek Orthodox Church v. Independent Bnay Abraham Sick Benefit
and Free Loan Assoc. and Riverside Cemetery, 29 W.W.R. (N.S.) 97, [1959]
20 D.L.R. (2d) 363 (Man. C.A.)..3.36
Vancouver Improvement Co., Re (1893), 3 B.C.R. 601 (C.A.).....................8.08
Van Koughnet v. Denison (1885), 11 O.A.R. 699..7.32
Watcham v. Att.-Gen. of East Africa Protectorate, [1919] A.C. 533, [1918-
19] All E.R. Rep. 455 (P.C.)..4.35, 8.46
Watson v. Jackson (1914), 31 O.L.R. 481, 19 D.L.R. 733 (C.A.)....................4.74
Weston v. Blackman (1917), 12 O.W.N. 96 (C.A.).....................8.69, 8.71, 11.63
Wilbur v. Tingley, 24 M.P.R. 175, [1949] 4 D.L.R. 113 (N.B. C.A.)..................8.39
Williams v. Pickard (1908), 17 O.L.R. 547 (C.A.)...........................6.44, 6.52
Williams v. Salter (1912), 23 O.W.R. 34 (C.A.)..................................6.13
Wilton v. Murray (1897), 12 Man. R. 35 (Q.B.).................................5.03
Winchester v. N. Rattenbury Ltd., [1953] 3 D.L.R. 660, 31 M.P.R. 69 (P.E.I.
C.A.)...3.14
Winter v. Keating & Gillis (1978), 24 N.S.R. (2d) 633, 35 A.P.R. 633 (C.A.).........3.13
Wood v. LeBlanc (1904), 34 S.C.R. 627..2.93
Woodberry v. Gates (1846), 3 N.S.R. 255 (C.A.)............................8.36, 8.43

Table of Statutes

References are to paragraphs of the text.

UNITED KINGDOM

Act of Union, 1840, 3 & 4 Vic., c. 35...10.07
Canada (Ontario Boundary) Act, 1889, 52 & 53 Vic., c. 28......................4.31
Commissioner of Sewers Act, 1531, 23 Hen. 8, c. 5.............................6.02
Land Transfer Act, 1875, 38 & 39 Vic., c. 87..................................3.96
Limitations Act, 1623, 21 Jac. 1, c. 16.......................................4.80
Prescription Act, 1832, 2 & 3 Will. 4, c. 71..................................2.55
Quebec Act, 1774, 14 Geo. 3, c. 83...............................6.07, 10.07, 10.307
Real Property Limitation Act, 1833, 3 & 4 Wil. 4, c. 27.......................4.80
Registry Act, 1865, 29 Vic., c. 24..3.92
Statute of Frauds, 1677, 29 Cha. 2, c. 3..................2.49, 4.25, 4.138, 8.04, 8.05
Statute of Uses, 1535, 27 Hen. 8, c. 10.......................................2.17
Statute of Westminister I, 1275, 3 Edw. 1.....................................2.54

CANADA

Access to Information Act, R.S.C. 1985, c. A-1...............................12.23
Act Respecting Registry Offices and Privileges and Hypothecs in Lower Canada, 1860, 23
    Vic., c. 59...10.138, 10.141, 10.148
Act respecting the Management of Public Lands adjoining non-navigable streams and
    lakes in the Provinces of Quebec, 1883, 46 Vic., c. 8......................10.310
Act Respecting the Profession of Land Surveyors in the Province of Manitoba, 1881, 44
    Vic., c. 29..11.14
Act respecting the sale and management of the Public Lands, S.U.C. 1860, 23 Vic., c. 2
    s. 38..4.08
Act to ascertain and establish on a permanent footing, the Boundary Lines of the different
    Townships of this Province, S.U.C. 1798, 38 Geo. 3, c. 1.....................4.24
Act to provide for the survey reservation and protection of lumber lands, 1884, 47 & 48
    Vic., c. 7
        s. 4...6.80
Act to provide for the disposal of the Public Lands in this Province, and for other purposes
    therein mentioned, 1837, 7 Wm. 4, c. 118....................................4.11
Act to repeal . . . "An Ordinance concerning Land Surveyors, and the Admeasurement of
    Lands" and also to extend the provisions of . . . "An Act to ascertain and establish on

Table of Statutes

a permanent footing the Boundary Lines of the different Townships of this Province",
S.U.C. 1818, 59 Geo. 3, c. 14 .... 4.24 4.57
Artic Waters Pollution Prevention Act, R.S.C. 1985, c. A-12 .... 5.96
British North America Act, 1867, 30 & 31 Vic.,
    c. 3 .... 1.18, 1.22, 1.55-1.60, 1.62, 1.66, 1.74, 4.18, 4.19, 10.07
        s. 91 .... 1.62, 1.63, 4.18, 10.283
            91(24) 2.23
            92 .... 4.18, 10.281, 10.282, 11.11
            92(13) 2.23, 20.124
            108 .... 4.18, 10.283
            109,117 .... 4.18, 10.280
            125 .... 10.313
Canada Act, (U.K.), 1982, c. 11 .... 1.66
Canada Evidence Act, R.S.C. 1985, c. C-5 .... 7.17
Canada Lands Surveys Act, S.C. 1976-77, c. 30 .... 11.49, 11.68
Canada Lands Surveys Act, R.S.C. 1985,
    c. L-6 .... 4.23, 4.47, 4.52, 5.68, 10.167, 11.15, 11.31
        ss. 24(1), (2), 26(2), 33(1), (2) 8.80
           48-60 .... 8.74
           49 .... 8.75
           50(2), (3), 51, 52(1), (2), (3), 53, 54 .... 8.75
           55(1)-(3), 56-59 .... 8.75
Canadian Bill of Rights, R.S.C. 1970, Appendix III 1.68
Canadian Charter of Rights and Freedoms, Part I of the *Constitution Act, 1982,* being
    Schedule B of the *Canada Act 1982* (U.K.), 1982, c. 11 ..... 1.05, 1.68, 1.69, 1.71, 2.14,
                                                                                        11.01, 11.44
        s. 24(1) 1.73
           25 .... 2.23
Constitution Act, 1867 (U.K.), 30 & 31 Vic., c. 3 .... 5.60, 5.63
Constitution Act, 1982 [en. by the *Canada Act 1982* (U.K.), c. 11] 1.69, 10.01
        s. 38 .... 1.67
           92 .... 11.11
Constitutional Act, 1791, 31 Geo. 3, c. 31 .... 6.07, 10.07, 10.286
Colonial Laws Validity Act, 1865, 28 & 29 Vic., c. 63 .... 1.56
Criminal Code, R.S.C. 1985, c. C-46 .... 11.56
Dominion Lands Act, S.C. 1872, 35 Vic., c. 23 .... 3.83, 4.23, 4.47, 6.17, 6.21
Dominion Lands Act, S.C. 1908, 7-8 Edw. 7, c. 20 .... 4.23, 4.118
        s. 57 .... 4.56
           62 .... 4.141
Income Tax Act, R.S.C. 1952, c. 148, as amended by S.C. 1970-72, c. 63 .... 1.19
Inquiries Act, R.S.C. 1985, c. I-11 .... 8.75
Law of England, S.U.C. 1792, 32 Geo. 3, c. 1 .... 6.07
Navigable Waters Protection Act, R.S.C. 1985, c. N-22 .... 10.167
North-west Irrigation Act, S.C. 1894, c. 30 .... 6.27, 6.83
        s. 5 .... 6.22
           7 .... 6.54
North-west Territories Act, S.C. 1886, c. 25
        s. 3 .... 6.20
Ordinance to Prescribe and Regulate the Registry of Titles, 1841, 4 Vic.,
    c. 30 .... 10.138, 10.139

Table of Statutes   xlvii

Privacy Act, R.S.C. 1985, c. P-21
    s. 2 .... 12.30
Railway Act, R.S.C. 1985, c. R-3 .... 10.167
Seigniorial Act, 1854, 18 Vic., c. 3 .... 10.138, 10.140
Statute of Westminister, 1931, 22 & 23 Geo. 5, c. 4 .... 1.56
Supreme Court Act, R.S.C. 1985, c. S-26 .... 1.74
Territorial Lands Act, R.S.C. 1985, C. T-7
    s. 9 .... 5.56
Territorial Sea and Fishing Zones Act, R.S.C. 1985, c. T-8 .... 5.96
Veterans Land Act, R.S.C. 1970, c. V-4 .... 8.10

ALBERTA

Arbitration Act, R.S.A. 1980, c. A-43
    s. 1 .... 11.07
Architects Act, R.S.A. 1980, c. A-44 .... 11.17
Boundary Surveys Act, R.S.A. 1980, c. B-10 .... 8.30
Guarantees Acknowledgment Act, R.S.A. 1980, c. G-12 .... 9.55
Land Surveyors Act, S.A. 1981, c. L-4.1 .... 11.09, 11.17, 11.27
Land Titles Act, R.S.A. 1980, c. L-5 .... 3.46
    s. 1(f) 3.56
        66 .... 3.45
        135.1 .... 3.79
        173(1)(e) 3.46
        177(1)(a) 3.46
        180 .... 3.60
        183 .... 8.58
Planning Act, R.S.A. 1980, c. P-9
    s. 120 .... 3.66
Public Lands Act, S.A. 1931, c. 43
    s. 8 .... 6.22
Public Lands Act, R.S.A. 1980, c. P-30
    s. 3 622
       4 .... 6.22
Surface Rights Act, S.A. 1983, c. S-27.1
    s. 14 .... 11.54
Surveys Act, R.S.A. 1980, c. S-29 .... 3.62
    ss. 22-25 .... 8.74
        31(1), 42(1) 8.17
        62-68 .... 8.74
        89 .... 11.58
Water Resources Act, R.S.A. 1980, c. W-5
    s. 3 .... 6.23

BRITISH COLUMBIA

Land Act, R.S.B.C. 1979, c-214 .... 5.56
    s. 1 .... 5.69, 5.70, 8.30
        52, 64(4), 76 .... 5.70
        78(1) 8.34
Land Survey Act, R.S.B.C. 1979, c. 216
    s. 5(1) 8.17

xlviii   Table of Statutes

Land Surveyors Act, R.S.B.C. 1979, c. 217
    s. 6..11.19
Land Title Act, R.S.B.C. 1979, c. 219
    s. 95..5.58
        322-364..8.74
        365..8.74, 11.57
Land Title Inquiry Act, R.S.B.C. 1979, c. 220........................8.10, 8.24, 8.77
Municipal Act, R.S.B.C. 1979, c. 290.....................................8.83
    ss. 887-890, 893, 899, 901, 908..8.83
Privacy Act, R.S.B.C. 1979, c. 336......................................12.30
Property Law Act, R.S.B.C. 1979, c. 340..................................8.60
Statute of Frauds, R.S.B.C. 1979, c. 393
    s. 1..8.04

MANITOBA

Boundary Lines and Line Fences Act, R.S.M. 1970, c. B-70.......................8.84
    s. 2...8.24, 8.84
       12..8.84
Crown Lands Act, R.S.M. 1970, c. C-340
    s. 5..6.22
Law of Property Act, R.S.M. 1970, c. L-90
    s. 27...8.58
       28..8.58, 8.60
Municipal Board Act, R.S.M. 1970, c. M-240................................8.83
Privacy Act, R.S.M., c. P-125..12.30
Real Property Act, R.S.M. 1970, c. R-30...................................8.58
    s. 2(1)(b)..3.56
       50.1..3.68
       59..3.45
       62(1)(d)..3.46
       66(1),(2)...3.36
       144.1...3.79
       161...3.60
Registry Act, R.S.M. 1987, c. R-50
    s. 16...3.17
       41..3.10
       55-56...3.12
Special Surveys Act, R.S.M. 1970, c. S-190..........................4.152, 8.74
Water Rights Act, S.M. 1982-83-84, c. 25
    s. 2..6.23

NEWFOUNDLAND

Limitation of Actions (Realty) Act, R.S.N. 1970, c. 207
    s. 16...8.10
Privacy Act, S.N. 1981, c. 6..12.30
Quieting of Titles Act, R.S.N. 1970, c. 324........................8.24, 8.77, 8.78
    s. 8(1)...8.86

NEW BRUNSWICK

Crown Lands and Forests Act, S.N.B. 1980, c. C-38.1
    s. 1 ....5.69, 6.46
    15-18 ....6.80
Expropriation Act, R.S.N.B. 1973, c. E-14
    s. 17(2) 2.102
Highways Act, R.S.N.B. 1973, c. H-5 ....2.57
    s. 30(4) 8.80
Land Surveyors Act, S.N.B. 1954, c. 97
    s. 40 ....11.55
    43 ....11.61
Land Titles Act, S.N.B. 1981, c. L-1.1 ....3.81, 8.74
    s. 24(21) 3.59
    30 ....3.36
    31 ....3.79
Limitations of Actions Act, R.S.N.B. 1973, c. L-8 ....2.91, 2.92
    s. 29 ....2.90
Property Act, R.S.N.B. 1973, c. P-19
    s. 20 ....2.108
Quieting of Titles Act, R.S.N.B. 1970, c. Q-4 ....8.24, 8.77
Registry Act, R.S.N.B. 1973, c. R-6
    s. 19(1) 3.13
    32 ....3.09
    50 ....3.09
    71(1)(a) 3.09
Statute of Frauds, R.S.N.B. 1973, c. S-14
    s. 8 ....2.49

NOVA SCOTIA

Angling Act, R.S.N.S. 1967, c. 9 ....8.11
Beaches and Foreshores Act, R.S.N.S. 1967, c. 19 ....5.48
Beaches Preservation and Protection Act, S.N.S. 1975, c. 6
    s. 3(a) 5.32
Crown Land Act, S.N.S. 1987, c. 5
    s. 15 ....8.34
Fences and Impounding of Animals Act, R.S.N.S. 1967, c. 104
    s. 12 ....8.42
Land Surveyors Act, S.N.S. 1977, c. 13 ....5.68
    s. 16 ....11.53
Land Titles Act, S.N.S. 1903-4, c. 47 ....8.74
Land Titles Clarification Act, R.S.N.S. 1967, c. 162 ....3.81, 8.10, 8.74
    s. 2(1),(3), 3(1)-(3), 4(1)-(4), (6), (7), 6(1), (2), (4), (5) 8.79
Public Highways Act, R.S.N.S. 1967, c. 248
    ss. 14(2), (3), 16 ....8.34
Public Records Disposal, R.S.N.S. 1976, c. 254 ....12.27
Quieting of Titles Act, R.S.N.S. 1967, c. 259 ....8.24, 8.77
Registry Act, R.S.N.S. 1967, c. 265
    ss. 10,26 ....3.09
    25 ....3.16

l Table of Statutes

Towns Act, R.S.N.S. 1967, c. 309
    s. 147 .... 8.34
Water Act, R.S.N.S. 1967, c. 335 .... 5.20, 5.58
    s. 1 .... 5.19
        1(k) 5.03
Water Amendment Act, S.N.S. 1972, c. 58
    s. 1(4) 5.03

ONTARIO

Bed of Navigable Waters Act, S.O. 1911, c.6 .... 6.12
Beds of Navigable Waters Act, S.O. 1951, c. 5 .... 6.32, 6.47
Beds of Navigable Waters Act, R.S.O. 1980, c. 40 .... 6.31, 6.89
    s. 1 .... 6.16
Boundaries Act, R.S.O. 1980, c. 47 .... 4.152, 8.23, 8.24, 8.74, 8.81, 8.82
    ss. 2, 3(1), (2), 5(1), (2), 6, 7, 8(1), (2), 9(1), 10, 12(1), (2), 13, 15, 16(1), 17 .. 8.81
Certification of Titles Act, S.O. 1958, c. 9 .... 3.95
Certification of Titles Act, R.S.O. 1980, c. 61 .... 4.81, 8.24, 8.77
Conveyancing and Law of Property Act, R.S.O. 1980, c. 90
    s. 37(1) 8.58
Courts of Justice Act, S.O. 1984, c. 11 .... 1.75
    s. 208 .... 4.81
Evidence Act, R.S.O. 1980, c. 145 .... 7.17
Investigation of Titles Act, S.O. 1929, c. 41 .... 3.93
Judicature Act, 1881 (Ont.), c. 5 .... 1.34
Land Registration Reform Act, S.O. 1984, c. 32 .... 3.96
Land Titles Act, R.S.O. 1980, c. 230 .... 3.17, 3.30, 3.91, 4.79, 4.87
    s. 141(2) 8.23
Limitations Act, R.S.O. 1980, c. 240 .... 4.80
Lord's Day Act, R.S.O. 1897, c. 246 .... 1.65
Municipal Boundary Negotiations Act, S.O. 1981, c. 70 .... 4.04
Navigable Waters Act, R.S.O. 1970, c. 40
    s. 1 .... 5.19
Planning Act, R.S.O. 1980, c. 379 .... 8.33, 8.81
Property and Civil Rights Act, R.S.O. 1980, c. 395 .... 6.07
Public Lands Act, R.S.O. 1980, c. 413 .... 4.08, 4.11, 6.13
    s. 7 .... 4.29
       8 .... 4.56
Quieting Titles Act, S.O. 1984, c. 427 .... 4.81, 8.24
Registry Act, R.S.O. 1980, c. 445 .... 4.79, 4.81
    ss. 51, 62 .... 3.94
       105 .... 3.93
       105(1) 8.10
       108 .... 3.17, 3.94
       109 .... 3.17
Road Access Act, R.S.O. 1980, c. 457 .... 4.77
Sale of Goods Act, R.S.O. 1980, c. 462 .... 9.55
Statute Law Amendment Act, S.O. 1940, c. 28 .... 6.47
    s. 3 .... 6.32, 6.37
Statute of Frauds, R.S.O. 1980, c. 481 .... 4.25, 4.121, 9.55

Table of Statutes li

Surveys Act, S.O. 1913, c. 33
    s. 2..6.13
Surveys Act, R.S.O. 1980, c. 493.......................4.117, 4.142, 4.144, 6.81, 8.82
    s. 1..4.62
        1(g)...4.03
        1(1)...4.144
        2..4.61
        9..674
        11..6.13, 6.23
        13(2), 17(2), 24(2), 31(2), 37(2), 44(1)..........................8.17
        58...4.62
Surveyors Act, R.S.O. 1960, c. 389
    s. 36(2)..11.45
Surveyors Act, R.S.O. 1980, c. 492..............................1.28, 1.41, 1.53
Surveyors Act, S.O. 1987, c. 6.....................................9.26
Territorial Division Act, R.S.O. 1980, c. 497..........................4.04
Vendors and Purchasers Act, R.S.O. 1980, c. 520
    s. 1...8.86

PRINCE EDWARD ISLAND

Investigation of Titles Act, R.S.P.E.I. 1974, c. I-7...........................8.77
    s. 2(1)..8.10
Land Titles Act, R.S.P.E.I. 1974, c. L-6.............................8.74
Quieting Titles Act, R.S.P.E.I. 1974, c. Q-2..........................8.24
Registry Act, R.S.P.E.I. 1974, c. R-11
    s. 10..3.09
        14..3.09
        43..3.13
Statute of Frauds, R.S.P.E.I. 1974, c. S-6............................8.04

QUEBEC

Act respecting the Lands in the Public Domain, R.S.Q., c. T-8.1
    art. 45..10.244
Act to amend and consolidate the laws relating to fisheries, S.Q. 1888, c. 17........10.310
Act to amend the Quebec Fish and Games Laws, S.Q. 1919, c. 31.................10.310
Act to Preserve Agricultural Lands, R.S.Q., c. P-41.1..............10.56, 10.233, 10.255
Act to promote the reform of the cadastre in Quebec, S.Q. 1985, c. 22.............10.151
Cadastre Act, R.S.Q., c. C-1...............10.143, 10.145, 10.146, 10.153, 10.155, 10.166
    art. 7...10.158
Cities and Towns Act, R.S.Q., c. C-19......................................10.167
Civil Code of Quebec
    art. 9...10.189
        13...10.23
        225..10.129
        304, 323...10.26
        374..10.49
        376-378..10.50
        379, 380..10.50, 10.51
        382..10.50
        383, 387, 388..10.51

400	10.288, 10.290, 10.291, 10.293, 10.304, 10.312
406	10.53
407	10.55
408	10.54, 10.300
414	10.54, 10.297
415	10.62, 10.76
420	10.300, 10.301
421	10.302
422	10.301
423	10.303
424, 425, 426	10.304
427	10.300, 10.305
4416-442p	10.61
443, 446	10.65
463	10.43
464	10.43
470	10.44
472	10.43
481	10.66
491, 493, 494	10.45
499	10.45, 10.72
500	10.73
501	10.74, 10.296
502	10.74, 10.298
503, 505	10.74
504	8.91, 10.74, 10.186, 10.190
506, 507	10.75
5.10	10.76, 10.78
511	10.78
514	10.114
518	10.46, 10.76, 10.77
519	10.46
520	10.46, 10.74, 10.248
523-527	10.78
528-531	10.79
532	10.79
533, 534	10.80, 10.81, 10.245
536, 537	10.80, 10.82, 10.245
538	10.80, 10.81, 10.82, 10.245
539	10.296
540, 543	10.83
541, 542, 544	10.83, 10.84
545	10.85, 10.90
546	10.85, 10.86
547	10.85, 10.87
548	10.85, 10.88
548	10.85
549	10.85, 10.89, 10.120
550, 551	10.85, 10.89
552, 553, 554, 555, 558	10.85, 10.90
556	10.85

Table of Statutes liii

557, 559, 560, 561	10.85, 10.91
562	10.85, 10.89, 10.91
563	10.85, 10.87, 10.91
564-566	10.85, 10.91
567	10.70
569, 570, 571, 580	10.71
579	10.70, 10.71
583	10.102
591	10.299
689	10.59
762	10.190, 10.192, 10.193
768	10.218
770	10.117, 10.129
771, 772	10.118
776	10.53, 10.126
805-807	10.131
983	10.14
984	10.25
985, 986, 987	10.26
989	10.33
991	10.28, 10.217
993	10.30
994-1000	10.31
1001, 1012	10.32
1007	10.39
1013-1022	10.34
1017	10.10
1040c	10.32
1053	10.38, 10.39
1057	10.14
1058	10.15
1069	10.21
1076	10.47
1079	10.16, 10.17
1089	10.18, 10.50
1100, 1105, 1106	10.19
1131	10.20
1207, 1208	10.169
1238, 1239	10.208
1245	10.205
1272, 1292	10.47
1425	10.47
1472	10.24
1980-1982	10.92
1983	10.92, 10.94
1984-1993	10.92
1994	10.92, 10.95
1995-2008	10.92
2009	10.92, 10.96, 10.184
2010-2015	10.92
2016, 2017	10.92, 10.97

2018, 2019	10.92, 10.98
2020-2029	10.92
2030	10.92, 10.99
2031-2036	10.92
2037, 2040	10.92, 10.101
2038, 2039	10.92
2041	10.92
2042	10.92, 10.101, 10.147, 10.162, 10.234
2043	10.92
2044	10.92, 10.101
2045-2081a	10.92
2082	10.166, 10.168, 10.182
2083, 2085	10.166, 10.183
2084	10.96, 10.166, 10.168, 10.184
2086-2091	10.166
2092	10.166, 10.178
2093-2097	10.166, 10.168, 10.183
2099	10.166, 10.183
2100-2116	10.166
2116a, 2116b	10.88, 10.89, 10.166
2117-2125	10.166
2126	10.166, 10.178
2127	10.166, 10.168
2128, 2129	10.166
2129a	10.166, 10.167
2130	10.101, 10.166, 10.184
2131	10.166, 10.169, 10.175, 10.176
2132-2135	10.166, 10.176
2136-2145	10.166, 10.179
2146, 2147	10.166
2148	10.166, 10.179
2149-2157a	10.166
2157b	10.166, 10.179
2158	10.166
2159	10.166, 10.179
2160	10.166
2161a, b	10.166, 10.170, 10.171, 10.172
2161c	10.166, 10.174
2162-2164	10.166
2164a	10.166, 10.174
2165	10.166
2166	10.145, 10.148, 10.166
2167	10.145, 10.166
2168	10.145, 10.147, 10.154, 10.161, 10.162, 10.166
2169	10.145, 10.147, 10.161, 10.166
2170	10.145, 10.161, 10.166, 10.173
2171	10.145, 10.147, 10.166
2172	10.145, 10.154, 10.161, 10.166
2173	10.145, 10.161, 10.166
2174	10.145, 10.158, 10.159, 10.163, 10.166, 10.207

2174a	10.145, 10.157, 10.158, 10.166
2174b	10.145, 10.150, 10.156, 10.166, 10.179, 10.181
2175	10.145, 10.154, 10.155, 10.166
2176	10.145, 10.150, 10.166
2176a	10.145, 10.166
2177-2182	10.166
2183	10.102, 10.119
2183a	10.102, 10.131
2184-2191	10.102
2192	10.102, 10.103, 10.105, 10.106
2193	10.102, 10.109, 10.123, 10.206
2194	10.102, 10.106, 10.109, 10.116
2195	10.102, 10.107
2196	10.102, 10.114
2197	10.102, 10.112
2198	10.102, 10.112, 10.113
2199	10.102, 10.110
2200	10.102, 10.127
2201	10.102, 10.121
2202	10.102, 10.125
2203	10.102, 10.105, 10.107, 10.120
2204	10.102
2205	10.102, 10.108
2206-2212	10.102
2213	10.102, 10.121, 10.312
2214, 2215	10.102
2216, 2217	10.102, 10.121
2218, 2219	10.102
2220, 2221	10.102, 10.121
2222, 2223, 2224	10.102, 10.111, 10.129
2225	10.102, 10.111
2226, 2227	10.102, 10.111, 10.129
2228-2231	10.102, 10.111
2232, 2233	10.102, 10.130
2234-2239	10.102
2240	10.102, 10.128
2241	10.102
2242	10.102, 10.123
2243-2250	10.102
2251	10.102, 10.124
2252	10.102
2253	10.102, 10.125
2254	10.102
2255	10.102, 10.128
2256-2270	10.102
Code of Civil Procedure, R.S.Q. 1977, c. C-25	10.09, 10.210, 10.230
art. 234	10.195
319	10.204
420	10.200
762	10.190, 10.192, 10.229

lvi   Table of Statutes

    763 .... 10.196, 10.197, 10.209, 10.210
    765 .... 10.212
    767 .... 10.212, 10.214
Consumer Protection Act, R.S.Q. 1977, c. P-40 .... 10.09, 10.32
Cultural Property Act, R.S.Q. 1977, c. B-4 .... 10.56
Land Surveyors Act, R.S.Q. 1977, c. A-23 ... 10.152, 10.232, 10.265, 10.267, 10.269, 10.272, 11.26
    art. 7 .... 11.40
    10 .... 11.46
    13(e), 56(2) 10.272
    24 .... 11.03, 11.22
    34 .... 10.220, 10.277
    47, 48 .... 10.201
    50 .... 10.204
    51 .... 10.212
    51(4) 10.215
    52 .... 10.216, 10.229
    52(1) g 10.214
    52(j) 10.215
    53 .... 10.229
    53(4) 8.97
    54 .... 10.229
    56 .... 10.257, 10.267
    57, 58 .... 10.273
Loi Sur l'aménagement et l'urbanisme, R.S.Q., c. A-19.1 .... 10.56
Mining Act, R.S.Q., c. M-13 .... 10.167
Railway Act, R.S.Q., c. C-14 .... 10.167
Régie du logement, R.S.Q., c. R-8.1 .... 10.56
    arts. 45-56 .... 10.256
Registry Office Act, R.S.Q., c. B-9 .... 10.151, 10.166
Survey Act, R.S.Q., c. A-22
    art. 1 .... 10.227
Watercourses Act, R.S.Q., c. R-13 .... 10.167
    art. 5 .... 10.306
    1524 .... 10.314

SASKATCHEWAN

Improvements Under Mistake of Title Act, R.S.S. 1978, c. I-1
    s. 2 .... 8.58
Land Surveys Act, R.S.S. 1978, c. L-4
    s. 24(1) 8.17
    60-94 .... 8.74
Land Surveyors Act, R.S.S. 1978, c. S-27
    s. 41(6) 11.45
Land Titles Act, R.S.S. 1978, c. L-5 .... 3.41, 11.40
Privacy Act, R.S.S. 1978, c. P-24 .... 12.30
Provincial Lands Act, R.S.S. 1978, c. P-31
    s. 12 .... 6.22
Water Rights Act, R.S.S. 1978, c. W-8
    s. 7 .... 6.23

1

The Canadian Legal System

*Brian Campbell, LL.B.**

§1.01 The surveying profession is one which, more than most other professions, comes into frequent contact with the law, from specific concepts such as property law, land registration systems and boundaries to the more philosophical legal concepts implicit in a self-governing profession. As such, this introductory chapter is intended to provide a general overview of the law in the context of the concepts which shape it, its basic structure, and principles of interpretation.

INTRODUCTION — THE NATURE OF LAW

§1.02 Laws are passed by the elected representatives of the public: Parliament and Legislature, both federally and provincially. However, it falls to the judiciary to interpret those laws and apply them to the particular facts of the case. In so doing, and in arriving at a decision which is, in the opinion of the judge, "fair", a judge is influenced to a greater or lesser extent by his or her view of the judge's role in the dispute. Legal analysts have attempted to classify the various schools of judicial thought reflected in judgments and to define them. Three such schools of thought or philosophies have generally been accepted: legal positivism, natural law, and realism.

---

* Ms. Carol A. Street was instrumental in putting together the information necessary for the first draft of the chapter. Her assistance is gratefully acknowledged.

§1.03 Positivism could perhaps be characterized as the purest application of the law. That is, those judges exercising a positivist philosophy seek to apply the law as derived from previous cases and statutes with no consideration for the law as it ought to be. In other words, this school of thought maintains that there must be a clear separation between the application of the law and influences of personal morality. Legal validity should depend upon legal criteria and not moral or ethical considerations. The rationale for such a view is that moral or ethical considerations which might affect a judgment are subjective; subjective criteria introduce uncertainty and inconsistency into legal decisions. If law is to provide a structure and order to society, society must be able to ascertain and know what behaviour is permissible.

§1.04 In reality, obviously, such a clear separation is impossible. As society becomes more complex, societal values arguably should be considered in a decision; if a judge's decision will have implications for society as a whole, then it may be beneficial for a judge in giving a decision to review those values as he or she interprets them and indicate how they are being considered. For example, in sentencing a convicted criminal a strict positivist view might simply review the sentence suggested by the statute, the cases where previous offenders had been sentenced, and provide a sentence which conformed with these two sources. However, if the judge took into account the particular offender's circumstances, prospects of rehabilitation, prior offences and any other factors deemed relevant, he or she would be deviating from a strictly positivist view.

§1.05 Legal realism, on the other hand, recognizes that there are other components intrinsic to judicial decision making. Legal realism extends positivism by including criteria patterns of judicial behaviour such as "the personality of the judge, the ability of the judge to distinguish, ignore and re-interpret precedent, the political or policy-making role of judges based upon a subjective perception of justice and equity by a particular judge, and the political, social and economic substructure that defines the context of a given case."[1] For example, journalists have recently reported concerns that a judge's political persuasion and philosophy will affect decisions rendered in *Charter of Rights* cases.[2] As a result, it has been suggested that hearings should be held prior to a judge's appointment, to ascertain his or her political beliefs — a practice common to the American system. To a legal positivist, such a hearing would be irrelevant; to a legal realist it would not.

---

1 Gerald L. Gall *The Canadian Legal System*, (Toronto: Carswell, 1983) at 11.
2 *Canadian Charter of Rights and Freedoms*, Part I of the *Constitution Act, 1982*, being Schedule B of the *Canada Act 1982* (U.K.), 1982, c.11.

§1.06 Realism and positivism are similar in that both look on law as the expression of the will of the state. However, according to positivists, the will of the state is conveyed through Parliament; according to realists, it is conveyed through the courts.[3]

§1.07 The third school of judicial thought is known as natural law. The proponents of natural law argue that justice cannot be based upon positive law alone. Rather, they believe that there are objective moral values which may be given a positive content. This amounts to a denial of the argument that there can be a rigid separation between what the law is and what it ought to be. Natural law has its roots in the concept that law consists of rules made in accordance with reason and nature.[4] The principles are couched in reason and common sense. By contrast, the principles of positivism can only be found in legal sources such as statutes.

§1.08 No one of these varying schools of judicial thought has managed to gain unchallenged acceptance as the pre-eminent modern legal philosophy. Each of the three has its share of proponents which in itself is a source of uncertainty in the law.

PRECEDENT AND STARE DECISIS

§1.09 As will be discussed further, the British common law tradition is based on judicial decisions in individual cases. As an example, over a period of time various cases will come before the courts which relate to real estate transactions which have failed to close. Each party will naturally blame the other. In deciding such cases, the courts will, by necessity, develop criteria for determining who was at fault: Was either party prepared to complete the transaction on the scheduled day? Was the property which the vendor tendered on closing the same property that the purchaser had actually agreed to buy? Was the vendor in a position to transfer a clear legal title to the purchaser on the scheduled day? Thus, over a number of cases, dealing with the same legal issue but different facts, principles will develop which guide not only the courts in making decisions but also society in that there is some certainty as to the acceptable boundaries of behaviour.

§1.10 Obviously, if such principles are to develop, each decision must in some way be binding on subsequent decisions. This concept is reflected in the legal doctrines of precedent and *stare decisis*. The doctrine of

---

3 D.J. Fitzgerald, ed., *Salmond on Jurisprudence*, 12th ed. (London: Sweet & Maxwell, 1966) at 35.
4 *Ibid.* at 15.

precedent requires a judge, when adjudicating upon a case, to adhere to the decision in a previous case which contains a fact situation similar to that in the case presently before the judge. *Stare decisis*, on the other hand, provides that a decision of a higher court is binding upon cases involving similar facts that arise in lower courts of the same jurisdiction. For example, a decision by the Ontario Court of Appeal is binding upon the courts below it, in a case where the issues before the lower court are the same as those which were before the higher court. Essentially, these two doctrines work to curb judicial law-making, which is arguably undesirable due to the fact that judges are not elected representatives of the public and are therefore unaccountable for their actions.

§1.11 As with legal positivism, however, the concept and the reality do not always correspond. While these principles are necessary to curb judicial law-making and to introduce some level of certainty into the law, judges have been ingenious in overcoming these rules when they deem it necessary. For example, the highest courts in various jurisdictions are inconsistent as to whether they consider themselves to be bound by their own past decisions. The Ontario Court of Appeal has held that it considers itself bound by its previous decisions, unless the decision in question was made *per incuriam* (that is, through inadvertence). At one time, the Supreme Court of Canada was of the same view with respect to the effect of its previous decisions.[5] In recent years, however, the Supreme Court of Canada seems to have altered this stance to the point where it is now possible that the court will feel free under certain circumstances to reconsider previous decisions, even if they were not made through inadvertence. In any event, *stare decisis* does not require a lower court judge to abide by a decision by a judge of equal jurisdiction. Therefore, until an issue is decided by a higher court, there is no certainty and no leash on judicial decision making.

§1.12 A more commonly used tool which also avoids the restrictions imposed by these doctrines is the process of distinguishing cases. Obviously, it is unlikely that any two cases will be totally identical in their facts. If one case can be distinguished in some way from a previous case, then the judge is not bound by the previous decision.

§1.13 There are generally four methods by which a judge can distinguish cases:

(1) First, a judge may compare the material facts of the precedent case with those of the present case and conclude that they are sufficiently different that the precedent case is not binding. This is the method

---

5 *Stuart v. Bank of Montreal* (1909), 41 S.C.R. 516.

most often relied upon by judges to remove an unwanted decision from their consideration.

(2) Second, a judge may circumvent a precedent by construing that part of it which is objectionable as *obiter dicta*. The principle of a previous decision or precedent being binding in subsequent cases provides that only the *ratio* of the case is what is binding on future courts. The *ratio* is the essential part of the decision. Comments which are made by the judge which are not essential to the decision are referred to as *obiter dicta* and they are not binding on other courts.

(3) Third, a judge may decide that a precedent is not to be followed because it was decided *per incuriam*; for example, a judge might note that a previous court had not been referred to a particular case or fact, which undermines the importance of the decision.

(4) Finally, in rare cases, a court may simply ignore a binding precedent.

§1.14 It is not to be inferred from this discussion that judges ignore previous decisions and simply render decisions at will. Judges are by and large conscientious about reviewing previous decisions and rationalizing why they do or do not apply. While the doctrines of *stare decisis* and precedent are necessary to provide some consistency and certainty to the law, techniques have necessarily developed to distinguish cases to prevent the law from becoming rigid and inflexible.

SOURCES OF LAW

§1.15 In this context, sources of law refer to the direct means by which the law comes into existence. Principally, this occurs through one of two major sources. As has already been mentioned, in the common law system, the law develops by way of judicial decisions rendered in particular situations. To determine what the law is on any particular issue, one must refer to the case law. In a pure common law state, this is the sole source of the law.

§1.16 Canada, however, is not a pure common law state. In Canada, the other and more significant source of law is statute law. Statutes are those codes enacted by bodies which have been given legislative authority to do so. To a certain extent, statutes have been enacted in order to codify branches of the law which have been formulated and developed over a number of years by the case law. Increasingly, however, statutes also attempt to set out an exhaustive and comprehensive structure of a specific area of law. In other words, statutes attempt to not only codify the law as it has developed, but also the law as it should be in the future. Statute law is therefore able to correct an injustice which may have been created by the common law system. For example, a judge faced with a particular situation may be unable to distinguish it from a previous decision. The application of the previous decision to the facts of the present case may result in an injustice. The judge, however, may feel that in overruling the previous decision and coming to a different conclusion in the present decision he would in effect be creating law. As already mentioned, judges have no authority to create new law and a judge in such a situation will often note that it is the responsibility of the Legislature to correct or change the law so that such an injustice will not result. Finally, statutes are used to carry out social policies which the common law cannot. For example, modern social services such as unemployment insurance, medicare, and workers compensation were all brought into existence by statute.

§1.17 As a codification of existing law, statutes obviously facilitate access to the law. However, it is understandably difficult for any legislative body to foresee all situations or disputes that may arise with regard to a particular area of law. To the extent that the statute is not comprehensive and fails to deal with as many foreseeable situations as possible, the courts must again fall back on the common law in rendering decisions. Also, the language of statutes is often convoluted and complex. As such, the interpretation to be given to any particular section in a statute is often controversial. Case law therefore develops on the interpretation which should be given to a particular section in a statute.

§1.18 The Canadian legal system relies on both statute and case law to provide sources of law. However, due to the doctrine of parliamentary sovereignty, statute and case law are not on an equal footing. Parliamentary sovereignty means that Canadian Legislative Assemblies, both federal and provincial, have unlimited power to create, alter and repeal law, so long as they do not overstep their constitutional mandate as set out in the *British North America Act*[6] of 1867, which will be discussed more fully below.

RULES AND PRINCIPLES OF STATUTORY INTERPRETATION

§1.19 When drafting statutes, the Legislature must attempt to foresee all possible circumstances in which the law will be applied. The Legislature must then attempt to set out in written language how the law is to be applied in particular cases. Often, the subject of a statute is not easily formulated in words. For example, it has often been said that the *Income Tax Act*[7] should be drafted in the form of mathematical formulas rather than words. In any event, the result is that statutes frequently require a great deal of interpretation.

---

6 (U.K.), 30 & 31 Vict., c. 3 (now *Constitution Act, 1867*).
7 R.S.C. 1952, c. 148 as amended by S.C. 1970-72, c. 63.

§1.20 Certain approaches to statutory interpretation have developed. The following is meant to be a brief review of the primary principles; complete texts have been written which may be referred to for more specific information on and analysis of statutory interpretation.[8]

§1.21 The "literal" or "plain meaning" rule requires that if the precise words used are plain and unambiguous, the judge is bound to construe them in their ordinary sense, even though such a construction might lead to an absurdity or a manifest injustice.[9] Under the "plain meaning" rule the judge will only consider the actual words of the statute; the context of the entire statute or the subject matter of the statute will not be referred to.

§1.22 As was discussed in the context of precedent, judges are often able to avoid such rules as the "plain meaning" rule to arrive at a particular decision. A famous example is the *persons* cases of the late 1920's. In those cases, section 24 of the *British North America Act* was considered, which provided that qualified *persons* could be appointed members of the Canadian Senate. Five women asked the Supreme Court of Canada to determine whether the word *persons* included women. Had the Court applied the literal or plain meaning rule of statutory interpretation, they would of necessity have come to the conclusion that women were included in the definition of *persons*. However, the Supreme Court of Canada applied a different rule of statutory interpretation: if a word in a statute is not plain and unambiguous, the meaning the word carries may only be ascertained by considering the context and the subject matter of the statute and then applying one of a number of presumptions of statutory interpretation. The Court looked at the date when the *British North America Act* had been passed, 1867, and concluded that at that time women were excluded by law and custom from holding office in a legislative body or indeed from voting in an election of such a body. Therefore, the Court concluded, the word *persons* in the section must be interpreted to exclude females.

§1.23 The decisions of the Supreme Court of Canada were at that time appealable to the British Privy Council, and an appeal was heard by that body. The Privy Council overruled the Supreme Court of Canada and effectively gave the word *persons* its literal interpretation.[10]

§1.24 The "golden" rule of statutory interpretation requires that "in interpreting an ambiguous statute, the grammatical and ordinary sense

---

8 See, for example, E.A. Dreidger *Construction of Statutes*, (Toronto: Butterworths, 1974).
9 *Abley v. Dale* (1851), 11 C.B. 378, per Jervis, C.J.
10 *Henrietta Edwards v. Att. Gen. of Can.*, [1930] A.C. 124.

of the word is to be adhered to, unless that would lead to some absurdity or some repugnance or inconsistency with the rest of the instrument, in which case the grammatical and ordinary sense of the words may be modified only so far as to avoid that absurdity and inconsistency".[11]

Must an absurdity which results from the language of the statute be objectively absurd or need it only be absurd in the opinion of the particular judge who is interpreting the statute? An example of an objective absurdity is seen in *R. v. City of London Court*.[12] There, the statute being considered provided that all actions commenced in a particular court could be transferred to another, but further provided that the other court had no jurisdiction to hear such matters. Obviously, an objective absurdity existed in the language of the statute.

§1.25 Cases which apply a subjective test of absurdity, not surprisingly, present more difficulty. An example of a case where the court considered whether an absurdity in language existed, and ultimately concluded it did not, is *R. v. Mojelski*.[13] There, the defendant was charged with being intoxicated in a public place. At the relevant time, he was in the back seat of his own car which was being driven along the street by another person. The words of the statute defined "public place", among other things, as "a conveyance while it is in any thoroughfare."

On appeal, the court concluded that the ordinary meaning of conveyance included the back seat of one's car and that no absurdity resulted from applying the ordinary meaning to the statute. It might be thought by some that a finding that the back seat of one's own car is a public place is an absurd conclusion; however, it is clear that the courts in this case applied a subjective test of absurdity.

§1.26 The "golden" rule is therefore victim to the same problems that beset the "plain meaning" rule; it has little application to statutes which are framed in wide or general terms. The initial question of whether an absurdity arises is an issue which in itself is open to interpretation. When an occasion arises in which this rule does find application, it may have the undesirable effect of allowing a judge to openly make exceptions to a statutory provision, which exceptions will naturally reflect the judge's personal views rather than the social policy which was intended by the Legislature to be followed.[14]

§1.27 The third approach to statutory interpretation is known as the "mischief" rule, also known as the "purpose" rule. It requires that in

---

11 *Grey v. Pearson*, 10 E.R. 1216 at 1234.
12 [1892] 1 Q.B. 273 (C.A.), as cited in Dreidger, *supra*, note 8 at 51.
13 (1968), 65 W.W.R. 565, reversing (1969), 60 W.W.R. 355 (Sask. C.A.).
14 J. Willis, "Statute Interpretation in a Nutshell" (1938), 16 Can. Bar Rev. 1.

interpreting an ambiguous statute the judge must consider four things: (1) what was the common law before the making of the Act; (2) what was the mischief and defect for which the common law did not provide; (3) what remedy did the Legislature choose to cure the mischief and defect; and (4) what was the true reason for the remedy. Once the court answers these questions, it is to make such construction as shall suppress the mischief and advance the remedy.[15] This rule was developed at a time when statutes were just starting to be used. Common law judges at that time frequently viewed statutes with suspicion and were prepared to change the effect of the statute having regard to the spirit or purpose of the Act, as they saw it.[16]

§1.28 Today, judges are less inclined to judicial law-making and the "mischief" rule is less likely to be applied unless the words of the statute are ambiguous. In such cases, the court will consider the purpose for which the legislation was passed; what mischief was the law meant to cure? Modern statutes are more likely to include sections which set out to some extent the objects or purposes of the legislation as does the *Surveyors Act*,[17] but the court may also choose to go beyond this and consider, for example, reports of a committee which drafted the legislation to determine the purpose and the proper interpretation of the wording of the statute.

§1.29 In addition to the above approaches to statutory interpretation there exist many other devices which aid interpretation. There are grammatical rules of construction, aids to statutory interpretation in the form of interpretation statutes, definition sections within the statute in question, treatises, legal dictionaries, and a set of presumptions. Despite all these aids, the interpretation of statutes is still, perhaps because of the nature of the exercise, a very inexact process.

THE ENGLISH LEGAL LEGACY

§1.30 At its inception, English law consisted solely of the common law which was a unification of the customs of the English under the guidance of the royal courts.[18] Judgments rendered in particular cases set out the principles upon which the common law is based and which had to be adhered to by subsequent judicial decisions.

§1.31 Originally, the common law provided that only certain remedies

---

15 *Heydon's Case* (1584), 3 Co. Rep. 7a, 76 E.R. 637.
16 *Supra*, note 8 at 75 and 82.
17 R.S.O. 1980, c. 492.
18 O.H. Phillips, *A First Book of English Law*, 6th ed. (London: Sweet & Maxwell, 1970) at 6.

would be available. In order to obtain a remedy under the common law, a writ had to be executed, a writ being simply the document which commenced an action. The common law developed to the point where there were specific writs for specific remedies and by the 13th century a check was imposed on the issue of new writs; if an individual sought a remedy for which the common law had no specific writ, that person could not proceed under the common law. With this limitation for actions in a developing social and economic world, unfairness eventually resulted. To circumvent this unfairness, a practice developed that a petition could be made directly to the king or queen for what came to be known as an extraordinary remedy, that is, one which was not available under the common law. The underlying principle, of course, was that the reigning monarch was the fountainhead of justice and therefore had the ability to respond to and grant requests for extraordinary remedies.

§1.32 Understandably, as the practice became more common, the monarch was unable to deal with all such requests and the jurisdiction for granting extraordinary remedies was given to the chancellor. The Courts of Chancery developed as the courts where the chancellor had ultimate authority to grant extraordinary remedies. Such remedies were also called equitable remedies, and the case law which developed from this court came to be known as the law of equity.

§1.33 The English courts, therefore, at one time consisted of the common law courts and the Courts of Chancery. Originally, the Courts of Chancery were intended to be flexible and responsive to new or unusual requests that in fairness and equity should be granted. There had to be no available remedy under the common law or the remedy of the common law, if it existed, had to be such as recognizably gave grave injustice to a litigant. As well, a petitioner would only be granted an extraordinary remedy in cases where the petitioner had behaved properly; that is, a remedy would not be granted to one who came to court with "unclean hands". Unfortunately, over time the Courts of Chancery developed their own rigid and inflexible and often unfair procedures. Charles Dickens' novel *Bleak House* chronicles the inefficiency and unfairness of chancery and the frustration of those caught in its web.

§1.34 This double-branched system of the English law became the basis of the Canadian legal system. It was not until well after the middle of the 19th century that the two courts were unified into one comprehensive court system. The *Judicature Act*[19] which structured this unification of the courts provided simply that both the common law and equity were to remain

---

19 1881 (Ont.), c. 5.

in force with equity to prevail in cases where the two conflicted. This principle applies to Canada today.

DIVISIONS OF LAW

Substantive Law

§1.35 Substantive law, generally the law which governs the interaction of persons as opposed to procedural law which governs such things as the conduct of trials, is divided into two basic areas: public law and private law. This division is loosely based upon whether or not the public interest is involved in the particular law. For example, constitutional, administrative and criminal law are generally categorized as public law, since they all affect the public interest in some way. Constitutional law is concerned with "the organization, powers and frame of government, the distribution of political and governmental authorities and functions, the fundamental principles which are to regulate the relations of government and citizen, and the plan and method according to which the public affairs of the state are to be administered."[20]

§1.36 Administrative law is the "body of law created by administrative agencies in the form of rules, regulations, orders and decision."[21] Examples of administrative agencies are the Ontario Labour Relations Board, the Ontario Censor Board, or the Discipline Committee of any professional group, such as surveyors, which is governed by statute. Even though decisions of administrative agencies usually deal with private interests, they are based upon guidelines which take into account and promote the public interest. As administrative law principles directly affect surveyors, a more complete review of this area of law follows at paragraph 1.41.

§1.37 Criminal law is regarded as an area of public law because a crime is considered an offence against the state and against the public interest.[22] Obviously, it is also an offence against an individual, but the overriding interest in dealing with crimes is the protection of society as a whole.

§1.38 Areas of law which do not affect the public interest include contract law, tort law and property law. Contract law is concerned with agreements which are binding and voluntary, which have legal objects, and which are entered into for mutual consideration (that is, each party has received something of value under the contract) by parties who intend to create a legal relationship. As contract law is only binding upon the individual

---

20 *Black's Law Dictionary*, 5th ed. (St. Paul, Minn.: West Publishing, 1979) at 282.
21 *Ibid.* at 43.
22 G. L. Gall, *The Canadian Legal System* (Toronto: Carswell, 1983) at 20 and 22.

parties to the contract or on those who have assumed obligations under the contract, it has no effect on the public interest. Surveyors, or other professionals, in carrying on their day-to-day practice, have one foot in private contract law and the other in public administrative law. The surveyor is free to contract with a client within certain defined parameters which have been put in place to protect the public interest.

§1.39 Tort law is that area of law concerned with intentional violations of the private rights of others and with the negligent performance of legally recognized duties of care owed to others.[23] Tort law is construed as private law as it is only concerned with the breach of a duty of care which one person owes to another, such as the duty of care which a land surveyor owes to third parties when he or she is performing his or her duties as a surveyor.

§1.40 Property law is concerned with the legally recognized rights and obligations which are attached to ownership and possession of real property, which refers to land, and personal property. Property law has little effect on the public interest as a whole.

Administrative Law

§1.41 As earlier noted, public law is that aspect of law which involves, in some way, the public interest, as opposed to private law which concerns and regulates a dispute between two or more individuals and has no immediate effect on the public in general. Administrative law is a specific type of public law which is now generally understood to mean that aspect of law which is outside of the traditional court system, at least at the first instance. Generally, a tribunal or other administrative body is created by a statute and this body is given strictly defined jurisdiction and authority to review and render decisions on a particular matter. The matter is always one in which the public has an interest, rather than one which affects only those parties or individuals immediately involved. Administrative law therefore directly concerns and affects surveyors.

§1.42 The *Surveyors Act* creates the Association of Ontario Land Surveyors and then creates the Council of the Association. The Council is an administrative body established to conduct the affairs of the Association. The statute strictly defines Council's authority: for example, sections 10 and 11 of the Act set out Council's authority for making regulations and by-laws. Council has no authority for taking any actions which do not come within a power set out in the Act.

---

23 *Ibid.* at 21.

§1.43 The public interest which is affected is specifically acknowledged in section 3 of the Act which provides that the objects of the Association are generally to maintain a certain standard of skill and behaviour among surveyors in Ontario.

§1.44 The Council itself is, therefore, an example of an administrative body. Council is further given the authority to hold hearings when disciplining members for professional misconduct and the panel which sits on these hearings is an administrative tribunal. As such, it is quasi-judicial in nature and must conduct its hearings according to the provisions of the statute regarding notice, examination, cross-examination, and the form of the decision. However, case law has also developed regarding the procedure of an administrative body or tribunal and the principles from these cases must similarly be followed. In particular, the concept of natural justice has developed, which will be discussed further.

§1.45 The reasons for the development of administrative tribunals are many. Their establishment is a recognition by the Legislature that certain matters can more efficiently and effectively be dealt with by a tribunal which has expertise in the area in question. For example, in the surveying profession, a discipline committee composed primarily of surveyors will comprehend, analyse and ultimately render a decision on the evidence much more quickly than could a court. In addition, a tribunal is more likely to have correctly assessed technical evidence, thus lessening the need for appeals. As a practical matter, the establishment of tribunals also lessens the burden on the courts and helps to ensure that matters are dealt with as expeditiously as possible. And finally, tribunals tend to be less formal than courts and hearings often proceed where one or both parties are unrepresented by lawyers. Tribunals are therefore more accessible to the ordinary person.

§1.46 Because tribunal decisions generally affect an individual's rights, the courts maintain a supervisory jurisdiction. Tribunals can therefore be characterized as quasi-judicial in nature; they do not deal solely with technical matters, but are also concerned with individual rights and obligations. As such, the courts, with their traditions of equity and fairness, have not relinquished their ultimate control over tribunal decisions.

§1.47 At common law, rules developed to protect the rights of individuals, which rules set out criteria regarding when it is appropriate for a court to interfere with the decision of a tribunal. Most fundamental of these rules is what has come to be known as the rule of natural justice. Natural justice is "a simple concept that may be defined completely in simple terms:

natural justice is fair play, nothing more."[24] Not surprisingly, this simple concept is less simple to apply to the facts of a particular case. However, as in other areas of the law, principles have developed to assist the courts in determining when there has been a denial of natural justice.

§1.48 Because the concept of natural justice is one which is based on fair play it is impossible to strictly define what the requirements of natural justice are. However, some examples can be given:[25]

(1) A person is entitled to a hearing before a decision is rendered; as a corollary, the person is entitled to notice of the hearing. But a person may waive his right to natural justice; having received notice of a hearing and having failed to attend at the hearing, the person affected cannot then say that there has been a denial of natural justice because a decision has been rendered without a hearing. If the hearing is to proceed on the basis of a charge, such as would occur before a discipline committee investigating the conduct of a surveyor, particulars of the charge must be given. If a tribunal is statutorily obliged to follow rules of evidence, a charge framed in general terms could be a denial of natural justice. For example, in the case of *Sen v. Sask. College of Physicians and Surgeons*,[26] Sen was charged with unbecoming, improper and unprofessional conduct. The charge was based on certain professional services which had allegedly been charged to the provincial insurance scheme and which, the College maintained, were unjustified. Sen asked for particulars of which services were not justified, including the type or description of the services, the name of the patient, the date, and why the services were not justifiable. The College was unable to provide the particulars requested and the court reviewing the College's actions held that if the rules of evidence apply to a particular tribunal, as they did in this case, then the particulars must be provided to enable a person charged to properly prepare a defence and to answer the charge. A failure to do so is a denial of natural justice.
(2) A refusal to disclose material evidence or a refusal to furnish copies will constitute a denial of natural justice.
(3) A refusal to grant an adjournment if there were good reasons for the request and if the result of the refusal was that the person was denied an opportunity to answer the case against him will constitute a denial of natural justice.

---

24 R.F. Reid & H. David, *Administrative Law and Practice*, 2d ed. (Butterworths: Toronto, 1978) at 213.
25 *Supra*, note 20 at 218-227.
26 (1969), 69 W.W.R. 201 (Sask. C.A.).

(4) The denial of the right to cross-examine a witness could be a denial of natural justice unless the tribunal has been given complete discretion in its enabling statute regarding the conduct of the hearing.
(5) A failure to base a decision on relevant evidence or to base a decision on evidence which was not presented at the hearing is a breach of the rules of natural justice, as is a finding of bias or prejudgment on the part of the tribunal. Even where a tribunal is given discretion regarding its procedure, if the discretion is exercised in an arbitrary manner the decision may be overturned.

§1.49 The above is not intended to be in any way exhaustive and a more thorough discussion can be found in administrative law texts.[27] However, it can be seen that natural justice encompasses a wide range of situations which all have in common the concept of fairness.

§1.50 A breach of the concepts of natural justice amounts to an "error of jurisdiction." In administrative law, this would occur where a tribunal has come to its decision without proper jurisdiction to do so. As such, the decision can be quashed and the reviewing court can either substitute its own decision or require the tribunal to re-hear the matter.

§1.51 As mentioned at the outset, the concept of natural justice is the most fundamental of the rules which will result in a court interfering with a tribunal's decision. However, a tribunal may also be overturned by a court on other grounds:

(1) First, a tribunal's decision must come within the ambit of the statute. A failure of it to do so will render a decision *ultra vires*, which literally means "beyond the powers", a concept more fully discussed in the context of constitutional law.
(2) Second, in certain situations a tribunal must establish a particular fact before it has any jurisdiction to proceed. If it proceeds without first establishing that fact, a tribunal is acting without jurisdiction in the decision subsequently rendered.
(3) Third, a court will intervene if a tribunal shows in its conduct of the hearing an abuse of power. For example, if it proceeds for an "improper purpose" such as not having regard to the policy objectives set out in the statute establishing the tribunal, its decision is liable to be quashed.
(4) Finally, if an error of law appears on the record of the proceedings, a supervisory court will intervene.

---

27 See, for example, Reid & David, *supra*, note 24 and S.A. DeSmith *Judicial Review of Administrative Action*, 4th ed. (London: Stevens & Sons Ltd., 1980).

§1.52 If an individual believes that a tribunal's decision is reviewable on any of these grounds, he may bring an application for judicial review. In Ontario such an application is brought to the Divisional Court. Should this court find that the tribunal has erred in any one of these ways, the usual remedy, as mentioned above, is to quash the decision. This remedy is called *certiorari*. The other remedies available on judicial review are: *prohibition*, in which the superior court can prohibit the hearing from proceeding; *mandamus*, in which the court can compel the performance of an act; *quo warranto*, which is the preventing of the continued exercise of an unlawfully asserted authority; and *habeas corpus*, (which literally means "you have the body") used for the release of a prisoner from unlawful custody. The latter two remedies arise in the criminal context.

§1.53 The *Surveyors Act* establishes a structure which is subject to the principles of administrative law.

CONSTITUTIONAL BASIS OF AUTHORITY

Structure of the Canadian Constitutional System

§1.54 The constitution of any country is most often embodied in a basic constitutional document setting out rules and procedures which allocate governmental power within a nation. The transformation of written rules to the practical daily realities of government and society must, by nature, encompass the various customs and mores of a society that may not be specifically included in the basic constitutional document.

§1.55 In Canada, the constitution is rooted in the principles and institutions of the United Kingdom. A collection of various British statutes, as well as recent innovative Canadian statutes, make up the written Canadian constitution. The prime constitutional statute of Canada is the *British North America Act*, enacted in 1867 by the British Parliament at Westminister. The Act created a "dominion" of Canada which united three colonies of British North America; the Act was drafted in a manner to allow for inclusion of additional colonies. The relationship between the Dominion of Canada and the United Kingdom was basically one of colonial dependence. There was no amending clause in the B.N.A. Act which could be exercised by the Dominion itself, but rather control was left in the hands of the British Parliament. Executive power vested with the Crown of Great Britain, which power would be exercised by a representative known as the Governor General.

§1.56 The written component of the early Canadian constitution consisted primarily of three British statutes: the *Colonial Laws Validity Act* of 1865, the *British North America Act* of 1867, and the *Statute of Westminister* of

1931.[28] When taken together, these three statutes provided the basic framework of the Canadian constitution. The *Colonial Laws Validity Act* and the *Statute of Westminister* provided the means of amending part of the B.N.A. Act and laid down some of the rules governing the relations between Canada and the Imperial Parliament.

The British North America Act

§1.57 The *British North America Act* (now called the *Constitution Act, 1867*) is Canada's principal constitutional document. The Act established the basis of a federal system in Canada. In a federal state, the power of government is distributed between a federal or central authority and several provincial or regional authorities. Thus, Canadians are governed by and subject to, the laws of two different authorities, the national or federal Parliament of Canada and the provincial Legislatures. This system contrasts with that of Britain which is a unitary state where the power of government is ascribed to one national authority and any local or municipal governments are subordinate to that national authority.

§1.58 From a law-making perspective, the B.N.A. Act is most important in its allocation of specified areas of power to the federal government and to the provincial authorities. Thus, the different branches of government are empowered to legislate and maintain dominance or jurisdiction over specific powers. For example, the federal government is given authority over criminal matters. In addition, and most importantly, the federal government has a residual power to pass laws relating to "peace, order and good government." The provinces are given power over "property and civil rights in the province," and thus, systems of land registration are provincial in nature.

§1.59 The various powers given to the federal and provincial governments under the B.N.A. Act are subject to the doctrine of *ultra vires* by which the B.N.A. Act possesses a supremacy over all statutes enacted by the federal or provincial governments. Therefore, any act or law passed by the federal or provincial government must comply with the jurisdictional constraints of the B.N.A. Act or, in other words, the federal Parliament cannot intrude on the law-making areas of the provincial Legislatures and *vice versa*. Any laws inconsistent with the powers given to the law-making body may be determined to be of no force and effect by the courts.

---

28 *Colonial Laws Validity Act*, 1865, 28 & 29 Vict., c. 63; *British North America Act*, supra, note 6; *Statute of Westminister*, 1931, 22 & 23 Geo. 5, c. 4.

18 Survey Law in Canada

The Unwritten Component of the Constitution

§1.60 Although the B.N.A. Act sets out the prime allocations of power between the provinces and federal government, a true understanding of the Canadian constitutional system must include some knowledge of conventions and usages derived from British constitutional history.

Conventions are rules of the constitution which are not written in any document and are, in effect, non-legal rules. These rules regulate the practical working conditions of the constitution and may actually assert or restrict the powers given to the various government bodies. For example, section 17 of the B.N.A. Act theoretically makes the Queen an essential party to the parliamentary system of Canada and section 55 confers upon the Queen and on the Governor General a power to refrain from giving royal assent to a bill enacted by the two Houses of Parliament. According to a strict interpretation of these rules, all legislation passed through the elected bodies of Canada would be subject to the discretion of the Queen or her representative. However, there has never been an instance in which royal assent was withheld, primarily due to a convention, or unwritten rule, which developed over time. It is therefore necessary to develop a knowledge of the history of Parliament and the various conventions when one deals with constitutional law.

§1.61 Usages are governmental practices which are ordinarily followed, although they are not considered to be obligatory. A usage may develop into a convention once sufficient time has passed and the ways of the usage have been continued. An example of usage may be a policy that a certain office appointed by the government should be occupied alternately by anglophone and francophone candidates. Such a usage may not be written into law, but continued practice over a lengthy period of time may result in a convention being established.

Federal and Provincial Powers

§1.62 The most important provisions of the *British North America Act* are sections 91 and 92 which allocate different legislative jurisdictions and powers between the Parliament of Canada and the provincial legislatures.

§1.63 *Power of the Federal Government: Section 91* According to section 91, the federal Parliament is able to make laws for the "peace, order and good government of Canada" in relation to any matters not listed in the classes of subjects enumerated in section 91. This provision has allowed the federal government to assume authority over matters which may have national dimensions and which could affect the entire country. In addition to this general power, the federal government is specifically given jurisdiction over several enumerated matters, most being national in scope such as criminal law, the regulation of trade and commerce, the postal service, the military, taxation, currency, laws pertaining to marriage and divorce, copyrights, and any matters which may not have been recognized in 1867 and which are not enumerated as being powers of the provinces. Thus, various matters created recently due to technological advances, particularly of a national scope, fall under the federal power; this includes airports, and radio and television communications.

§1.64 *Powers of the Provincial Government: Section 92* The provincial governments are also given jurisdictional authority over several enumerated subjects. These include such matters as establishment and maintenance of public and reformatory prisons, the establishment and maintenance of hospitals, and the incorporation of companies with provincial objects. Also listed are the important areas of "property and civil rights in the province" and all matters of a "local or private nature in the province", such as the law of property, local transportation systems, schools, and licensing of drivers.

§1.65 In some cases there tends to be an overlap of areas of power given to the federal Parliament and provincial Legislatures. For example, the federal Parliament has jurisdiction over the matter of divorce, while the provincial government legislates with respect to the division of family property; therefore, in the family law context federal and provincial laws mesh. In other areas, however, conflicts may develop between federal and provincial jurisdiction. Criminal law is within the federal government's jurisdiction, while property and civil rights are provincial. In the 1903 case of the *Att. Gen. of Ont. v. Hamilton Street Railway*,[29] the Ontario *Lord's Day Act*[30] being a law requiring shops to close for Sunday observance, was struck down as being a criminal law, thus being within federal jurisdiction. Though the determination of whether shops should close may have been considered to be a matter of property and civil rights, the particular law was found by the courts to so overlap the area of criminal law as to be rendered invalid as inconsistent with provincial jurisdiction. In some cases, the courts try to reconcile laws that cover multi-jurisdictional areas, and an entire textbook could be devoted to discussion of those methods. Suffice to say, the courts have been naturally hesitant to strike down laws unless there is a complete inconsistency between the federal and provincial laws. Where some overlap between the inconsistent laws is inevitable, the court may reconcile the inconsistency and allow the laws to stand.

---

29 [1903] A.C. 524 (P.C.).
30 R.S.O. 1897, c. 246.

20   Survey Law in Canada

The Constitution Act, 1982

§1.66 In 1982 changes were made to the basic structure of the Canadian constitutional system. Historical developments had altered the dependence of Canada on the United Kingdom and consequently the total independence and right to amend its constitution, was a fact of Canadian nationhood. It was legally the final step in making Canada a truly independent state. The *Canada Act*[31] incorporated and enacted the *Constitution Act, 1982* — the earlier *British North America Act* of 1867 renamed; it excluded any future Act of the United Kingdom Parliament from extending to Canada as part of Canadian law. The *Constitution Act, 1982* also introduced the *Canadian Charter of Rights and Freedoms* and it set up a procedure whereby the constitution of Canada may be amended by a formula requiring the support of the federal government and a specified number of provinces.

Amending Formula

§1.67 Section 38 of the *Constitution Act, 1982* provides an amending formula whereby the constitution of Canada may be amended by resolutions of the Senate, of the House of Commons, and of the Legislative Assemblies of at least two-thirds of the provinces that have at least 50 per cent of the population of all the provinces. This formula represented a substantial compromise based on various federal and provincial dialogues. Any amendment that lessens the powers of any provincial government requires resolutions supported by a majority of the members of the Senate, the House of Commons, and the Legislative Assemblies mentioned above, but such amendment shall not have effect in any province which has expressed its dissent by resolution of a majority of its members. The unanimous support of the provinces is required for certain specified amendments, particularly concerning the use of the English or French language as guaranteed in the constitution, and any amendment to the amending procedure.

The Charter of Rights

§1.68 The introduction of the *Canadian Charter of Rights and Freedoms* is a milestone in the social and legal history of Canada. The Charter declares the existence of basic fundamental freedoms and civil liberties for all Canadians. The Charter is entrenched in the constitution, meaning that it forms part of the constitution and can be amended only by the formula used to amend the constitution. Prior to enactment of the Charter

---

31 (U.K.), 1982, c. 11.

the federal Parliament, in 1960, had enacted the *Canadian Bill of Rights*[32] as a statute designed to recognize and declare that various human rights and fundamental freedoms existed in Canada. The question consistently arose as to what would result if a later statute contravened the Bill of Rights. That is, what would happen if a law enacted by the federal Parliament infringed one of the rights specified in the Bill of Rights, such as freedom of speech? A leading case considering the subject, *R. v. Drybones*,[33] a decision of the Supreme Court of Canada, inferred that legislation contravening the Bill of Rights would be of no force and effect, but there was a general hesitancy by the courts to hold that the Bill of Rights had overriding authority over all federal legislation because the Bill was not entrenched; it was merely a federal statute. As a consequence, the Bill of Rights failed to live up to what was presumably its drafter's intent, that is, the protection of fundamental freedoms.

§1.69 Enactment of the *Charter of Rights* changed all this. Section 32 declares that the Charter applies to the laws of the Parliament of Canada and the Legislatures of the provinces. Section 52 of the *Constitution Act, 1982* proclaims that the constitution of Canada (which includes the *Charter of Rights*) is the supreme law of Canada and any law that is inconsistent with its provisions is, to the extent of the inconsistency, of no force or effect. Therefore, any law of Parliament or the provinces which is inconsistent with the Charter is to be declared void.

§1.70 There are exceptions to this absolute position. First, Parliament or the Legislature of a province may expressly declare that a law shall operate notwithstanding a breach of the Charter provisions found in section 2 and sections 7 to 15 inclusive. These sections refer to the fundamental freedoms of religion, thought, belief, opinion and expression, freedom of the press, freedom of peaceful assembly and freedom of association, as well as various legal rights, including the right to life, liberty and security of the person. Notwithstanding that there is a real possibility that a Legislature may pass laws which contravene the Charter in these important areas, it seems unlikely that this will occur on a regular basis. The existence of the Charter provisions alone puts political pressure on the governing bodies in Canada to enact legislation which conforms with philosophical precepts set out in the Charter.

§1.71 Another abridgment of the absolute rights of the Charter, *vis-à-vis* ordinary legislation, is that the Charter guarantees the rights and freedoms set out therein subject only to "such reasonable limits prescribed

---

32 R.S.C. 1970, Appendix III.
33 [1970] S.C.R. 282.

by law as can be demonstrably justified in a free and democratic society".[34] The importance of this clause is that a law properly passed by the Parliament of Canada or a Legislature of the provinces for a *bona fide* purpose may still be valid even though it literally infringes a guaranteed right, so long as the infringement of that right can be justified.

§1.72 A question currently before the courts is, to which bodies, governmental or private, does the Charter apply? For instance, section 32 provides that the Charter is applicable to the Parliament of Canada and to the Legislature of each province. However, does the Charter apply to the acts of a university or a professional governing body? Both universities and professional governing bodies are created by statute. In this nexus with Parliament sufficiently close, are the actions of such bodies subject to the Charter? This question is not easily answered. A decision was recently rendered regarding mandatory retirement policies in force at various universities in Ontario. Certain employees maintained that such policies were a breach of the *Charter of Rights*; in defence, the universities pleaded, among other things, that the Charter provisions do not apply to them. The universities prevailed: the court concluded that the provisions of the Charter did not apply to the policies of the universities.[35]

§1.73 On April 17, 1985, a section protecting equality rights came into force. Section 15(1) of the Charter provides that

> 15(1) [E]very individual is equal before and under the law and has the right to the equal protection and equal benefit of the law without discrimination and, in particular, without discrimination based on race, national or ethnic origin, colour, religion, sex, age or mental or physical disability.

Section 24(1) of the Charter provides that anyone whose rights or freedoms, as guaranteed under the Charter, have been infringed or denied may apply to a court of competent jurisdiction to obtain a remedy that the court considers appropriate and just in the circumstances. Thus, many court cases have developed in which individuals or groups have sought a remedy for an alleged infringement of their rights. Such remedy may include a declaration that the law is to be of no force or effect, or a request for injunctive relief or damages. The enactment of the Charter has caused the reassessment of all of our laws and many of our fundamental institutions. Many challenges to the *status quo* are made daily.

---

34 *Supra*, note 2, s. 1.
35 *McKinney v. Bd. of Governors of the University of Guelph* (1986), 57 O.R. (2d) 1 (H.C.).

CONSTITUTIONAL BASIS OF JUDICIAL AUTHORITY

§1.74 In Canada, there are both federal and provincial court systems, each established pursuant to governing sections in the B.N.A. Act. The Supreme Court of Canada is created by the *Supreme Court Act*,[36] which was enacted pursuant to section 101 of the B.N.A. Act. This section allowed the Parliament of Canada to maintain a set of courts for the better administration of laws. The Federal Court of Canada was also created pursuant to section 101 of the B.N.A. Act. The Supreme Court of Canada is generally the court of final recourse and it may consider private disputes, provincial disputes, and constitutional matters referred to it by the various levels of government. The Federal Court of Canada is established to resolve disputes involving matters relating exclusively to the jurisdiction of the Parliament of Canada, that is, matters within section 91 of the B.N.A. Act. Various federal administrative tribunals have also been created pursuant to section 101 of the B.N.A. Act and these may resolve disputes arising from policy decisions of the various federal departments.

§1.75 Section 92(14) of the B.N.A. Act allows provincial Legislatures to set up and administer provincial courts. Thus, each province generally enacts statutes which create or regulate the court system. The Supreme Court of Ontario, for example, is regulated by the *Courts of Justice Act*,[37] as is the District or County Court. Various provincial courts are maintained as a result of a variety of provincial statutes. Similarly, the province has created many administrative tribunals which often resolve disputes arising from the interpretation and application of provincial legislation and regulations. Such tribunals may also be asked to consider disputes arising from agencies or bodies created by statutes of the province, such as professional organizations or quasi-governmental bodies.

System of Courts

§1.76 Following is a general representation of the Canadian judicial hierarchy. The court system within a particular province may depart somewhat from the general model.[38]

---

36 R.S.C. 1985, c. S-26.
37 S.O. 1984, c. 11.
38 This chart is from G. Gall, *The Canadian Legal System*, 2d ed. (Toronto: Carswell, 1983) at 117 and 118.

CHART 1
The System of Courts in Canada Generally

A. *Federal Courts — Courts constituted under federal statutes with judges federally appointed.*

Supreme Court of Canada

- The Chief Justice of the Supreme Court is also the Chief Justice of Canada
- Eight Puisne Justices

Federal Court of Canada

- Appellate Division
- Trial Division

B. *Provincial Courts — Courts constituted under provincial statutes with judges federally appointed.*

**Courts of Superior Jurisdiction of a Province
or the
Supreme Court of a Province**

Apellate Division
- This court is often referred to as the Court of Appeal of the province
- The Chief Justice of the Appellate Division is also the Chief Justice of the province

Trial Division
- In some provinces, such as Manitoba, for example, the two divisions here are separate courts constituted by separate statutes, with the trial court known as the Court of Queen's Bench
- Often this court is simply referred to as the Supreme Court of the province
- The Chief Justice of this court is properly referred to as the Chief Justice of the Trial Division

Note: In the province of Ontario, there is a further subdivision with the creation of the Divisional Court. The Divisional Court has an administrative law jurisdiction in respect of the granting of prerogative remedies. In addition, in December of 1976, the Supreme Court of Ontario was further subdivided to create a family law division.

County or District Courts

- In certain circumstances, the District or County Court judges exercise the jurisdiction of local judges of the Supreme Court of a province

Surrogate Courts

- Usually, judges of the County or District Court serve in the capacity of Surrogate Court Judges

Note: As indicated elsewhere, many of the provinces have or are about to merge their county or district courts with their superior courts. The result of such an amalgamation is the elimination of an intermediate court of trial jurisdiction with judges who are federally appointed.

C. *Provincial Courts — Courts constituted under provincial statutes with judges provincially appointed.*

Provincial Courts

| Juvenile Court | Family Court | Provincial Court (Criminal Jurisdiction) | Small Claims Court |

Note: In some provinces, by the operation of various enabling statutes, two or more of the above courts are combined into a single court, with various divisions. For example, in Ontario the Provincial Court is divided into Family and Criminal Divisions, whereas the Small Claims Court is established under a separate statute. Alternatively, some provinces provide concurrent jurisdiction for judges serving on one or more of the above courts to deal with matters arising in another of the above courts. For example, in Alberta, under the Provincial Court Act, R.S.A. 1980, c. P-20, a judge of the Provincial Court has jurisdiction to sit in either, some or all of the criminal, small claims, family or juvenile divisions.

CONCLUSION

§1.77 This introductory chapter is meant to provide a broad overview of the law. As such, it has not been possible to go into great detail regarding any specific area of it. It is to be hoped, however, that it has provided the reader with a basic understanding of the structures of the law and how they have developed.

2

Real Property Law in Common Law Provinces

James S. Dobbin, B.Sc. Eng., LL.B.

INTRODUCTION

§2.01 This chapter will introduce the reader to real property law in common law provinces. The subject will be presented as follows:

(1) History and evolution of real property law — some background into where the law is today and why and how it got to be that way.
(2) Theories of ownership — an introduction to the how of owning real property and rights therein.
(3) Conveyancing — a brief look at transferring title to real property.
(4) Legal principles affecting ownership — a collection of principles.
(5) Future development — a brief look at possible directions of the evolution of the law of real property and the forces behind those changes.

§2.02 The chapter, as the title implies, relates only to common law provinces. The scope has been kept as broad as possible. As a result, much of the finer detail relating to the law of each individual province, particularly with regard to statute law, must be left to the reader's research. Because of the volume of the material which should be considered in a review of this nature, many topics have been dealt with in an introductory fashion only. Many others have been omitted altogether. It is hoped that these decisions on emphasis have not diminished the value of the chapter to the reader.

28 Survey Law in Canada

§2.03 The law is stated as of January 1, 1988.

HISTORY AND EVOLUTION OF REAL PROPERTY LAW
Some Background

§2.04 Real property law is the set of legal principles which govern, firstly, the use of real property by individuals and, secondly, the interaction between individuals in that use.

§2.05 Land is an economic commodity. As a culture develops, its use of land develops. The nature of that use depends upon many variables such as, for example, the economic and social needs of the people, the physical characteristics of the land and external forces such as drought or invasion. Each culture faces a different set of variables and will respond to them in a unique fashion. The law of real property of a culture is a reflection of that response.

§2.06 Canadian real property law is based on that of England. To fully appreciate the present law it is therefore important to understand the cultural forces existing in both countries over an extended period of time.

Norman Conquest

§2.07 All studies of English property law seem to begin with the Norman Conquest of 1066. History tells us that prior to that point real property was subject to absolute or allodial ownership.[1] The changes which followed the Conquest set the foundation for modern real property law and thus make it a most convenient starting point.

§2.08 William the Conqueror, with the help of some fifteen hundred barons invaded and conquered England and set himself up as King. Reasoning that the inhabitants had resisted him and had been conquered, William claimed all the land in the country as his personally. The land and culture were primarily agrarian. William need to do two things: consolidate power and derive some income from all of his efforts. He devised a scheme whereby in return for what he needed he granted certain rights in the land to the barons who had helped him. Their rights were restricted to the use or possession of the land, or both, but did not extend to ownership. In return for the rights granted by the king, the barons would provide money, food or army service to the king. Normally, therefore, those already living

---
1 See Halsbury's *Laws of England*, 4th ed., Vol. 39 (London: Butterworths, 1982) par. 304, footnote 1 and works therein cited.

on the land would continue to do so while furnishing money, food or military service to their masters.

§**2.09** It is critical to recall that the king still owned the land. All that had been granted to the barons were rights to use or possess the land and as a result all that could be granted by the barons was what they had — the right to use or possess the land. The farmer working the field derived his right to do so from his lord who derived his right from his lord and so on up to the king from whom all rights were derived. The flow of rights down this feudal pyramid was matched by an equal flow of money, loyalty, troops, and even prayer, up the pyramid. In addition, control over the land was maintained by a system which gave the lords rights to determine (among other things) who would marry whom and to whom property rights could be transferred.

§**2.10** Further detail on the feudal system can be found in any comprehensive text on English property law.[2] Suffice for the purpose of this chapter to say that over the following centuries the flow of goods and services up the pyramid gradually stopped and the intermediate lords disappeared. This evolution left the owner of land in a position where he derived his rights directly from the king. The modern landowner cannot rightly claim to actually own the land. Such absolute ownership is referred to as allodial ownership. All he can claim are rights to use and possession of the land which rights are subject to the Crown's ultimate ownership. The concept of ownership of rights to land is critical to an understanding of English property law.

Early English Developments

§**2.11** This section will not examine the finer detail of the evolution of English property law, but will concentrate on the development of and relationship between common law, equity, and statute law.

§**2.12** Immediately following the Norman Conquest a system of dispute resolution developed which was based on the adversary system. Each party to a dispute presented his position to a judge. It was the judge's responsibility to resolve any dispute as to facts, either alone or by use of a jury, and to apply the law to the facts in coming to a decision. In the course of time the law came to consist principally of previously decided cases of which some record had been kept. The judge's application of the law to the facts was based on the doctrines of precedent and *stare decisis*. Precedent

---

[2] See, for example, Anger and Honsberger, *Law of Real Property*, 2d ed. (Aurora: Canada Law Book, 1985) at Chapter 2.

30   Survey Law in Canada

required the judge to come to the same decision as had been reached in identical circumstances by a court whose decisions were binding on that judge. *Stare decisis* set up a hierarchy which determined which court's decisions were binding on that judge. This is the essence of the common law. A body of decided cases grew and enabled individuals to have some degree of certainty as to what decision a judge might reach on a particular point. This process also imparted a certain degree of inflexibility on the common law.[3] From this starting point of pure common law the development of equity and statute law cut across all levels of the law.

§2.13 The rigidity of the common law system often resulted in situations where strict application of common law rules led to decisions which were in some aspect unfair or unjust. In those situations the only recourse of the aggrieved party was to petition the king himself for an equitable or fair solution. The king, being all powerful, had the capacity to impose the "proper" decision. As time progressed the king delegated this authority to his chancellor and the chancellor further delegated it to a set of courts which became known as the Chancery Courts. Thus, a parallel system of courts evolved where, in theory at least, decisions were based more on the merits of the claims rather than a strict application of precedent and *stare decisis*. In Canada, this dual court system survived until this century when the common law and chancery courts were merged. The body of law generated by the chancery courts can be termed the law of equity.

§2.14 Statute law is founded in the concept of supremacy of Parliament. Under the English parliamentary system, any law Parliament wishes to pass it may pass, and upon passage that law overrides any existing common law to the contrary.[4] There are two general ways in which statute law may be applicable. Firstly, where some problem is seen to exist in the common law, a statute may be passed to remedy that specific situation. Several statutes from the Middle Ages fall into this category. Secondly, an entire scheme or system may be artificially created in an area not yet fully considered by the common law. Examples include land use regulation and condominium statutes. It is statute law which has become more and more important in present day Canada. This reflects an ever increasing complexity in society and the uses to which it puts land. But while statutes fairly freely create new law, there appears to be a fairly consistent position

---

[3] For a more complete discussion of the role and effect of precedent and *stare decisis* see *supra*, chapter 1, "The Canadian Legal System".

[4] The supremacy of Parliament has been somewhat tempered in Canada by the adoption of the *Canadian Charter of Rights and Freedoms,* Part I of the *Constitution Act, 1982,* being Schedule B of the *Canada Act 1982* (U.K.), 1982, c. 11.

that a statute does not overrule common law unless it is expressly stated to do so.

Transfer to Canada

§2.15 During the development of an English colony the laws to be applied would vary over time. At the earliest stages of settlement the entire body of English law would have been enforced as the colony would have been considered as a part of the mother country. As the colony developed a degree of independence, it would be entitled to elect its own legislature and pass its own laws. This process posed an interesting problem: should the new Legislature be forced to start from scratch and create an entire body of law sufficient to run the colony or should some short-cut be adopted. The expedient solution was adopted and the concept of reception was developed. At some point (termed the date of reception) the entire body of existing English law — common, equitable and statue — became the law of the colony. Even this approach caused some problems as some laws would not have been applicable to circumstances existing in the colony. In such cases the non-applicable laws were considered not to have been transferred.

§2.16 The reception date applicable to a colony depends on several factors. In some cases the date is fixed by legislation but often there was to be no formal adoption of a date and it will have to be determined by examination of surrounding circumstances such as whether the colony was conquered, settled or acquired by treaty. Each province has a different reception date and reference will have to be made to local authorities as in some cases the date may still be in dispute. It is sufficient for the purposes of this chapter to indicate that the real property law received from England was for the most part uniform across the common law provinces.

§2.17 It is important to emphasize the effect of the reception of English law. The whole body of law, whether common, equitable or statute (with the exception noted above) became the common law of the colony. Thus, important English statutory reforms (e.g., the *Statute of Uses*)[5] are part of the common law of the receiving colony and are not usually to be found in its statutes.

§2.18 From the starting point of reception, the colony would proceed to develop its own laws. Statutes would be passed within the framework of the legislative authority given to the colony. Common law and equity would develop on precedent and *stare decisis* and the conditions existing in the colony.

---

5 1535, 27 Hen. 8, c. 10.

Modern Developments

§2.19 Over the last century developments in real property law have been almost exclusively statute law developments. If, as claimed earlier, all law reflects the society which generates it, the present situation reveals a need to have more flexibility in real property laws to accommodate an ever greater diversity in the economic and social uses to which society wishes to put the land. Examples of this economic and social complexity are condominium and matrimonial property legislation.

§2.20 As society continues to evolve and its needs change, no doubt further legislative schemes will be adopted to accommodate those needs.

Aboriginal Title

§2.21 Aboriginal title is used here to denote the complex set of disputes which presently exists over claims made by the native populations in Canada. It is presented here as a topic more as a caveat to the reader than to provide any comprehensive or authoritative analysis.[6]

§2.22 The issues involved in this area often seem better suited to the political than the legal arena. Some of the matters which must be considered are:

(1) the nature of both the aboriginal use of the land and the European settlement;
(2) the effect of such Crown charters as that granted to the Hudson's Bay Company;
(3) the effect of a number of treaties between the Europeans and the aboriginal peoples; and
(4) the *Royal Proclamation of 1763*.

The combination of these thorny legal issues in the political arena has led to great uncertainty. Native peoples have presented well documented and well-argued claims based on aboriginal title in many areas of Canada. Each claim presents a unique set of issues and facts.

§2.23 Under the *British North America Act*, the federal government has responsibility for native peoples[7] while the provincial governments have

---

[6] One of the most helpful works on the subject is Slattery, *The Land Rights of Indigenous Canadian Peoples, as Affected by the Crown's Acquisition of Their Territories* (Oxford: Oxford University Press, 1979).

[7] *British North America Act 1867, now the Constitution Act, 1867*, (U.K.), 30 & 31 Vict., c. 3, s. 91(24).

jurisdiction over property rights.[8] The *Canadian Charter of Rights and Freedoms* purports to maintain existing aboriginal rights.[9] It is apparent that any overall solution will have to involve both the federal and provincial governments.

§2.24 At this point it appears that the federal government has adopted a policy of recognizing some form of aboriginal title as legitimate and a long and slow process of negotiations, punctuated by more frequent resort to the courts, continues.

THEORIES OF OWNERSHIP

Introduction

§2.25 As described in the previous section, one of the basic principles of English real property law dictates that only the Crown can own land absolutely.[10] All that an individual can own are rights to the land. This section will develop that theme, setting out firstly how rights to land can be owned and secondly what form those rights may take.

Tenure

§2.26 Tenure is the how of owning rights to land. Much of the evolution of the various types of tenure is taken up in the history of the Middle Ages. A full review of that evolution is beyond the scope of this work.[11] Briefly, as seen in the previous section, the Crown granted rights to large parcels of land to great lords who made grants in turn to lesser lords and so on down to the lowest level — usually the farmer working the land. In return for each of these grants, some form of compensation flowed from the grantee upward to the grantor. The type of compensation flowing upward was critical to defining the quality of the right held.

§2.27 For the purposes of this work, tenure can be divided into three categories: (1) free tenure; (2) unfree tenure; and (3) leasehold tenure.

§2.28 Freehold tenure, that is, the holding or tenure of land by a free man was any tenure under which the owner of a right in real property could enforce that right in a common law court by way of a real action — that is, an action for return of his land if his right to possession had been challenged. This can be contrasted with unfree tenure which generally

---

8 *Ibid.*, s. 92 (13).
9 *Supra*, note 4, s. 25.
10 See §§2.08 to 2.10.
11 See Halsbury's, *supra*, note 1 at 215 *et seq.*

existed only at or near the bottom of the feudal pyramid. At that low level, the local lord had almost complete control over those underneath him. All disputes had to be resolved in his local or manor court rather than before a common law court. The remedies available there did not include return of the land. With subsequent reform, the holders of unfree tenure eventually acquired equal access to the common law courts and the distinction disappeared. The only continuing relevance of this distinction is the retention of the term freehold. With the dying out of feudal incidents and the disappearance of the middle men from the feudal pyramid, all holders of free tenure eventually held their rights directly of the Crown. That is the situation which was brought to Canada. Individuals who own land hold that land by free tenure. They hold their rights directly of the Crown and those rights can be enforced in a common law court in a real action.

§2.29 Leasehold tenure grew up in a different fashion. The relationship between the owner of the land and the tenant is defined by a contractual agreement between them termed a lease. The quality of the rights of the tenant (the tenure) is therefore as varied as there are forms of leases. In addition, the tenant was originally incapable of maintaining a real action for return of the land. All the tenant could do was maintain an action for damages.

Estates

§2.30 If tenure is the how of owning rights to land, the estate is the measure of the quantity of those rights, in the sense of time particularly; the quantity in the sense of area is the parcel. As with tenure, it is convenient to classify estates as freehold and leasehold.

§2.31 Freehold estates can be fee simple estates, fee tail estates, or life estates.[12] The use of the word "fee" indicates that the estate can survive beyond the death of the holder and may potentially last forever.

§2.32 A fee simple estate is the maximum estate possible. The owner of that estate holds all of the rights which it is possible to own. Almost all of the freehold estates which will be encountered in present day Canada are fee simple estates.

§2.33 A fee tail estate was almost as broad in scope as a fee simple estate, but its continuance was dependant upon the production of heirs by the holder of the estate as set out in the document creating the estate. The

---

12 See Sinclair, *Introduction to Real Property Law*, (Toronto: Butterworths, 1969) at 17 to 31 for an excellent discussion of fee simple, fee tail and life estates.

fee tail has been effectively abolished in most of the provinces and it is highly unlikely that such an estate will be encountered.

§2.34 The life estate is simply that — an estate in land which lasts only as long as the life of some named person. That person might be the owner of the estate himself or some other person (a life estate *pur autre vie*). Life estates can often be found where an individual has died leaving his or her real property to the surviving spouse for his or her life and then to the children of the marriage.

§2.35 All of the above freehold estates might be modified so that they may terminate on the happening or non-happening of some specific event. This termination of the estate might be automatic (the determinable fee) or might require some overt act on the part of the original grantor (the fee upon condition subsequent).[13]

§2.36 Leasehold estates are defined by the lease agreement between the lessor and lessee much as leasehold tenure described above.[14]

§2.37 Two or more estates can exist in regard to the same parcel of land at the same time. A might have a life estate with a reversion to B in fee simple on A's death. At the same time, A might grant a leasehold estate to C. A, B, and C all have estates in the parcel. All own an interest in it. The sum total of all of the estates existing in a parcel of land must add up to a fee simple estate.

Seisin

§2.38 Seisin is often described as the connecting link between tenure and estates.[15] Seisin is the fact of having the immediate right to possession of a freehold estate in land held in freehold tenure. Thus, the owner of a freehold estate in land held in freehold tenure who has an immediate right to possession of the land is said to be seized of the land. It is clear from this that a tenant could not have seisin — he has no freehold estate. In addition, even though the owner of the freehold estate (the lessor) has given up possession to the tenant, he still has seisin. This can be understood in two ways: firstly, the possession of the tenant may be considered to be possession on behalf of the lessor; secondly, it may be remembered that if the lessor removed the tenant, the tenant could not take a real action to return himself into possession but could take only a personal action in damages — thus the lessor has an immediate right to possession.

---

13 See Anger and Honsberger, *supra*, note 2 at 124-127 for a more complete discussion.
14 See Sinclair, *supra*, note 12 at 31-36 for a discussion of the leasehold estate.
15 See Anger and Honsberger, *supra*, note 2 at 25.

Legal and Beneficial Ownership

§2.39 One final categorization of ownership is necessary to complete the picture and that is to distinguish legal ownership from beneficial or equitable ownership.

§2.40 The distinction between common law courts and courts of chancery or equity has already been discussed.[16] If an owner had a right which was enforceable in the common law court, he had a legal interest. If the owner's remedy was restricted to courts of equity, he had an equitable or beneficial interest. Even with the merger of the courts of common law and equity this distinction remains.

§2.41 The most common example of the separation of legal and beneficial ownership is the mortgage.[17] The mortgage document conveys the legal ownership of the estate in question to the lender (the mortgagee). The borrower (the mortgagor) is left with the equitable ownership of the estate. He has the right to regain the legal ownership of the estate on payment of the mortgage debt and performance of all other obligations contained in the mortgage contract. This right was originally enforceable only in the courts of equity and is still referred to as the equitable or beneficial ownership of the estate.

OTHER RIGHTS

Introduction

§2.42 As seen above, ownership of a fee simple estate in land involves the ownership of all of the rights to that parcel which English law recognizes. There are several rights to land which can be carved out of the fee simple estate and bought and sold independently of ownership of the rest of the estate. The following is a listing of some of the more common rights which may be encountered.

Easements

§2.43 In its simplest terms, an easement[18] is a right for the owner of one parcel of land (known as the dominant tenement) to use another parcel (known as the servient tenement) for the benefit of the dominant tenement. In the negative sense, it might also involve the right to require the occupier

---

16 See §2.13.
17 For more detail on mortgages see §§2.84 to 2.86.
18 For a more complete discussion of the topic of easements see *Gale on Easements*, 15th ed. (London: Sweet & Maxwell, 1986).

of the servient tenement to refrain from doing something with that land, again for the benefit of the dominant tenement. In both cases, the right is not a right to possession but a right to use.

§**2.44** An easement is greater than a personal right in the owner of the dominant tenement; it benefits not the owner but the parcel itself. The right may pass with the conveyance of the dominant tenement. The law terms such a right as hereditable or as a hereditament — it will not die with the death of the original grantee of the right.

§**2.45** There are several types of rights which may be easements. The most common would probably be a right of passage over the land of the servient tenement to the dominant tenement. Other possibilities might be easements for supports of structures like the right to require the owner of the servient tenement not to excavate his parcel so as to cause structures on the dominant tenement to be damaged; easements for light or air such as the right to require the owner of the servient tenement not to interfere with windows or vents on the dominant tenement; easements for drainage like the right to cause surface waters to drain from the dominant tenement to the servient tenement; or an easement to use a well such as the right to draw water from a well on the servient tenement for the benefit of the dominant tenement.

§**2.46** In order for an easement to exist there must be two separate parcels under different ownership. Although the most common examples would involve two contiguous parcels, cases indicate that this may not be necessary.[19]

§**2.47** Some situations which look like easements are more difficult to classify. This is the situation where there is no specific parcel which can be identified as the dominant tenement. These cases are common in provision of utilities or in certain classes of highways where a utility line or roadway crosses a parcel (the servient tenement) and yet there is no specific dominant tenement. These are termed easements in gross and English and Canadian law do not recognize them as true easements. They have been termed public easements.[20]

§**2.48** For the purposes of this chapter, easements can be created in three ways — by grant, by statute, or by prescription.[21]

---

19 *Adamson v. Bell Telephone Co.* (1920), 55 D.L.R. 157 (C.A.).
20 Sinclair, *supra*, note 12 at 48. See §2.57 below for a discussion of highways.
21 A fourth method — creation by proprietary estoppel — has been noted but it is a rare situation and beyond the scope of this work. See *Crabb v. Arun District Council*, [1976] Ch. 179 (C.A.).

38   Survey Law in Canada

§2.49 Generally, to create an easement by grant, some form of written document must be involved.[22] In these cases, the terms of the instrument must be examined to define the rights involved. Disputes may arise as to whether the right granted is a true easement or some lesser right or as to the extent of the use permitted under the terms of the document.

§2.50 In contrast to creation by express grant, easements may also be created by implied grant or reservation. As seen above,[23] in order for an easement to exist, there must be two adjacent (or nearly adjacent) parcels under different ownership. Where two parcels are under one ownership and a right is being exercised for the benefit of one to the burden of the second, a quasi-easement exists. If the owner of the parcels transfers the quasi-dominant tenement without specific reference to the easement, it may be possible for the new owner to claim that the right remains for the benefit of his parcel. This process of transfer without express reference to the easement is termed implied grant.

§2.51 As a general rule, the reverse is not true. The law will not generally allow the owner of the two above parcels to transfer the quasi-servient tenement without reference to the easement and then claim the benefit of the easement for the quasi-dominant tenement he has retained. The phrase used by the courts to express this is "a man cannot derogate from his grant." There are exceptions to this rule however, the principal one being where the easement is one of necessity. Where it would be impossible to use the retained parcel without the benefit of the easement, it is said to be an easement of necessity and the owner of the dominant tenement will be permitted to use it. The most common example would be where the conveyance of the quasi-servient tenement cuts off all access to the quasi-dominant tenement.

§2.52 Easements may be created by federal, provincial, local or municipal legislation. In each case, reference must be made to the statute creating the easement to determine the characteristics and extent of the right involved.

§2.53 Easements may also be created by prescription. A prescriptive claim may be based on common law principles or on statutory provisions.

---

22 The English *Statute of Frauds*, 1677, 29 Cha. 2, c. 3, was received as common law by the provinces and in some cases it has been re-enacted. The statute forbids the transfer of an interest in land unless it is accomplished by some written document. See for example, *Statute of Frauds*, R.S.N.B. 1973, c. S-14, s. 8 which provides: "No interest in lands shall be assigned, granted, or surrendered, unless it is by deed or note in writing, signed by the party assigning, granting, or surrendering the same, or by his agent thereunto lawfully authorized by writing, or by act and operation of law."

23 See §2.46.

Prescription

§2.54 Prescription under common law principles had a difficult beginning. Under the original doctrine developed by common law courts, a claimant had to show use of the right "from time immemorial." Time immemorial was first defined by the Statute of Westminister I[24] to run from the year A.D. 1189. Of course a Canadian claimant would find it impossible to show use from that date. Over the years the courts adopted more and more liberal rules, eventually settling on the principles of "lost modern grant." If use of the right could be proven for a period of 20 years, the courts would conveniently presume that at some point earlier in time the owner of the servient tenement had made an express grant of the easement. The lack of any proof of an actual grant did not affect the process as the court simply made the further presumption that the document had been lost.[25]

§2.55 Statutory prescription can be traced to 1832 when the British Parliament enacted the *Prescription Act*.[26] The provisions of the applicable statutes vary across Canada. Some have retained the thrust of the original British legislation which codified the lost modern grant principles with minor modification. Some eliminate prescriptive easements. Reference must be had to the appropriate legislation in each instance.

§2.56 Easements can be terminated in the following ways:

(1) by release — the owner of the dominant tenement may, by writing, expressly relinquish his right to the easement;
(2) by natural termination — if a time limit is expressed on creation of the easement and the time expires. Also, in some circumstances, the reason for the existence of the easement may disappear leading the courts to hold that the easement has been terminated;
(3) by unity of ownership — where dominant and servient tenements fall under one ownership, the easement ceases. Refer, however, to the above discussion of quasi-easements;[27]
(4) by abandonment — if the owner of the dominant tenement conducts himself in such a fashion as to imply that he no longer requires the easement. Non-use over a long period of time is evidence from which a court may infer abandonment. Prescriptive periods may come into play here;

---

24 1275, 3 Edw. 1.
25 See *Gale on Easements*, 14th ed. (London: Sweet & Maxwell, 1972) at 137-143 for a more complete view.
26 1832, 2 & 3 Will. 4, c. 71.
27 See §§2.50 and 2.51.

(5) by destruction — this applies generally to support easements and party or common walls. Where the dominant tenement is destroyed by any means, a support easement will continue. If the servient tenement is destroyed through no fault of the owner of it, the support tenement will be terminated as the law will not force the owner to reconstruct; or
(6) by expropriation — some expropriation statutes provide that where the servient tenement is expropriated, the easement will be terminated and the owner of the dominant tenement may claim compensation.

Highways

§2.57 As noted above, highways are a special class of easement and thus some special rules apply to them. A highway may be created by the owner of land by the process of dedication and acceptance. This is a two step process whereunder the owner must establish or dedicate a highway and the public must accept and use it. The dedication may be deliberate and formal as in the filing of a subdivision plan showing intent that the highway is to be public or it may be informal as in the clearing or improving of a physical roadway over the property of the owner. The acceptance may also be formal such as in the approval of a subdivision plan by the appropriate authorities as public representatives or informal such as actual use by members of the public. The actual status of a roadway may thus be in doubt as the fact of dedication or acceptance may be questioned. Provincial legislation may contain provisions whereunder the status of a roadway may be designated.[28]

§2.58 It has been said that the doctrine of lost modern grant cannot apply to the creation of a highway as one cannot make a document in favour of the public at large. Courts have avoided this difficulty by holding that where there has been long use of a roadway there must have been a dedication by the landowner and thus have developed a sort of lost modern dedication.[29]

Licences

§2.59 A licence is a personal right to certain privileges in or over the land of another. This must be distinguished from an easement which is a right for the benefit of a parcel of land not an individual. As the licence is personal, it is not hereditable — it dies with the death of the licensee.

---

28 See for example, The *Highways Act*, R.S.N.B. 1973, c. H-5.
29 See Anger and Honsberger, *supra*, note 2 at s. 2008 for a fuller discussion of the topic.

§2.60 A common example of a licence might be guests to one's home. The guests (licensees) have the right to be on the property of the owner (the licensor) as a result of the express permission of the licensor. In some circumstances, the permission may be implied, as in the delivery of mail, for example.

§2.61 As a general rule, a bare licence is revocable by the owner at any time. More complex situations can arise however. If consideration has been paid in order to obtain the licence, as in, for example, a patron at a theatre, more restrictions may be placed on the owner's right to revoke. It might, therefore, be improper for the owner to revoke the licence without some good cause, as perhaps if the patron were disturbing others. In addition, licences may be accompanied by some other contractual right which may restrict the owner's right to revoke. A common example of this would be the right of a utility company to enter on property of the owner in order to service its equipment. The contract between the parties may stipulate that the utility will have a non-revokable licence to enter on the property of the owner to service or remove equipment.

Profits à Prendre

§2.62 A *profit à prendre* is a right to go onto land of another and take something from that land.[30] The *profit à prendre* has many of the characteristics of the easement. The two distinguishing features are:

(1) there is no requirement that the owner of the *profit à prendre* own any parcel which the right will benefit; and
(2) the *profit à prendre* is restricted to the right to come and take away some part of the product of the soil or the soil itself of the subject parcel.

§2.63 The *profit à prendre* is a hereditament — it is an interest in the subject property and may be dealt with as such. It may be assigned (sold) or released and does not die with the death of the grantee. The *profit à prendre* may be limited as to time in the original grant.

§2.64 The subject matter of the *profit à prendre* must be capable of severance from the soil and must be capable of being reduced to possession by the taker. Thus, such items as soil, minerals, and timber are easily identifiable as potential subjects. Less apparent would be fish, animals, or ice. By an interesting quirk of English law, water is not capable of

---

30 For a general discussion of *profit à prendre* see Anger and Honsberger, *supra*, note 2 at 973-977.

being owned although ice is. A right to take water therefore, cannot be a *profit à prendre* but must be an easement.[31]

Options and Rights of First Refusal

§2.65 Where a potential purchaser of a parcel of land wishes to establish a situation where the owner of the land is restricted as to whom the land may be sold, there are two devices which may be employed — the option and the right of first refusal.

§2.66 An option is a contract between an owner and a potential purchaser whereunder the owner agrees that for a specified period of time the potential purchaser has the right to decide whether or not to buy the land on specified terms. The potential purchaser may do nothing and the option will lapse at the end of the specified term. The potential purchaser may exercise the option by some overt act specified in the contract. On exercise of the option, the parties become obligated to one another — the seller to sell and the purchaser to buy.

§2.67 A right of first refusal is a contract whereunder the owner of the land agrees with the potential purchaser that before entering into an agreement with any other party for the sale of the subject property he must first offer the property for sale to the holder of the right of first refusal on identical terms as offered by the third party. The holder of the right of first refusal may accept the offer and thus enter into an agreement for purchase of the property, or he may decline in which case the owner will be free to sell to the third party.

Covenants

§2.68 The owner of a parcel of land desiring to sell all or part of that land may have a legitimate interest in restricting the activity which takes place on the parcel sold. If he sells part of the whole parcel, he may wish to restrict the purchaser from carrying on some noisy or obnoxious activity which will diminish his enjoyment of the part retained. If he sells the entire parcel, he may wish to restrict what types of business may be carried on if he intends to re-establish in the vicinity. If he wishes to subdivide the entire parcel, he may wish to set up standards of construction or use which purchasers must meet. These restrictions may make the lots more valuable to potential purchasers and increase the profit of the subdivider. In all of the above situations, the vendor may employ covenants in the conveyance

---

31 *Grand Hotel Co. v. Cross* (1879), 44 U.C.R. 153.

to attempt to bring about the desired result.[32]

§2.69 The covenant is a personal contractual obligation between seller and buyer concerning use of the parcel sold. Breach of the covenant will give rise to a personal action for restraint or damages or both. The law has no difficulty with this. The parties have entered into a contract and one party has failed to meet his obligations. Difficulties do arise where the original seller or buyer or both no longer own the parcels in question. In this situation, the eventual owners of the parcels no longer are the original parties to the contract and the law says that there is no privity of contract between them. The general rule is that a contract cannot be enforced by a party where there is no privity. As with all general rules, there are exceptions.

§2.70 Where there is no privity, the rules of enforceability of a contract can be looked at in two ways: first, is the benefit of a covenant enforceable; or second, is the burden of a covenant enforceable. The owner of a parcel of land is said to have the benefit of a covenant if the owner of the other parcel involved in the covenant is required to do or not do something which enhances the value of the first parcel. The owner of the other parcel will have the burden in these circumstances. Thus, where an owner sells a portion of his property he may exact from the purchaser a covenant to the effect that no business may be undertaken on the parcel sold. The parcel sold has the burden of the covenant while the parcel retained by the original owner has the benefit.

§2.71 The benefit of a covenant will run with the land (that is, be enforceable by a successor in title to the original party to the covenant) so long as the covenant "touches and concerns the land." To touch and concern the land is to affect the land as regards its mode of occupation or its value.[33] In the above example of a covenant restricting use of the burdened parcel to non-commercial uses, a breach of the covenant would in all probability lower the value of the benefitted parcel. A successor in title to the originally benefitted parcel may successfully maintain an action for restraint or damages or both even though he has no privity of contract with the originally burdened owner.

§2.72 The burden of a covenant will run with the land if the covenant is of a negative nature (that is, it does not require some positive act or money outlay), if it relates directly to the use or value of the land and if the benefit accrues to some identifiable parcel of land. In the situation

---

32 See Anger and Honsberger, *supra*, note 2 at 899-918 for a more complete discussion of covenants.
33 *Rogers v. Hosegood*, [1900] 2 Ch. 388 (C.A.).

above, the covenant concerning commercial use of the parcel would probably be enforceable against a successor in title to the originally burdened owner so long as the successor purchased with knowledge of the covenant. The covenant is negative in nature; it relates to use of the land and the benefit accrues to the parcel originally retained.

§2.73 A final example of enforceability arises in building schemes. As seen above, an owner wishing to subdivide and sell lots may impose restrictions on individual purchasers in hopes of increasing profit. He may have little interest in enforcing the covenants once all of the lots are sold. In this type of arrangement, where a common owner creates a series of parcels and imposes a set of restrictions intended to apply to and be for the benefit of all lot owners, any owner may maintain an action based on breach of the restrictions by any of the other owners.

CONVEYANCING

Introduction

§2.74 In this section, important aspects of real estate transactions will be examined. The reader is cautioned that much of the finer detail involves individual practice which will have developed not only in particular provinces, but often in particular cities or areas. As a result, much of that detail must be omitted in favour of discussions of general principles.

Preliminaries

§2.75 The real estate transaction will begin when a land owner decides to sell. At that point, he may attempt to find a purchaser on his own or he may engage the services of a real estate sales agency. In the latter case, the owner and the real estate broker will enter into a contract commonly known as a listing agreement. The listing agreement will provide that for a specified period of time the broker may seek out potential purchasers for the property and the owner will entertain offers to purchase presented only from that broker. If a sale is successfully completed, the listing agreement will provide that the broker will be entitled to a commission. This commission typically runs from five to seven percent of sales price on a residential home and up to ten percent for vacant land or mobile homes.

§2.76 Potential purchasers will view the property and if one decides to make an offer to purchase that offer will be presented to the owner by the broker. Such offers are usually set out on a printed form of agreement of purchase and sale. The owner may ignore it altogether, accept it, or counter it — that is, modify some of the terms (generally the price) and have it referred to the purchaser for his acceptance or rejection.

§2.77 If the original offer is accepted by the owner or if the counter offer is accepted by the purchaser, the document will constitute an effective agreement of purchase and sale.

§2.78 Often the agreement will be contingent on certain other events, such as the selling of the buyer's present home or the buyer obtaining satisfactory financing for the purchase, all within a specified time period. If these events do not occur, the agreement will be considered never to have existed and the seller must start again.

Representation by Solicitors

§2.79 Once an effective agreement of purchase and sale is in place both buyer and seller will seek legal representation. In addition, the lending agency which is financing the purchaser may engage its own solicitor or may appoint the buyer's solicitor to act on its behalf in the transaction.

§2.80 The seller's solicitor is responsible for preparing closing documents. These will generally consist of a deed or transfer to be executed by the seller and a statement of adjustments or balance sheet setting out deductions from or credits to the contract selling price for such items as property taxes. In addition, other documentation may be required in some jurisdictions. The seller's solicitor will also contact the holder of the existing mortgage to obtain details on paying out the mortgage or having it assumed by the purchaser.

§2.81 The purchaser's solicitor will examine the vendor's title to the property. In a land title system, this will entail acquisition of a certificate of title for the property from the appropriate authority as well as examining any possible charges or encumbrances not covered by the certificate. Any encumbrance will be brought to the attention of the vendor's solicitor who will arrange for its payout and discharge. In a deed registry system, the purchaser's solicitor will examine the records at the local land registry office and will formulate an abstract of title for the property based on those records. The abstract will record the history of all transactions concerning the property arranged sequentially and starting at some appropriate point. The time period may be set out in applicable provincial legislation or it will be governed by practice in the area. Each document in the abstract will be examined to determine its effectiveness both in content and from the point of view of its having been executed by the proper parties. The solicitor will then formulate an opinion as to the marketability of the vendor's title. Any defects uncovered by the title examination will be presented to the vendor's solicitor and must either be corrected by the vendor or waived by the purchaser and purchaser's lender.

§2.82 The lender's solicitor will prepare mortgage documentation for execution by the purchaser. The lender's instructions to its solicitor will define the lender's requirements. These will generally include the acquisition of a plan prepared by a licensed surveyor showing the boundaries of the parcel involved and any structures thereon. This will enable the lender's solicitor to establish that there are no encroachments and that applicable set-back requirements have been met.

§2.83 On the date established for completion of the transaction, the solicitors will meet. The purchaser's solicitor will have accumulated funds from the purchaser and the lender sufficient to pay the adjusted closing price. The vendor's solicitor will have in his possession an executed deed or transfer and the keys to the property. The solicitors trade the above items and purchase funds are held in escrow until the purchaser's solicitor has advised the vendor's solicitor that he may take the funds out of escrow. The vendor's solicitor will pay out any existing mortgage on the property as well as the real estate commission owing and any other obligations which he has undertaken to satisfy. The remainder will go to the owner and the transfer is complete.

Mortgages

§2.84 The mortgage is a conveyancing device used to facilitate lending. The property mortgaged or charged acts as security to the lender that the monies lent will be repaid. The word mortgage is said to have been derived from the Norman-French *mort* (dead) *gage* (pledge). In its earliest forms, the mortgagor (borrower) actually gave up possession, and thus profits, to the mortgagee (the lender). The land had thus been pledged and was useless (dead) to the borrower.[34]

§2.85 The development of the mortgage is very much involved with the courts of equity.[35] At common law a conveyance of land was just that — a conveyance. Once the words of grant were used in a document, the common law allowed no retreat from that position. The land had been conveyed and no limitation could be placed on that — no one would be entitled to derogate from his grant. The need for an effective security

---

34 See *Falconbridge on Mortgages*, 4th ed. (Agincourt: Canada Law Book, 1977) at 3 and 4.

35 For a general discussion of the development of the mortgage see *Falconbridge on Mortgages, ibid.* at chapters 1 and 2.

document was so great that by the end of the 15th century, the common law courts had come to accept as valid a document developed by conveyancers which was, in effect, a deed "subject to a provision for disfeasance." This disfeasance, or undoing, was very strictly limited by the common law court. A borrower was required to make payments at the appointed times or he lost any chance of getting the property back and he still owed the balance of the debt. The courts of equity soon stepped in and under their approach, the right of the mortgagor developed into a right to redeem. Simply put, the mortgagee had legal title to the property under the mortgage subject to the equitable rights of the mortgagor to redeem the mortgage and get the property back. In order to perfect his legal title on default by the mortgagor, the mortgagee had to go through a process of eliminating the mortgagor's right to redeem. The mortgagee could generally only commence proceedings to perfect his title if the mortgagor was in default under some provision of the mortgage — generally the provision for payment. At any time before completion of the perfection process, the mortgagor could remedy the default, bring his payments up to date, and the perfection process would have to stop. On complete payment of the debt, the mortgage was said to be redeemed. This has the effect of rendering the mortgage a nullity and conveying legal title back to the mortgagor.

§2.86 This is the situation which exists under present day rules in some provinces. Many of the respective rights and obligations of mortgagor and mortgagee have been codified under legislation and reference should be made to that legislation for details in each province. Those areas having systems of title registration have generally replaced the mortgage with a document called a charge. The charge does not have the effect of conveying the legal title to the mortgagee, but instead, gives the mortgagee a method of instituting such a conveyance on default.

LEGAL PRINCIPLES AFFECTING OWNERSHIP

Introduction

§2.87 The following section is intended to give an introduction to a number of concepts which affect real property law. It cannot be too frequently repeated that space constraints preclude all but the most rudimentary discussions of these principles. For more detail, reference should be made to more specialized works.

Limitations of Actions

§2.88 The common law has long held to the notion that if an individual is wronged in some fashion, the courts will only hear an action within a reasonable time after the cause of action arose. At common law, this principle became known as the doctrine of laches. Legal theorists argue that the principle is based on some sort of basic unfairness in allowing a plaintiff to pursue a claim following a long delay — a delay that might have caused the defendant to do something to his own detriment.

§2.89 It is the effect of the principle which is important here — the court makes no ruling whatsoever on the merits of the plaintiff's claim. All that happens is that the court refuses to hear the case at all.

§2.90 The principle of laches has been incorporated into the statute law of all common law provinces by way of *Limitations of Actions Acts*. These statutes set forth a number of situations where after a certain specified period of time has elapsed, any action will be barred and a court will refuse to hear it.[36] It is important to note that in some situations, time periods in the Acts may be extended, as for example, if the potential plaintiff was an infant or under a mental disability at the point time would normally have begun to run.

§2.91 The implications of the above on real property law are reflected in the principle of adverse possession. In general terms, if B is trespassing on the lands of A, and A desires to have the courts help him get rid of B, A will commence an action for ejection. If B can establish before the court that A's cause of action has been in existence for longer than the appropriate statutory period as outlined in the *Limitations of Actions Act*, the court will refuse to go any further. The court will make no pronouncement on the title of B to the land in question — it will only refuse to hear A's claim. The statute is said to operate in a negative fashion in that it only extinguishes rights. It does not create any title in the trespasser. B will have to follow some other course in order to have the court confirm title to the land in him.

§2.92 A set of requirements has arisen which a trespasser will be asked by the court to meet in order to invoke the protection of a *Limitations of Actions Act*.[37] These are:

(1) possession must have been actual and exclusive. The trespasser (or his agent) must be physically occupying the land to the exclusion of all others. Courts will consider the type of land involved in deciding on what level of possession is to be required. If, for example, the land

---

36 See, for example, *Limitations of Actions Act*, R.S.N.B. 1973, c. L-8, s. 29 which provides "No person shall take proceedings to recover any land but within twenty years next after the time at which . . . the right first accrued to the person taking proceedings. . . ."
37 See *Sherren v. Pearson* (1887), 14 S.C.R. 581.

is arable and the trespasser has planted and harvested it, even though he has not actually resided on it, the courts may consider those acts to be possession to the extent to which the land was capable;
(2) possession must have been open and notorious. The acts of possession must have been free for all the world to see and must have been without the permission of the true owner; and
(3) possession must have been continuous. Occasional adverse use of the property will not be sufficient to result in a successful claim of protection under a *Limitations of Actions Act*.

Colour of Title

§2.93 Very similar to adverse possession discussed above, colour of title arises when a trespasser occupies land by virtue of a deed or other title document which is defective in some aspect. Even though there might be some individual with a better documentary title to the property, the possession of the trespasser will run in the same fashion as in adverse possession. The essential difference is that under adverse possession, the true owner's claim will only be barred to those lands actually possessed, while under colour of title, possession by the trespasser of any part will be extended and be considered possession of all of the lands described in the title documents.[38]

Prescription

§2.94 The above description of adverse possession and colour of title outlined how rights to possession of property might be obtained. Prescription is a parallel process whereby rights to use, as opposed to possession, are obtained. In adverse possession it is necessary for a trespasser to establish possession of the property. Under the doctrine of prescription, it is only necessary to establish use of the property.

§2.95 As with adverse possession requirements concerning the nature of the possession, in order to invoke the doctrine of prescription the claimant must be able to establish a certain quality of use. The quality of the use required has been variously described as *nec vi, nec clam, nec precario* (neither as the result of force, nor secrecy or evasion, nor as dependant upon the consent of the owner)[39] or " . . .enjoyment had not been secretly or by stealth, or by tacit sufferance or by permission".[40] Many of the issues

---

38 See *Wood v. LeBlanc* (1904), 34 S.C.R. 627.
39 Sinclair, *supra*, note 12 at 45.
40 *Tickle v. Brown* (1836), 111 E.R. 826.

50 Survey Law in Canada

involved in an action concerning adverse possession will arise in prescription cases. The quality of use will be a factual issue to be determined in each instance.

§2.96 Another critical issue in prescription cases involves the duration of the use. At common law, the use had to be proven from time immemorial. In England, through some ancient judicial reasoning, that time was fixed at 1189, the beginning of the reign of Richard I. As noted above, proof of use from that date would be impossible in Canada.

§2.97 In any event, the common law developed a convenient fiction to circumvent the onerous requirements of proving use from time immemorial. The doctrine of lost modern grant[41] has developed so that where a user can establish use for a 20 year period, the court can invoke a presumption that the owner or some predecessor in title to him had actually made a grant of the right and that the document had been lost. Thus the situation was no longer one of prescription but one of grant.

§2.98 Finally, many common law provinces have addressed the situation by introducing legislation. In several provinces, a uniform *Easements Act* is in force which sets down time periods required to establish prescription and to define the resultant relationship between the owner and the user.

Expropriation

§2.99 The term expropriation is derived from the Latin *ex* (out) and *proprium* (property). It is used in the sense of real property law to describe a taking away of some property right from an individual by the state or some other individual given that power by the state. As discussed earlier, ownership of real property in common law provinces actually vests in the Crown.[42] That ultimate ownership is often used as a theoretical basis for the power of the state to expropriate. It would appear, however, that this foundation is not necessary as most, if not all, jurisdictions which have allodial systems of ownership of real property (as, for example, some American states) recognize some form of expropriation power.

§2.100 All of the expropriation mechanisms existing in Canada today are based on statutory schemes. Each of these statutes must be examined in its own context for specific provisions. The most that can be attempted here is a discussion of the common approaches to be found.

§2.101 Expropriation statutes generally break down into three parts —

---
41 See §2.54.
42 See §§2.07 — 2.10.

requirements which the expropriating authority must meet in order to be able to invoke the power of expropriation, mechanics of the actual taking, and compensation to the owner and any others who have suffered loss.

§2.102 The expropriation statutes attempt to strike a balance between the interests of the expropriating authority and the affected land owner. Some sort of public hearing is generally required where the expropriating authority presents its case and land owners may question that case and present other views. After all of the presentations are in, the adjudicator[43] at the hearing must make some decision as to whether or not to allow the expropriation to take place. In that decision, the adjudicator must consider whether or nor the expropriation

> (a) is reasonably necessary to accomplish the objectives of the expropriating authority or applicant,
> (b) is fair, balancing the objectives of the expropriating authority or applicant against the interests of the owner that would be extinguished by the expropriation, and
> (c) in the case of an application made under section 7, [an expropriation by a private individual] is consistent with the public interest.[44]

If the adjudicator decides that the expropriating authority has not established that the proposed expropriation meets the requirements of the Act, the application will be turned down. Otherwise, the application will be permitted, subject to some right to appeal that decision.

§2.103 Each individual Act must be consulted to determine what procedure is to be followed in the expropriation process.

§2.104 All expropriation schemes require that the expropriating authority compensate a landowner whose interests are affected either directly by loss of some or all of his land or indirectly by a diminution of the value of his land as a result of the process although none of his land was taken. In order to arrive at an appropriate compensation level, the Expropriation Acts generally set out three phases which the parties go through — negotiation, arbitration, and litigation.

Concurrent Ownership

§2.105 Where title to real property is vested in more than one individual, that communal ownership must be characterized as one of:

---

43 The adjudicator is generally an appointed official whose work is restricted to expropriation hearings only.
44 To use as an example, the words of the New Brunswick *Expropriation Act*, R.S.N.B. 1973, c. E-14, s. 17(2).

(1) joint tenancy,
(2) tenancy in common,
(3) co-parcenary, or
(4) tenancy by the entireties.

The last two categories appear to be of purely historical interest and will not be dealt with here.[45]

§2.106 Whether a communal ownership is a joint tenancy or a tenancy in common is important principally because of what happens on the death of one of the owners. In a joint tenancy, the interest of the dying owner disappears. If the joint tenancy was originally between only two co-owners, the survivor will hold the entire estate. If it was between more than two co-owners, the survivors continue as joint tenants. Eventually, of course, only one of the original co-owners will remain and that person will take the entire estate. This principle is known as the right of survivorship. In contrast, under a tenancy in common, the interest of the dying co-owner will pass under the provisions of his will or the appropriate Devolution of Estates Act if there is no will. The tenancy in common will continue with the original tenants in common and the successor(s) to the dying co-owner.

§2.107 As a prerequisite to a discussion of how to tell a joint tenancy from a tenancy in common, it is necessary to examine four conditions which must exist in a joint tenancy. These conditions have been termed the four unities and they are:

(1) unity of possession — each joint owner must be entitled to possession of all of the property in question subject to identical rights of possession in all of the other co-owners;
(2) unity of title — all co-owners must have derived title from the same instrument — (be it a will or a conveyance) or act (as in the case of title acquired by adverse possession);
(3) unity of time — the interest of each co-owner must vest at the same instant in time; and
(4) unity of interest — all co-owners must have the same bundle of rights to the property.

§2.108 In order to determine whether a given situation is a joint tenancy or a tenancy in common, it is first necessary to look for the four unities. If they are not present, the co-ownership cannot be a joint tenancy. Even if the four unities are present, it is possible that the co-ownership is a

---

45 See Anger and Honsberger, *Law of Real Property*, 2d ed. (Aurora: Canada Law Book, 1985) at 827 — 830.

tenancy in common. The nature of a co-ownership depends on the intent of the creator of the co-ownership. That intent is to be found in the document creating the co-ownership or, if that document is silent or equivocal on the point, in the statutory provision found in all common law provinces. The statutory preference is toward tenancy in common.[46] This is apparently based on some notion of fairness — that is, that it is more equitable to allow a co-owner's heirs to take his interest than to have it disappear altogether on his death.

§2.109 If a joint tenancy exists, there are several methods available to sever it — that is, to convert it into a tenancy in common. This severance may be unilateral, as for example, where one joint tenant conveys his interest to a third party thus destroying the unities of time and title. The severance may be mutual, as for example, by agreement of all the joint tenants. The severance may also be as a result of some act of a third party, as for example, where the interest of one joint tenant is seized by a creditor and sold.

Mechanics Liens

§2.110 Mechanics liens exist by virtue of specific legislation enacted in all common law provinces. The general thrust of all of the provincial statutes is to provide protection in the form of a lien to contractors, sub-contractors, wage earners and suppliers who have been involved in the construction of something (generally some sort of structure) which is affixed to a definable parcel of land. The Acts provide that a member of the above classes becomes possessed of a lien on provision of material or service and provides for the preservation and enforcement of the lien against the property should the lienholder not be paid.

§2.111 The various Acts also make provision for the owner of the property to protect himself from claims (by sub-contractors, wage earners, or suppliers who have not been paid) by holding back from payment to the general contractor a specified percentage of the contract price (or the value of the work) for a specified time period beyond the date of completion of the work. If no claims are filed by the lienholders within the specified time periods, the liens cease to exist and no claim can be made against the property. At that point, the owner can release the payments which were held back to the general contractor. If claims are filed within the

---

46 See for example New Brunswick's *Property Act* (R.S.N.B. 1973, c. P-19, s. 20) which provides "An estate hereafter created, granted or devised to two or more persons in their own right, shall be a tenancy in common, unless expressly declared to be a joint tenancy. . . ."

specified time periods, the owner will be protected if he has held back the specified percentage in that lienholders will not be able to claim against the owner's property, but only against monies held back. This is the case even if the total of the claims far exceeds the amount which was held back. The unpaid claimants will have to pursue their actions against the party who should have paid them for any amount not covered by the held back monies.

§2.112 Perhaps the most convenient method of making the above more clear would be to follow a simplified example of the construction of a residential home. The owner of the land will typically enter into a contract with a builder (the general contractor) for construction of the home. That contract will generally provide for progress payments to be made to the builder at various stages of completion of the project. The builder will employ labourers (wage earners), will enter into contracts with sub-contractors for such items as electrical or plumbing systems and will purchase materials from a variety of suppliers. In addition, the sub-contractors will purchase materials or employ wage earners or both. All of these individuals will have a lien on the property on supply of material or effort. That lien is created by the Act and will remain in existence for a time period specified in the Act. If the owner pays the builder and the builder pays his suppliers, sub-contractors and wage earners and the sub-contractors pay their sub-sub-contractors, wage earners and suppliers (and so on down the line), there will be no difficulties and the liens will all expire on the lapse of the lien period. If, however, the chain breaks down at some point and someone does not get paid, a claim will be filed following the procedures outlined in the Act. The filing of the claim will preserve the lien beyond the time period in the Act. The ultimate remedy of a claimant under the Act is sale of the property and payment of the claimant(s) out of the proceeds of that sale. In order to avoid that possibility, the owner will hold back a percentage, as defined under the Act, from each progress payment made to the builder. If the chain of payment breaks down and claims are filed, the owner may ask the courts to distribute the monies held back to the claimants and no further action can be taken against the property. If the owner has neglected to hold back the appropriate amounts, the property might very well be sold to satisfy the claims.

Trusts

§2.113 Where one individual holds the legal title to property, but holds it for the benefit of another, a trust exists. Legal title is held by the trustee and beneficial or equitable title is held by the *cestui qui trust*, more commonly called the beneficiary. The person who sets up the trust by conveying legal title to the trustee subject to the trust is termed the settlor. The distinction between the legal and equitable title was most important at a time when there were separate courts of equity. At that time, common law courts would only recognize the conveyance to the trustee but not the restrictions imposed on the conveyance. The beneficiary would have had to resort to the courts of equity to have his rights recognized. Following merger of the common law and equity courts, this distinction became less critical but for a complete understanding of some of the finer aspects of the law of trusts, it is still important.

§2.114 Trusts are most often created voluntarily by the settlor. He will convey title to the trustee and in some fashion establish what the trustee is to do with the property for the benefit of the beneficiary. In each case, the document or agreement creating the trust must be examined in order to determine the rights of the respective parties. It might be, for example, that under a will a parent devises real property to a trustee to hold for the benefit of minor children. The trustee may be empowered to allow the children to reside on the property, to rent it and apply the rental income for the benefit of the children or to sell it and hold the proceeds of sale for the benefit of the children. As well as the obligations imposed on the trustee by the settlor, all provinces have legislation which will more completely set out a trustee's duties and responsibilities as, for example, what investments a trustee may make with trust monies and guidelines for avoidance of conflict of interest.

§2.115 While many trusts are voluntarily created, courts will often impose a trust in a situation where that does not appear to be the intention of the parties involved. These circumstances are almost always cases where it would be unfair, in the court's mind, to allow the holder of the legal title of real property to reap all of the benefits of that ownership. Before the introduction of the various *Matrimonial Property Acts*, courts often held that even though legal title to, for example, a family home was held by one of the spouses, on marriage breakup it would have been unfair to allow the holder of the legal title to disregard the other spouse. A trust would be imposed by the court whereunder the holder of the legal title would be deemed to be holding on behalf of both spouses. In addition to this family situation, there are many other circumstances where a court will exercise this capacity to impose a fair resolution.

Matrimonial Legislation

§2.116 As marital breakdown became more and more prevalent in the 1970's and 80's, provincial legislatures turned their attention to the division of assets of a couple on failure of marriage. The result has been a series of statutes (referred to herein as the *Matrimonial Property Acts*) which address this complex issue.

§2.117 Common law had long considered a husband and wife as one entity.[47] All of the rights of the wife, including property rights, were abated during the marriage. Through a long process of evolution, the courts of equity had developed principles which afforded some protection to the wife by the 1800's. In addition, the introduction of various Married Women's Property Acts in the late 1800's allowed a wife to acquire, hold, and dispose of real property independently from her husband. This situation existed until the recent introduction of the various *Matrimonial Property Acts*. The Acts must each be examined for specific provisions. However, some generalizations can be made.

§2.118 Typically, the *Matrimonial Property Acts* do not interfere with the existing property rights of spouses until some particular event occurs. That trigger may be marriage breakdown or, in some cases, death of one spouse. On the occurrence of this event, spouses will be entitled to make application to a court for a division of some of the assets of the couple. The affected assets are variously defined but the marital home is always dealt with. The legislation establishes guidelines which the court must consider in coming to a division of the assets. As a rule of thumb, an equal split of property is indicated, however some flexibility is retained by the courts which may be reflected in an unequal split of assets based on conduct of the parties.

§2.119 It must be noted that this division of property can be undertaken without regard to the status of the registered title to the property in question. In addition, it is possible in all jurisdictions for spouses to contract out of the provisions of the applicable Act in certain circumstances and within certain limits by use of marital contracts or cohabitation agreements.

Planning and Land Use Control

§2.120 The time has long passed when an individual landowner could do what he pleased with his property.

§2.121 The common law has always been reasonably effective in governing the relationship between two specific landowners. Thus, where one owner is using his land in a fashion which is injurious in some way to his neighbour, the neighbour would have the right to commence an action for damages and, more importantly, for an order restraining the offending

---

47 For a review of the status of a married woman see Mendes da Costa, *Studies in Canadian Family Law* (Toronto: Butterworths, 1972) at chapter 5.

Real Property Law   57

use. Such situations are very much a part of modern life, as evidenced by a growing series of actions by homeowners against farmers claiming that odours created by farming amount to a nuisance and should be curtailed. In some provinces, the legislatures have passed right to farm legislation which restricts, in some circumstances, the right to commence such a nuisance action.

§2.122 The common law has not been very successful in addressing a situation involving a balance between one landowner's right to use his land and the public good. The response to this difficulty has been the enactment of legislation which is aimed at defining and protecting the public good by establishing a framework of rules governing the use of property. For present purposes, these statutes can be broadly classified into two groups — zoning legislation and indirect legislation.

§2.123 Zoning legislation is used here to denote a series of direct controls over the use and development of land. Typically, these controls include:

(1) the establishment of some form of plan for development of a community;
(2) the establishment of a series of zones or areas within the community and a determination of what uses will be permitted in each zone;
(3) the regulation of the process of development of raw land; and
(4) the regulation of improvements or alterations to existing parcels of land.

§2.124 The responsibility for establishment and enforcement of these controls may fall to the federal, provincial or municipal levels of government, depending on the land in question or the use involved. The federal government will be responsible for controls in fields where its jurisdictional responsibilities lie, as for example in federal parks, on Indian reserves or in the vicinity of airports. The provincial governments play a major role in land use control pursuant to their responsibilities under section 92(13) of the *British North America Act*[48] (property and civil rights). These responsibilities are often delegated by the provinces to municipalities so that control will exist at the local level.

§2.125 Indirect legislation is intended here to cover all controls, other than zoning legislation, over the use of land. Typical of indirect legislation governing uses of land are:

(1) pollution control regulations;
(2) hunting and fishing controls;
(3) prospecting and mining controls;

---

48 1867, now the *Constitution Act, 1867*, (U.K.), 30 & 31 Vict., c. 3, s. 92(13).

(4) statutes prohibiting or regulating criminal acts (such as prostitution or gambling); and
(5) taxing statutes.

There are several other examples. All of these statutes, while dealing principally with some subject other than land use, can indirectly affect the use a landowner can make of his land.

Condominiums and Air Space Parcels

§2.126 One of the fundamental principles of real property law is that a building constructed on a parcel becomes part of that parcel. Any conveyance of the parcel will automatically include buildings or other improvements. A corollary to this principle is expressed in the Latin *dominus soli est coeli et inferorum vel usque ad infero* (property rights extend up to the heavens and down into the earth). The combination of these two principles would seem to render it impossible for the common law to recognize ownership of, for example, an apartment on an upper floor of a large building. The owner of the ground floor apartment would, in theory, own all above him.

§2.127 Economic forces tending to favour multi-level construction and social forces tending to favour ownership, as opposed to rental, run counter to these difficulties. The result has been the development of the condominium and air space parcel legislation.

§2.128 In its simplest form, a condominium is made up of:

(1) division of real property, horizontally and vertically, into individual freehold units;
(2) allocation of the remaining real property as common areas owned in tenancy in common by all the unit owners; and
(3) establishment of an administrative framework to manage the scheme.

§2.129 All jurisdictions in Canada now have statutes enabling development of condominium projects and this section will deal almost exclusively with that legislation. In contrast, however, there are some situations existing outside of any statutory framework which entails forms of strata ownership. Notable among these are flat ownership in Great Britain and co-operative ownership in some American states. A full examination of these common law condominiums is beyond the scope of this work.[49]

§2.130 The condominium statutes in the common law provinces all provide for establishment of a condominium scheme by some positive act on the

---

49 See Anger and Honsberger, *supra*, note 45 at chapter 38.

part of the owner of the subject parcel. All require some form of plan to be filed with the appropriate authority delimiting the boundaries of the individual units and the common areas. In all but Alberta, Saskatchewan, and British Columbia, some form of declaration or constitution must also be filed. This declaration or constitution will contain much of the administrative framework for the management of the project. In the above-mentioned western provinces, the statutes provide that framework. All statutes provide for the creation of a condominium corporation as part of the scheme, individual taxation of the units, enforcement of common area charges, deregistration of the condominium, what happens on partial or total destruction of the building(s) housing the units, and insurance of the units and the common areas.

§2.131 The result of all of this is that upon establishment of a condominium, an individual is free to purchase what is essentially a cube of space. The cube is defined by reference, through filed plans, to the building surrounding it. It is important to note that the ownership of the cube of space is entirely independent of any ownership of some definable parcel of the surface of the earth. The cube may be dealt with in exactly the same fashion as any traditional parcel of land, subject to the rules of the condominium.

§2.132 As an extension of the condominium concept, some jurisdictions are introducing legislation permitting owners of a parcel to deal in an economic fashion with the air space above the ground level. That economic use may take a number of forms from absolute conveyance to splitting off individual rights associated with the air space parcel. The former is most often encountered in large, multi-level construction projects. It would be possible under an air space parcel statute to separate ownership of a mall from ownership of an underground parking lot below the mall. The latter type of use (splitting off individual rights) can occur in several situations. A common example often arises in large urban areas. Cities often place height restrictions on central areas. These restrictions tend to apply to the total height of all the buildings in a central area and owners who wish to exceed an average maximum will be entitled to purchase extra height credits from owners of lower buildings.

Wills and Estates

§2.133 A person may die testate (with an effective will), intestate (with no effective will), or partly testate and partly intestate (with a will which does not provide for disposal of some of his property). On the death of an individual some person or corporation may be appointed by the appropriate authority to represent the deceased in dealing with his assets.

This individual is termed the personal representative. If the deceased was testate, he will in all probability have named a personal representative in his will. In those circumstances, the personal representative will be called the executor. If the deceased was intestate, or if his will did not name a personal representative, the individual appointed will be called the administrator. The personal representative is responsible for managing the deceased person's assets (his estate) in compliance with the terms of the will, or if there is no will, in compliance with the applicable *Devolution of Estates Act*.

§2.134 A will may provide for several things to happen in regard to real property. It might instruct that the property be held for the benefit of minor children, be sold and the proceeds divided, or be conveyed directly to one or more individuals.

§2.135 A Devolution of Estates Act is often referred to as a statutory will. The various statutes dictate how an intestate's estate will be divided among his relatives.

Escheat and Forfeiture

§2.136 At common law, if a person died intestate with no heirs, all of his property reverted or escheated to the Crown. As discussed above, under English common law, the Crown held the absolute ownership of all real property. This was therefore, the logical result of failure of heirs. In Canada today, all common law provinces have enacted legislation which retains the Crown's rights under the principle of escheat at least to some degree. Reference should be made to the applicable statute in each province.

§2.137 In addition to escheat for lack of heirs, English law provided that where a person was convicted of treason or some other serious criminal offence, all his property would escheat to the Crown. These latter provisions have now been overruled by criminal legislation and are no longer applicable in Canada.

FUTURE DEVELOPMENTS

§2.138 In discussing the history and evolution of real property law earlier in this chapter, it was asserted that any law is a reflection of the society in which it exists. When speculating on future development in real property law it is therefore necessary to examine the forces in society which tend toward change in the relationship between individuals and in the use of land. What follows is a review of some of the areas where change can be expected.

Real Property Law 61

§2.139 One of the most interesting developments in land use in the past few decades has been the re-emergence of the downtown core of cities. This has been reflected by a re-working of the use patterns of large areas including renovation of residential buildings, redevelopment of commercial buildings into residential or retail space, and interconnection of commercial spaces with underground or aerial walkways. Some projects extend over the banks of rivers or harbours and may require multi-level division of ownership. All of these uses, and no doubt several yet to be created, require innovative approaches to the real property law issues involved. Some of the problems encountered can be solved using existing common or statute law, but many will require new legislation.

§2.140 As the population increases and land use becomes more competitive, it can be expected that conflicts between neighbouring owners over the use of land will increase. This will be especially apparent in areas where residential use borders commercial use. It has already been seen in conflicts between farmers and neighbouring residents over noises and odours typical of some farming operations and in concerns raised over pollution of ground water by service stations or other suburban commercial uses. Further conflicts have arisen as a result of the recent trend to attempt to integrate mentally handicapped persons or paroled offenders into neighbourhood settings. These conflicting interests have political and social aspects; they will certainly involve the introduction of legislative schemes in an attempt to establish appropriate priorities.

§2.141 The changing demographics of society also place strains on the law of real property. The aging of the population has been reflected by an increasing number of elderly persons who wish to retain their independence. Two developments which assist are the reverse mortgage and the granny flat. A reverse mortgage can be helpful where a retired person or couple do not have sufficient income to maintain a desired standard of living but do have a home which is free of mortgage. In that case, the homeowner can put a reverse mortgage in place which allows a monthly income to be drawn which is based on the equity of the home. On death, the home will be sold to pay off the accrued debt. In circumstances where an older person does not wish to maintain or is not capable of maintaining a separate home, a granny flat may be an ideal solution. The granny flat is a self-contained apartment-like structure which is placed on the property of a relative. The older person may live independently but still be close enough to relatives for security.

§2.142 Another demographic change resulting in changes to real property law has been the increase in divorce and separation. This issue has been

62   Survey Law in Canada

discussed above.[50] Further changes in the law to accommodate these pressures can certainly be expected.

§2.143   The issue of aboriginal title claims has also been discussed above.[51] This especially is an area where extensive change can be expected in the laws relating to ownership and use of land, even to the fundamental concept of Crown ownership of land.

§2.144   All the above and many other issues illustrate how societal pressures lead to changes in the law of real property. Any predictions relating to future developments in the law must, therefore, be related to predictions of the needs of society.

REFERENCES

Anger and Honsberger. *Laws of Real Property*, 2d ed., (Aurora: Canada Law Book, 1985).

Falconbridge on Mortgages, 4th ed., (Agincourt: Canada Law Book, 1977).

Gale on Easements, 14th ed., (London: Sweet & Maxwell, 1972).

Halsbury's *Laws of England*, 4th ed., (London: Butterworths, 1982).

Mendes da Costa. *Studies in Canadian Family Law*, (Toronto: Butterworths, 1972).

Sinclair. *Introduction to Real Property Law*, (Toronto: Butterworths, 1969).

Slattery. *The Land Right of Indigenous Canadian Peoples, as Affected by the Crown's Acquisition of Their Territories*, (Oxford: University of Oxford, 1979).

---

50   See §§2.116 to 2.119.
51   See §§2.21 to 2.24.

3

Land Registration Systems

Donald McLaurin Lamont, Q.C.*

INTRODUCTION

§3.01 This chapter will be general in nature, devoted more to the promotion of an understanding of the basic principles governing the title registration systems in the common law provinces of Canada, than as a legal reference for resolving disputes. Adequate treatment of the statutory laws and regulations, and the case law applicable to the system in any one province would require a book, not a chapter. Add the diversity between provinces, and it is readily apparent that this chapter can be no more than an introduction.

§3.02 One of the primary purposes of a legal system is to provide a set of rules to enforce at least a minimum level of morality. The situations which create legal problems in a land registration system are few compared to the number of registered transactions because most people exceed the minimum level of morality in their dealings with each other and take precautions to protect themselves from those who don't. One measure of the effectiveness of a title system is the ready availability of accurate information to enable these precautions to be taken.

---

* C.W. MacIntosh Q.C., Director of Land Titles & Legal Services, Land Registration and Information Service, Halifax, Nova Scotia provided the information about the New Brunswick Land Titles Act and reviewed the manuscript in its final stage. His assistance is gratefully acknowledged.

DEED REGISTRY SYSTEM

Basic Principles

§3.03 The system provides for the registration of deeds or assurances of title and is dependent for its efficacy upon the registration in a public office of all instruments affecting title to a parcel of land. The legal effect of each instrument is derived from common law or statutory rules.

§3.04 The registered documents form a chain, and each instrument in the chain must be examined, usually by the solicitor who is acting for an intending purchaser or lender, to ensure that the owner shown as grantee in the last deed in the chain has good title, subject only to agreed encumbrances, easements or interests affecting the parcel. The various provinces have enacted statutes which provide a limitation period for the recovery of title to land. It may not be necessary to search back more than the number of years specified in the statute to a deed which would form a good root of title. In some cases it is wise to search back to the original grant of the land from the Crown in order to determine whether or not there are reservations in the grant itself or in subsequent deeds, which have been omitted from the land description in later deeds. The decision as to how far back to search would be influenced by the history and use of the land; for example, a missed mineral reservation would have less impact upon the purchaser of a small city residential lot than upon the purchaser of a large rural parcel.

§3.05 A deed registry system works well if the boundaries of the land parcels are properly defined in the registered instruments and can be ascertained with reasonable certainty from monumentation or natural features; if the instruments are carefully recorded in a public office using an indexing system which enables every registered instrument which affects a specific land parcel to be retrieved quickly; and most importantly, if the risk assumed by a person who fails to register is such that registration under the system is essentially compulsory.

Overriding Interests

§3.06 It is safe to state that in all provinces there are interests in land which may be enforced against a registered owner without being registered under the system. These are generally referred to as overriding interests. The fewer the number of overriding interests permitted under provincial laws, the easier it is for an intending purchaser to ascertain the state of the vendor's title.

Land Registration Systems 65

Defective or Missing Deeds

§3.07 There may also be defective documents in the chain of title, or there may be an essential deed missing from the chain. In case the defect cannot be remedied or the missing deed cannot be found or replaced by a new deed, it may be necessary to have resort to the courts to obtain an order under a Quieting of Titles statute.

Indexing Methods

§3.08 Which option our ancestors selected when the indexing system was established has a major effect upon the ease with which registered documents can be retrieved today for the purpose of searching or copying. To change an indexing system that has been in place for generations to another system which is perceived to be better is an enormous undertaking.

§3.09 The Maritime Provinces have been faced with the task, not only of changing the indexing system but, of converting to a torrens system based upon a comprehensive surveying and mapping program. The indexing system which has been in use in those provinces since the inception of the registration system is what is commonly known as a grantor/grantee index. With certain exceptions permitted by the statute, documents are copied in full in separate registers for each county.[1] An alphabetical index is maintained under which the names of both grantor and grantee are entered under the proper letter.[2] It is obvious that with the passage of time, such a system could become awkward to use. Some of the problems which might have arisen have been avoided due to collections of particulars of the registered records being kept outside of the registry offices, notably by long-established law offices.

§3.10 The other system of indexing which is in use in Canada under the deed registry system is based on land parcels. It is the method adopted in Ontario and copied by Manitoba when the *Registry Act* was passed in that Province. Under that Act particulars of a document are entered in a register under a separate head for each lot or part of a lot as originally patented by the Crown or as defined on a subsequent plan of subdivision.[3] Hindsight indicates that the system would have been improved to some extent if new headings had been established each time a parcel was subdivided by any method.

---

1 *Registry Act*, R.S.P.E.I. 1974, c. R-11, s. 14; *Registry Act* R.S.N.B. 1973, c. R-6, s. 50; *Registry Act*, R.S.N.S. 1967, c. 265, s. 26.
2 *Ibid.*, P.E.I., s. 10; N.B., ss. 71(1)a, 32; N.S., s. 10.
3 *The Registry Act*, R.S.M. 1987, c. R 50, s. 41.

§3.11 The parcel method of indexing, outlined above, applies to documents which contain a description of the land parcel affected. There are other instruments which may be registered that for various reasons do not contain a land description, but instead create an interest in or affect dealings with all of the lands of a person. Examples are government liens, writs of execution or certificates of judgment, and grants of probate or letters of administration. These documents are indexed under alphabetical headings against the name of the person whose interest in land is charged or otherwise affected. The indexed particulars of the registered instruments and the instruments themselves constitute the general register. Registration in the general register is permitted under the deed registration system and, in most of the provinces which have a torrens system, it is also permitted under that system.

Priority

§3.12 Priority of registered over unregistered instruments is, as mentioned earlier, fundamental to the deed registration system. Sections 55 and 56 of the *Registry Act* of Manitoba illustrate the statutory establishment of this priority.

> **55**. Except as mentioned in sections 57 and 58, priority of registration under this Act in all cases prevails, unless, before any such prior registration, there has been actual notice of the prior instrument to the party claiming under the prior registration.
>
> **56**. Except as mentioned in sections 57 and 58, any instrument that may be registered in pursuance of this Act, affecting any lands whatsoever situated in Manitoba, whether there has been any grant from the Crown of those lands or not, shall be adjudged fraudulent and void against any subsequent purchaser or mortgagee for valuable consideration, without actual notice, unless the instrument is registered in the manner in this Act directed before the registering of the instrument under which the subsequent purchaser or mortgagee may claim.

It must be noted that in order for a subsequent purchaser or mortgagee to establish priority over a person claiming under a competing instrument, executed prior to the instrument under which he claims, he must have paid valuable consideration; in other words, the section does not protect the recipient of a gift. Also, he must not have had actual notice of the prior instrument.

§3.13 Similar provisions are found in subsection 19(1) of the *Registry Act* of New Brunswick and section 43 of the *Registry Act* of Prince Edward Island which read as follows:

> **19**(1) All instruments may be registered in the registry office for the county

where the lands lie, and if not so registered, shall, subject to the provisions of subsections (3), (4) and (5), be deemed fraudulent and void against subsequent purchasers for valuable consideration whose conveyances are previously registered.

43 No constructive or other notice of any unregistered deed or mortgage shall defeat, impeach, or affect, or be construed to affect, any deed or mortgage relating to all or any part of the same lands, tenements or hereditaments, which has been registered under this Act, but every such unregistered deed or mortgage shall be deemed to be fraudulent and void against subsequent purchasers or incumbrancers for valuable consideration, whose deeds or mortgages are previously registered, whether the purchasers or incumbrancers had notice thereof or not, but nothing in this Act affects, or is construed to impeach any will, or security for a debt due or to become due to the Crown, although it or a memorial or entry thereof is not recorded in the office of a registrar.

The exceptions to the application of subsection (1) set out in subsections (3), (4) and (5), of section 19 of the New Brunswick *Registry Act* involve leases, sales at public auction pursuant to law, and wills.

§3.14 Unlike the Manitoba sections, these sections do not specify that a person who acquires a registered interest with actual knowledge of a prior unregistered interest is not entitled to the priority established by the respective sections. Nevertheless, interpretation of these sections by the courts has established a similar result., The following quotation is taken from the headnote in the case of *Winchester v. N. Rattenbury Ltd.*,[4] a decision of the Prince Edward Island Court of Appeal in Equity:

Held, although on the wording of s. 34 it is not clear whether actual notice is included in the expression "no constructive or other notice", the section must be construed to cover only constructive or other implied notice in a case where the actual notice is such as would render the conduct of the subsequent purchaser fraudulent or unconscionable. A *Registry Act* should be given a reasonable rather than a literal interpretation so as not to become an instrument of fraud by way of defeating an equitable title or one arising under an unregistered instrument of which a subsequent purchaser is fully aware. "Purchasers or incumbrancers" in s. 34 must be limited to those who become such in good faith, acquiring their title *bona fide* on the strength of the register.

§3.15 The New Brunswick courts have given a similar interpretation to their section and in that province actual notice will defeat a later conveyance.[5]

§3.16 Section 25 of the *Registry Act* of Nova Scotia abolishes the doctrine

---

4 [1953] 3 D.L.R. 660 (P.E.I. C.A.).
5 See *Thomas v. Thomas*, [1938] 4 D.L.R. 566 (N.B.S.C.).

of constructive notice and leaves actual notice as the only kind of notice which will defeat a registered interest of a purchaser. It reads:

> **25.** No equitable lien, charge or interest affecting land shall be valid as against a registered instrument executed by the same person, his heirs or assigns.[6]

Functions of the Registrar

§3.17 Apart from the administrative functions and the indexing and recording functions, the registrar has a responsibility to ensure that the instruments which are registered meet certain statutory requirements. If a deed must be accompanied by an affidavit of a subscribing witness, as is the case in Manitoba, then it is his duty not to register a deed which does not meet the requirement, although if the defective deed is accidentally registered it is not thereby made void.[7] Basically, his staff screen the documents for the minimum requirements, but they may also point out other defects to the registrant which could give rise to problems at a later date. This courtesy is generally appreciated. In the event that the registrant still wishes to complete registration of a document which meets the minimum requirements, it is the duty of the registrar to register it. The registration of plans of survey is governed by special rules in each jurisdiction. Although the registrar requires that the formalities of legal execution are met, he takes no responsibility for the validity or effectiveness of a registered instrument. While his is more than a mere recording function, the examination is for the purpose of avoiding registration of obviously incomplete instruments or those for which registration is prohibited by law. If the registrar makes an error in the recording of instruments, or commits some other act or omission while carrying out his duties resulting in damages, he is not personally liable, but any successful claim will be paid from the general revenue of the province. The foregoing reference is to the Manitoba *Registry Act*.[8] In addition to the right to proceed against the registrar for negligence, the Ontario Act[9] provides for compensation from the assurance fund set up under the *Land Titles Act*[10] in certain situations, e.g. an error in recording an instrument.

---

6 For the interpretation by the courts, the following cases apply: *Ross v. Hunter* (1882), 7 S.C.R. 289; *Winter v. Keating & Gillis* (1978), 24 N.S.R. (2d) 633 (C.A.); and *O'Toole v. Walters* (1979), 34 N.S.R. (2d) 246 (T.D.).
7 *Supra*, note 3 at s. 33. But this may not be the case in Nova Scotia; see *Burchell v. Bigelow* (1904), 40 N.S.R. 493 (C.A.).
8 *Supra*, note 3 at s. 16.
9 *Registry Act*, R.S.O. 1980, c. 445, ss. 108 & 109.
10 R.S.O. 1980, c. 230.

Instruments which may be Registered

§3.18 A description of the deed registration system would not be complete without some explanation of the most important documents which operate to convey ownership or encumber title to land. Transfer of ownership is usually accomplished by a deed of grant, simply referred to as a deed. This instrument contains recitals of the names and particulars of the grantor and the grantee, the consideration paid, a clause vesting title in the grantee, and another clause called the habendum clause which commences "to have and to hold" and sets out the interest being conveyed, the tenancy, and in some cases a reservation in favour of the grantor. There are also covenants by the grantor which are assurances to the grantee that, for example, the grantor has good title, that the grantee shall have quiet possession and that, if necessary, the grantor will execute further documents. These covenants are in a short form covered by the *Short Forms Act* of the particular province. The benefit of these covenants to the grantee and his successors in title is that they operate as an estoppel against the grantor; in other words, the law will not permit the grantor to later deny the truth of the covenants. As a result, after the passage of the number of years sufficient to establish a marketable title in the province, such a deed to a purchaser for value will form a good root of title.[11] A quit claim deed, which may also be used as a conveyance but is most commonly used to release equitable interests, or to rectify errors in land description, does not contain these covenants for title. In a normal property purchase, unless otherwise agreed, a purchaser is entitled to insist upon a deed with the usual covenants.

§3.19 A first mortgage under the deed registry system also operates as a conveyance of the legal estate in the land to the mortgagee, subject to defeasance in the event that the money lent is repaid in accordance with the terms of the mortgage. It is essential to understand the nature of equitable rights in order to understand the nature of a mortgage under one of the Registry Acts. The mortgagee has the legal estate but the mortgagor has the equity of redemption. All of the usual incidents of ownership are attached to the equity of redemption, including the right to possession, and to the rents and profits of the land. There are so many equities in favour of the mortgagor that the mortgagee is strictly limited to well-defined methods of realizing the security in the event of default.

§3.20 There are numerous other types of instruments which may be

---

11 For further information see the article by C.W. MacIntosh Q.C., "How Far Back Do You Have To Search" *Nova Scotia Law News*, Vol. 14, No. 3, December 1987.

registered including leases, agreements for sale, easements, plans of survey, discharges of mortgages, to name a few of the most common.

Advantages and Disadvantages

§3.21 How a particular system is viewed by different individuals depends upon numerous subjective factors. An attempt will be made to point out the differences which may be regarded as an advantage or a disadvantage of a particular system. It is unlikely that either system will be looked upon as being ideal. The torrens system will be examined in some detail later in this chapter. Its introduction was the result of difficulties with the deed registration system experienced in the last century. Pointing out the arguments used at that time to support the introduction of the torrens system is not an assertion that such problems exist to the same extent in the deed registration systems in Canada today as they did then.

§3.22 Under a deed registration system, it is easier to have the interest of an innocent purchaser for value set aside by the courts in favour of a person who was deprived of the land through fraud or error. This would be seen as an advantage by someone who had been deprived of an interest in land under a torrens system without knowing about it until after the limitation period for an action against the registrar (through a claim against the assurance fund) had expired.

§3.23 A deed registration system allows more flexibility in terms of the form and the range of documents which may be registered. Torrens statutes generally provide forms for use under the system and also establish the instruments which may be registered under the system.

§3.24 In some jurisdictions a transaction can be closed more quickly under the deed registration system than under the torrens system. The reason for this is that when dealing under the deed registration system, the solicitors for the parties have completed their evaluation of title prior to attending at the registry office to register documents, and the only remaining step is to ensure by current searches that there has been no registration since their previous searches which would prevent the transaction from being closed. Under a torrens system, with perhaps the exception of Ontario and New Brunswick for various reasons, most transactions are not completed by the solicitors for the parties until registration of the documents has been completed by the registrar and, in the case of a conveyance, a certificate of title has been issued by the registrar. The registrar's staff examine the documents and prepare the certificate of title. A number of factors may contribute to delays between the time an instrument is presented for registration and the time that registration is completed. A substantial delay can result in financial loss to the parties as well as inconvenience. The use of technology to speed up processing of documents goes far to resolve this problem. Measures for its implementation are under way in New Brunswick, Ontario, Alberta and Manitoba. British Columbia has an operational automated system.

§3.25 Many beneficial features of the torrens system will become apparent as the system is examined in some detail later in this chapter. Those who pressed for the introduction of the torrens system in the latter part of the last century were eloquent in their support and pointed out the flaws in the deed registration system as it was at that time. The following statement was made by Mr. George S. Holmestead, then Inspector of Titles in the High Court of Justice of Ontario, at a public meeting in Toronto in 1890 under the auspices of the Canadian Land Law Amendment Association, and was quoted in *Thom's Canadian Torrens System*:[12]

> Let us, in imagination, go for a moment to one of our registry offices. We want to ascertain the title to a lot of land. We are shown what is called an 'Abstract Index,' in which is set forth a list of all the instruments affecting the land we are inquiring about. From this it may appear that there is a perfect chain of titles, but no one is safe in relying on this index alone; he must examine each instrument indexed, and carefully consider whether it is made by the proper parties, in proper form, and is duly executed. This he must do at his own risk. He must determine for himself the legal effect of each instrument at the risk that if he, or his legal adviser make a mistake, he may lose the land. And the great trouble is that after exercising all the care and caution possible, and procuring the best legal assistance, some unexpected flaw may be discovered. Let me give you a few practical instances of the way in which our present registry system works.
>
> There was a case of *Nex v. Seddon* before Mr. Justice Robertson lately. The action was brought to enforce a contract for the sale of a parcel of land. In this action it appeared that the vendor had a good registered title — no links in the chain were wanting, no flaws were apparent. It appeared that the vendor claimed as devisee under his uncle's will and had effected loans on the property, and I believe had made one or two mortgages to loan companies. Now very few persons, I venture to think, would guess what was the objection to this title. The will under which the vendor claimed was in due form, duly executed, duly registered, and yet *it was of no more value than a piece of waste paper.* The difficulty was that after the testator had made his will, he married, which had the effect of revoking the will (see R.S.O., c. 109, s.20), so that it was absolutely null and void; and yet, as you have seen, it was nevertheless registered.
>
> There was another case not long ago before the courts of *Munsie v. Lindsay* [(1882), 1 O.R. 164, affirmed (1885), 11 O.R. 520 (Ch.D.)], in which the facts were as follows: In 1854 a man named Munsie died leaving a will whereby

---

12 D.J. Thom, *Canadian Torrens System* (Calgary: Burroughs & Co., 1912) 21.

he devised his farm to his widow for her life, and then after her death to his son Robert. The will was duly registered. Robert shortly afterwards purchased his mother's life interest, and then thinking himself owner of the lot absolutely, he sold it to his brother James,who subsequently sold it to the defendant Lindsay, who bought on the faith of his vendor having a good registered title. Lindsay lived on the lot and worked and improved it for a good many years, thinking, no doubt, to leave it as a provision for his family on his death. In 1874, however, the widow of the testator died, and within ten years afterwards the heirs at law instituted a suit against Lindsay and recovered the land from him, on the ground that the devise to Robert was void, he happening to have been a witness to the will. All the hard-earned fruits of Lindsay's labor were thus taken from him without any compensation, except a lien for the value of the permanent improvements he had made, against which was set off an occupation rent for the premises since the widow's death. I think that you will confess that a registration system which leads to such results is, to say the least, not a very perfect or efficient system.

I will give you another instance of its effects. Some little time ago a case was before the courts, *Beaty v. Shaw* [(1888), 14 O.A.R. 600], in which the following state of facts appeared: There were two persons, executors and trustees of an estate. One of the trustees owed a sum of money to the estate, and executed a mortgage on a parcel of land to his co-trustee to secure its payment. The co-trustee died, and the other who owed the money survived him, and he then (without paying the debt to the trust estate) as surviving trustee and executor, dishonestly executed a discharge of his mortgage, which was duly registered. He then sold the land to the defendants in the action, who bought it in good faith, relying on the mortgage having been discharged. The purchasers worked it and made valuable improvements on it, and were somewhat astonished after the lapse of two years to learn that the discharge of mortgage was utterly worthless. The result of the suit was to take the land away from them, with all their improvements, without any compensation whatever; and yet you will be careful to observe that the discharge, or what purported to be a discharge, was duly registered.

§3.26 It is to be noted that in all of the cases mentioned, there was no difficulty apparent from an examination of the Abstract Index, or an examination of the registered instruments. This, of course, is not always the case as mentioned earlier.

§3.27 These are the shortcomings of the deed registration system which the torrens system was designed to overcome by providing for the issuance of a certificate of title to the owner, upon which is endorsed a memorial of every registered interest affecting the parcel. The interest of a *bona fide* purchaser for value named in the certificate is indefeasible with certain exceptions and any person suffering loss through the operation of the system is entitled to compensation from a fund set up under the torrens statute.

Land Surveys

§3.28 Mention has not been made to this point of the role of the land surveyor. It is obvious that without accurate boundary surveys and the registration of plans and other evidence to perpetuate a record of those surveys, a fundamental foundation would be missing from any title registration system. When the prairies were being opened for settlement, the survey of the township system was a monumental achievement, enabling the early settlers to receive a grant from the Crown for an easily described parcel, monumented on the ground, (at least at that time), and forming part of a huge grid to enable the geographic location of any parcel to be located on a topographic map with a township overlay. In the earlier settled parts of Canada, the role of the surveyor has been of equal importance, although in many cases the job was made more difficult as a result of very ambiguous land descriptions in early deeds. The physical difficulties involved with surveys in the mountainous terrain of British Columbia and western Alberta; the rock, lake, and muskeg of the Canadian Shield; and the rugged and often mountainous areas of Ontario, Quebec, and the Atlantic Provinces, which have been surmounted by surveyors from the earliest days of exploration in what is now Canada, command the admiration and respect of every Canadian. In modern times, the work has gone forward using the marvels of electronic technology. The land survey is a key component of any title registration system.

§3.29 It has been the writer's privilege to work with professional land surveyors and survey technicians in many different capacities because of the close interrelationship which exists between the land surveyor and the registration system. This applies, not only to the surveyor in private practice, but to those involved with government in many departments and in Crown corporations. The closest involvement was with those employed in the survey section of a land titles office. One of the features of the torrens system which appeared to be appreciated was the fact that care was taken when land was brought under the system and in case of subdivision thereafter to ensure that the land description was tied to a survey and was clear and unambiguous, or was shown as a parcel on a plan of survey. Experience indicated that faulty descriptions were perpetuated under the deed registration system. This is not a necessary flaw of the system but stemmed from the fact that the deeds were accepted for registration after minimal scrutiny and the faulty descriptions were carried forward from one deed to the next until a boundary problem was discovered.

TITLE REGISTRATION SYSTEMS

§3.30 The term title registration system is used to describe any system under which the title to land is certified, as opposed to a system of registration of deeds or assurances of title for what they are worth. Title registration systems were being formulated in South Australia and in England at approximately the same time. The system adopted in the western provinces, the Yukon, and the Northwest Territories is modeled after the system introduced in South Australia in June 1858. The Ontario title registration system under the *Land Titles Act* was modeled after the English Act of 1875. They are both commonly referred to as torrens systems, but there are some differences which will be mentioned later.

TORRENS LAND TITLES SYSTEM

Historical Background

§3.31 The torrens system originated in South Australia through the efforts of Sir Robert Torrens. He was the collector of customs at the port of Adelaide, and he stated that the system was modeled after the registers used under the Shipping Acts. In his view, the greatest flaw of the deed registration system was that registration in a public office did not override the common law principle *nemo dat quod non habet* (no one can give that which he does not own). In effect, a registered deed in favour of a purchaser for value could always be set aside as a result of a flaw in the vendor's title. He was a man of great tenacity, and mainly through his efforts the first torrens statute became law in South Australia in June 1858.

The Certificate of Title

§3.32 The certificate of title issued by the registrar and entered in the register under a torrens statute is the central document in the system. In itself it constitutes an index of the document from which it was created, (the transfer of land, etc.); the plan of subdivision or other survey upon which the legal description of the land is based; and all registered encumbrances, liens, easements, and interests which affect the land described in the certificate. It is, of course, much more than that because of the protection afforded to the registered owner of land or of a registered interest in the land, and to all third parties dealing on the strength of the register.

Powers and Duties of the Registrar

§3.33 The registrar under the torrens system is a statutory official who assumes the legal responsibility for the acts, omissions, mistakes, or misfeasances of those who are employed to carry out those functions of the registrar which may be delegated. There is a registrar for each district created under the authority of the torrens statute. There is also a senior official, a Registrar General or Master of Titles, who establishes policy and practice for all districts, and who may exercise any of the powers of the registrar in each district.

§3.34 The primary functions of the registrar, described in broad terms, are to register instruments, retain them in safe custody, issue certificates of title, and provide title information to the public. This entails developing a system for numbering, filing, copying, and retrieving instruments. The process of registration involves a careful examination of the instruments presented for registration for proper execution and compliance with the laws governing registrability. The registrar must maintain the integrity of the system by ensuring that only authorized instruments are registered and that they meet accepted standards. This includes ensuring that all plans of survey registered have been prepared in accordance with the statutes and regulations governing their preparation and registration and that they are compatible with the boundaries of adjoining land parcels. Under the scheme of the torrens statutes, a conveyance or other instrument does not become operative until the registrar issues the certificate of title, or executes the certificate of registration of the instrument, as the case requires. (In Ontario and New Brunswick an unregistered instrument may be effective for some purposes between the parties to it).

Correction of the Register

§3.35 The registrar has power to correct errors in the register which do not affect the rights of purchasers for value. The exercise of this power has given rise to some serious difficulties, particularly with respect to mineral rights in Alberta. In the early days of the system, mineral reservations had not been the focus of attention and a substantial number of certificates of title were issued without a proper exception of mines and minerals which had been reserved in an earlier instrument. When the omissions were later discovered corrections were made, but care was not always taken to ensure that there was no intervening *bona fide* purchaser for value. With the discovery of major oil fields a number of law suits resulted. The leading case known as the *Turta* case went to the Supreme Court of Canada.[13] It was held that the registrar had no power to make the corrections because they affected rights conferred for value, and therefore the purported corrections were nullities. There were other

---

13 *Turta v. C.P.R. Co. and Imperial Oil Ltd.* (1952), 5 W.W.R. (N.S.) 529, affirmed [1953] 4 D.L.R. 87 (Alta. C.A.), which was affirmed [1954] S.C.R. 427.

important torrens issues decided in the case. The registrar can always correct a certificate or instrument under the authority of an order of the court, and need not be a party to the court proceedings unless a claim against the registrar is involved. In Manitoba, the Registrar General has power to correct "any omission, clerical error, or other defect" in registered plans of survey.[14] Whether specifically authorized by statute or not, it would be unreasonable not to allow corrections of minor defects in plans in cases where no one would be prejudiced.

Rejection of Instruments

§3.36 The torrens statutes give the registrar power to reject an instrument which appears to be unfit for registration. The Manitoba *Real Property Act* prohibits the registrar from registering an instrument not in accordance with the provisions of the Act.[15] In order that persons may rely upon the register, a rejected instrument loses the priority which it would have had if it had not been rejected. This is so, even in situations where the grounds for rejection might be found to be erroneous or insufficient. In this context, mention must be made of the Manitoba Court of Appeal decision in the case of *Ukrainian Greek Orthodox Church v. Independent Bnay Abraham Sick Benefit and Free Loan Assoc. and Riverside Cemetery*.[16] In this case the registrar had rejected a caveat on the grounds that the land description was insufficient. The agreement, which was the subject matter of the caveat, was based upon the previous certificate of title which contained a metes and bounds description of the land parcel, and this description was used in the caveat. The certificate of title had been cancelled and a new certificate issued as the result of the registration of a plan of survey. The examiner applied the principle that although the boundaries might well be coincidental, it was now the survey posts which governed, and therefore a caveat claiming under the metes and bounds description might cover more or less land than described in the certificate of title. This was held by the court to be irrelevant as long as the description was sufficient to enable the registrar to determine the certificate of title upon which the memorial of filing of the caveat should be entered. The facts of the case were such that the court was faced with a serious dilemma involving consecrated ground in a cemetery. It was held that the priority of the rejected caveat should date back to the time that it was originally submitted. To apply this as a general principle in the torrens system would raise many practical difficulties. It would require the entry upon the certificate of title

---

14 *Supra*, note 4 at s. 51.
15 R.S.M. 1988, c. R30, s. 66 (1), (2).
16 (1959), 29 W.W.R. (N.S.) 97 (Man. C.A.).

of a memorial of all caveats presented for filing with a notation of the rejection of those which appeared to the registrar to be unfit for filing. It would then be the responsibility of the person intending to acquire an interest to find out whether or not the caveator intended to dispute the reason for rejection and apply to have the caveat registered retaining priority from the time that it was originally presented to the registrar for registration. The Manitoba *Real Property Act* has since been amended to provide for priority based upon a serial number assigned at the time that the instrument is presented for registration. Perhaps this amendment is sufficient to establish that when an instrument, including a caveat is rejected it loses its priority on the register. In New Brunswick this problem is dealt with by registering all caveats; the Registrar General then processes them and may strike off those which are not proper.[17]

Appeals from the Decisions of the Registrar

§3.37 The system of internal appeals in the various jurisdictions may vary to some extent, but basically the appeal goes from the examiner who rejected the document, or his supervisor, to the registrar of the district. An appeal from the registrar may be made to the senior official responsible for all land titles districts. A further appeal lies to the court.

First Registration

§3.38 Lands in the public domain, that is, surveyed, unpatented Crown lands, may be brought under the torrens system by whatever method the particular province has decided to use. It may be by order-in-council, ministerial order, or application to the registrar. Crown grants (letters patent) issued by a province are generally forwarded directly to the land title office. In cases where the land has been dealt with under the deed registration system, an application to bring the land under the operation of the torrens statute is required. The registrar obtains an abstract of the title including any deeds and copies of registered deeds held by the applicant.[18] He acts in much the same manner as a solicitor for a purchaser would, perusing every deed or other instrument in the chain of title. He will requisition missing deeds or other evidence to satisfy himself that the

---

17 *Land Titles Act*, S.N.B. 1981, c. L-1.1, s. 30.
18 There are local variations of this general principle. In New Brunswick, for example, he would obtain an abstract from the solicitor; in Manitoba the abstract is the register under the *Registry Act* and the registrar only requisitions copies of deeds, etc. in the possession of the owner. These are usually limited to copies of registered deeds and perhaps the latest deed from a previous owner to the applicant for registration.

applicant has a safeholding title. A safeholding title is one which, while the proofs may not be such that it could be forced upon an unwilling purchaser, nevertheless, the owner is unlikely to be disturbed in the possession and enjoyment of the land. From the survey aspect, the land description will be examined, and it may be that a survey and registered plan will be required. Inconsistencies with adjoining boundaries may be resolved by quit claim deeds.

Fundamentals of the System

§3.39 *Indefeasibility* The purpose behind the development of the torrens system and its counterparts in England and elsewhere was to avoid some of the uncertainties facing a purchaser under the system which it was intended to replace, one of them being the need to verify, to the extent possible, the validity of the predecessor's title. To this end, the torrens statutes provide that the certificate of title shall be conclusive at law and in equity against all persons and the Crown and that the person named therein is the owner of the estate or interest specified. This indefeasibility is subject to the exceptions stated in the statute and no others. It would be difficult to find a better summary of the principles of the torrens system than that set out by Turgeon J.A. of the Saskatchewan Court of Appeal in *Hackworth v. Baker:*[19]

> Some of the cardinal principles of the Torrens registration system are embodied in these sections, and in one respect or another they are designed to do away with some of the rules of the old law of real property, and consequently with some of the difficulties and controversies to which the old rules gave rise. These sections establish: (1) That estates and interests pass upon the registration and not upon the execution of an instrument; exception being made only in the case of certain leasehold interests; (2) That priority dates from the time of registration and not from the time of execution; (3) That the registered owner, except in the case of his own fraud, holds his land free from all estates or interests not noted on the register, saving certain leasehold interests already mentioned, and subject to the reservations and incidents provided by the statute; (4) That possession by another shall not derogate from the registered owner's right; (5) That a person taking a transfer from the registered owner shall not, except in the case of his own fraud, be affected by any notice given him of another's equity or unregistered interest in the land; that, further on this point, knowledge of such equity or unregistered interest shall not be considered a fraud; and it is expressly set out that this protection is to be given to the purchaser "any rule of law or equity to the contrary notwithstanding."

§3.40 One of the situations which has given rise to differences in interpretation by the courts is the situation of a *bona fide* purchaser for

---

19 [1936] 1 W.W.R. 321 at 332-333 (Sask. C.A.).

value from a forger. Until the year 1967, the law in Canada appeared to be clear that the title acquired by a person under a forged instrument could be set aside in favour of the previous owner, but if the land had been further transferred to an innocent purchaser, then that purchaser's title would be indefeasible. In one sense, this seems to be logical because the purchaser is in a much better position than the registrar to ensure that he is dealing with the registered owner and not with a forger. This principle is often referred to as deferred indefeasibility. The leading case which established this principle is *Gibbs v. Messer*,[20] a decision of the Privy Council on appeal from the courts of Victoria, Australia. The following quotations are taken from the judgment of Lord Watson:

> Their Lordships do not propose to criticize in detail the various enactments of the statute relating to the validity of registered rights. The main object of the Act, and the legislative scheme for the attainment of that object, appear to them to be equally plain. The object is to save persons dealing with registered proprietors from the trouble and expense of going behind the register, in order to investigate the history of their author's title, and to satisfy themselves of its validity. That end is accomplished by providing that every one who purchases in bona fide and for value, from a registered proprietor, and enters his deed of transfer or mortgage on the register, shall thereby acquire an indefeasible right, notwithstanding the infirmity of his author's title. In the present case, if Hugh Cameron had been a real person whose name was fraudulently registered by Cresswell, his certificates of title, so long as he remained undivested by the issue of new certificates to a bona fide transferee, would have been liable to cancellation at the instance of Mrs. Messer; but a mortgage executed by Cameron himself, in the knowledge of Cresswell's fraud, would have constituted a valid encumbrance in favor of a bona fide mortgagee.
>
> . . . .
>
> Although a forged transfer or mortgage, which is void at common law, will, when duly entered in the register, become the root of a valid title, in a bona fide purchaser by force of the statute, there is no enactment which makes indefeasible the registered right of the transferee or mortgagee under a null deed. The McIntyres cannot bring themselves within the protection of the statute, because the mortgage which they put upon the register is a nullity. The result is unfortunate, but it is due to their having dealt, not with the registered proprietor, but with an agent and forger, whose name was not on the register, in reliance upon his honesty. In the opinion of their Lordships, the duty of ascertaining the identity of the principal for whom an agent professes to act with the person who stands on the register as proprietor, and of seeing that they get a genuine deed executed by that principal, rests with the mortgagees themselves; and if they accept a forgery they must bear the consequences.

---

20 [1891] A.C. 248 at 254-258 (P.C.).

§3.41 In 1967, the Privy Council, in the case of *Frazer v. Walker*[21] on appeal from New Zealand, distinguished the case of *Gibbs v. Messer* on the facts and held that the person who had acquired a certificate of title through a forged instrument held an indefeasible title. Appeals to the Privy Council had been abolished in Canada prior to the decision in *Frazer v. Walker*, and therefore, the case is not binding upon Canadian courts. It was followed in the case of *Hermanson v. Martin* in the Court of Queen's Bench of Saskatchewan, in September, 1982.[22] In that case, the transfer to the defendant had been executed by only one of two registered owners, the other signature being a forgery. It was held that the defendant was a *bona fide* purchaser for value with no knowledge of the fraudulent execution of the transfer, and entitled to the protection of section 213(1) of the *Land Titles Act*.[23] The reasoning in *Frazer v. Walker* was adopted.

§3.42 It is too early to assess whether or not the future trend in Canada will be to follow the reasoning in *Frazer v. Walker*, and by doing so, eliminate the principle of deferred indefeasibility. The time is ripe for clarification of the legislation to at least establish the degree of care which must be exercised by a purchaser or mortgagee to avoid dealing with a forger, before being entitled to an indefeasible title or compensation from the assurance fund. As mentioned earlier, the registrar is not in a position to determine whether or not an instrument has been executed by the registered owner, or a forger, or has been otherwise fraudulently obtained.

§3.43 *Nature of fraud which would invalidate a certificate of title* There are numerous cases on this subject from which the leading case of *Assets Co. v. Mere Roihi*,[24] a decision of the Judicial Committee of the Privy Council on appeal from New Zealand, has been selected to illustrate the general principle. The following is an extract from the reasons for decision expressed by Lord Lindley:

> [B]y fraud in these Acts is meant actual fraud, i.e., dishonesty of some sort, not what was called constructive or equitable fraud — an unfortunate expression and one very apt to mislead, but often used, for want of a better term, to denote transactions having consequences in equity similar to those which flow from fraud. Further, it appears to their Lordships that the fraud which must be proved in order to invalidate the title of a registered purchaser for value, whether he buys from a prior registered owner or from a person claiming under a title certified under the Native Lands Acts, must be brought home to the person whose registered title is impeached or to his agents. Fraud by persons from whom he claims did not affect him unless knowledge of

---

21 [1967] 1 A.C. 569 (P.C.).
22 [1982] 6 W.W.R. 312 (Sask. Q.B.).
23 R.S.S. 1978, c. L-5.
24 [1905] A.C. 176 at 210 (P.C.).

it is brought home to him or his agents. The mere fact that he might have found out fraud if he had been more vigilant, and had made further inquiries which he omitted to make, does not of itself prove fraud on his part. But if it be shown that his suspicions were aroused, and that he abstained from making inquiries for fear of learning the truth, the case is very different, and fraud may be properly ascribed to him. A person who presents for registration a document which is forged or has been fraudulently or improperly obtained is not guilty of fraud if he honestly believes it to be a genuine document which could be properly acted upon.

§3.44 Thomas W. Mapp, Professor of Law at the University of Alberta, in his article entitled *Torrens' Elusive Title*,[25] made these observations:

> In an early New Zealand case, *National Bank of New Zealand v. National Mortgage and Agency Co.*,[26] Richmond J. discussed the problem as follows:
>
>> In many instances the rule of equity that notice is fraud must be recognized as consentaneous with the principles of common morality; for it may be an act of downright dishonesty knowingly to accept from the registered owner a transfer of property which he has no right to dispose of.
>>
>> . . . .
>
> The Act does not go so far as to shelter a purchaser who takes with full knowledge that the transfer to himself will unjustly deprive the true owner of his property without adequate compensation.
>
> The author believes that the statement of Richmond J. in the subsequent case of *Locher v. Howlett*,[27] and the emphasized language in particular, identifies the basic principle of law accurately:
>
>> It may be considered as the settled construction of this enactment that a purchaser is not affected by knowledge of the mere existence of a trust or unregistered interest, but that *he is affected by knowledge that the trust is being broken*, or that that owner of the unregistered interest is being improperly deprived of it by the transfer under which the purchaser himself is taking.
>
> It may be that unnecessary confusion has been created by the tendency of courts to use a variety of different expressions to articulate a single fundamental principle. It is said to be fraud if the subsequent purchaser "knows that the registered owner *has no authority* to dispose of the property", or "knows that the transfer *will unjustly deprive* the true owner of his property", or "intends to deprive another of his *just rights*." Are these expressions [not] all formulations of a basic rule that if C enters into a transaction with B with actual knowledge that it involves a breach of B's trust obligations to E, then C has knowingly participated in B's fraudulent conduct?

---

25 *Alberta Law Review, Book Series*, Vol. 1 at 115.
26 (1895), N.Z.L.R. 3 S.C. 257 at 263, 265.
27 (1894), 13 N.Z.L.R. 584 at 585 (emphasis added).

§3.45 ***Two certificates of title for same land*** In the situation where there is more than one certificate of title in existence for the same land, the earliest certificate or a certificate derived from the earliest certificate shall prevail over the latter.[28] This exception to indefeasibility does not apply to the situation where the earlier certificate has been improperly cancelled. In that case, the general rule for correction applies, and the cancelled certificate of title may be re-instated by correction, but the correction must be made prior to the acquisition of an interest by a *bona fide* purchaser for value. This rule is applied in the situations caused by the use of poorly worded land descriptions in some of the early certificates of title. For example, a lot shown on a plan as being 100 feet in width may have been transferred at different times as the north 50 feet and the south 50 feet. In the opinion of the writer, in the event of a shortage, it would be the owner under the latest certificate who would have less than 50 feet.

§3.46 ***Misdescription*** In a situation where the boundaries of land are incorrectly described in a certificate of title, the erroneous description may be corrected against the registered owner, at any time prior to registration of title in the name of a *bona fide* transferee for value. Correction may be made against a *bona fide* transferee for value in cases where there are two certificates of title issued for the same land. It has been held by the Supreme Court that the misdescription exception applies to the description of the surface boundaries and not to a missed exception of mines and minerals.[29] In Alberta, it is not clearly set out in the legislation that the exception to conclusiveness of the register, in case of misdescription, does not apply against a *bona fide* purchaser for value. It would appear to defeat the scheme of a torrens statute if the relevant sections of the *Land Titles Act* of Alberta[30] were applied differently than similar sections have been in New Zealand or in Manitoba, where it is specifically set out that the exception to conclusiveness on this ground does not apply to a *bona fide* purchaser for value.[31] There is a dimension which applies to a purchaser of land which has been misdescribed in a certificate of title which may not apply in other situations where the *bona fides* of the transaction is at issue. A purchaser may be unaware of the error in his predecessor's certificate of title, but thoroughly familiar with the location and dimensions of the parcel which it was his intention to buy and for which the purchase price was established. If the error in his predecessor's title is carried forward without detection and the innocent purchaser is later advised that his certificate of title contains a windfall, then it would

---

28 *Land Titles Act*, R.S.A. 1980, c. L-5, s. 66; and *supra*, note 15 at s. 59.
29 *Supra*, note 13. *Turta v. Canadian Pacific Ry. Co. and Imperial Oil Ltd.*, *supra*, note 13.
30 *Supra*, note 28 at ss. 66(1), 173(1)(*e*) and 177(1)(*a*).
31 *Supra*, note 15 at s. 62(1)(*d*).

appear that if these facts are established, the certificate of title could be corrected.

§3.47 *Historical searches* The principle, which is one of the fundamentals of the system, that the certificate of title is conclusive and there is no need to search behind the register, still receives general public acceptance when dealing with land except mines and minerals, which is usually thought of as the surface. It has become established practice to search behind the current certificate of title in the case of mineral interests. This is, in part, a result of the fact that the Assurance Fund would not fully compensate those who might lose an interest in mines and minerals through the operation of one of the exceptions to indefeasibility. Of particular importance is the fact that in some cases there have been improper entries made in the register resulting in their being two certificates of title for the same interest. Another reason, unrelated to the principle of indefeasibility, is that the expression "mines and minerals" is not a fixed term, varying to some extent in different time periods. It is therefore, in some cases, important to examine the instrument which severed title to the mines and minerals from the title to the surface.

§3.48 *Completeness of the Register — overriding interests* Registration of certain interests in land is not necessary in order to maintain priority of that interest over a registered interest in the land under either the torrens system or the deed registration system. These interests are called overriding interests. Ideally, the list of overriding interests should be kept as short as possible. Under the torrens system, the statute lists the overriding interests in a separate section as exceptions to the conclusiveness of the register, stating that the certificate of title is subject by implication to the interests listed. In order to determine what unregistered interests to guard against, the torrens statute for the province must be examined. Some of the most common of the overriding interests are listed below.

(1) Reservations in the original grant from the Crown.
(2) Municipal taxes.
(3) A lease for less than three years where there is actual occupation under the lease.
(4) Orders of attachment, writs of execution, judgments or orders for the payment of money etc.
(5) Planning schemes and by-laws.
(6) Any right of expropriation by statute.

§3.49 The above list is by no means exhaustive, but it is illustrative of the types of inquiries outside the land titles office which must be made by or on behalf of an intending purchaser.[32] With reference to orders of attachment, writs, judgments etc., these may be ascertained from a search of the general register index in the land titles office. In British Columbia and New Brunswick, this type of charge is no longer an overriding interest, but instead, must be registered against the land and will appear as an encumbrance in the certificate of title.

§3.50 *Assurance fund* The protection afforded to the *bona fide* purchaser for value under the torrens system cannot be provided without loss to equally innocent persons who would have been able to recover their interest in the land under the deed registration system which it replaced. It was therefore necessary to establish a fund from which any person who suffered loss through the operation of the torrens system would be compensated. This fund is called the assurance fund. A claim for compensation from the assurance fund is made against the registrar in his name or office. The various torrens statutes provide a procedure for payment of claims on the recommendation of the registrar without the necessity of an action in court. The criteria for payment without an action vary to some extent between jurisdictions, but the basic purpose is to permit the payment of just claims without requiring a court proceeding. In other situations, the applicant will be required to establish the right to compensation in court.

§3.51 The liability of the fund is divided into two categories referred to as primary and secondary. The primary liability of the fund is to compensate for errors, omissions or misfeasances of the registrar in the course of carrying out his duties under the statute. This can be a fairly onerous liability due to the emphasis which the system places upon certification by the registrar and the numerous opportunities for both clerical and legal errors which are inherent in the system. The secondary liability of the fund is for loss resulting from the fraud or wrongful act of some person other than the registrar, or resulting from the fraud or wrongful act of that person and the error, omission or misfeasance of the registrar. In cases where there is a defendant other than the registrar and the court decides that the other defendant should pay the claim, then judgment is not entered against the registrar until it has been determined that the money cannot be realized from the assets of the other defendant.

Registration of Instruments (except plans)

§3.52 *Transfers and leases* The instrument most commonly used under the torrens system to convey title is the transfer of land, a simple form of conveyance provided by the statute. The execution and delivery of a

---

32 See also §9.11.

transfer of land to a purchaser with the intent that it should be registered confers upon the transferee the right to register it, but the actual conveyance does not take place until the transfer has been registered. Registration occurs when the registrar signs the certificate of registration of the transfer and issues the certificate of title. The formality of signing completes registration, but does not determine priority, which is established at the time that the transfer is received, either by date and time or by date and serial number. The most common estate for which a certificate of title is issued is an estate in fee simple. This may be held in different tenancies such as joint tenants or as tenants in common for example. The fee simple may be in possession or in remainder expectant upon the decease of the owner of a life estate.

§3.53 A lease is a conveyance of an estate for a "term certain" which may be for a term of years or for a life or lives. It is not necessary that the registration of a lease will result in the issuance of a certificate of title to the lessee. In most jurisdictions a certificate of title for a leasehold estate will be issued if it is requested. Many valuable properties are the subject of long term leases and it simplifies future registrations against the leasehold estate to have a separate certificate of title. Short term leases, where there is occupation, do not require registration to preserve their priority, but are included in the section providing for priority of overriding interests in the statute.

§3.54 *Transmission of title* In certain situations, the right to deal with the interest of the registered owner has been vested in another person and, in other situations, title to the interest has vested in another person upon the occurrence of an event. Examples of the first situation are:

(1) The death of the owner, and the appointment of an executor or administrator to administer the estate.
(2) The bankruptcy of the owner and the appointment of a trustee to administer the bankrupt estate.

Examples of the second situation are:

(1) Vesting of title in the survivor upon the death of a joint tenant.
(2) Vesting of title as the result of sale or confiscation of the land for arrears of municipal taxes; mortgage foreclosure; or expropriation proceedings.

§3.55 In all of these situations, the title is transmitted. In the first situation, the transmission is to the person appointed to administer the estate of the deceased person in the representative capacity or to the trustee in bankruptcy. In the second situation, the transmission is to the person in whom title has been vested. Whether or not a formal transmission application is required varies with the situation and with the jurisdiction.

§3.56 *Mortgages and encumbrances* Under the torrens system, a mortgage derives its force from the statute and operates as security only, in contrast to a mortgage under the deed registration system which conveys the legal estate in the land, subject to defeasance in the event the mortgage is paid in accordance with its terms. A special statutory form is provided, but fairly substantial deviations from the form are permitted. As with all instruments, registration is not complete until the registrar signs the certificate of registration. The word encumbrance has been used in a number of different senses, particularly in Alberta, the Northwest Territories and Manitoba. The first usage covers any registered instrument, for which a memorial has been endorsed upon the certificate of title, including, for example, mortgages, easements, caveats, and mechanics' liens. The second usage refers only to those instruments which create a charge or lien on land. This is in accordance with the definition used in the Acts.[33] The third usage is to describe the statutory encumbrance used to secure an annuity, rent charge or sum of money. The context in which the word is used is, on occasion, the only guide to its intended meaning.

§3.57 *Restrictive covenants* Prior to the widespread use of zoning regulations, registration of a restrictive covenant was the means used to protect the residents of an area from unregulated use of the land in that area, usually a large subdivision. They are still popular for certain purposes and their use is often promoted by planning departments. The common law rule is that to be enforceable, the covenants must be negative in nature, that is, they can be restrictive as to use, but they cannot require any positive action by the covenantor such as clearing snow from the sidewalk. To state that the covenantor will not allow snow to accumulate on the sidewalk would not convert such a covenant to a negative one. If there are any negative covenants in an instrument which also contains positive covenants, it may be registered, but the positive covenants will not run with the land. They might be enforceable by an action against the person who executed the instrument, but could not be enforced against a subsequent owner. The practice was established from the inception of the system in Manitoba, requiring these restrictive covenants to be filed as caveats called building restriction caveats. All restrictive covenants are accorded a special status whether filed as a caveat or a restrictive covenant instrument. For example, they are not wiped out by sale or confiscation for arrears of taxes by a municipal corporation. Also, they cannot be withdrawn or released by the person who filed the caveat or registered the instrument, unless that person is the only beneficiary of the covenant.

---

33 *Supra*, note 28 (Alta.) at s. 1(*f*); *supra*, note 15 at s. 2(1)(*b*).

§3.58 *Easements* An easement is a right annexed to a parcel of land over an adjoining parcel. The parcel to which the right is annexed is called the dominant tenement and the parcel which is subject to the right is called the servient tenement. The most common rights which are the subject of easement agreements or grants are rights of way for various purposes, for example, a driveway for access to another parcel owned by the owner of the dominant tenement. With the introduction of "zero lot line" housing, an easement is required in order to provide access for maintenance of the wall of the house which is on the line.

§3.59 *Statutory rights of way for utilities* Instruments of this nature are often referred to as easements, although they do not meet the requirement of a dominant and servient tenement. Instead, they draw their validity from the torrens statute and from various statutes which govern the public and private corporations which are entitled to carry out the functions which give rise to the need for a legally enforceable right of way. These include *Public Works Acts, Pipelines Acts*, etc. These acts set the standards for the instruments which are registered. These instruments are generally based upon a plan of survey of the right of way. In Ontario, under the *Land Titles Act*, a separate register has been set up for pipeline registrations. In the opinion of the writer, it is unfortunate that a similar register has not been established, at least in Alberta and Manitoba, where the writer is familiar with the difficulties caused by the method in use in those provinces. The pipeline and the right of way of major systems are mortgaged by a deed of trust and mortgage in favour of a large financial institution for a very large sum. A memorial of a right of way agreement is endorsed on every certificate of title for land which is crossed by the right of way. There is no difficulty with this, being the standard procedure for all registrations. The procedure was established of also entering a memorial of the mortgage on each certificate of title affected by the right of way and this has caused considerable inconvenience. These rights of way are usually mortgaged again, as the result of refinancing, every five years. This results in an onerous task for the solicitor for the pipeline corporation, the solicitor for the mortgagee, and for the registrar each time the right of way is mortgaged, or an earlier mortgage is discharged. It also causes concern to the owner when he sees these large mortgages endorsed against the title, requiring assurance that it is only the pipeline right of way which is affected. It is a very major undertaking to change a system of this type once it has been in use for a number of years. (Under the New Brunswick Act an easement may be granted a separate certificate of title, avoiding the problems set out above.)[34]

---

34 *Supra*, note 17 at s. 24(21).

§3.60 Removal of interests from the register With few exceptions, interests which have been registered or filed against the certificate of title of the registered owner, may be withdrawn or discharged by the registration of an instrument executed by the person appearing on the register to be the owner of the interest or by the authorized agent of that person. The court has authority in any proceeding to direct the registrar to make any entry or do any act necessary to give effect to a judgment or order of the court.[35] This would include discharging an interest. In the case of caveats, the person claiming the interest may be served with a notice to take proceedings in court and file a certificate of *lis pendens* within the time limited in the statute. At the expiration of the period specified, if no certificate of *lis pendens* has been filed, the registrar may remove the caveat as lapsed. In the case of mechanics' liens or builders liens, the statute provides that after the expiration of a specified period of time the lien shall cease to exist unless an action is taken in which the lien may be realized and a certificate of *lis pendens* is filed. A single certificate of *lis pendens* will protect all liens filed because all lien claimants may join in the action. There is also provision for service of a notice to require a lien claimant to commence proceedings within the time specified in the statute and file a certificate of *lis pendens* or the lien will cease to exist.

Registration of Plans

§3.61 The plan registration function requires close cooperation between the registrar and his staff in the survey section of a land titles office and the land surveyor, whether in private practice or as an administrator or working professional in a government department. Guidelines governing the registration of plans should be prepared through a consultative process, recognizing that the registrar is responsible for maintaining the integrity of the title system. These comments will be limited to the requirements for registration in the land titles office. These requirements are not isolated from the legal framework governing surveys, and the preparation and registration of survey plans. There is a different set of laws, and a correspondingly different set of requirements for registration under the *Land Titles Acts* or *Land Registry Acts* of the various jurisdictions. There are, however, certain principles of general application which will apply to registration requirements under torrens statutes and the discussion will be limited to these.

§3.62 Plans of subdivision Although the registrar does not assume responsibility for the area or the extent of a lot shown on a plan of

---

35 *Supra*, note 28 (Alta.) at s. 180; *supra*, note 15 at s. 161.

subdivision, he is given powers to enforce certain survey standards. This power may be given by the statute to the senior official, such as the Master of Titles in Saskatchewan or the Registrar General in Manitoba. The daily exercise of the function will be carried out by the official appointed to manage the survey section of a land titles office, who is a professional land surveyor. The situation is somewhat different in Alberta, where for historical reasons some of the plan examination functions of the registrar have been delegated to the Director of Surveys appointed under the *Surveys Act*,[36] and the enforcement of survey standards for patented as well as unpatented land is also vested in the same official. It is obviously important that a high standard of accuracy should be maintained because of the interdependence of surveys and the difficulty involved with correction at a later date. One of the powers is to require a survey to be verified on the ground before a plan is registered in a land titles office. To enable the practising land surveyor to conduct the survey and prepare a plan which will meet the requirements for registration, practice guidelines or regulations are issued by the survey official in the land titles office, prepared after consultation with representatives of the practising land surveyors. Other statutory functions of the registrar relate to the registration process. A plan of subdivision, when registered, may not only cause new certificates of title to issue to the registered owner of the subdivided land, but in cases where areas on the plan are shown as public roads, streets, lanes, parks, reserves, etc., there is a statutory vesting of these areas in the Crown or a municipal corporation as specified in the statute and certificates of title will issue accordingly. That is the major reason why the statutes require that all persons having registered interests shall consent to the registration of the plan.

§3.63 The subdivision control provisions of the provincial *Planning Acts* place the onus upon the registrar to enforce a number of the requirements of the legislation, one of them being to ensure that subdivision does not take place without the approval required under the relevant *Planning Act*. It is simple for the registrar to check to ensure that the proper endorsement of approval by the planning authority appears on a plan of subdivision presented for registration, but there are other situations which are more complex, such as a request for separate certificates of title for parcels in the same quarter section, but on opposite sides of a public highway. It is probably impossible to cover every situation by unambiguous legislation or regulation.

§3.64 A subdivision plan will be carefully examined prior to registration to ensure that it complies with the statutory requirements and the guidelines

---

36 R.S.A. 1980, c. S-29.

or regulations. The examiner will check to ensure that sufficient boundaries of the township or other framework survey are shown to enable the location of the subdivided area to be determined, and that there are sufficient ties to previously surveyed lines and corner monuments. Title to adjoining parcels and plans of survey of adjoining parcels will be examined to ensure that there are no boundary discrepancies.

§3.65 A subdivision of an area which has not been previously subdivided in any manner, while it may require a similar exercise of surveying expertise, does not give rise to the same legal and administrative problems as those which are often encountered when subdividing an area which has in the past been divided into a number of allotments with different owners. One of the common difficulties is with different mineral exceptions based upon the boundary definitions which will no longer apply when the subdivision plan is registered. It would severely limit subdivision design if some of the lot boundaries had to coincide with those of previous parcels in order that the exceptions may be accurately shown on the new certificates of title which will issue following registration of the plan. It complicates a certificate of title to show two exceptions for minerals for undefined portions of a lot. In some cases, the excepted minerals may be acquired by the subdivider and consolidated with the surface to eliminate all mineral exceptions. They could be conveyed again to the mineral owner in accordance with the new lot boundaries. In many cases they cannot be acquired and the only solution is to limit the subdivision to the surface only. While this solves the immediate problem, it does result in an unsatisfactory situation because records must be maintained of plans and legal descriptions which would otherwise be superseded by the new plan, and certificates of title maintained for the minerals according to those plans and legal descriptions. Mineral exceptions are not the only interests which may cause difficulties in this situation. There may be an easement for a right of way along the boundary of one of the earlier parcels. It may be necessary to provide the owner of the dominant tenement with an alternate means of access in order to obtain a surrender of the easement, or it may be necessary to show the area which is subject to the easement as a separate lot on the plan. These two examples do not exhaust the list of registered interests which can complicate the subdivision process. For the complication to apply, it is not necessary that the area being subdivided shall have been previously subdivided since the original survey of the area may, for example, straddle the boundary between two quarter sections.

§3.66 Where land covered by a plan of subdivision is to be re-subdivided, the best course may be to apply to the proper tribunal for an order cancelling the existing plan in whole or in part prior to the registration of the new plan. In Alberta, the proper authority is the Planning Board appointed under the *Planning Act*. If all interested parties consent, the Board may make an order and the registrar upon registration of the order, shall cancel the plan of subdivision in accordance with the order and do all things necessary to give effect to the order.[37]

§3.67 The registration of plans of subdivision provides a basis for simple record keeping, not only in a land titles office, but in other branches of government, notably, municipal assessment and taxation. This simplicity does not apply to the same extent with respect to mines and minerals which have been severed from the surface. Mention has been made earlier of the complexities which may arise in connection with subdivision of land where there are previously established internal boundaries in the area being subdivided. The record keeping duties of the registrar can be carried out most effectively when, for every exception from the land described in a certificate of title, there is another certificate of title for the exception based upon the same description of the boundaries. In the case of mines and minerals, this is not possible because they have been severed from the surface by reservation at various periods of time. The best that can be hoped for is that the exception for mines and minerals will be uniform over the entire area of the subdivision. In effect, there is a double mosaic created which requires, in some cases, retention of boundary references which no longer apply to the surface.

§3.68 Where one of the boundaries of a parcel which is to be subdivided is a natural water boundary, a difficulty may arise in situations where the present location of the boundary differs substantially from location at the time when the boundary upon which the certificate of title is based was established. The nature of accretion is discussed in Chapter 5. It has been the experience of the writer that a small accretion presents little difficulty and can be safely included in the subdivision. It is quite common for large areas to be accreted, particularly to parcels adjoining shallow lakes. The registrar would have no authority to make a determination that title to the additional land had been acquired by the riparian owner through lawful accretion. He would refuse to register a plan of subdivision which showed lots, streets, and reserves on land which was at one time part of the bed of a water body owned by the Crown, which is the owner of the bed in most situations. The Crown may be prepared to agree that title has passed to the riparian owner. In that case, the only remaining hurdle is to ensure that the addition of the accretion to the owner is properly done so that adjoining riparian owners are not prejudiced. This may be accomplished by obtaining the consents of adjoining owners. The registrar may require a consent order or other order of the court. In Manitoba, there is a special

---

[37] *Planning Act*, R.S.A. 1980, c. P-9, s. 120.

section in the *Real Property Act* which permits the registrar to act on the consent of the Crown and of any adjoining owners who might be prejudiced.[38]

§3.69 Expropriation plans Governments at all levels, and many Crown corporations and other corporations created for certain purposes, such as constructing railways, transmission lines, pipelines, etc., have power to take the land which they need to enable them to carry out their objects without the consent of the registered owner or other persons who have registered interests in the land. This is the power to expropriate. Even in the case of a superior government, it must be exercised in the furtherance of some legitimate government purpose and not indiscriminately. The limits upon the exercise of the power, the procedures to be followed, and the provisions for compensation are set out in the federal and provincial *Expropriation Acts* and in the Acts of incorporation of the respective corporations. Where land is being acquired for a purpose for which the land may be expropriated, it is still a compulsory taking if a price is negotiated with an owner who agrees to convey the land without the formal expropriation procedures being followed. It is considered that the parties are not negotiating on an even footing when one party can say, "If you don't sell, I will be required to take expropriation proceedings." The negotiations simply establish the amount of compensation. Municipal expropriation acts usually provide that upon the registration of the plan, title vests in the Crown or in the municipal corporation, as the statute provides. If all of the persons having registered interests in the land have agreed to the amount of compensation, then a statement to that effect is endorsed on the plan. If the formal expropriation procedures have been followed, then the proper certification will accompany the plan. In both cases, title will issue clear of registered encumbrances, the compensation taking the place of the interest expropriated.

§3.70 Condominium plans The condominium method of ownership has gained wide public acceptance. Under the scheme of a *Condominium Act*, a building may be subdivided into areas called units which may be owned separately from the rest of the building or from the land upon which the building stands. Other parts of the building and the land are owned by the owners of the units in the percentage or fractional interest assigned to each unit. These interests are called common interests, and the areas not part of the units are called common elements or common areas. The *Condominium Act* of the local jurisdiction may permit the establishment of a bare land condominium. In that case, the land is divided into lots which are units, and the rest of the area is common elements or common areas. The entire condominium is governed by a corporation which comes

---

38 *Supra*, note 15 at s. 50.1.

into existence at the time that the condominium documents, including the plan, are registered. The land titles office becomes a repository for information about the constitution and by-laws of the condominium corporation, in addition to the registered plan and the certificates of title for the units and the common interests.

Search Facilities

§3.71 Historically, a function of the land title systems has been to provide safe custody for the registered instruments and the certificates of title and to enable the public to have access to search any instrument or certificate of title upon the payment of a small fee. In addition, the registrar was required to issue a certificate of search (called an abstract in some jurisdictions) upon request, for a slightly larger fee. Provision was also made for certifying copies of registered instruments. The system was geared to the needs of an intending purchaser of an interest in land. With the introduction of copying machines and microfilm, the need for personal searching of certificates of title and documents has been dramatically reduced. In the case of survey information, copies of registered plans are readily available, but the land surveyor preparing to make a survey will find that it is advantageous to examine the plans in the land titles office before deciding upon the need for prints. The introduction of a computer data base into the land titles system opens the door to serving the needs or desires of those who require massive amounts of specialized information for purposes totally unrelated to land transactions of any sort. It also opens the door to providing an enhanced service to those requiring information for land transactions and to the branches of government which carry out regulatory or taxing functions related to land.

Caveats

§3.72 The main use of the caveat under the torrens system is to enable equitable or other legitimate interests in land, to obtain priority for the interest claimed from the time of registration of the caveat. The word caveat is derived from the Latin *caveo* which means beware or take care. The registration of a caveat does not add any validity to the interest claimed. The following comments of Duff J. in *Alexander v. McKillop & Benjafield*[39] explain the principle involved.

> The fundamental principle of the system of conveyancing established by this and like enactments is that title to land and interests in land is to

---

39 [1912] 1 W.W.R. 871 at 886-887 (S.C.C.).

depend upon registration by a public officer and not upon the effect of transactions *inter partes*. The act at the same time recognizes unregistered rights respecting land, confirms the jurisdiction of the courts in respect of such rights and furthermore makes provision — by the machinery of the caveat — for protecting such rights without resort to the courts. This machinery, however, was designed for the protection of rights and not for the creation of rights. A caveat prevents any disposition of his title by the registered proprietor in derogation of the caveator's claim until that claim has been satisfied or disposed of; but the caveator's claim must stand or fall on its own merits. If the caveator has no right enforceable against the registered owner which entitles him to restrain the alienation of the owner's title, then the caveat itself cannot and does not impose any burden on the registered title.

§3.73 After the registration of a caveat, any person acquiring an interest in the land through the registration of an instrument will be required to register the instrument, subject to the claim of the caveator. This is not an admission that the claim of the caveator is valid, and it is still open to a person acquiring an interest, subject to the caveat, to take proceedings to have the caveat removed.

§3.74 There are variations in forms and procedures between jurisdictions, but basically, a caveator will be required to state the nature of the claim and the grounds upon which it is based. Also, the caveat will be supported by an affidavit in which it is asserted that the deponent believes that the caveator has a good and valid claim and that the caveat is not being filed for the purpose of delaying or embarrassing the owner or any person claiming through him.

§3.75 A caveator may, in lieu of or after filing a caveat, take proceedings in court and file a certificate of *lis pendens*. A certificate of *lis pendens* is issued by the court to give notice that an action has been commenced in which some right, title, or interest in the land is called into question. A person acquiring an interest after the certificate of *lis pendens* has been filed would be bound by the decision of the court. It would, therefore, be in the best interest of a person who intends to acquire an interest in land, described in a certificate of title which is subject to a certificate of *lis pendens*, to investigate the court proceedings to find out if the decision of the court might adversely affect that interest.

§3.76 Although it is not necessary to attach a copy of the agreement or other document upon which the claim of the caveator is based, it is common practice to do so. This enables anyone searching the caveat to discover the full basis for the claim. It is also a good way to ensure that the caveat protects all of the interests of the caveator under the document. The grounds of the claim should still be set out in the caveat but a "catch all" phrase may be inserted to cover all other rights and interests in favour of the caveator set out in the document. The importance of claiming all interests granted by an agreement or other document was highlighted by the decision of the Supreme Court in *Ruptash v. Zawick*[40] which held that interests in the agreement which were not claimed in the caveat were not protected by the caveat.

§3.77 It is a person claiming an interest in land who is entitled to file a caveat. The registrar has a duty to require that the nature and the grounds of the claim are set out, but there have been different views expressed about the responsibility of the registrar to refuse to register a caveat unless the interest claimed, would, if established in court, be an interest in land recognized by law. The argument of those who say that if the form of the caveat is properly completed and the nature and grounds of the claim are set out, it is not the duty of the registrar to question whether or not the nature of the claim is such that it would, if established, be an interest in land, is based upon the language of Tritschler J. in *C.P.R. v. District Registrar of Dauphin Land Titles Office*.[41]

> Applicant does so claim and it is not disputed that its caveat is in the correct statutory form. A caveat is merely a notice of a *claim* which may or may not be a valid one. The validity of the claim must be determined after and not before the filing of the caveat. The purpose of caveats is to warn the registered owner and, what is more important, all persons who might deal on the faith of the certificate of title, that the caveator claims an interest which is not disclosed on the certificate of title. This purpose would be frustrated if a claim to an interest had to be established by the decision of a Court or of the District Registrar before a caveat could be filed. It is trite law that caveats are to be used for the protection of alleged as well as of proved interests and that a caveat is merely a warning which creates no new rights but protects existing rights, if any.

§3.78 It is the writer's view that if the interest claimed in the caveat is one which has been held by the courts not to be an interest in land and there is no special statutory requirement to register a caveat to protect such an interest, then the caveat should not be registered. In case there is doubt as to whether or not the interest, if established, would be an interest in land, then it should be resolved in favour of accepting the caveat. For example, to accept a caveat to protect a promissory note, where there is no statement in the caveat that the land has been charged as security, would be unfair to the registered owner. Although remedies are available to the registered owner, they can be both costly and time consuming. The

---

40 [1956] S.C.R. 357.
41 (1956), 4 D.L.R. (2d) 518 at 521 (Man. Q.B.).

following comment appears in *Thom's Canadian Torrens System*:[42]

> However, while respectful agreement may be accorded to the following statement by Tritschler J. at pp. 243-4 in *C.P.R. v. District Registrar of Dauphin Land Titles Office* it is submitted that there is nothing therein which suggests that the interest necessary to found a caveat may be anything less than an interest in land:
>
>> The Act does not require that the caveator shall claim to have an estate or interest in land which is derived from the registered owner nor does it empower the district registrar to decide whether a claim asserted by a caveator is binding upon or enforceable against the registered owner.... It is trite law that caveats are to be used for the protection of alleged as well as of proved interests and that a caveat is merely a warning which creates no new rights but protects existing rights, if any.

§3.79 One of the imaginative concepts originated by the Land Registration and Information Service (L.R.I.S.) project in the Maritimes, and subsequently adopted in Alberta and Manitoba,[43] is the transfer of caveat. This recognizes that the caveat is a useful means of achieving status on the register for many interests of a long term nature and that these interests are frequently assigned to another person, with the result that a series of caveats are filed to protect the same interest. The original caveat must be maintained in order to retain priority back to the time that it was registered. The original caveator no longer has an interest and, with the passage of time, it may be difficult to locate the caveator when a withdrawal of the caveat is required or the consent of the caveator is needed for the registration of a plan of subdivision. The transfer of caveat eliminates this problem because there is only one caveat registered and the last transferee is entitled to withdraw the caveat or consent to the registration of the subdivision plan.

§3.80 The methods of requiring the caveator to take proceedings in court to establish the right to maintain the caveat and the withdrawal procedures were discussed briefly earlier in this chapter. In view of the differences in the various jurisdictions, it would serve no useful purpose to add to that brief generalization.

Registration of Mineral Interests

§3.81 Under the torrens statutes, the definition of land includes mines and minerals, unless the same are expressly excepted. This is not interpreted

---

42 V. Di Castri, *Thom's Canadian Torrens System*, 2d ed. (Calgary: Burroughs & Co., 1962) at 608.
43 *Supra*, note 28 (Alta.) at s. 135.1; *supra*, note 15 at s. 144.1. See also, *supra*, note 17 at s. 31.

to mean that mines and minerals cease to be land if they are owned separately from the surface. A mineral ceases to be land when it is physically extracted from the land, but while it is in place it retains its character as land.[44]

§**3.82** The Crown is by far the largest owner of minerals in Canada. A set of records for keeping track of dealings with these interests is maintained by the department of government responsible for the administration of Crown lands. Although in some cases, the Crown may have a certificate of title under the torrens system which covers or includes mines and minerals, the leases or other dealings by the Crown are recorded in the department concerned. There will be no attempt to describe any of the record keeping functions of a department which administers Crown lands.

§**3.83** One of the overriding interests in the torrens statutes is *any subsisting reservation contained in the original grant of the land from the Crown*. The earlier grants in the prairie provinces and the territories were issued by the Government of Canada, under the provisions of the *Dominion Lands Act*.[45] Until about the year 1887, there were no reservations of mines and minerals in the grants, but, depending upon the date of the homestead entry, grants after that date reserved mines and minerals. The grant must be checked in order to be certain what reservations, if any, were contained in it. At the time of the transfer of natural resources to the provinces by the federal government in 1930, these mineral reservations passed to the Crown in right of the respective provinces. After that date, grants of land in the public domain were issued by the province and mines and minerals were usually reserved. Both the Hudson's Bay Company and the Canadian Pacific Railway Company were early land owners who sold some of their land without reserving mines and minerals and later reserved mines and minerals in their deeds or transfers. In the case of the Crown reservations, it is possible for a certificate of title to contain a land description without mention of any exception or reservation of mines and minerals in favour of the Crown, yet the reservation might apply under the section of the Act which makes reservations in the original grant an exception to the conclusiveness of the register.

§**3.84** It is clear that if mines and minerals were not reserved in favour of the Crown in the original grant, they became land for which a certificate of title could be issued under the torrens system. This was done in a number of ways:

---

44 The New Brunswick and Nova Scotia Land Titles Acts (*supra*, note 17 and R.S.N.S. 1967, c. 162) specifically exclude mines and minerals from the definition of "land".
45 1872, 35 Vict., c. 23.

(1) by issuing certificates of title without mention of any exception for mines and minerals;
(2) by stating in a certificate of title that mines and minerals were included; and
(3) By issuing separate certificates of titles for mines and minerals.

§3.85 In the first case, to be satisfied about ownership of the mines and minerals, it is necessary to search the original grant from the Crown. It is also wise to make an historical search of the intervening certificates of title to ensure that there are no situations under which a person could claim under a prior certificate of title. Where the certificate of title states that minerals are included, the situation is little different from the first, except that the assurance fund might be resorted to for compensation in cases where the facts are such that another person is entitled to the mines and minerals. It is still wise to make the historical search and check the reservations in the original grant. In the third case, where a separate certificate of title was issued, there is still some risk involved in relying on the certificate alone, and the historical search, including a search of the grant from the Crown, is a wise precaution.

§3.86 It is unfortunate that it is necessary to look behind the certificate of title in the case of mineral ownership, but the fact that there were more opportunities for error introduced by the methods of recording mineral interests than applied to the routine for issuing certificates of title to the surface has created an air of uncertainty. In addition, the value of a mineral interest is unknown until a marketable mineral has been discovered, at which time it may become extremely valuable. Concerned about the possibility of huge claims, the provinces placed an upper limit on claims against the assurance fund arising from errors in certificates of title for mines and minerals. This has led those leasing mineral interests and intending to invest substantial capital in exploration and drilling or mining to take every possible precaution, including the historical searches.

§3.87 The problems relating to mineral title are complicated by the fact that there is a wide divergence in the language used in the reservation clauses in the deeds or transfers which create the exceptions. The registrar may be called upon to issue a certificate of title for a specific substance. If mines and minerals are included in the certificate of title for the surface, there would be no difficulty involved with the registration of a transfer of a specific substance. However, if there is separate ownership of the surface and mines and minerals, then it might be beyond the scope of the responsibilities of the registrar to make a determination as to whether or not the substance is included in the certificate of title to the mines and minerals, or the certificate of title to the surface. When confronted with the task of interpreting the meaning of the reservation clauses, the courts have applied the common law rules of interpretation. For example, a clause reserving "all mines and minerals, including without limiting the generality of the foregoing, all petroleum" would cover less than a reservation of "all mines and minerals". This is because the courts would apply the *ejusdem generis* rule. Under this rule, the enumeration of one or more substances of a particular nature, limits the meaning of the general description to similar substances. Even the expression "all mines and minerals" may be interpreted differently depending upon the date of severance. This principle is outlined in the decision in *Borys v. C.P.R.*[46] The following quotation is from the judgment delivered by Lord Porter:

> When endeavouring to ascertain the meaning which is to be attributed to "petroleum" in the original reservation, it becomes necessary to decide who are the persons whose use of the word is to determine the sense in which it is employed in the relevant document, inasmuch as the chemist and laboratory expert may attribute to it a meaning different from that which the lay mind would adopt. As has been said, the chemical contents of the petroleum and the natural gas found in the field are the same and regarded scientifically the substances are therefore the same. But a scientific similarity of substance does not establish that the materials are themselves rightly to be described by the same name. The proper approach, says the appellant, is to ascertain the meaning of the word in the mouths of those non-scientific persons who are concerned with its use, such as landowners, business men and engineers, and to be guided by them as to the true construction of the reservation. The vernacular not the scientific meaning is, he maintains, the true one, and in support of this contention he calls attention to the observations of Lord Halsbury in *Glasgow Corpn. v. Farie* (1888) 13 App Cas 657, at 669, 58 LJPC 33, when he says of mines and minerals that in construing the expression it has to be determined what these words mean in the vernacular of the mining world, the commercial world and landowners at the time when the grant is made. This method of interpretation has been repeated and accepted more than once and their Lordships agree that, where it can be ascertained that a particular vernacular meaning is attributed to words under circumstances similar to those in which the expression to be construed is found, the vernacular meaning must prevail over the scientific. But the distinction is not a rigid one to be applied without regard to the circumstances in which the word is used. It was said by Lord Watson in the same case at p. 675:
>
>> "Mines" and "minerals" are not definite terms: they are susceptible of limitation or expansion, according to the intention with which they are used.
>
> In their Lordships' view the same observations are true of the meaning of petroleum. It may vary according to the circumstances in which it is used.

---

46 (1953), 7 W.W.R. 546 at 552-553 (P.C.).

§3.88 The commonly used method of protecting petroleum and natural leases under the torrens system is by the registration of a caveat. Those obtaining sub-leases or assignments also file caveats. In the case of a full assignment of the caveator's interest, a transfer of the caveat may be registered in those provinces where provision has been made for the registration of transfers. The lease operates as a conveyance of a profit *a prendre*, that is, the ownership of the substance when extracted from the land.

§3.89 The lessees generally regard themselves as the owner of the leased minerals and speak of the registered owner as the owner of the "royalty interest". The lease may be limited to specific substances between defined strata below the surface, but the caveat registered to protect such a lease must use the land description in the certificate of title.

§3.90 Substances such as petroleum and natural gas are ambulatory in nature, and, therefore, it is possible for a well on one owner's land to draw these substances from beneath surrounding lands. This has made it necessary to legislate special rules for sharing underground deposits.

LAND REGISTRATION SYSTEMS IN THE PROVINCE OF ONTARIO

§3.91 There are two systems of registration in the Province of Ontario. The first is the deed registration system under the *Registry Act*[47] and the second is the land titles system under the *Land Titles Act*.[48] Both systems have been modified to make improvements over the years and major changes are taking place at the present time, stemming from the Province of Ontario Land Registration and Information System (POLARIS) project and the Land Registration Improvement project.

§3.92 The original registration system was purely voluntary and an alphabetical system of indexing was used. The *Registry Act*[49] of 1865 forms the basis for the present system. It introduced: (1) the parcel based index (abstract index); (2) the requirement to register original executed documents; and (3) provisions which effectively made the use of the system compulsory. The main features of this system were outlined earlier in the section dealing with deed registration systems.

§3.93 It is important to note that in Ontario, it is not necessary to search back more than 40 years unless there has been no deed registered within

---

47 R.S.O. 1980, c. 445.
48 R.S.O. 1980, c. 230.
49 1865, 29 Vict., c. 24.

the 40-year period. The original enactment limiting the search period to a root deed outside the 40-year period was the *Investigation of Titles Act*[50] of 1929 which came into force in 1930. This was revised as Part III of the *Registry Act* to eliminate the requirement to search back to a root deed outside the 40-year period, unless there was no deed registered within the 40-year period. The latest revision which came into force in August 1981[51] eliminated the preservative effect of acknowledgments of instruments registered outside the 40-year period contained in instruments registered within the 40-year period. When a search of the abstract is requested, it will cover only a 40-year period unless a longer period is specified in the request.

§3.94 Another feature of the Ontario system requires the involvement of the registrar. This is the provision for deletion from the abstract of mortgages which appear to have been validly discharged by a discharge of mortgage. Where a discharge of mortgage has been registered for more than ten years but the mortgage has not been deleted by the registrar, the mortgage ceases to affect subsequently acquired interests.[52] If the registrar removes a mortgage which has not been effectively discharged, the assurance fund set up under section 57 of the *Land Titles Act* is liable to pay compensation.[53] The registrar may also delete from the abstract references to certain other instruments which have been discharged for more than two years. These interests cease to affect the land after the expiration of the two year period. Examples are a claim for a mechanic's lien, a registered notice of a conditional sales contract, and a registered gas or oil lease.[54] The assurance fund is also liable to pay compensation in the event of deletion of an interest which has not been properly discharged.[55]

§3.95 The registry system has also been enhanced by the *Certification of Titles Act*,[56] under which a procedure was available after September 1, 1958, for certifying the state of title at the time of registration of a plan of subdivision. Certification areas were designated by regulations and before the registration of a plan of subdivision, the owner's title had to be certified under that Act or registered under the *Land Titles Act* (in an area where available). The requirement has been extended by subsequent amendments to require certification of all plans of subdivision before

---

50 S.O. 1929, c. 41.
51 *Supra*, note 47 at s. 105.
52 *Ibid.* s. 51.
53 *Ibid.* s. 108.
54 *Ibid.* s. 62.
55 *Ibid.* s. 108.
56 S.O. 1958, c. 9.

registration, or registration under the *Land Titles Act*, and the Director of Titles has been given authority to retrospectively, without application or consent, certify the titles of subdividers. A significant number of subdividers' titles have been certified under the project undertaken by the Ministry of Consumer and Commercial Relations. It is readily apparent that it is advantageous to persons dealing with lots in a plan of subdivision not to be required to search back of the registered plan. The *Registry Act* continues to apply to the certified land and a new chain of title begins for each subdivision lot.

§3.96 The system under the *Land Titles Act* was introduced in Ontario in 1885. As mentioned earlier, it was not derived from the original torrens statute enacted in New South Wales, but from the English *Land Transfer Act*.[57] The basic principles of the systems are similar, the differences being more procedural than substantive. Under the Ontario system, the caution takes the place of the caveat, but it is not used freely to protect interests under instruments which could be registered under the system. The emphasis is upon providing for the registration of the instrument. Unlike the system in the prairie provinces, a new certificate of title is not issued each time there is a new registered owner. Instead, the entry in the parcel register showing the ownership of the former owner is deleted and an entry showing the ownership by the new owner is added and certified by the registrar. At one time, the registrar issued a certificate of ownership, but this practice has been discontinued and a copy of the parcel register is acceptable proof of ownership and of the state of the title. Registration under the *Land Titles Act* is still not available in all areas of Ontario.

§3.97 The *Land Registration Reform Act*[58] contains provisions which introduce major changes to the registration system. Among these changes are the following:

(1) to be registered under either the *Registry Act* or the *Land Titles Act*, a document must be in one of four prescribed forms: Transfer/Deed of Land; Charge/Mortgage of Land; Discharge of Charge/Mortgage; Document General. The document general is used to cover any instrument other than one of the first three. Additional particulars of the documents may be set out in the schedules to the forms;
(2) Most of the commonly required affidavits have been replaced by statements in the documents;
(3) Statutory covenants will be implied rather than set out in the forms;

---

57 1875, 38 & 39 Vict., c. 87.
58 S.O. 1984, c. 32.

the covenants may be expressly excluded or waived in a schedule; and
(4) Standard charge terms may be filed with the Director of Land Registration. This will avoid the repetitious registration by major lending institutions of the same terms in numerous charges.

§3.98 One of the objectives of the POLARIS project is to create property maps and provide unique identifiers to each land parcel. It will then be possible to obtain a computer-generated search of title to any parcel in the register.

§3.99 Most of the information used to prepare this section on the registration system in Ontario has been kindly provided by Richard E. Priddle Q.C., until recently the Director of Land Registration for the Province of Ontario. The following information concerning registration of plans of survey is a direct quotation from his letter to the writer dated March 6, 1986:

> Plan examination began in the Land Titles Office in Toronto in the mid 1950's. The first "Code of Standards" was embodied in a regulation under the Land Titles Act in 1958, and a similar version was included in regulations made under the Certification of Titles Act, 1958 later the same year. The first regulation dealing with "Surveys, Plans and Descriptions of Land" under the Registry Act came into force on July 1, 1964. About seven years ago we combined several survey related regulations into one, which is now under the Registry Act as R.R.O. 1980, Reg. 898.

§3.100 Mr. Priddle also advises that some of the authority for establishing standards for surveys and plans is to be transferred to the Association of Ontario Land Surveyors, and that most of the responsibility for plan examination was recently transferred from the Ministry of Consumer and Commercial relations to the Association.

THE LAND REGISTRATION SYSTEM IN NEWFOUNDLAND

§3.101 Although the Province of Newfoundland has a deed registration system similar to those found in the Maritime Provinces, the historical development of the legal framework governing land ownership is unique. The writer was directed by Mr. Neil MacNaughton, Director of Surveying for the Department of Forest Resources and Lands in Newfoundland, to an article by Dr. A.C. McEwen which contains a concise outline of the historical development of land ownership law in Newfoundland. With reference to the current basic registration system, the following statement appears:[59]

---

59 A.C. McEwen, "Land Titles in Newfoundland", *The Canadian Surveyor*, Vol. 31, No. 2, June 1977, 151 at 158.

The present Registry of Deeds is essentially a repository of documents, and its staff has neither the legal authority nor the technical resources to check or scrutinize submissions for other than matters of form. Provided the instrument is properly executed and proved for registration, it must be accepted upon payment of the prescribed fees. Title searches are assisted by an alphabetical index of parties to registered transactions, and all documents are reproduced in bound volumes. In Newfoundland, unlike some other jurisdictions, a vendor's obligation to show a good and sufficient chain of title is not limited by any statutory minimum period, and this frequently requires the conveyancer to trace a root of title into the distant past.

§3.102 The system is essentially compulsory because an unregistered instrument is void against a subsequent purchaser or mortgagee for valuable consideration who registered his deed or mortgage before the other instrument.[60]

LAND REGISTRATION IN THE MARITIME PROVINCES

§3.103 The registry system is currently being used in the provinces of New Brunswick, Nova Scotia and Prince Edward Island where it has been in place since these provinces were established two hundred or more years ago. Land titles legislation has been developed by the Land Registration and Information Service, Council of Maritime Premiers. Land titles statutes have been enacted in New Brunswick and in Nova Scotia; the New Brunswick Act is in force in Albert County.

§3.104 Some special features of this legislation are outlined below:[61]

(1) Intake Triggered by Conveyance of Property Under Old System
A deed of land may not, by regulation, be registered in the Registry of Deeds without the approval of the Registrar General who will grant this if an application is made to bring the land under the *Land Titles Act*, or special circumstances exist.

(2) Records can be Electronic
The format for documents has been designed to accommodate electronic storage, but hard copy retention is also possible.

(3) General Register Abolished
All instruments, including executions and judgments, must be filed against specific parcels.

---

60 *Ibid.*, at 157.
61 C.W. MacIntosh, "Special Features of the Maritime Land Titles Act enacted in N.B. and in N.S. and in force in Albert County" (L.R.I.S., 1987).

(4) Registration Retroactive to Time of Acceptance
Documents are date and time stamped as they are paid for and if found to be in good order obtain their priority from the time they were handed in.

(5) Caveats Assignable
A claim supported by a caveat may be assigned like any other encumbrance, and it does not require discharge and refiling. In this way there are no problems with priority.

(6) Certificate Shows Entire Title
The Certificate of Registered Ownership shows on the face of it the names of the present owners of the property and lists all encumbrances against title in sufficient detail that original documents need not be consulted.

(7) Mortgages are in Short Form and Contain Lengthy Covenants by Implication
Parties may delete statutory covenants and insert optional covenants by use of assigned numbers. In this way documents can be reduced to one page and those which depart from the standard can be immediately identified.

(8) Debentures in Short Form
A standard form of debenture consisting of one page with implied covenants has been developed. These can be modified in the same manner as mortgages.

(9) Leases
A standard form of lease consisting of one page with implied covenants has been developed. These can be modified in the same manner as mortgages.

(10) Stop Order
This is a document issued by the Registrar General which can immediately freeze a title pending investigation and correction.

(11) Lots on Plans
When an approved plan is filed new certificates of registered ownership are issued to the owner with respect to each parcel shown on the plan.

(12) Miscellaneous Documents
When a document comes in that does not conform to the standard format, the Registrar General may grant an order giving it effect as if it were a transfer, easement or whatever legal interest is intended.

CONCLUSION

§3.105 Only principles and the cases which illustrate those principles which are of universal importance in each of the two major registration systems have been dealt with in this chapter. Regional differences have been pointed out in some, but not all situations. There is sufficient information to promote a basic understanding of the main components of the registration systems.

§3.106 In all jurisdictions, the daily functioning of the system is complicated by numerous statutes which impose duties upon the registrar, some based upon regulatory functions of government not directly related to land ownership. When it is understood that the size of a land registry increases constantly, not only in the number of separate land parcels as subdivision and condominium construction takes place, but also as a repository for documents which are registered in increasing numbers, and that the systems in recent times must accommodate the regulatory functions as well, then the need to find the most efficient means of handling the increased volume and the added duties becomes apparent. Computer data processing systems are being developed or implemented in many provinces. British Columbia has an automated torrens system in place at the time of writing, and the largest land titles offices in the province have been converted to the system. Ontario, Alberta and Manitoba are all developing automated systems. By virtue of being able to avoid re-typing unchanged information when issuing new certificates of title, and the capability of built-in reasonability tests, an automated system should reduce the number of manual or oversight errors in a torrens system. In time, much of the registered plan information will be in automated graphic form. Those who are developing the systems with all of the attendant problems will pave the way for an easier to manage and simpler registration service to the public in the future.

4

Boundaries

David W. Lambden, B.Sc.F., F.R.I.C.S., F.I.S. Aust., C.L.S., O.L.S.
and
*Izaak de Rijcke, B. Sc., LL.B., O.L.S.**

BOUNDARY DEFINED — NATURE AND FUNCTION

§**4.01** A boundary is the line of division between two parcels of land. It is a limiting line; by it is ascertained the extent of parcels in separate ownership or subject to different rights.

§**4.02** Halsbury defines a boundary as "an imaginary line which marks the confines or line of division of two contiguous parcels of land".[1] On the whole, surveyors would be happier if the definition used the word 'invisible' rather than 'imaginary', and Rowton Simpson was careful to point out this distinction in *Land Law and Registration*.[2] A secondary definition by Halsbury is that 'boundary' denotes the physical features or objects by reference to which the line of division is described; and used in this sense the boundary is a visible feature which, by the common law, may be ascribed in ownership to one or to both of the adjoining owners.

---

* The authors wish to record their thanks to W.W. Shepherd, M.L.S., J.N. Gardiner, O.L.S. and M. Viminitz, S.L.S. for their efforts in reading the manuscript of this chapter (a request of very short notice) and giving their comments and valued criticism.
1 *Halsbury's Laws of England*, 4th ed., Lord Hailsham of St. Marylebone (London: Butterworths, 1973), Vol. IV, at 356.
2 S.R. Simpson, *Land Law and Registration* (Cambridge: Cambridge University Press, 1976).

There is, of course, sometimes a need to find the precise property line within or beside the boundary feature and such a fixing of the boundary creates an invisible (or imaginary) line through and along the physical feature of a ditch, a wall, a fence, or other artificial boundary erected by man. A surveyed boundary would fall midway of these two definitions as it produces an invisible line between artificial marks. A natural boundary appears, in first instance, to be a tangible boundary, and this is true if it is at the visible interface of land and water which is a water's edge boundary on inland non-tidal waters; it is less easy to see in the case of the ordinary mean high water line applicable along tidal waters. And if the line of division is the middle thread of a stream, the limit under water is certainly invisible, although its position is dependent on the observable line of the banks. These are natural monuments and are properly not staked, marked or monumented by the work of surveyors; they should, of course, be located at the time of survey of a parcel by measurements connected to other monumented lines of survey since, at later times, the change in position of natural ambulatory boundaries may be evidenced by the measurements.

§4.03 Defining 'boundary' requires that the subject — the land parcel — also be defined.[3] Except as it may be varied by any definition unique to a particular statute, 'land' is deemed, at common law, to include land under the waters of swamps, bogs, ponds and other areas of superficial water within the boundaries and also under those streams that cross the property that have not been excluded by special provisions.[4] The maxim *cujus est solum, ejus est usque ad coelum et ad inferos* is the common law proposition that he who holds the surface of the earth — in modern terms, 'owns' the parcel — holds also the airspace directly above and the subsoil below to the centre of the earth. However, this common law proposition has been varied of practical necessity; above the parcel a somewhat uncertain sense of boundary pertains, especially in consideration of aircraft passage and air pollution. Mineral and other subsurface rights are almost always determinable from the words of the grant.

§4.04 These concepts of boundaries apply to the division of land into individual units to which terms of title are applied establishing the rights and interests in the parcels. The same principles apply to international

---

3 *Supra*, note 1, (1982), Vol. 39 at 262.
4 *Surveys Act*, R.S.O. 1980, c. 493, s. 1(*g*). 'Land' includes land covered with water. The exclusion of land under water may be by the words of the grant if sufficient to rebut the presumption that the upland parcel includes all or some part of the bed, or the exclusion may be by virtue of statutory provisions. See Chapters 5 and 6 in respect of tidal and inland non-tidal waters.

and inter-provincial boundaries and also within a province to the boundaries of local government jurisdiction or legislative constituencies.[5]

§4.05 Boundaries must be created, or delimited, by the proper actions of owners, and they also must be brought into physical existence by being identified and marked on the ground. The latter is, generally, the work of land surveyors in the measurement and marking of ground points and in the drafting of plans of survey. These activities distinguish the property limits and hence demonstrate the parcels on the ground by survey monuments and in documents, such as a plan or other graphic representation, or in some other statement of record such as a written description.

§4.06 This principle is simply illustrated. As a first example, the boundary between the provinces of Alberta and British Columbia was specified and created by statutes. The height of land portion had to be found by the observation of a physical feature on the ground and the monuments are useful ancillaries. The meridian portion had to be found by measurement for longitude, then run as a line on the ground with survey monuments set as necessary identifiers. In this case, the survey follows the creation of the boundary. A second example, a plan of subdivision, seems at first instance to follow a different route in that the surveyor's work on the ground comes first. However, it is the combined result of the owner's statement on the plan, the registration of the plan, and the dealing with land parcels represented on the plan that creates the boundaries; this fulfills the owner's intent to divide the larger parcel. Before these combined actions, the marks on the ground are merely pieces of iron or wood; after the actions, the marks have become 'monuments' and the lines have become 'boundaries'. The boundaries so monumented are sometimes said to have become 'final and conclusive' or 'true and unalterable'. Boundaries are the result of creation, of marking or identifying and of the sanction of both recognition and adoption; they are then *the* boundaries. The statutes are restating the common law principles and the qualifiers are, in fact, tautology.

HISTORY AND GEOGRAPHY — CANADIAN LAND RESOURCES

§4.07 In Canada, we speak in the traditional manner of the Crown as the apex of state power, as the outright owner of ungranted lands and as the ultimate owner of all lands of which individuals are, in feudal theory, mere tenants. By the ancient rules, the feudal theme applied to a colonial territory, placed the ownership of landed property, from which came the royal revenues, in the monarch personally; the modern scheme settles

---

[5] *Territorial Division Act*, R.S.O. 1980, c. 497; *Municipal Boundary Negotiations Act*, S.O. 1981, c. 70.

110 Survey Law in Canada

ownership in the sovereign state. In this abstract concept, the monarch, or his appointed representative in public capacity as a body politic, is the Crown and the head of the executive power of government from which the right of grant arises.

§4.08 Ontario treats its Crown lands under the *Public Lands Act*.[6] It is more a matter of semantics than realism. However, the notion of public lands was the death knell to the Establishment of 150 years ago: "a 'public lands' includes lands heretofore designated as Crown lands, school lands and clergy lands."

§4.09 The British North American colonies which united in 1867 to form Canada first came into existence by conquest or by settlement. They were Crown colonies and the monarch personally had acceded to the ownership of all lands and resources and the revenue from these as previously and thereafter granted.

§4.10 Quebec was conquered territory and out of Quebec came Ontario — in part. Both the English and the old French laws may affect the determination of ownership of the resources of Ontario and the position of certain boundaries. Between the Crown and a private party the rights of ownership have seldom been in contest and are usually easily determined from the Crown grant as the origins of private title.

§4.11 The revenue raised on private land is the property tax and in English law it has long been settled that Parliament controls taxation. Upon the sale of Crown land, the rule of old English law and of the old French law, held that the Crown as owner received the revenues directly. The *Public Lands Act*, as a modern measure, treats these revenues as public property and receipts go to the consolidated revenue fund of the province. This has been the situation for Ontario since 1837[7] and analogous administrative structures can be identified in the other provinces.

§4.12 It is the revenues that are of prime interest. The sovereign enjoyed the prerogative rights of revenue from Crown land and resources in the Canadian colonies years after they had in England been converted into

---

6 R.S.O. 1980, c. 413; and see *An Act respecting the sale and management of the Public Lands*, S.U.C. 1860, 23 Vic., c. 2., s. 38.

7 And see, for an historical account of some elements, *Niagara Falls Park v. Howard* (1892), 23 O.R. 1, affirmed (1896), 23 O.A.R. 355. Per Chancellor Boyd in 23 O.R. 1 at 23: "An important date is 1837, when the custody, control, and ownership of all public lands in Upper Canada was transferred to the Provincial Government by the Act 7 Wm. IV., ch. 118, to which the Royal assent was given." This was *An Act to provide for the disposal of the Public Lands in this Province, and for other purposes therein mentioned*.

specific payments under the various acts that created the civil list system to provide the funds to maintain the royal household and other traditional royal functions and responsibilities. This conversion from hereditary revenues to annual grants of the British Parliament was effected between 1760, the accession of George III, and 1831, in the reign of William IV.

§4.13 The governments of Upper and Lower Canada in the early 1800's — the Establishment or Family Compact governments — were representative but not responsible. Only the monies from taxation were, as a matter of course, paid into the consolidated revenue fund of the colony. Crown surveys were funded from the military purse in Whitehall and profits from the sale of Crown lands remained with the executive which was, therefore, financially independent from any real need for tax revenues that were refused by the legislatures. The elected legislatures could exercise only some of the usual controls.

§4.14 These issues were major ones in the 1837 Rebellions in Upper and Lower Canada. The end result for all the Canadian colonies, whether by force of rebellion or by negotiation, was a strong sovereign position in 1867 with each colony holding an independent destiny as regards the whole of its assets and resources: in order of time — New Brunswick, 1837; Canada (as predecessor to Ontario and Quebec at Confederation), 1840; Nova Scotia, 1849. Prince Edward Island, which joined the confederacy in 1873, had been granted this same control of territorial and casual revenues in 1851.

§4.15 British Columbia, which joined in 1871, has provided a more complex case in determining provincial rights of control, but several decisions have stated that its position, in respect of its resources, is the same as for the other provinces. In 1949, Newfoundland also brought its control of provincial resources into the federation and has done so with some rather cogent arguments as to continuing control and jurisdiction.

§4.16 As stated by Lord Watson in *St. Catherine's Milling and Lbr. Co. v. R.*,[8] in dealing with the Ontario lands and resources:

> By an Imperial statute passed in the year 1840 (3 & 4 Vict. c. 35), the provinces of Ontario and Quebec, then known as Upper and Lower Canada, were united under the name of the Province of Canada, and it was, inter alia, enacted that, in consideration of certain annual payments which Her Majesty had agreed to accept by way of civil list, the produce of all territorial and other revenues at the disposal of the Crown arising in either of the united Provinces should be paid into the consolidated fund of the new Province. There was no transfer to the Province of any legal estate in the Crown lands, which

---

8 (1883), 13 S.C.R. 577, affirmed 14 App. Cas. 46 at 55-56 (P.C.).

continued to be vested in the Sovereign; but all monies realized by sales or in any other manner became the property of the Province. In other words, all beneficial interest in such lands within the provincial boundaries belonging to the Queen, and either producing or capable of producing revenue, passed to the Province, the title still remaining in the Crown. That continued to be the right of the Province until the passing of the British North America Act, 1867.

And further:

In construing these enactments, it must always be kept in view that, wherever public land with its incidents is described as 'the property of' or as 'belonging to' the Dominion or a Province, these expressions merely import that the right to its beneficial use, or to its proceeds, has been appropriated to the Dominion or the Province, as the case may be, and is subject to the control of its legislature, the land itself being vested in the Crown.

§4.17 Title to property that has not been granted by Letters Patent of the Crown is, then, vested in Her Majesty; the revenue — the beneficial use — accrues to Her Majesty subject to legislative control, that is, to the provincial treasury; the granting of land is by Her Majesty, under seal, on the advice and signed endorsement of the minister of the Crown designated by statute as the administrator for the granting of Crown lands.

§4.18 The Crown-owned property of the colonial provinces was brought forward at Confederation in 1867 for allocation according to the agreed formula set out in the *British North America Act, 1867*[9] much in the same manner as the agreed division of legislative powers in sections 91 and 92. Certain property was to be vested in the Crown in right of Canada (section 108) and the balance was to remain vested in the Crown in right of the province (sections 109 and 117) "subject to the Right of Canada to assume any lands or Public Property required for Fortifications or for the Defence of the Country." What must have seemed eminently simple at that time became complex all too quickly and this has been especially true for the category of Public Works and Property which, by section 108, was to be part of the federal jurisdiction. Legally, the two Crowns are quite different when issues are tried in the courts on contests of jurisdiction over and ownership of land and resources.

§4.19 In precise terms or not, the *British North America Act* of 1867 gave the Crown in right of Canada the power to do those tasks it had been created to handle, namely:

---

9 Now the *Constitution Act, 1867* (U.K.), 30 & 31 Vict., c. 3 as enacted by the *Canada Act 1982* (U.K.), 1982, c. 11.

(a) — For trade and commerce, lands as follows —

    (i) Canals with Lands and Water Power connected therewith
    (ii) Public Harbours
    (iii) Lighthouses and Piers, and Sable Island
    (iv) Rivers and Lake Improvements.

(b) — For trade and commerce, equipment as follows —

    (i) Steamboats, Dredges and Public Vessels
    (ii) Railway and ancillary items.

(c) — For communication and general administration — Custom Houses, Post Offices and Public Buildings.

(d) — For defence purposes — Armories, Drill Sheds, and the like.

§4.20 In fact, however, the federal Crown received what existed as to place and extent of a particular facility in 1867. For example, quite apart from the questions that arise in determining what constitutes a particular harbour, there is the further problem of finding out what that was on 1 July, 1867, or at the date the province joined.

§4.21 The harbours have provided interesting illustrations of the legal struggles and political processes. There are two ways to settle whether or not a public harbour is federal property: firstly, by litigation as in the case of the harbours at Summerside, Sydney, Halifax, Montreal, Toronto, Vancouver, Quebec, and others; and secondly, by legislation. In 1924, the British Columbia and federal governments, by orders-in-council, settled the federal public harbour status of Victoria, Esquimault, Nanaimo, Alberni, Burrard Inlet and New Westminster. In 1963, an agreement was achieved between Ontario and the federal government. This resolved at least part of the problem by setting forth the 27 public harbours mutually agreed to (of some 60 that had existed in 1867) and with legal descriptions of the properties. The remaining harbours are the property of Ontario, at least to the extent that provincial Crown ownership is determinable and has not been conveyed to private holding. Similar agreements do not exist for the rest of Canada. They would certainly be of benefit in resolving issues that are costly to settle in the courts, especially where there are questions of fact as to the extent and scope of a harbour as federal public property.

§4.22 This bare mention of federal/provincial relations in respect of land, and the illustrations of harbours, is given to suggest that Crown rights and responsibilities are a complex matter that merits serious thought by the surveying profession. There is an effect on certain boundaries and surveys. There are many other instances, such as the determination of navigability and the lands of Native Peoples and Metis, which have led to particular repositories for records, the authority of provincial and Canada

114 Survey Law in Canada

Lands Surveyors to conduct surveys and the appropriate forums or tribunals for the resolution of title and boundary issues.[10]

ORIGINAL SURVEYS OF THE CROWN

§4.23 From the beginning of the government, surveys in the present provinces of Manitoba, Saskatchewan, Alberta, part of British Columbia and the remaining portions of Canada's original North West Territory (now called the Yukon Territory and the Northwest Territories), the instructions issued under the statutory provisions of the *Dominion Lands Act*[11] specifically defined 'monuments' as certain described physical things and 'boundary' as the straight line between these monuments. The boundaries were, therefore, the physical creations of the original federal Crown surveys, an action which the Crown could undertake under the Act or probably without it as a prerogative of the Crown to deal with and dispose of Crown lands; the point is that the procedures were wholly statutory in origin. By contrast, in the Atlantic provinces there was never a system of survey.

§4.24 In Ontario, a rudimentary concept of system was applied in the earliest surveys of 1783, later to become increasingly rigorous,[12] but never to the extent of the Dominion system.[13] The present controlling statute for Ontario is the *Surveys Act* of 1980, which had its beginning in the 1783 Crown instructions for settlement surveys and the ordinance of 1785. The origins of the details, however, are more properly ascribed to the 1798 statutes of Upper Canada[14] and, most notably, to the 1818 statutes.[15] Section 2 of the present Ontario statute provides that:

---

10 As one recent reference see G.V. La Forest, *Natural Resources and Public Property under the Canadian Constitution.* (Toronto: University of Toronto Press, 1969).
11 *Dominion Lands Act*, S.C. 1872, 35 Vic., c. 23, succeeding statutes, and the *Manual of Instructions*, various editions, of which a catalogue is appended. In 1908, the *Dominion Lands Act* was divided and enacted as *Dominion Lands Act*, S.C. 1908, 7-8 Edw. 7, c. 20, and *Dominion Lands Surveys Act*, S.C. 1908, 7-8 Edw. 7, c. 21. The present provisions are in the *Canada Lands Surveys Act*, R.S.C. 1985, c. L-6, and the *Manual of Instructions for the Survey of Canada Lands* (2d ed.) 1979.
12 R.M. Anderson, "The Development of Township Surveys in Ontario." *The Canadian Surveyor*, April, 1936, Vol. 5, No. 8. See also R.L. Gentilcore, "Lines on the Land", *Ontario History*, Vol. LXI, No. 2, (1969) at 55-73.
13 E.M. Dennis, "The History of the System of Dominion Land Surveys" in *Annual Report, A.O.L.S.* (Toronto: Association of Ontario Land Surveyors, 1915). Also published in the *Report of the 9th Annual Meeting of the Association of Dominion Land Surveyors.* (Ottawa: Association of Dominion Land Surveyors, 1915) at 26. See also J.K. Benner, "The New System of Survey of Dominion Lands" in *Annual Report, A.O.L.S.* (Toronto: Association of Ontario Land Surveyors, 1918).
14 Statutes of Upper Canada, 1798, 38 Geo. 3, c. 1, *An Act to ascertain and establish on a permanent footing, the Boundary Lines of the different Townships of this Province.*
15 Statutes of Upper Canada, 1818, 59 Geo. 3, c. 14, *An Act to repeal an Ordinance of*

Boundaries 115

No survey of land for the purpose of defining, locating or describing any line, boundary or corner of a parcel of land is valid unless made by a surveyor or under the personal supervision of a surveyor.

However, neither here, nor in any other statutory provision, is it provided that the survey makes or creates the boundary. The common law provisions prevail and it is the actions of people, land 'owners', including the Crown, that create boundaries. Boundaries, in the legal sense, come into existence only by some means that has legal validity or standing, or otherwise stated, boundaries must have a character fixed by law. Also the absence of any differentiation in the legal interest held in each of two abutting parcels of land would, if one unique party owned both parcels, deny the existence of a legal boundary. Real property law recognizes this in the doctrine of merger of title.

§4.25 The right to give, sell or otherwise transfer a parcel of land, to which a certain right or estate is attached, is a fact of English law. This was not so in 1066, but became the right of English people in the course of time. Most importantly, there had to be some definite action recognized at law in order to effect the conveyance. As between free men this was originally a publicly witnessed ceremony, a deliberate, open and observable act that did not need documentation. The necessity for the agreement for a transfer to be expressed in writing became part of our laws after the enactment of the *Statute of Frauds* in 1677,[16] though charters of transfer were much used before that time.

§4.26 The law has always recognized that there has to be some legal sanction to make a transfer of land enforceable. Among several meanings, a sanction is an authorization and express permission, a binding and confirming influence, and it is these meanings of sanction that pertain to the legal creation of boundaries.

§4.27 When an estate is divided, and this includes the Crown estate, a new title is created and a new boundary is established. It need not require such a straightforward action as a conveyance by sale or gift, and there

---

the Province of Quebec, passed in the twenty-fifth year of His Majesty's Reign, entitled, "An Ordinance concerning Land Surveyors, and the Admeasurement of Lands," and also to extend the provisions of an Act passed in the thirty-eighth year of His Majesty's Reign, entitled, "An Act to ascertain and establish on a permanent footing the Boundary Lines of the different Townships of this Province," and further to regulate the manner in which Lands are hereafter to be Surveyed.

16 *Statute of Frauds*, 1677, 29 Car. 2, c. 3, and *Statute of Frauds*, R.S.O. 1980, c. 481. The date of reception of English law into the Canadian provinces must be considered in many instances respecting property and, therefore, boundaries. See B. Laskin, *Cases and Notes on Land Law*. rev'd ed. (Toronto: University of Toronto Press, 1964).

are other legal sanctions besides those that pertain to written documents of transfer.

§4.28 For instance, an owner who neglects to exercise his rights of ownership and possession over all his estate, thus maintaining the boundaries of his parcel, may lose his right by the adverse action of another. The man who takes over in the role of a trespasser will then hold the part or the whole of the parcel in possession but will not himself have the full legal title to the area under the claim until such time as a legal process bestows this on him by an order that settles or 'quiets' the title. In this instance, it is the judicial order that is the sanction that establishes the boundary. Until that order is given, the boundary of the claim is not known; that is, the order may not vest full legal title to the whole or part of the parcel that has been the subject of possession. The order may confer title only to that portion of the possession which the court is prepared to recognize and sanction as having satisfied necessary and prerequisite conditions.

§4.29 Again, a boundary may be created by an official act of survey by the Crown even though there is no transfer of any portion. This is the situation in areas such as the western provinces after 1870 and in Ontario from the earliest times[17] where the Crown estate was divided according to preconceived schemes. The basic authority for this status of surveyed lines as boundaries is in the Crown or public lands Acts of the particular jurisdictions. That these limits are in existence as real boundaries is further evidenced, for example in Ontario, by the fact that the Crown may annul the division of a tract of land; but this cannot be done in respect of the rights to any parcel that has been granted by the Crown.[18] The schemes serve as methods of unique parcel descriptions and, if the scheme is annulled, the parcel is redescribed in some other fashion.

THE ORIGIN OF BOUNDARIES

§4.30 The general proposition is summarized as follows:[19]

Boundaries are fixed either (1) by proved acts of the respective owners, or

---

17 The first instructions of civil government for surveys in the future Upper Canada were given by General Frederick Haldimand, Governor of the Province of Quebec, on 11 September, 1783, to Deputy Surveyor John Collins. Some surveys prior to this date had been ordered by the military authority at Niagara: see R.M. Anderson, "The Development of Township Surveys in Ontario", *The Canadian Surveyor*, April, 1936, Vol. 5, No. 8.
18 *Public Lands Act, supra,* note 6 at s. 7.
19 *Halsbury's Laws of England,* 4th ed. (London: Butterworths, 1973) Vol. IV at 356.

(2) by statutes or by orders of the authorities having jurisdiction, or (3), in the absence of such acts, statutes, or orders, by legal presumption.

If the word 'sanctioned' is used in place of the word 'fixed', the legal concept is perhaps more clearly expressed: a severance must have a legal sanction to be enforceable and it is the severance that creates the boundary. In the case of the Crown division of townships, the sanction lies in anticipation of the severance and the progression to fulfillment of procedures under statute.

§4.31 We have had some notable boundary cases in Canada and reasonably sufficient that we need probably not look too often to other common law jurisdictions for precedents. For cases taken to the level of the Supreme Court of Canada, and before 1949 to the Judicial Committee of the Privy Council, there is the Ontario-Manitoba boundary dispute,[20] the Labrador boundary,[21] and others. But for one reason or another, having to do with the legal issues involved, some decisions at these high levels are not all that helpful in establishing concepts, except by rather involved analysis.

§4.32 For a clear statement of the law about boundaries that have been surveyed and marked on the ground, few judgments provide so clear an instruction as the Privy Council decision of *South Australia State v. Victoria State*,[22] upholding the 1911 judgment of the High Court of Australia. There were many issues and there was not unanimity in the High Court decision. It was a long drawn out dispute between the two States.

§4.33 The bare essential details are these: (1) South Australia was created as a Province in 1836, with its eastern boundary to be the 141st meridian of east longitude; (2) within a decade the necessity arose of defining the boundary between the Province and that part of New South Wales which became Victoria in 1850; (3) joint and common arrangements were made and the survey conducted between 1845 and 1847 in which 123 miles of the line were run northward from the sea towards the Murray River, clearly marked and monumented and shown on official maps and officially proclaimed; (4) under a further agreement, the boundary line was extended north to the Murray River in 1850; (5) in 1868, a joint commission between New South Wales and South Australia made longitude observations at the Murray River preliminary to running the line further to the north and it was then found that the earlier line was two miles, 19 chains west of the

---

20 Report of the Judicial Committee of the Privy Council, ordered 11 August, 1884. See also, *Canada (Ontario Boundary) Act, 1889*, (Imp.), 52 & 53 Vic., c. 28, reprinted in R.S.O. 1980, App. B.
21 *Re Labrador Boundary*, [1927] 2 D.L.R. 401 (P.C.).
22 [1914] A.C. 283 (P.C.), affirming (1911), 12 C.L.R. 667 (H.C.).

118   Survey Law in Canada

new fix of the 141st east meridian; and (6) after much correspondence and debate, the State of South Australia brought the case before the High Court in 1911 and, losing the action, appealed to the Privy Council.

§4.34 The Privy Council held that the legal sanction creating the boundary was fulfilled by official actions of the two states; in the words of the headnote:[23]

> [U]pon the true construction of the letters patent, it was contemplated that the 141st degree of east longitude should be ascertained and represented upon the surface of the earth, and that there was implied authority given to the Executives of the two Colonies to do such acts as were necessary to that end; that, upon the facts, the Executives of the two Colonies had acted within that implied authority, and that the line agreed and marked became and was the boundary between the States.

The last clause states what actually constitutes the boundary: the line as first marked and sanctioned was the actual boundary and not the theoretic 141st meridian of east longitude.

§4.35 The *South Australia* case is well worth the reading, as much for the discussion of survey techniques and the concepts of accuracy as for the law that is stated. It is a reasoned expression of the common law principle that the conveyance of a parcel of land, followed by endorsement of the line of decision by the usage of the abutting parties, establishes the boundary as a fact on the ground. It may not be the described parcel of the deed; hence the recourse to the courts to settle the issue and the necessary consideration of the evidence, intrinsic and extrinsic, to resolve the latent and patent ambiguities.[24]

§4.36 Propositions of uniform application in common law may now be stated:[25]

(1) The position of a boundary is primarily governed by the expressed intention of the originating party or parties or, where the intention is uncertain, by the subsequent behaviour of the parties (or beneficiary of the severance) pursuant to any such expressed or implied intention.
(2) That which is intended to set a limit to the extent of a boundary is referred to as its 'bound'. *What constitutes the bound for any particular boundary is a question of law.*

---

23 *Ibid.* at 284.
24 *Watcham v. Att.-Gen. of East Africa Protectorate*, [1919] A.C. 533 (P.C.). See also *Schuler A.G. v. Wickman Machine Tool Sales Ltd.*, [1974] A.C. 235 at 261 and 269 (H.L.). See also *Grasett v. Carter* (1884), 10 S.C.R. 105.
25 Particular acknowledgement is here, and for the next paragraph, made to a late colleague and friend Frank M. Hallmann, *Legal Aspects of Boundary Surveying as Apply in New South Wales.* (Sydney: Institution of Surveyors, 1973) at 174-175.

(3) *Where the bounds are to be located (found) on the ground is a question of fact* to be determined in the light of the best admissible evidence.

§4.37 The following tabulation, which may not be all inclusive, summarizes the various forms of title severance that create boundaries.

(a) Documentary
- (i) Where the intention to sever is unilateral:
  - (a) deed of gift
  - (b) devise
  - (c) expropriation or compulsory acquisition
  - (d) dedication to the public
  - (e) proclamation of vesting
  - (f) statutory vesting
  - (g) order of court
  - (h) plan registration (followed by conveyance/mortgage/charge of one lot therein)
- (ii) Where the intention is bilateral or multilateral:
  - (a) conveyance by purchase and sale; and the earlier, now outmoded, bargain and sale, and lease and release
  - (b) partition
  - (c) agreement as to position of a lost or confused boundary (Theoretically this is not a title severance but factually it is. Replacement of a lost line by statutory rules falls in this same category. One must qualify this categorization by recognizing that certain judicial views of the matter are founded on the presumption that the agreement or statutory prescription for the boundary's position is that the true location, in any event, is based on the best available evidence.)
  - (d) quit claim
- (iii) By statute law for the division of Crown lands — the systems established by the surveys acts, or under Crown or public lands acts or by royal prerogative.

(b) Non-documentary
- (i) Presumption of common law
  - (a) *ad medium filum* rule (rivers and roads)
  - (b) doctrine of accretion (includes erosion)
- (ii) Statutory impositions
  - (a) adverse possession against the Crown
  - (b) adverse possession against a municipality in respect of a road allowance

(c) adverse possession against a freeholder
(d) prescription and user

SURVEYS, PLANS AND EVIDENCE

§4.38 Preceding comments have addressed the meaning of boundary, in the basic sense of a dividing line between title rights and the legal process of creation. The notion of what constitutes a boundary is more than a little complex because what may be seen as a real thing on the ground may also be seen as a representation on the paper record in the form of words or diagrams or both. Differences in the perceived images are fruitful sources of litigation.

§4.39 Modern administrative and legal practices clearly show that recourse must inevitably be had to survey records. In some systems, such as in the United Kingdom, there are large-scale mapping records. In others, for example, South Africa and most European countries, there are intensive ground surveys completely tied to control from which come numerically coordinated values, in theory at least, expressing the position of boundary corners. In yet others, including Canada generally, there is the individual parcel, with an individual survey, by an individual surveyor, forming an individual record, increasingly proving that individuality is, in these things, somewhat of a handicap since, as in most other activities, individuals have neighbours.

§4.40 The locations of boundaries are a matter of fact in the legal sense and the surveying profession acts as the carrier of these facts from the ground to the courts in cases of dispute that are brought to action. In day-to-day operations (as there is actually very little litigation about boundaries), the surveyor creates the record of things on the ground to be thereafter used by people far-distant from the ground location.

§4.41 In one instance, the surveyor is the creator of fresh evidence as in the original Crown surveys of wilderness lands for settlement or in the marking of lots in a new subdivision for a private owner. In another instance, he is the interpreter of old evidence to locate the position of an original boundary, but he is not the authority of final decision; that remains the function of the courts.

§4.42 It is worth noting here that the surveyor's work in placing new ground marks before boundaries are legally created and, the weight given his opinion expressed on the redefining of original boundaries according to physical and recorded evidence, place a heavy responsibility upon him. This responsibility does not end with the completion of a job, nor is the onus shifted to a land registration office on deposit and registration of survey records or necessarily to a solicitor certifying a common law title for deeds registry.[26]

§4.43 At all times, the location of a boundary of a parcel is a question of fact to be based on evidence. The orders of reliability of evidence that are definitive of a boundary reflect those things which the courts have found least likely of error, namely, first preference to the natural boundaries of parcels; second preference to original monuments placed or recognized by survey; third preference to features of possessory evidence that can be related in time to the original survey (this is not adverse possession); and fourth preference to measurements.[27]

§4.44 This placement of linear and angular measurements, as evidence, in a position of least weight is not a mark of discredit to the surveyor. Rather than that, it is the realization that there is more to fixing a boundary than using measured dimensions from other points which may be equally uncertain in position or deducing that a given line is straight simply because a single direction is recorded for it. Modern techniques can reproduce stated dimensions within very fine limits and lines can be run straight, but old lines and dimensions, set out and measured with quite different equipment such as the Gunter's chain, with its actual links, for distances and the magnetic compass for directions, were almost invariably inaccurate and imprecise. Alternatively, the marks may be considered to be incorrectly placed and in error from the recorded dimensions; this may give grounds for resurvey, that is, for a new definition. Obviously, long-accepted features, to which property owners have lived quietly and to which they have acquiesced as their boundaries, cannot be lightly upset.[28]

§4.45 Once an area is covered by control survey markers which are not disturbed, the measurements connecting these points and parcel boundaries may assume significance as reliable evidence. (The writers hesitate to presume the views of the courts on the true significance.) This situation barely exists in Canada and will not be a strong feature of the survey of parcels for many years to come until the full effect of properly administered survey integration policies are secured and, presumably, the whole concept of numerical definition of boundaries has been tested in litigation. In fairness to proponents of co-ordination, and to themselves,

---

26 See, for example, *LeBlanc v. DeWitt* (1984), 34 R.P.R. 196 (N.B. Q.B.).
27 The sources in *Greenleaf on Evidence*, *infra*, note 45, and in the Canadian decisions are given in D.W. Lambden and I. de Rijcke, *Boundaries and Surveys*, (Canadian Encyclopedic Digest (Ont. 3rd), Title 19) (Toronto: Carswell, 1985).
28 *Palmer v. Thornbeck* (1877), 27 U.C.C.P. 291 (Ont. C.A.). See also *South Australia State v. Victoria State*, *supra*, note 22.

122 Survey Law in Canada

the writers leave unanswered the whole question of "properly administered survey integration policies".

§4.46 There are two possible situations that may emerge in examining a parcel on the ground and the title records of that same parcel: (1) that the fences or other occupational evidence may or may not be on the boundaries; and (2) that the stated dimensions of an original title plan may or may not be accurate (and the older the plan, the less accurate it will probably be as a reflection of the true dimensions). This points up the need for the surveyor to address these critical situations in his analysis of the evidence and in his subsequent written report to his client in which he expresses his opinion.

SURVEY SYSTEMS AND STATUTES

§4.47 Statutory provisions for surveys vary considerably from one Canadian common law province to another. The differences are a main cause of differing concepts and philosophies about surveys and between surveyors and are reflected in education and the actualities of reciprocity. Manitoba, Saskatchewan, Alberta and British Columbia have derivative Acts of the original *Dominion Lands Act* of 1872, while the present federal statute, the *Canada Lands Surveys Act*,[29] applies to the Yukon and Northwest Territories, some federal lands in the provinces and the offshore. The Ontario *Surveys Act* has yet older origins. The Atlantic provinces are characteristically common law in having no general schemes for Crown lands division and for the execution of surveys.

§4.48 There seems little doubt that, in writing the 1871 *Manual shewing the System of Survey adopted for the Public Lands of Canada in Manitoba and the North-West Territories with Instructions to Surveyors*,[30] John Stoughton Dennis, first Surveyor General, was reflecting a favourable view of the American system and manual of instructions — this he acknowledged — and, in the writers' opinion, a strong aversion to the Ontario systems and procedures with which he was familiar due to several decades of surveying as a provincial land surveyor. This chapter is not the place to recite the Dominion Lands Survey System or other provincial systems and

---

29 R.S.C. 1985, c. L-6.
30 A catalogue of the ten editions of the *Manual of Instructions for the Survey of Dominion Lands* and of the two editions for the *Survey of Canada Lands* is given at the end of this chapter.

the reader is referred to the *Manuals* and to various papers.[31] What Dennis implemented was a rigorous scheme on a designed theoretical basis to locate uniform and fully-surveyed blocks a mile square — in fact quartered by lines subsequently to be run to create the quarter-section of 160 acres more or less as the homestead unit for a family. This was the area which notions of land use — and some political views, undoubtedly — ordained as appropriate. Dennis' own surveys in Ontario, made as a provincial land surveyor, are typical of those of the mid 1800's; perhaps his original surveys for the Crown should be credited as somewhat better than most for the standard of care he gave to measurement. For the new territory, Dennis rejected the procedures for surveys of the older provinces. Lines would be run by theodolite and direction controlled by frequent astronomic observations; the compass was not to be used, except for some minor surveys. For distance measurement at that time there was only the Gunter's chain of actual links; painstaking care was needed for the error would quickly show up on the closure of each section. The steel band did not come into general use until 1878 or 1879.[32] This was also the time (1881) when steel markers were first required to be set and the use of wooden posts, cut at the site as the corner monuments, came to an end for the Dominion surveys. Wooden markers were thereafter used only as visual guides to the corners. In the prairie country, the wooden posts were mostly poplar which has short life as a monument, but, it appears, they lasted long enough for the settlers to take up the land and to be subsequently found as solid evidence during early retracement surveys in Manitoba. It is not uncommon to still find the pits and mounds erected at the time of the first survey, but the wooden posts have long since disappeared.

§4.49 The system, of course, was not controlled in the geodetic sense through the use of extended control and the breaking down of a survey from the whole to the part. The Special Survey after 1876, with telegraphic time signals and other devices, reasonably succeeded in injecting a much better control over the extension and propagation of errors.

§4.50 In Ontario, it was not until the first decade of this century that Crown surveyors generally set iron markers; only nine were required for the large townships of the 1800-acre section system in the northern Clay Belt from Cochrane west. It would also appear that the compass continued in use

---

31 R. Thistlethwaite, *Systems of Survey in Canada*. (Ottawa: Dept. of Mines and Technical Surveys, now Energy, Mines and Resources Canada, 1962) and L.M. Sebert, "The History of the Rectangular Townships of the Canadian Prairies", *The Canadian Surveyor*, December, 1963, Vol. 17, No. 5.

32 E. Deville, "*Standards of Length and Errors in Dominion Lands Surveys*" Sessional Paper no. 25. Annual report of the Department of the Interior for the Fiscal Year ending 31 March, 1913, Vol. 1, 1914.

as a line instrument well into the present century as it did until after the Second World War in the Atlantic provinces.

§4.51 These original surveys of the Crown were settlement surveys and accuracy was secondary to progress. In comparison with the system for the western provinces, however, more criticism can be lodged against the Ontario systems for the lack of control, methodology and procedure. What purports to be a system, of course, is a matter of interpretation.

§4.52 The point must be made that once the original surveyed lines, faulty for distance and crooked from compass error as they might be, were recognized by the settlers and converted to physical features and once a lot line, not run in the original Crown survey, emerged as a physical feature, whether from a known survey of the lot line or not, then it would appear that the courts are reluctant to decide for any other line than that which comes from the long-standing acceptance of the adjoining owners.[33] This is entirely logical. The argument as to whether this is the lot line by the *Surveys Act* or some other type of lot line (e.g., lot line by occupation or established lot line) becomes nonsensical. There is inherent and common sense logic against changing the title records for each failure of the ground-based facts to conform to the theory of the system for corners and lines and numerical values as there are stated in records and intended to express the position of boundaries on the ground.

§4.53 The emphasis shifts from survey to boundary. Theobald in *Law of Land* expressed the rule that 'the law bounds every man's property and is his fence' and that title to land — to the whole or any part — is not lost merely because the boundaries are lost or presumed lost, or because a subsequent survey shows that the first demarcation of a line is incorrect in its placement. Mr. Justice O'Connor, in his judgment in the High Court decision in the *South Australia* case,[34] added a further notion for the definition of a boundary, " . . . the very term 'boundary' connotes in its ordinary natural meaning a line of division capable of being permanently fixed." He found that:

> [A] reading of the expression 'fix the boundaries' which could confine its operation to a fixing on paper could hardly have been within the contemplation of the legislature. In its ordinary everyday meaning the words are wide enough to cover a fixing and marking on the ground as well as a fixing by written

---

33 See §4.43 respecting the priorities of evidence recognized by judicial decisions. See also *Palmer v. Thornbeck, supra,* note 28; and numerous cases cited in D.W. Lambden and I. de Rijcke, *Boundaries and Surveys,* (Canadian Encyclopedic Digest (Ont. 3rd), Title 19), (Toronto: Carswell, 1985). Also Canadian Abridgement — Title *Boundaries* (Toronto: Carswell 1966) and supplements.

34 *Supra,* note 22 (H.C.), at 712-714.

description, and in that wide sense they ought in my opinion to be construed. ... The power to fix territorial boundaries must imply the power to fix them permanently.

§4.54 Mr. Justice Isaacs noted that the Imperial authorities making the charter grant of the province had not undertaken the demarcation on the ground; it had been left to the local authorities as the interested parties. "And, once done, it was of the essence of the matter that it should be permanent."[35] As a direct parallel, it must be noted that the Crown of Ontario did not undertake the survey of the lot lines and there is no rule that says the adjoining owners cannot settle their boundaries by their own first demarcation; the inference is that the task had been left to the adjoining owners and, "once done . . . it should be permanent."

> Nor is it a reasonable supposition that the line was fixed with an implied reservation that it was to be subsequently altered if found incorrect. The notion is scarcely conceivable that titles were to be taken and paid for, and homes built, political ties and institutions formed and established, all with the consciousness that the boundary once fixed might at any moment be altered, and at indefinite intervals swing backwards and forwards. The only reasonable idea is permanency of a boundary delimitation in accordance with the law so far as that was then practically possible, having regard to the circumstances of the country, the state of science when the operation was undertaken, and the urgency of the occasion.[36]

Stated in another way, the two and a quarter mile offset of that meridian boundary run in 1833 fell within the range of 'more or less'. The views expressed by O'Connor and Isaacs JJ. were fully endorsed by the Privy Council.

§4.55 In the case of *Bell v. Howard*,[37] Mr. Justice Draper, Chief Justice of the Court of Appeal of Upper Canada, stated:

> [I]t was a question proper to be submitted to the jury upon the evidence whether a division line had been adopted and agreed upon between the owners and occupiers of these two lots, according to which they had used, occupied, and enjoyed, respectively, for more than twenty years before the commencement of this action; and that if there had been a division line so agreed upon, and the occupation of the respective proprietors had been so mutually limited thereby for twenty years or upwards, the parties would be bound by it, though on an accurate survey it should be found to vary from the true division line ascertained according to the original plan of survey. It seems to have been conceded all round that the division line between these lots had not been run in the original survey of the township; but there is strong evidence to show that a line had been run a good many years ago dividing these lots,

---

35 *Ibid.* (H.C.), at 724.
36 *Ibid.*
37 (1857), 6 U.C.C.P. 292 at 296 (C.A.).

the marks of which blazed on the trees are still to be traced, and at the north end of which there was a stake standing more than twenty years ago, and considerably within that period; that the occupants of both lots, in cutting timber or disposing of timber to others, had asserted their own rights up to this line, claiming nothing beyond it, and giving directions to those employed by, or acting under, them to observe and not to cut the trees which were marked to designate the line. I think this is evidence of occupation on the one hand and of acquiescence on the other, of mutual agreement as to the boundary and of waiver of any right to set up or claim any other boundary; and if believed by the jury, sufficient to warrant their verdict, which was in fact rested on this ground.

And this renders it, in my opinion, unnecessary to consider the question as to the mode of the original survey, whether by section or otherwise; for the verdict is not affected by the assumption that the survey proved by the plaintiff was correct in principle and in execution, which at present I see no sufficient reason to question.

But I think also that the plaintiff's action, based on the correctness of this survey, is one of strict legal right; and that when it is admitted, as his surveyor proved, that he has his full complement of land, though he fails in this action, and that to establish this survey will have the effect of disturbing all the boundaries in the immediate neighbourhood, we ought not to set aside a verdict which avoids that consequence, and which is consistent with long possession undisputed till the bringing of this action, and apparently acquiesced in by those interested in it, unless upon very clear and satisfactory grounds; and that if the verdict can be upheld as not being contrary to law, nor unsupported by evidence we are bound to uphold it.

§4.56 Section 8 of the Ontario *Public Lands Act* provides for resurveys and for altering and amending plans but the effect on entitlement seems unclear. Section 57 of the *Dominion Lands Surveys Act* was most explicit about resurveys.

> Whenever through an error in the survey, a boundary monument is not at the place where it should have been erected, the Minister may order that such monument be removed and that a new monument be erected at the proper place; but no monument defining the boundary of land for which letters patent have issued shall be displaced without the consent in writing of the owner thereof; nor shall a monument defining the boundary of land held as a homestead or under lease, licence or agreement of sale be displaced without the consent in writing of the holder thereof, unless the error in the position of the monument is at least five chains, in which event the Minister may, without the consent of the holder, authorize the correction of the error, but the person or persons acquiring through such correction any improvements on the land shall be required to pay the owner of such improvements therefore such an amount as may be fixed by the Minister [or as determined by an arbitrator].

Instances of actual use of this section are not known to the writers, but the tenor of its provisions is cavalier. A resurvey so defined is not a redefinition of original boundaries; it is the creation and marking of wholly new ones.

§4.57 In the Upper Canada decision of *Dennison v. Chew*,[38] the view is quite different. The essence is that the boundaries, once settled and acquiesced in by the adjoining proprietors, must not be upset by any statutory provision that might present a different position, whether by greater accuracy or change of rules:

> I do not conceive [that] the [Surveys] act[39] can be resorted to to undo what mutual arrangements, or the lapse of time and long continued actual possession, may have accomplished. The legislature would not, I think, intentionally make a law with such an object, or to produce such an effect.

§4.58 The 1818 Act certainly reads to an effect that the court would not accept. It was not amended, nor were subsequent derivative sections changed in wording. Administrators and survey personnel, then as now, tend to limit their reading to the statutes and dodge the decisions of the courts which expound on the meaning of the statutes. The misconceptions that were perpetuated by successive generations of surveyors could have been anticipated. The courts have the last word where the issues go to litigation and they were consistent in that where the evidence shows the acceptance (that is, the sanction) of a boundary by the adjoining owners, then that boundary will be held to be the dividing line and also the lot line if that was the intent of the grant. This is recognizably true if that line was run by a surveyor; where it cannot be proven that the line was not run by a surveyor, a presumption appears to prevail that it was, indeed, the work of a surveyor fulfilling, by presumption, the intent of the statute.

§4.59 The Privy Council decision in *James v. Stevenson*[40] held "that a long line of old fencing of unknown origin, erected in a position where one might expect the boundary to be, would be presumed by the law to be erected on the true line of the boundary in the absence of very cogent evidence to the contrary." As the length of the fencing, some two miles, was a significant part of the evidence, it would seem that this presumption would be appropriate only to rural surveys.

§4.60 In the Upper Canada case of *Palmer v. Thornbeck*[41] the line of division between two lots was disputed. Every second lot in the concessions of the township was abutted by road allowances which throughout the whole township had been retraced on the evidence and the survey confirmed by special statute. Using this evidence to locate by proportion the starting corner and the required direction placed a new line of survey at odds with an old fence located on a line and position of unknown origin.

---

38 (1836), 5 O.S. 161 (Ont. C.A.).
39 S.U.C. 1818, 59 Geo. 3, c. 14, *supra*, note 15.
40 [1893] A.C. 162 (P.C.).
41 *Supra*, note 28.

The court held that, on the evidence, the lot line boundary was the line of fence and that the onus of proof was on the neighbour who attempted to upset settled possession.

§4.61 In summary, then, we have these propositions, valid certainly for Ontario and the Atlantic provinces, perhaps restricted somewhat for the western provinces:

(1) Where a boundary is first to be placed is, initially, purely conceptual, but once used in a document of title it is nonetheless a boundary. If a feature on the ground is not adopted, then some fixing on the ground must be undertaken in accordance with the concept.
(2) The boundary is then placed on the ground for the first time and, while the concept may envisage that a surveyor will do this, it is not an absolute necessity. The Ontario *Surveys Act*,[42] for example, does not say that a *boundary* is invalid for want of a surveyor; it says only that the *survey* is not valid.
(3) It is the operation of the sanction of law that establishes the boundary as fixed. A line does not become a boundary nor a survey mark a monument until they appear as items in a document of conveyance of some interest in land. A contest in the courts must first of all upset the sanction and the onus of proof is on the litigant who wishes to upset the *status quo*. "The burden of proof lies on him who affirms a fact, not on him who denies it."[43]

TRUE AND UNALTERABLE BOUNDARIES: PRIORITIES OF EVIDENCE

§4.62 In the use of the term 'true and unalterable', the statutes are adding an unnecessary gloss to a restricted group of lines. Any boundary (the 'original' by Crown survey,[44] registered plans of subdivision and other 'first' surveys, and redefinitions that have been adjudicated) as a line of division that has been duly sanctioned is a fixture and true and unalterable. The courts use the term 'true line' quite often in order to distinguish a particular line from some other that is being put forward in a dispute. The survey monuments are the markers of positions of boundaries and are only valid so long as they are in their original positions. It is the boundaries that

---

[42] R.S.O. 1980, c. 493, s. 2.
[43] *Ei qui affirmat, non ei qui negat, incumbit probatio*. See, for example, *Palmer v. Thornbeck*, supra, note 28, per Gwynne J., at 294: " . . . as to the true boundary line between lots, the *onus probandi* lies upon [him] who seeks to change the possession."
[44] *Supra*, note 42 at s. 1, provides a restricted meaning of 'original survey' for the purposes of the Act. See also 'original post' defined in s. 1, and the provisions of s. 58 respecting highway lands.

are (if we must use the words) true and unalterable: it is not the monuments, if such monuments are anything but the original or first-placed monuments in original positions defining the boundary for the first time.

§4.63 This concept of the inviolable *status and position* of a boundary as first set on the ground is the foundation for all but one of the rules for extent of property. There is one class of boundaries where the *status* may hold eternal but the *position* will change: natural boundaries are discussed in Chapters 5 and 6.

§4.64 It is by the many and consistent decisions that the American legal writer Greenleaf[45] compiled the statement for the order of priorities for the evidence of the true boundaries where there is ambiguity in a grant. These are common law principles long recognized in Canadian decisions. In current phraseology,[46] Greenleaf's statement is:

> Where there is ambiguity in a grant, the object is to interpret the instrument by ascertaining the intent of the parties; and the rule to find the intent is to give most effect to those things about which men are least liable to mistake. On this principle, the things by which the land granted is described are thus ranked according to the regard which is to be given them: (1) natural boundaries; (2) lines actually run and corners actually marked at the time of the grant; (3) the lines and courses of an adjoining tract, if these are called for and if they are sufficiently established, to which the lines will be extended; and (4) the courses and distances, giving preference to one or the other according to circumstances.

PROPERTY RIGHTS AND BOUNDARIES

§4.65 The ordinary rights of property are determined for the parcel set forth in the title documents, that is, within the specified boundaries. However, there are some variations on this principle. The first to be noted is that which arises from contact of an upland parcel with water. This gives rise to the inherent rights of a riparian proprietor and these are rights of property. The subject is discussed in Chaperts 5 and 6. Discussions of other variations — easements and possessory claims — follow below.

Easements

§4.66 There are certain other rights that may be attached to property as incorporeal interests and extend beyond the boundaries of the parcel. The principal one to note in the context of surveying activities is easements.

---

45 *Greenleaf on the Law of Evidence*, 16th ed., W.D. Lewis ed. (Philadelphia: R. Welsh & Co., 1896), Vol. I, sec. 301, at 488, n. 3.
46 Lambden and de Rijcke, *supra*, note 27 at §90.

These are privileges enjoyed by the owner of one parcel over the land of another person.

§4.67 The English Court of Appeal case *Re Ellenborough Park*[47] provides an excellent introduction. The judgment of the Court was delivered by the Master of the Rolls, Lord Evershed.

> [I]t will be proper, as a foundation for all that follows in this judgment, to attempt a brief account of the emergence in the course of the history of our law, of the rights known to us as "easements," and thereafter, so far as relevant for present purposes, to formulate what can now be taken to be the essential qualities of those rights. For the former purpose we cannot do better than cite a considerable passage from the late Sir William Holdworth's Historical Introduction to the Land Law (Clarendon Press, 1927, p. 265). The author states: "Both the term 'easement' and the thing itself were known to the mediaeval common law. At the latter part of the sixteenth century it was described in Kitchin's book on courts, and defined in the later edition of the 'Termes de la Ley.' " After stating the definition and observing its obvious defects from the point of view of modern law, Sir William proceeds:
>
>> "But these defects in the definition are instructive, because they indicate that the law as to easements was as yet rudimentary.
>>
>> It was still rudimentary when Blackstone wrote. In fact, right down to the beginning of the nineteenth century, there was but little authority on many parts of this subject. Gale, writing in 1839, said: 'The difficulties which arise from the abstruseness and refinements incident to the subject have been increased by the comparatively small number of decided cases affording matter for defining and systematising this branch of the law. Upon some points indeed there is no authority at all in English law.'
>>
>> The industrial revolution, which caused the growth of large towns and manufacturing industries, naturally brought into prominence such easements as ways, watercourses, light, and support; and so Gale's book became the starting-point of the modern law, which rests largely upon comparatively recent decisions.
>>
>> But, though the law of easements is comparatively modern, some of its rules have ancient roots. There is a basis of Roman rules introduced into English law by Bracton, and acclimatized by Coke . . . . The law, as thus developed, sufficed for the needs of the country in the eighteenth century. But, as it was no longer sufficient for the new economic needs of the nineteenth century, an expansion and an elaboration of this branch of the law became necessary. It was expanded and elaborated partly on the basis of the old rules, which had been evolved by the working of the assize of nuisance, and its successor the action on the case; partly by the help of Bracton's Roman rules; and partly, as Gale's book shows, by the help of the Roman rules taken from the Digest, which he frequently

---

47 *Re Ellenborough Park; Re Davies; Powell v. Madison*, [1956] Ch. 131 at 161-163 (C.A.).

and continuously uses to illustrate and to supplement the existing rules of law."

. . . .

The passage which we have read from Sir William Holdsworth sufficiently serves to explain the appearance and the prominence of Roman dicta in the English law of easements, commonly called, indeed, by the Latin name of "servitudes" . . . .

For the purposes of the argument before us Mr. Cross and Mr. Goff were content to adopt, as correct, the four characteristics formulated in Dr. Cheshire's Modern Real Property, 7th ed., pp. 456 et seq. They are (1) there must be a dominant and a servient tenement: (2) an easement must "accommodate" the dominant tenement: (3) dominant and servient owners must be different persons, and (4) a right over land cannot amount to an easement, unless it is capable of forming the subject-matter of a grant.

§4.68 Easements go with or against the fee simple of the land, not with or against the owners. For example, an easement for a sewer line may be granted by the owner of a burdened parcel (the servient tenement) to the owner of benefitted lands (the dominant tenement). This is an affirmative form of easement: the servient permits the dominant. Another group of easements is characterized as negative; here, the owner of the dominant tenement may compel the owner of the servient tenement to do something, such as not to block by construction as easement of light. Upon the sale of either or both parcels the easement goes as an attachment to one and a restriction to the other, but if one owner buys the other property the easement is erased.

§4.69 There are many different kinds of easements.[48] Those most commonly found are easements for human or vehicle passage, for drainage and sewage, for party wall purposes, for light and air, for electrical reticulation and other services, for buttresses or footings of supporting walls, for overhanging eaves and down-pipes of roof drainage. There are also unusual ones for extending the bowsprit of a ship over a wharf, creating a nuisance such as pollution,[49] or using a toilet. The provision of corridors for electrical power lines, oil and gas pipelines and municipal water-supply systems are sufficiently important that they are often created as rights of

---

48 Sir Robert Megarry and H.W.R. Wade, *The Law of Real Property*, 5th ed. (London: Stevens, 1984). In Part 4 of 'Easements and Profits' a comprehensive treatment is given of the law and the many forms of easements are listed. See also V. DiCastri, *Registration of Title to Land*. (Toronto: Carswell, 1987) Chapter 18:9, Easements.
49 *McKie v. K.V.P. Co.*, [1948] O.R. 398, affirmed [1948] O.W.N. 812, affirmed [1949] S.C.R. 698. A prescriptive right to pollute was refused, but see the *K.V.P. Company Limited Act*, 1950, S.O. 1950, c. 33.

way in fee simple, in the same sense as railway rights of way; they are then not easements.

§4.70 The number of different forms of easements increased markedly in the decades of the industrial revolution. Conveyancers readily expressed the intentions and the courts were not reluctant to recognize and enforce the rights. Some easement rights are undefined, and undefinable, as to position and dimension. The ones of special concern to the land surveyor are those of dimensionable extent.

§4.71 As noted above, an easement belongs to or runs with a parcel of land, that is, it is appurtenant to a parcel. Those that are laid down over second and successive parcels beyond the first or immediately abutting parcel to the dominant tenement are therefore attached by the preceding appurtenant easement of the same nature. If not attached, the easement is severed from the dominant tenement, and it is only in more recent times that there have been statute amendments to permit otherwise severed easements to be acquired as legal rights. The problem of severed rights is readily imagined if one considers the task of acquiring a long easement for services crossing many parcels. If severed easements are not legally admitted, then the acquiring authority is obliged to acquire each segment of the easement one after the other in sequence, a situation of practical impossibility.

§4.72 The definition of the extent of an easement may be by dimensioning the whole of the easement area or by locating a centre line or one side line or an offset line and noting the widths to each side. Regulations may require the survey and monumentation of the easement area and its separate treatment in title records; thus while several parts may comprise the extent of the whole parcel in fee simple, one specified part may be subject to an easement. In any event, the limits of easements are boundaries in every sense of the word.

§4.73 The creation of an easement is most generally by express grant registered under deed registry or land titles acts; or it may be by statute; or by prescription (for a right, less than freehold, which matures after a term of years of undisputed enjoyment under common law unless the lands are registered under a land titles system); or by necessity (such as a right of access to a public highway arising from failure to specify any right of way for an otherwise landlocked parcel). Records of the first two are, or should be, readily obtained, but the surveyor, as an agent for a client, would be lax in his duty not to record the observation made and information learned about any potential easement rights.

Prescriptive Rights — Lost Modern Grant

§4.74 The expressions, 'prescriptive rights', 'possessory rights', 'adverse possession'; and 'prescriptive title' are often heard and sometimes wrongly used. Qualifiers are also applied to prescription when used as a term of the civil law for the principle of the creation, or the extinguishment, of rights by the lapse of time; respectively, these are deemed to be positive and negative. The positive or creative form is also the ancient action of the common law that may give the right of easement if used and enjoyed *nec vi, nec clam, nec precario*[50] and undisputed from time immemorial. Negative or extinctive prescription of the civil law is seldom a term of use in the common law; it is adverse possession in the common law, that is, the occupation, use and enjoyment of real property leading to freehold rights contrary to the right of the true owner who is the holder of the title.[51] Prescription for rights of easement became a common law doctrine whereas adverse possession, through limitation periods that extinguish paper title and protect a possessor, is statutory. Laskin provides the following note on prescription.[52]

> The basic assumption of prescription ... is that the use claimed as an easement had a valid (or legal) origin, which may be presumed in case of long-continued enjoyment. The presumption, which would be of a grant, operated originally under the rules of common law prescription, by which a continuous use ... back to 1189 (specifically to the date of accession of Richard I, the date of "legal memory") was conclusively presumed to be founded upon a grant of the alleged easement prior to that time. This meant, of course, that any claim of an easement by common law prescription would be defeated on a showing that the alleged easement could not or did not exist at some particular time after 1189. Moreover, the farther 1189 receded the more difficult it became to prove continuous use back to that time, and in consequence the Courts modified the required proof by accepting proof of use for as long as living witnesses could remember .... As before, however, this did not help where the evidence showed that the easement did not or could not have existed at some time after 1189; for example, the "dominant" and "servient" tenements might have come into common ownership since that time, or, in the case of an easement of support of buildings, the buildings were not put up until after 1189.
>
> We need not be concerned in this country with common law prescription. It is inapplicable because ... "the Court may take judicial notice of the fact that America was discovered in 1492" ....[53]

---

50 *nec vi, nec clam, nec precario*: neither as the result of force, secrecy, or evasion.
51 See §§4.78-4.88.
52 B. Laskin, *Cases and Notes on Land Law*, rev'd ed. (Toronto: University of Toronto Press, 1964) at 576.
53 *Abell v. Woodbridge and York* (1917), 39 O.L.R. 382, reversed on other grounds 45 O.L.R. 79, but restored 61 S.C.R. 345.

134 Survey Law in Canada

The absurdity of the rule of "legal memory" led to the development of the doctrine of lost (modern) grant, based on a presumption of grant since 1189 but before the commencement of the long-continued use on which a claimant relied. Thus, a period of no less than 20 years' unexplained continuous use (adopted by analogy to Statutes of Limitation) would suffice to justify a finding that a grant, now lost, had been made; and further, the period of use did not have to be immediately before action.

The defects of lost modern grant were not quite as formidable as those of common law prescription. However, the presumed grant would have to be by a capable grantor to a capable grantee .... the Courts apparently would not listen to an argument that no grant was in fact made. The "lost grant" doctrine was accepted in the common law provinces as part of the applicable common law ....[54]

§4.75 In the fifth Volume of *Re-statement of Property*[55] published by the American Law Institute, the elements of prescription are clearly set forth:

S. 457. Creation of Easements by Prescription.

An easement is created by such use of land, for the period of prescription, as would be privileged if an easement existed, provided the use is
 (a) adverse, and
 (b) for the period of prescription, continuous and uninterrupted.

S. 458. Adverse Use

A use of land is adverse to the owner of an interest in land which is or may become possessory when it is
 (a) not made in subordination to him, and
 (b) wrongful, or may be made by him wrongful, as to him, and
 (c) open and notorious.

S. 459. Continuous and Uninterrupted Adverse Use

 1) An adverse use is continuous when it is made without a break in the essential attitude of mind required for use.
 2) An adverse use is uninterrupted when those against whom the use is adverse do not
 (a) bring and pursue to judgment legal proceedings in which the use is determined to be without legal justification, or
 (b) cause a cessation of the use without the aid of legal proceedings.

§4.76 Laskin also quotes from the second edition of *Halsbury's Laws of England* for the definition of user as of right, that is, the basis for a prescriptive claim of an easement. The relevant paragraphs read:[56]

---

54 *Watson v. Jackson* (1914), 31 O.L.R. 481 (C.A.); *Jones v. Jones* (1843), 4 N.B.R. 265 (C.A.).
55 *Restatement of the Law of Property*. (St. Paul: American Law Institute, 1944). Cited by Laskin, *supra*, note 52 at 582.
56 *Supra*, note 19, Vol. XIV at 41-42, §§83 and 84.

83. *User must be as of right.* In order to support a prescriptive claim under the doctrine of prescription at common law, the user or enjoyment of an alleged right must be shown to have been user "as of right", having been enjoyed nec vi, nec clam, nec precario (neither as the result of force, secrecy, or evasion), nor as dependent upon the consent of the owner of the servient tenement. Consent or acquiescence on the part of the servient owner lies at the root of prescription, and a grant cannot be presumed from long use without his having had knowledge or at least the means of knowledge. He cannot be said to acquiesce in an act enforced by mere violence, or in an act which fear on his part hinders him from preventing, or in an act of which he has no knowledge actual or constructive, or which he contests and endeavours to interrupt, or which he sanctions only for temporary purposes or in return for recurrent consideration.

84. *Ignorance of servient owner.* The servient owner's actual ignorance of the exercise or enjoyment of the alleged right will not in every case prevent the enjoyment from being as of right. There are some things which every man ought to be presumed to know, at any rate while in occupation. Very slight circumstances may put the servient owner upon inquiry, and if he neglects to make inquiry it may be that knowledge must be imputed to him. Where an ordinary owner of land, diligent in the protection of his interests, would have a reasonable opportunity of becoming aware of the enjoyment by another person of a right over his land, he cannot allege that it was secret. If, however, the enjoyment be fraudulent or surreptitious, it cannot support a prescriptive claim.

§4.77 In Ontario, the *Road Access Act*[57] does not purport to create an easement nor a public right. Its effect on property rights is important, although too few issues have been litigated to clearly assess the likely effects of this statute. A surveyor is advised to show any roadways or trails crossing a client's lands. The limits are probably undeterminable but a reasonable centre line by straights and deflection points would constitute a good depiction.

Adverse Possession

§4.78 Adverse possession combines the abstract idea of rights in land with the real fact of occupation on the ground in a manner that is inconsistent with the rights of the true owner. In common law, the rights or interests in land are expressed in terms of estates and of these it is the freehold estate in fee simple with which adverse possession is primarily linked.

§4.79 The principles of adverse possession are those of the law of property; the administrative systems for recording rights have responded variously. For instance, adverse possession is admitted as a part of land titles systems, that is, of title by registration laws, in England while it is firmly rejected in Ontario.

---

57 R.S.O. 1980, c. 457, first enacted by S.O. 1978, c. 61.

136 Survey Law in Canada

South Australia, Victoria and New Zealand have admitted its place in the scheme of the acts during the last 30 odd years, and while New South Wales had been as adamantly opposed as Ontario, the principle has more recently been admitted in a restricted form that does not operate in any sense to defeat the statute. The New Brunswick statute, which is the most recent, rejects adverse possessory claims against registered title, while in Alberta, it is, with certain constraints, a long accepted feature. In Ontario, there are far more titles recorded as deeds under the *Registry Act*[58] than as titles under the *Land Titles Act*[59] and, as a consequence, the adverse possessory claim against the fee simple ownership is a very real issue.

§4.80 In 1623, the English statute[60] shifted the historic emphasis in actions for possession "from a time limitation on tracing back seisin to a time limitation on asserting a right of entry."[61] By the *Real Property Limitation Act*,[62] the next step was taken that extinguished the title of the dispossessed former owner upon the proof of the adverse claim for the time specified.

§4.81 Thus, the modern *Limitations Act* specifies the time limit within which the claimant to a better title, who will usually be the holder of the paper title, must act against an adverse possessor. If that time passes without an effort being made to recover the land, the act bars the present owner from ever doing so and extinguishes his title. The occupant then holds a title by the fact of possession, but he does not hold a documentary title that is evidence of his rights: *Brown v. Phillips*.[63] The vesting of title in the possessor can only occur as a result of pleadings in the litigation seeking a declaratory order to this effect (which, it appears, a court would be wisely loath to do as parties other than the litigants may have rights); or by an administrative law process brought under a statute with vesting provisions (such as the procedures for a land titles first application[64] or certification,[65] or in some provinces, quieting titles[66]); or by the common law process of registering the judgment under a Registry Act[67] and using and claiming this as the root of title.

---

58 R.S.O. 1980, c. 445.
59 R.S.O. 1980, c. 230.
60 *Limitations Act*, 1623, 21 Jac. 1, c. 16.
61 *Supra*, note 52 at 613.
62 1833, 3 & 4 Wil. 4, c. 27. For example, the Ontario *Limitations Act*, R.S.O. 1980, c. 240.
63 [1964] 1 O.R. 292 (C.A.).
64 *Supra*, note 59.
65 *Certification of Titles Act*, R.S.O. 1980, c. 61. This statute appears to be unique in Canadian law.
66 The Ontario *Quieting Titles Act*, S.O. 1984, c. 427, with its origins in S.O. 1865, c. 25, was repealed by the *Courts of Justice Act*, S.O. 1984, c. 11, s. 208.
67 For example, the Ontario *Registry Act*, *supra*, note 58.

§4.82 The conditions which must be fulfilled to perfect a possessory claim are that there is actual, open and visible, notorious, exclusive and continuous possession and enjoyment of the use of the land in a fashion which is adverse or hostile to and in derogation of the title of another person, by the claimant and those through whom he claims. The land must be under a claim of title as the whole or a part of a parcel, or, in other words, *the possessory claim must be over a boundary.* There has to be a defined, or definable, entity of land. Where part only of a parcel is claimed, adjoining other land of an adverse claimant, both the line of original boundary and the line of extent of the possessory claim must be shown by a survey. At this stage, the latter line is not a boundary; it is not definitive of extent of title until the court determines the validity of the claim and it is the court[68] and the court alone, that will create the new boundary and state its position. A surveyor would be unwise to set marks as monuments prior to the decision; whatever he sets can only be reference or recovery marks for the survey of features until such time as the boundary is settled by the court.

§4.83 As owner is deemed by law to have constructive possession of the whole of the parcel of his title although he may not use it all and therefore may not be in actual physical possession of it all. This principle of constructive possession does not operate to the benefit of a claimant for adverse possession; hence the claim can only be for the extent of the actual use and occupation, but if no defence is raised the extent could be of the whole parcel.

§4.84 The proof of the preceding conditions is essential to the claim. The proof is strengthened by evidence showing that there is the receipt of profits as part of the enjoyment of the property, that there is the discharge of the burdens attached to the property such as the payment of taxes, and that there is the repair and maintenance of the property whether of buildings or of fences or of the land. The quality and the extent of possession are considered: an adverse claimant must show the conduct of a legitimate owner in possession, that is, *animus possidendi* or intention to possess in such a manner that prevents the owner from enjoying the land to the extent of which it is capable.

§4.85 There are certain conditions that operate against claims to title by adverse possession: for example, if there has been an acknowledgement in writing by the adverse claimant, the time will date from that acknowl-

---

68 'Court' is here used in a generic sense to include an administrative court hearing applications for first registration under a land titles act or under the Ontario *Certification of Titles Act, supra,* note 65.

edgement; and a claim will not succeed unless the ousted owner was *sui juris*, that is, of full legal capacity under no disability of infancy or mental incompetency. (Absence, as for instance on military service, is cause for extension of the limitation period in some jurisdictions but not, it appears, in Canada.)

§4.86 Surveyors are cautioned against concluding adverse possession without proper consideration of the legal aspects which must be satisfied. These are quite apart from visible features on the ground which may stir the fancy as an immediate solution to a complex boundary problem. The perfection of a claim of adverse possession for title good against all the world including the dispossessed holder of the paper title is no small matter. This is not the place to discuss the many legal points which are open to testing in the courts; rather, it is directed to the surveyor's role of providing evidence of relevant factual things.

§4.87 The materials needed to support a claim of adverse possession can be assembled in four categories, as follows:

(1) The claim is brought before a court by satisfying the procedural prerequisites in commencing an action. (In Ontario, this is done under Rule 14 of the Rules of Civil Procedure,[69] or on application to the Director of Titles under the *Certification of Titles Act*,[70] or to a land registrar under the *Land Titles Act*.[71]) The basis of the claim must be presented in all necessary detail, usually by a statutory declaration of facts confirmed by disinterested parties stating their means or sources of knowledge of the alleged facts.

(2) Evidence must be provided as to when the land under claim was first enclosed to exclude the ousted owner; by whom the enclosure was made and the circumstances; the materials of fences or other structures of enclosure; the conditions of these enclosing structures from time to time and the repairs and maintenance carried out and by whom; the means of ingress to the land; the improvements that have been made to land and buildings and the dates of improvements and made by whom; and the purpose for which the applicant uses the land.

(3) If the applicant claims through predecessors in title, the assignment of the possessory rights of the predecessors is needed.

(4) A plan of survey is so much an essential part of the application that it will seldom be dispensed with except for whole parcels already

---

69 O. Reg. 560/84 under the *Courts of Justice Act*.
70 See Note 65 above.
71 *Land Titles Act*, R.S.O. 1980, c. 230.

clearly defined.[72] The surveyor must ascertain the history of the possession sufficiently to be able to show the enclosures on which the owner relies for his claim; the relation of these enclosures to the original boundaries of parcels; the roads, road allowances, lanes, paths, gates, doors or other means of entry to the land; the location and description of all buildings; the land-use such as pasture, cultivated field, orchard, market garden, storeyard, car park, etc.; any apparent easements, rights of way or encroachments.

§4.88 While the onus is on the applicant to prove valid possession of the property (that is, reliance is placed on the strength of the claim, not the weakness of the other party's possession), the surveyor's role is one of affording information. Generally, there will need to be disclosure of all the information obtained, including discrepancies noted, difficulties in reconciling facts and features, uncertainties and even doubts. The court or tribunal seeks accuracy and completeness of the certification it is being asked to give. Consequently, the surveyor in these instances should see himself as an agent and advisor of the tribunal and never as a protagonist for the applicant.

STANDARDS OF MEASUREMENTS

§4.89 The present role of the land surveyor has been competently addressed in many publications from professional associations. However, mention is seldom made of the basic cause for the present role, that the need for repeated surveys of boundaries lies with individual desires, business needs and social legislation that call for dimensions and areas and things on the parcels. Severances, including extensive subdivisions, are now probably controlled more by legislation for planning than for title record purposes. A person may live happily and with no boundary dispute with neighbours on a parcel that is undersized, in fact, on the ground *vis-a-vis* its record of title. If an owner wishes to divide the parcel, minimum dimension standards may prohibit a severance or at least require special approval. The task of setting a new division line is at best a high-class technical operation: measure correctly, set good markers, abide by the

---

72 Some acts, and their regulations if any, have not spelt out a requirement for plans of survey. Per J.F. Doig: "There is no statutory requirement for a plan of survey under the New Brunswick *Quieting of Titles Act* [Revised Statutes, c. Q-4, consolidated to 31/08/87]. There it is the abstract of title which is certified, though plans of survey are used during proceedings." This apparent ignoring of surveys and plans should be seen in proper perspective. It is title that can be and is certified; the bounds are to be found on the ground by the evidence; the representation by the numerical values of a survey should be certified by the surveyor but numerics are not certifiable by a court.

standards and establish a proper record of the evidence created. That is not the subject here. Our concern is with the identification of the previously located bounds, and with measurements as an integral part of the total evidence which must be considered.

§4.90 By present standards, the old, original surveys of the Crown were grossly inaccurate.[73] The purpose of the surveys has to be considered, however; these were settlement surveys and, for the settlement of the heavily forested wilderness of eastern Canada, the work of those surveyors was to their credit; they showed resourcefulness and mastery of the logistics of the field work in trying conditions. By contrast, there can be little excuse for modern surveys not being numerically correct. The purpose is different and the equipment is superb. Where a survey is now found faulty, the cause lies chiefly in the failure to, firstly, search, find, and evaluate evidence, and secondly, by logical deduction within certain broad principles of law, arrive at a sound opinion of boundary position, that is, a conclusion founded on facts and law. As pointed out by I. de Rijcke in the chapter on Evidence, it is vital to this operation that the surveyor understand the legal process in the courts and how evidence is treated in the search for facts.

§4.91 In the long evolution of principles of the law of evidence about boundaries, succinctly summarized by Greenleaf[74] in the last century, measurements of record were relegated to low standing. A careful analysis of judgments given down the years shows that the courts were not often critical of the survey standards and expertise. The courts were, and still are, rarely forced to be critical of surveyors testifying to their own good work. What they were doing was rejecting the thoughtless use of numerical values of old surveys to reset corners and lines. This is at the very root of the objection to re-establishing boundaries by setting out recorded directions and distances if any other method can be used — firstly and always, evidence; then perhaps proportioning, which, of course, is not always the legally, or logically, correct procedure.

§4.92 The description in the parcel clause of the document of transfer

---

73 E.C. Frome, *Outline of the Method of Conducting a Trigonometrical Survey*, 2d ed., rev'd (London: John Weale, 1850) at Chapter IX: Colonial Surveying, footnote at 121. "The rude and inaccurate mode in which land has been marked out in Canada by the chain and compass . . . renders the survey of that country not a fair point of comparison with that of more modern colonies." Nevertheless, seen in retrospect, no deduction should be made from Frome's comment that the lands of other colonies were much better surveyed, and by personal experience we can vouch for the fact they were not. Frome was writing of what should be done. When he became the third surveyor general of South Australia, he had the usual problem and had to live with the practicalities arising from political and administrative indifference to surveying needs.

74 Greenleaf, *supra*, note 45 Vol. 1, s. 301, at 488, n. 3.

or in the title register is an attempt to define the entity of land on the ground. It is the best statement that responsible conveyancers can articulate in words and diagrams to describe the actual parcel; and that actual parcel on which people live, build houses, work and invest money may not agree with the representation. The difficulties and ambiguities encountered here have resulted in the subject of Interpretation of Descriptions. Where the bounds of a parcel are identifiable by ancient features of known repute, there is no need for measurement: this is the legal parcel known as Blackacre, the ancient manor. Blackacre can even be divided with no potential for dispute, cutting off, for example, the south field, provided it is the well-known and only south field lying beyond the old stone wall. There is still no need for measurements to demonstrate the entity. But once there is a need for knowledge of the acreage or frontage to be paid for, there is a need for survey. This is dimensional division of land and is now virtually the rule to satisfy primarily the planning authorities, but also investors and the titles administrations.

§4.93 The measurements should be correct. Measurements made at an earlier time and by others never are. It is not the author's purpose to write about measurement techniques and statistical analysis of results, nor modern methods of network design for error control.[75] (Indeed, we are entirely persuaded that the measurements of the last-made survey, especially if modern, are correct to the nth degree.) Our concern is with all prior surveys used as the basis for titles, especially surveys by other people, a combination of events that gives a warning to be cautious. The error analysis of present surveys, if applicable and used, can tell the surveyor the likely range of error, and it will be very fine; this is not the 'more or less' of the courts which is a rough measure of what people will, or should, accept as discrepancies between the facts on the ground and the numerical values in the record.

§4.94 'Accuracy' and 'precision' are words often used interchangeably but they warrant distinction: precision is the fineness of measurement, accuracy is its correctness. Thus a distance between two ground markers stated to three decimal places is precise, but it is not accurate if it is incorrect. This applies also to directions shown to the finest units. The mathematical rules of significant figures are not fairly applied in the analysis of the records of early surveyors as they either did not know them or did not respect them. They are also improperly applied in modern surveys where the rules

---

75 T.J. Blachut, A. Chrzanowski and J.H. Saastamoinen, *Urban Surveying and Mapping*. (New York: Springer-Verlag, 1979). See also E.J. Krakiwsky, ed., *Papers for the C.I.S.M. Adjustment and Analysis Seminars*, 2d ed. (Ottawa: Canadian Institute of Surveying and Mapping, 1987).

are substantially ignored by some practitioners or overruled by regulations, as, for example, where the introduction of metric units had led to the practice of expressing all metric values to three places of decimals.[76] This is regulation verging on the ridiculous and does not advance the prestige of the profession; the public, as the users of surveying services, are prone to believe the results, only later to be disillusioned. The change to metric units did not increase precision or accuracy.

§4.95 No system of title by registration guarantees the boundaries in the sense of parcel dimensions; for two lots abutting, it is a certainty that the division line is the lot line for that is the basis of the two titles, but the dimensions to and along the lot line are not guaranteed. The corner markers as physical things, including surveyors' monuments, are matters for evidence evaluation in the field.

§4.96 As earlier noted the original surveys were seldom reliable for accuracy of dimensions. In ratio terms, misclosures of blocks by field measurements of the order of one part in 300 were accepted by the administrators for settlement surveys, if indeed a figure could be calculated. In the running of concession lines some nine miles across townships of the single front system of Ontario, the recorded values are as measured or, more correctly, as thought to be measured. There was no connection between concessions. Even when a check system was introduced in 1829 by the double-front sectional system of lots in concessions, many township plans were returned with theoretic or scheme measurements. There was no valid checking process as corner marks were not marks of an interrelated system. The Ontario patterns should be recognized as the orderly disposition of settlers on the ground, slotting them nicely into the pigeonholes of lots in the tiers of concessions. Methodology of survey it was not.

§4.97 The Dominion Lands Survey system was a considerable advance on the Ontario method for orderly settlement and substantially improved on the surveying technique. The Dominion Lands Survey *Manual of Instructions* of 1903 contained a clause[77] that one would consider redundant

---

76 Two older references are still pertinent to Canada which has half-heartedly adopted the metric systems in property surveying: A.P.H. Werner, "Problems and Effects of the Conversion of Weights and Measures to S.I. in Surveying", *The Australian Surveyor*, March 1972, Vol. 24, No. 1; and M.D. Kellock, "The Quest for the Millimetre", *The Australian Surveyor*, March 1973, Vol. 24, No. 1.

77 Dr. E. Deville, Surveyor General, *Manual of Instructions for the Survey of Dominion Lands* (First Part) (Ottawa: Government Printing Bureau, 1903) at Clause 201 which follows (in the text) appeared first in the fifth edition of 1903, that is, 30 years after the start of the surveys. The experience with the Manitoba surveys alone could have led to the new clause as retracements showed the discrepancies with the record values.

if it were not for the fact that it was needed to revoke some faulty practices that had emerged and which the writers of the *Manual* were not going to tolerate.

> The bearings, distances, and other data must be entered in the field notes as actually found on the ground by the surveyor's own measurements, whether the same do or do not agree with previous surveys or with the provisions of law or of the Manual of Surveys. The entry of conventional, theoretic or assumed or supposed data is absolutely forbidden.

§4.98 The mathematical closure of sections provides a guard against blunders and some errors of measurement. It can show directional error by the bearing misclosure of each section, or any block. Azimuth observations are an even better guide. Linear measurements are harder to isolate — a chain too short or too long will still give a numerically good closure and will not reveal that the parcel is outsized. The real check occurs when lines are closed on control positions.

§4.99 The *Manual* required horizontal distance to be actually measured by the surveyor and "[i]n chaining over uneven ground, should the same be so broken as not to permit of the full chain being levelled, the measurement should be made with such portion thereof as may be easily levelled ...."[78] The Gunter's chain used in the early Dominion surveys, as everywhere else in those days, was notoriously inaccurate by its very form and weight, but more serious, the constant pulling progressively lengthened the linkages. This was recognized by the *Manual*.[79] "The chain is to be tested and corrected, at least every other day during use, by a standard measure which shall have been previously compared and approved by the Surveyor General." The standards were calibrated wooden poles. The correction was usually done by removing a link, shortening it with a new end loop and replacing it.

§4.100 In his history "From Compass to Satellite",[80] W.D. Stretton wrote:

> In the Annual Report of the Department of the Interior for the year 1913, Surveyor General Edouard Deville, reported tremendous errors found in the meticulous survey made by surveyor Milner Hart in 1871 along the Principal Meridian. Township 35 was found to be one third of a mile too far north; A.G. Stuart was sent to resurvey the meridian and found that every mile was 13 feet too long, indicating that Hart's chain had been two inches too long. If Hart had verified his chain by stretching it out and measuring it with the

---

78 J.S. Dennis, Surveyor General, *Manual showing the System Adopted for the Public Lands of Canada in Manitoba and the North-West Territories with Instructions to Surveyors*, (Ottawa: Queen's Printer, 1871), clause 6.
79 *Ibid.*, clause 5.
80 W.D. Stretton, "From Compass to Satellite", *The Canadian Surveyor*, December 1982, Vol. 36, No. 4, at 21.

wooden yardstick furnished as a standard, the error on each yard length would have been less than one tenth of an inch, and perhaps not more could be expected for this method.

§4.101 Three years after the start of the survey, awareness of creeping errors led to the Special Survey of meridians and baselines by an order-in-council of February 1874, to provide a net of extended control as a master grid of the main meridians and base lines. The lack of correct time for observations had faulted the longitude results, but, nevertheless, the main monuments of positions were in place on the principle of breakdown from the whole to the part (albeit with some error) avoiding the extreme accumulation of error that would have otherwise been inevitable. Comparing the smallest unit surveys of township sections with the new extended control showed a pattern of overrun, positional error, of about nine and one half feet to the mile[81] or one part in 570, roughly 0.2%. It could mean an overrun in the area of a section of about 0.4%. The full use of the steel band tapes (of four chains and 300 feet lengths) probably dates from about 1878 or 1879, and thereafter the surveys attained a dramatic increase in accuracy.[82]

§4.102 This is not the same as the relative error of closure noted previously (§4.96) in respect of the early sectional systems in Ontario; the latter, of course, provided closures of sections only for those surveys where the surveyor had recorded his actual measurements. A relative closure of one part in 570 in the perimeter of a square mile (to use the same area unit) means a block misclosure in lineal terms of 37 feet which might be the result of errors in either directional or linear measurements or both. Comparing original against modern accurate measurements, discrepancies of 3% are not exceptional in Ontario and the Atlantic provinces and it may be overrun or shortage. The area difference is then 6%.[83]

§4.103 Misclosure of a surround is only a crude test of survey accuracy. Checking directions against the meridian by astronomic observation is a nominal improvement as also is checking against independent astronom-

---

[81] W. Pearce, in The Proceedings of the Association of Manitoba Lands Surveyors 34th Annual Meeting, 1914, cited in J.G. McGregor, *Vision of an Ordered Land*, (Saskatoon: Western Producer Prairie Books, 1981) at 40.

[82] E. Deville, *Standards of Length and Errors in Dominion Lands Surveys*, Sessional Paper No. 25 in Annual Report of the Department of the Interior for the Fiscal Year ending 31 March 1913, Vol. 1, 1914.

[83] For an analysis of the Dominion Lands Survey System, see S.H. de Jong, *Analysis of a Sample Area of the Dominion Lands Survey System*. Report to the Surveys and Mapping Branch, Department of Energy, Mines and Resources, Canada, 1966. See also the 1953 *Supplement to the Manual of Instructions for the Survey of Canada Lands*, (Ottawa: Queen's Printer) at 52 for geodetic survey positioning of numerous section corners.

ically determined positions. The proper test is closure against positions coordinated by full geodetic survey, a facility that was not available for the greater part of all the original Canadian land surveys and is still not today. The Special Surveys of the West came as close as possible to a proper control system as could be reasonably achieved at that time. This probably explains the present general indifference to control surveys; there is no tradition of use. Reputed accuracy is nearly meaningless in uncontrolled land surveys.

§4.104 In modern property surveys, reference of directions to the astronomic meridian is fairly general. It avoids the instability of the magnetic compass. However, reference to the magnetic north at a place and date, as used in the earliest Ontario surveys and generally in the Atlantic provinces, led to a factually true statement. On the other hand, in Ontario usage more often than not, the wording in reference to astronomic direction is not true and the bearing statement on the plan is a charade. The reason for this is historical. In the last century, when so much of the habitable land was being surveyed by the Crown, surveyors were instructed in both Upper and Lower Canada to determine a meridian for their township survey, probably at the start of the survey, and on that meridian determine the compass variation. This was set off on the compass dial and the compass would then, at that point of observation, read zero degrees for the astronomic north. The place of their observation was not always noted. It was the compass that was then used as the line instrument; the variation changes from east to west and local attraction plays its own part in producing an erroneous record. What is really important, of course, is that some line is being adopted as the datum of bearings for particular surveys.

§4.105 For correct usage, the correlation of direction within the township unit of the Dominion Lands Survey system is the best demonstration short of a mapping grid. The latter has the central median of the projection; grid direction is related to this control meridian by the values of the convergence of meridians and the difference between the arc of the line of sight and the chord on the projection (arc-to-chord factor). The township, while not functioning as a mapping grid, also has a central meridian of reference. At latitude 50 degrees, the township is 1/44th of the width of the six-degree wide mapping zone of the Universal Transverse Mercator Projection. The convergence is significant at about one minute of arc to the mile of departure east and west; the arc-to-chord factor is negligible at three miles from the central meridian and ignored for these boundary surveys. The directions of all surveys within a township are interrelated by convergence to the central meridian; they are correctly designated

146   Survey Law in Canada

'bearings' and are given as full circle values from north.[84] (A quadrant bearing is merely a species of bearing in a survey developed as if made on a plane surface.) Even where derived by measurements from prior established lines, the relationship is sound and a bearing statement in the following form on the plan is correct: bearings are astronomic derived from the bearing of the line (defined) and referred to the central meridian. If the original survey was in error in its direction, then it could be argued that the derived bearing is not astronomic; it is a matter of degree. Furthermore, it is orientation that is involved and this has nothing to do with survey accuracy unless position is an added feature. It has a lot to do with boundary accuracy. Taking different datums for surveys can, and does, lead to different parcel boundaries as set by two surveyors. In view of the accuracy of field astronomy for general surveys, an orientation within a minute of arc seems reasonable.

§4.106 By such a standard, however, the majority of survey records of the Atlantic provinces and Ontario (southern Ontario, certainly, and generally most surveys before 1900) cannot be relied upon for orientation. Where orientation was not directly by the compass, but by the practice of the time described above, the result was an attractive record but an unreliable one that cannot be unraveled unless the field notes record the point of meridian determination. The equipment available and the method used also made the meridian determination suspect. It will be readily seen that this situation in Ontario is in fact worse than the general situation in the Atlantic provinces because for whole townships, some as large as a hundred square miles, myths of astronomic direction were developed, whereas numerous variations of direction of five to eight degrees are known. Knowing the erratic behaviour of the compass, that it does not run a straight line, experienced surveyors will often turn this variability of the compass to advantage; allowing for the annual change of the declination since the original survey, it is likely that evidence of the original line will be found with a compass whereas running a stated astronomic course with a theodolite will just as frequently produce nothing except frustration.

§4.107 An examiner of surveys is primarily interested in the evidence found of prior surveys and that which is placed for new lines of severances. The

---

84   Bearing defined for the purposes of boundary or cadastral surveys; the bearing of a line at a point is equal to the observed astronomical azimuth at the same point, less the convergence of meridians if east of the reference meridian, and plus the convergence if to the west. Convergence equals the difference in longitude multiplied by the sine of the middle latitude of the line. For the limited extent of cadastral surveys, the additional terms of the convergence formula, the arc-to-chord factor and the deviation of the vertical are not significant.

angles between lines are far more important than the stated directions, and orientation is only mildly of interest. In Manitoba, indeed, the land titles administration began to use angles in descriptions in preference to directions early. If bearings are to be adopted (assumed?) from prior records, the truth of the whole practice is adhered to by the simple statement of what is in fact being done: "The datum of bearings for this survey is ... [and state the proven line]."

§4.108 The progression of the charade is evident from the following statements found on plans of the last few decades:

> Bearings are astronomic, referred to [or related to] the bearing of [line on plan].
> Bearings are astronomic, assumed from the line [on plan].
> Bearings are assumed astronomic, derived from the line [on plan].
> Bearings are astronomic derived from the compass observations of [surveyor] as shown on plan ...

§4.109 A bearing used as a datum and taken from a control survey on a geodetic base is not properly called 'astronomic'. The usage is, unfortunately, frequent; it is suggestive of sound survey practice defeated by concern for the temper of plan examiners (or pedantic adherence to a misinterpretation of faulty regulations).

COORDINATION AND INTEGRATION OF SURVEYS

§4.110 These two concepts are in fact the same, one word supplanting the other with renewed attempts to establish a systematic approach to the total survey operation within a region. Over-emphasis on numerical coordination of position missed the associated components of coordination of effort and of record.[85] The co-operative nature of such projects has had little success without the administrative regulation of some authority and the contribution (induced or seduced) of private practitioners.

§4.111 Coordination of monument positions can add greatly to the efficiency of all survey work in a region provided that it is recognized for what it is. Grid co-ordinates are but the two-dimensional (x and y, or E and N) expression of the bearing and distance vectors of conventional survey, but the process adds position and hence controls orientation. Legislation or regulations, providing that coordinate survey control may be considered as evidence of positions, is no more nor less than statutory recognition of the common law principles recognizing measurements. Coordination merely sweeps in a greater range of witnesses to a particular

---

85 T.J. Blachut, "The Integration of Surveys and Its Requirements" *The Canadian Surveyor*, March 1960. Vol. XV, No. 2.

position. The system invokes proper checks and tests of measurement accuracy and its use is better considered as sound common sense which, obviously, still requires intelligent and proper usage. For example, if three points lie on a boundary that is a straight line and the replacement of the intermediate point is required, restructuring by any control that produces a point of deflection is immediately suspect.

§4.112 The great advantage of coordination of boundary positions is for the implementation of a parcel-based land information system and the so-called multipurpose cadastre, provided that the record of a feature (including a boundary monument) and its position by coordinates is recognized as descriptive and not definitive.[86]

LOST BOUNDARIES

§4.113 The fact that boundaries are creations of man is emphasized in numerous places in these chapters. It is a natural inclination in lands developed from colonial settlement to think first of artificial markings by surveys and monumentation and secondly of physical structures such as walls and fences adopted as dividing *features*. The latter objects have width and can be wholly owned by one of the adjoining parties or divided in ownership depending on the location of the dividing *line*, since boundary, in the ultimate fine sense, is a line without width. Obviously the feature, the monuments and the line, can disappear and in that sense be lost. However, the rules of evidence, also creations of man, do not appear to label simple disappearance as lost; since a boundary line is defined as imaginary or invisible it does not lend itself to the general notion of lost as applied to an entity. This abtruse aspect of the problem of boundary replacement needs the legal concepts of evidence, and the rules of evidence, to be understandable. Evidence is in many respects a slippery notion to define. Generally, it is the encompassing facts which tend to convince a tribunal or an enquirer — a surveyor — that some subject of enquiry, an event or a thing — such as a boundary — that is unknown or uncertain of status, is determinable. So that this process is not left to personal prejudice or whims, but is rational and certain in real terms, a number of guiding principles have been settled. A discussion of standard, admissibility, relevancy, weight and burden of proof is given in the chapter on Evidence.

§4.114 Statutory provisions for surveys have generally stated that a surveyor shall apply the best evidence obtainable, that is, for example, direct evidence

---

[86] The most recent reference on this matter is K.N. Toms, J.T. Major, T.W. Hughes and E.R. Robinson, "Towards Legal Co-ordinates for Boundaries" in *The Australian Surveyor*, March 1988, Vol. 34, No. 1.

preferred to hearsay, an original document preferred to a copy and, it follows, an original post preferred to a replacement; and, of course, the post must be in its original position. This best evidence must not be misconstrued as to meaning. *Wigmore on Evidence*[87] notes that from its first formal statement about 200 years ago, "[t]he phrase about 'producing the best evidence'... has long been nothing more than a loose and shifting name for various specific rules", and concludes the section (with citations): "[t]he sooner the phrase is wholly abandoned, the better." To arrive at a judicious opinion on a boundary replacement this would appear to substantially widen the range of evidence that is generally considered by surveyors.

§4.115 From time to time, the belief appears that when a land surveyor, authorized and registered by statute, has replaced an old boundary, that is the end of the matter; it has been tried and decided. This is too far from the truth. The surveyor has expressed an opinion and shown it by replacement marks on the ground as well as by records. If his treatment of the evidence is sound his opinion will be sound, but it does not follow that the location of the boundary is settled.

§4.116 "Where there is ambiguity in a deed or other written instrument, the construction to be placed on the instrument is for the court."[88] Whenever any matter is up for decision, especially where this calls for the admission of evidence extrinsic to the record, the *decision* rests with the court. It is precisely this that makes the surveyor's solution an opinion and not a decision. Obviously, if the surveyor is to do his work correctly, that is, as if a judge and jury were looking over his shoulder, he must consider the evidence in the same manner as would a court. He must consider all the evidence that may be adduced and advise accordingly, and this includes a duty to state that the boundary is finally resolvable only by a court and that the use of some particular evidence he has considered for his opinion may be a use reserved only to a court for final decision. His opinion is vital to any issue and it must be remembered that a surveyor is paid for his opinion, not for his doubts.

§4.117 This is especially the case with lost boundaries. The cases clearly show that a surveyor's opinion and an affirming court decision will replace a boundary by the use of secondary evidence such as bearing trees and other witness marks and by testimony given under oath. At some stage

---

87 J.H. Chadbourne,*Wigmore on Evidence*, 4th ed. (Boston: Little Brown & Co., 1972), Vol. 4 *"Evidence in Trials at Common Law"* s. 1174. See also P. Atkinson, *Evidence* (Canadian Encyclopedic Digest (Ont. 3rd), Title 57), Paragraph 1118 *et seq.*

88 D.W. Lambden and I. de Rijcke *Boundaries and Surveys.* (Canadian Encyclopedic Digest (Ont. 3rd), Title 19), Part III: Interpretation of Descriptions.

or distance, from the missing point or line, the evidence fails to be local, and the corner or line is deemed lost. The statutes for surveys, such as in Ontario and the western provinces,[89] recite the methods of re-establishing the missing element.

Conventional Lines

§4.118 However, there is also the common law principle of conventional, consensual, or agreed boundaries.

> While in a sense all boundaries arise from acceptance or agreement or convention, the expression "conventional boundary" has a special meaning. The concept of a conventional boundary, as between adjoining property owners, rests on the prerequisite that another, the true boundary line of division, cannot be found, that it is uncertain and undeterminable, that it is lost, and not merely that it is unknown because sufficient enquiries have not been made or surveys performed. The Surveys Act defines "lost corner" and "obliterated boundary" for original surveys of the Crown and for surveys for registered plans of subdivision and provides specific rules for their re-establishment. It would appear that there is no place for the admission of a conventional line where the true line is redeterminable on the evidence or the prescribed rules of the Act. But there are many lines to which the concept would be applicable, as in the determination of boundaries created by descriptions in which ambiguities are present and not reconcilable other than by recognition of the agreement between adjoining owners.[90]

§4.119 To go deeper into the matter than the above quotation in respect of the statement: "[I]t would appear that there is no place for the admission of a conventional line where the true line is redeterminable on the evidence or the prescribed rules of the Act", it has to be noted that the acts of Ontario and of the western provinces and federal lands do not state that the acts overrule the common law. The systems are for the layout of the Crown estate; they also prescribe rules for redefinition of lost corners and lines, and it may be argued that with a royal or Crown prerogative for the division and disposal of Crown property, the common law rules would be overridden whether so stated or not., Thus it may be a point of nice law whether two valid methods, by statute and by convention, still apply for replacement of a lost line if that line is one for which rules are set forth in the surveys acts. As Madame Justice Boland noted in *Bea v.*

---

89 See *Surveys Act*, R.S.O. 1980, c. 493. See also the surveys acts of the western provinces generally which are derivatives in greater or lesser degree of the *Dominion Lands Acts*, or *Dominion Lands Surveys Acts* after S.C. 1908, 7-8 Edw. 8, c. 21; the applicability of all parts appears to depend to some extent on the date of assumption of control of surveys from the federal authority.

90 *Supra*, note 88 at §42.

Robinson,[91] the conventional line principle has great appeal; it is interesting to note that in her decision the Ontario *Surveys Act* was not part of the statute law considered. If the line does not fall within the ambit of the acts, the conventional line solution would appear to be acceptable if the line is indeed lost.

§4.120 "In order to establish a conventional line as the line dividing two lots of land it must be shown that there was an agreement, perhaps not necessarily in writing, between the respective owners of the lots and that the line was adopted and lived up to by them."[92] In *Bernard v. Gibson*,[93] it was noted by Vice-Chancellor Strong that he could not find any case where a parole agreement had been held conclusive without the addition of long-continued possession sufficient to give a title, or such standing-by and acquiescence in the acts of the other party as would constitute equitable estoppel. The possession referred to is not, of course, adverse possession, but a logical application of the limitation period.

§4.121 It is interesting to note that there is an abundance of decisions from the Atlantic provinces determining boundaries on the grounds of conventional lines, a limited number from Ontario, and none (to the writers' knowledge) from the western provinces. The concept appears to warrant intensive study as an unsettled aspect of boundary law in the common law provinces where there is no parallel to the concept of *bornage* in the civil law of Quebec, which province, it should be noted, does not have a *Statute of Frauds*.[94]

Estoppel

§4.122 Any treatment of conventional lines as boundaries raises the doctrine of estoppel which, as a doctrine or principle of law, is seldom considered directly relevant to the surveyor's professional activity. However, in boundary retracement, the potential for the application of estoppel cannot be ignored. It requires special understanding and treatment.

§4.123 Estoppel is a doctrine of law that has evolved through the common law. It stands alone and distinct from equity. Some decisions have called estoppel a mere rule of evidence and have said that estoppel cannot create

---

91 (1977), 18 O.R. (2d) 12 (H.C.).
92 *Supra*, note 88 at §45.
93 (1874), 21 Gr. 195 at 203 (C.A.).
94 Bornage is discussed in chapter 10.

a substantive right in law which does not, or would not, otherwise exist.[95] However, from a review of the law of estoppel below, one will note that estoppel operates very much in giving the benefit of a substantive law right to a party successfully alleging estoppel against another party.

§4.124 Estoppel is a relatively complex legal notion which draws upon and is founded in the law of contract, the law of property and certain equitable principles. The application has been found useful in these areas of law as well as in many others. As a legal notion, estoppel involves certain essential elements in combination, such as a statement made by one party and intended to be acted upon, the action by another party on the faith of that statement and the resulting detriment to the latter.

§4.125 Estoppel has been defined by di Castri in *The Law of Vendor and Purchaser*[96] as follows:

> A person making a false representation as to a present fact, for the purpose of fraudulently influencing the conduct of another person who acts upon the representation, is estopped from denying its truth and may be compelled by the court to give effect to the representation. It is the very essence of justice that between those two parties their rights should be regulated, not by the real state of the facts, but by that conventional state of facts which the two parties agree to make the basis of their action, and this is what is meant by estoppel.

§4.126 As between two parties where estoppel is to operate or exist, it is the voluntary assumption or acquiescence in a fictional set of facts which, in the absence of truth, is accepted in lieu of the truth. Once this fiction has been accepted and subsequently acted upon by one of the parties to his detriment, the party who might stand to receive the benefit of that person's detriment, although the benefit need not be directly the opposite of the detriment, may not thereupon rely on the falsehood as a defence to the consequences of the fictional set of facts being imposed by or held in a court of law as binding on him. Invoking the principle of estoppel is possible only as a means of protection for the action made in good faith on the basis of the statement of the other party; that is, estoppel is a shield and not a sword to use as the basis of a cause of action.[97] The essence of estoppel, then, is that the party who has previously asserted a statement of fact is precluded (that is, estopped) from denying the truth of that statement. He is, by the origins of the words, 'closed up', 'plugged up', 'shut up'.

---

95 A. Jayne, *Estoppel.* (Canadian Encyclopedic Digest (Ont. 3rd), Title 56) §6.
96 V. DiCastri, *The Law of Vendor and Purchaser.* (Toronto: Carswell, 1976) at 314.
97 *Supra*, note 95, §§7, 8.

§4.127 The courts have held that four basic factors are principal elements in the doctrine of estoppel. These four elements are:[98]

> (1) there must be acts or conduct on the part of the party to be estopped which amount to a representation or concealment of material facts, made either with knowledge or culpable negligence; (2) he must intend that his conduct shall be acted upon or must so act as to cause the other party reasonably to believe it was so intended; (3) the other party must be ignorant of the true situation; and (4) he must rely upon that conduct to his prejudice or injury.

§4.128 The first element includes the need for a representation of facts or a concealment of material facts. On this point, di Castri notes the distinction between representation as an element of contract and representation as the basis of an estoppel. The first may give rise to a cause of action. However, representation as the basis of an estoppel permits a rule of evidence to be invoked. Di Castri gives the example[99] of a situation where a purchaser alleged that he bought certain resort lots on the representation (by the vendor's agents) that he would be permitted the use of a certain park which the vendor owned close by. After ownership of the lots and enjoyment of the park had continued for approximately 13 years, the vendor then decided to fence off the park and begin demanding an admission fee. The courts held that since the park had been enjoyed by the purchaser, as well as by others, without any charge or hindrance whatsoever for the 13 year period, the vendor had now precluded himself from denying that this right was one that had accrued to the purchaser. Estoppel worked in this situation so as to entitle the purchaser to invoke a rule of evidence. It did not have the operation it would in the law of contract for a remedy in the nature of damages for breach of representation or collateral warranty.

§4.129 Estoppel will never establish title or a right.[100] The trial decision was slightly varied to make it 'technically correct'; the purchaser was not given an easement or any other form of legal right. Instead, the court awarded a declaration, in the purchaser's favour, to the effect that the vendor was estopped from denying that the plaintiff purchaser and all persons claiming through or under him are entitled to free passage to the park. The effect on the land titles certificate was discussed:[101]

> It must be borne in mind . . . that the registered owner is in absolute control of his land. There is nothing in the . . . Act to protect him from the consequences

---

98 *Supra*, note 96 at 314. Also see note 95, *supra*, at §5.
99 *Supra*, note 96 at 314, citing *Hugg v. Low*, [1929] 3 D.L.R. 725 (Sask. C.A.).
100 *Ibid. Hugg v. Low* at 731.
101 *Ibid.* at 730.

of doing something himself, if he chooses, which will derogate from his own title. Thus, if he permits himself to become responsible for the making of representations which are conceived at the time to be to his advantage, and thereby leads another to change his position, as the defendant [purchaser] has done here, he can have no right to complain if it be held that he has stepped beyond the protection afforded to him by the statute, and must make such representations good.

§4.130 The element of negligence is not one that all writers agree as to being necessary. For example, the Canadian Encyclopedic Digest[102] specifically points out that an estoppel may arise in circumstances which are devoid of both fraud and negligence. To that end, estoppel operates as a shield and never a sword.

§4.131 The representation in question must be one of an existing fact. It is apparently not absolutely necessary, in the opinion of some, for the representation to be knowingly untrue to the person against whom estoppel works. Therefore, if a person represents himself as having authority to do an act when in fact he has no such authority to do it, and another is thereby drawn into a contract with him and that contract becomes void because of a lack of authority, he is consequently liable for damages which result to the party who had accepted that representation, whether the party making it acted with knowledge of its falsity or not for he undertakes the truth of its representation; that is, the validity of the authority is the maker's responsibility.

§4.132 Expanding further on the three last formal requirements set out in paragraph 4.127, it is necessary that the party invoking the doctrine of estoppel must have been mistaken as to his legal rights. He must also have acted to his potential detriment on the faith of his mistaken rights. On the other hand, the party against whom the estoppel is set up should know of the existence of his own rights and of the other party's mistaken belief. He must actually have gone so far as to encourage the latter to spend money either directly or indirectly or otherwise have induced detrimental reliance by abstaining from asserting his legal rights. While this usually involves money, the expenditure may be indirect, as the labour on the planting of crops, and labour generally, both of which can be expressed in terms of money. It reasonably also includes the cost and labour of a logging operation when undertaken in good faith and reliance on an adjoining owner's clear assertion of the boundary position which under other circumstances might constitute a trespass.[103]

§4.133 Conduct or action which is misleading may give rise to estoppel

---

102 *Supra*, note 95 at §10.
103 See, for example, *Bell v. Howard* (1857), 6 U.C.C.P. 292 (C.A.).

in certain circumstances, as where there is ignorance on the part of the person being misled about his legal rights; but mere silence or inaction is not a strong ground for inferring acquiescence unless the party was fully aware of his right to do otherwise. Generally speaking, however, silence can be effective to create an estoppel. Silence is innocent and safe where there exists no duty to speak. For estoppel to arise, it is necessary that the truth should not be known to the party relying upon the estoppel. Very simply stated, if a person has kept silent when he should have spoken, he will not then be permitted to speak when he should keep silent. "Silence will not create an estoppel where the person who has remained silent was ignorant of the facts. 'Silence' or 'deliberate silence', to amount to a misrepresentation or to conduct of omission giving rise to an estoppel must mean the withholding of facts or circumstances known to the party which it is his duty to reveal."[104] Failure to reply to letters, correspondence or telephone calls may therefore give rise to an estoppel.

§4.134 Estoppel, as a principle with relevance to activities in boundary retracement, will therefore require the surveyor to conduct inquiries of the persons having knowledge or memory of the circumstances surrounding the placement of boundary delimiting features. As these features will most often be of an origin other than of survey, specific skills and requirements in making records of such evidence would be expected. The use of affidavits, photographs and receipts for money spent on materials for fencing are only a few of the types of evidence that should be considered in preparing a complete record. Again, the paramount role of the surveyor is collection of the evidence, although much of the evidence relative to an issue that sets up an estoppel may be distinctly outside of the ambit of a surveyor's proper engagement. Finally, the role of the courts in making the final determination of where a lost boundary is to be located should never be overlooked.

§4.135 The clear application of the defence of estoppel appears in *Rollings v. Smith*[105] where there was no suggestion of what are generally conceived as conventional lines. Events surrounding the construction of a house and the fencing of the immediate curtilage brought an issue to trial and further to appeal. The court set forth the applicable principles of estoppel[106] and, citing the facts of the case relative to each, stated them as follows, citing Anger and Honsberger:[107]

---

104 *Supra*, note 95 at §§35-38.
105 (1978), 15 Nfld. & P.E.I.R. 128 (P.E.I. C.A.).
106 See §4.127.
107 H.D. Anger and J.D. Honsberger, *Canadian Law of Real Property* (Toronto: Canada Law Book, 1959) at 311.

156 Survey Law in Canada

[I]n order that A may be estopped in equity from complaining of the violation of his rights by B, the following must be established:
(1) A must know of his legal rights because the rule is founded on his conduct in the light of that knowledge.
(2) B must be mistaken as to his own legal rights because, if he is aware he is infringing on A's rights, he takes the risk of A later asserting them.
(3) B must spend money or do some act to his prejudice because otherwise he would not suffer by A subsequently asserting his rights.
(4) A must know of B's mistaken belief so as to make it inequitable of him to keep silent and allow B to proceed.

§4.136 On the other hand, estoppel applies in the concept of boundaries by agreement or convention in the fashion noted by Vice-Chancellor Strong in his decision in *Bernard v. Gibson*.[108]

RE-DEFINITION OF BOUNDARIES: RESURVEY, RETRACEMENT, RESTORATION

§4.137 In the first demarcation or marking on the ground of the boundaries of a parcel of land by a surveyor, the work is mainly an applied technology to ensure that the limits are correctly placed by a trail of monuments of some special and (one hopes) permanent character that are recognizable to those creating and taking interests as occupiers on the ground. These aspects are entrusted to surveyors to ensure a proper record is made of the boundary features as evidence for consideration in restoring the original lines at some later time. While the principle is that surveyors in the Canadian provinces are chartered by statute for the making of surveys, not for the making of boundaries, the two events virtually coincide in the majority of instances as a matter of process. Thereafter, the real test of the professional skill of land surveying is the re-definition of boundaries that, through the ravages of time or bulldozers or neglect to maintain recognizable physical features, have deteriorated to the extent that uncertainty arises as to location.

§4.138 The common law has always seen boundary re-definition as a matter of evidence and not of preconceived rules of where, in theory, a boundary should be placed.[109] In the old English context, boundaries existed

---

108 *Supra*, note 93 at 203. See also §4.120.
109 One of the notable expositions on this aspect is to be found in the essay "*The Judicial Function of Surveyors*" by Thomas M. Cooley, Chief Justice of the Michigan Supreme Court from 1864 to 1885. Written about 1885, the paper appeared first in the Michigan Engineering Society Journal published by the University of Michigan. See J.B. Johnson,

as physical features of the landscape. If a new division line was created for a fee simple or other estate, it was made as a fact on the ground by the open actions of the parties. The *Statute of Frauds* of 1677 brought written charters to the fore as evidence of land dealings because the developing economy had changed from the old practice of feoffment with livery of seisin on the land. Charters require a parcel clause to identify the entity of the dealing.

§**4.139** The initial definition, and the subdivision, of the ancient manor of Blackacre was earlier discussed. In marked contrast, colonial development of total wilderness lands needed survey definition on the ground and this had to precede the granting of title (and, admittedly, this did not always occur): the parcel came into physical existence by the defining survey marks. In some instances, all bounds of a parcel were marked by survey and it followed that such marks of the first survey had to assume the status of inviolable boundary determiners. Frequently, and unfortunately, not all parcel limits were marked on the ground at the time of grant. For example, a quarter-section in the Dominion Lands Survey system would be granted after the original Crown survey of only two sides with three corners set in place. And, in Ontario, grants were made with only one side surveyed and two markers in place and even of lands with no markers. The statutes of the survey systems ordain the boundaries of the original Crown surveys as true and unalterable. The common law prescribes the same status for all first definitions of parcels.

§**4.140** To complete the boundary definition, recourse was therefore taken to statutory provisions that set forth the manner of locating the lines not run and monumented in the original Crown survey, but the Ontario statute in fact left such lines as then first run unspecified as to status. There is no problem, however; the courts have almost consistently solved the problem by common law principles and their solutions have been for clear reasons.

§**4.141** In the Dominion Lands Survey system, and in another of the same nature in Ontario, likewise a derivative of the United States Public Lands Survey system, the unrun interior lines of a section are prescribed to connect monuments directly across the section. The Dominion system went further, however, and defined a corner monument uniquely as a certain physical thing set by a surveyor and a boundary as the straight line between these

---

The Theory and Practice of Surveying, 16th ed. (New York: John Wiley & Sons, 1909), Appendix A; also see *Surveying and Mapping*, April-June, 1954, Vol. XIV, No. 2, at 161-168, where reprinted. A converter to apply the expressed comments and principles to the Canadian scene a century later is written as accompanying notes to studies in survey law at the Centre for Surveying Science, University of Toronto, Erindale Campus.

monuments; furthermore, blazed lines are solely indicators and not boundary features.[110] This is essentially a point-to-point rule that is entirely reasonable in open prairie country and especially under a fairly strict system of good accuracy standards and the exclusion of the use of the magnetic compass.

§4.142 By contrast, in Ontario, after a short use of this concept (still, of course, applicable for the first survey of a line where the townships were so designed), logic dictated a return to the point-line system, much more suited to a land heavily forested. The directions of the side lines of parcels — the lots in tiers of concessions — were to be run parallel to (otherwise generally expressed as "on the course of") some governing line that was to control directions, sometimes as that line was reinstated on the ground and sometimes as the numerical value of direction stated in the original Crown survey records.[111]

§4.143 These lot boundaries by first survey of the lines of the scheme, which were not run in the original survey by the Crown-appointed surveyor, were also marked along their length by the blazing of trees on each side of the cut line. By the common law operation of the sanctions that settle boundaries, the lines so marked are the true boundaries, crooked and out of place though they may be.

§4.144 The Ontario *Surveys Act* defines an obliterated boundary as one where the corners or blazed trees cannot be found; that is, the status of blazed trees as evidence is endorsed in the statute in section 1(l). The result is that Ontario lot lines are almost inevitably not straight lines between the terminal markers. The decisions of the Ontario courts have logically called for any boundary in dispute to be re-defined on the evidence of the first running of the line. Thus it is the priorities of evidence which strictly control the character of the Dominion system and not preconceived notions of a theoretic scheme and the strict rules of boundary. Unfortunately, this has led to much argument as shown by incidents of directly opposed survey practices: lot lines restructured by the Act and lot lines relocated by the evidence of things on the ground.

§4.145 In the Atlantic provinces, the common law rules prevail for re-definition of boundaries strictly on the evidence of the first demarcation

---

110 See, for example, as to original boundary lines, the *Manual of Instructions for the Survey of Dominion Lands*, 9th ed. (Ottawa: King's Printer, 1918) and the *Dominion Lands Surveys Act*, S.C. 1908, 7-8 Edw. 7, c. 21, s. 62; and as to blazed lines, see clause 98 of the *Manual* "Blazed lines are not intended to mark the boundaries or limits of the parcels of land laid out; they are opened and blazed for the sole purpose of assisting in finding the corner monuments."
111 The methods are explicitly stated in the *Surveys Act*, R.S.O. 1980, c. 493.

or on the presumption drawn from the real things that exist and the usage of neighbours. As in Ontario, the standards of survey were lax and, indeed, even more so and well into recent times. Many practitioners used the compass for line work until this mid-century.

§4.146 The parcel clause of the transfer deed defines the boundaries in quite variable fashion depending on time and place, but in the final analysis judicial decisions settle the problem with the nicety of the common law. Present social and economic demands require good parcel descriptions, and modern instrumentation and practice can readily provide this. But this applies to present and future work. There still remain the problems of reconciling the numerics of old surveys with the facts on the ground as well as from the documentary records. The differences can be disturbing to a layman.

MODERN SURVEYS, TITLES AND LAND DEALINGS

§4.147 The settled lands of Canada are fully surveyed only in the sense of the incomplete original Crown surveys. This does *not* mean: (1) that all the corners are still marked by the original monuments — extremely rare in Ontario and the Atlantic provinces which were divided before the age of the relatively permanent iron monument; (2) that in the east, where lines are evidenced by blazed trees, that any of the original line evidence remains; (3) that original survey lines and corners are evidenced by a chain of survey records: that lines have in fact been maintained or re-defined by survey is more often than not speculative because it is not a tradition to keep public records of survey, and later private divisions of land may be in no better situation.

§4.148 The ancient manor of Blackacre was secure for its bounds by local knowledge and custom, possibly the extent better assured than the title. One of the inspirations to Robert Torrens was the disaster of title that befell his Irish cousin; he made no mention of boundary problems. The registered title concept is good for guarantee by the state, or the deed by the solicitor, if the bounds are unique in words — theoretically, that is. Thus the title to "Lot 1 in Concession IV" may be absolute; the extent is perfect in words and, in the absence of conflicting evidence on the ground, the extent is probably perfect on the ground. A title system based entirely on words, with adjoining parcel descriptions using the same words to describe the common boundary and with the rejection of adverse possession, is virtually faultless. The assurance fund is likely, on this score,

160   Survey Law in Canada

to be characterized by what Theo Ruoff called 'indecent solvency.'[112] The one flaw is the unpredictable nature of people who put buildings, fences, services and pools in wrong places in relation to the bounds which are, in the classic words, imaginary, which is only slightly worse than invisible. The loss of survey marks is a serious matter.

§4.149 This analysis is almost historical. There are few sharp divisions in history and this is not one of them. For the new lines of land divisions now to be drawn on the earth for homes, work places and routes of travel, the measurements and mathematics of survey are taking a new place. In a new development there are no ancient occupations and the loss of survey marks makes measurements (inevitably now coordinated on a local if not national system) the only sure means of replacement of these boundaries. There is still a line of distinction to be drawn: the significance of measurements as evidence does not make the resulting coordinates the corners.

§4.150 To make a mathematical system work, it needs checks and more checks, thus incorporating mathematical analysis and adjustment routines formerly regarded only as techniques for geodetic and project control surveys. This, in our opinion, is the true meaning of coordination and integration. It is a logical accompaniment to electronic measurement and computer technology. There is nothing unusual about this development in countries where there have been waves of rapid development. At first settlement, the settler looked to the survey marks to show his limits. Decades later he would treat the fence as his boundary, even if that fence did not lie along a surveyed line.

§4.151 It is not settled, of course, that re-definition by coordinates is the right way. It is convenient for surveyors and administrators but it will be seen by many as technology dictating human relations: as earlier stated, boundaries are 'people things'. The priorities of evidence are still valid because of the original surveys and all other prior surveys of poor mathematical structure and insubstantial evidence. This will remain the case until the opinions of surveyors are tested by adjudication.

§4.152 The Ontario *Boundaries Act*[113] has provided admirable solutions for nearly three decades, in large numbers and province-wide. It is surprising that equally simple approaches to boundary adjudication have not sprung up in the other provinces but it is, of course, a matter of demand.

---

112 T.B.F. Ruoff, *An Englishman Looks at the Torrens System: being some provocative essays on the operation of the system after one hundred years.* (Sydney: Law Book Co. of Australasia Pty., 1957).
113 R.S.O. 1980, c. 47.

In Ontario in the 1950's, the pace of development outstripped the existing methods of finalizing re-definitions of boundaries. The Act was framed on the same premises as stated for land titles registration: efficient, expeditious and economical. Manitoba utilizes the provisions of the *Special Survey Act*[114] to resolve problems with boundaries which are actually title problems. Once the Plan of Special Survey is registered, it is supposedly binding from that time forward, but it is noted that if errors are found in a special survey, this may be corrected by another. The "binding" and "fixing" character is uncertain.

§4.153 The early history of surveys under land title regimes is an interesting episode. From his writings, it would appear that Robert Torrens had little concern about boundaries, undoubtedly picturing the creation of walls and other substantial boundary features or relying on the marks and numerics of surveys with no feeling for the problems of re-definition. Reliance was certainly placed on the wording of descriptions and the correlation between diagrams and parcels. In Australasia, the setting of survey standards and the checking of surveys for private land dealings had barely started 30 to 40 years after inception of guaranteed title systems; in Ontario, it was more than 70 years. A clear picture of earlier practice emerges by comparing the modern plan to accompany a first application with those done prior to 1958 when standards were first set down.

§4.154 The main element that was missing was the old evidence for re-definition and the new evidence created. It was felt that the public record for evidence of title should also do something to ensure against wandering boundaries where the evidence was not respected for re-definition. Thirty years later these first rules, to be enforced before plan deposit, are now the standards of the Association of Ontario Land Surveyors fulfulling its professional mandate by post-deposit examination.

§4.155 The modern survey, and the survey problems, are different because of standards, techniques, equipment, computing ease, education, performance, responsibility and liability; also because of more people, more demand for and on land, higher land prices, more facilities from highways to swimming pools, condominiums and office buildings, more social planning, more issues to be resolved and, probably, less tolerance for discrepancies. It may seem like a new world, but the same basic problem remains of re-defining old boundaries to which new boundaries are tied and doing this with a high degree of certainty that the surveyor's opinion will be upheld by some adjudication tribunal.

§4.156 Modern practice, which has developed dramatically since the

---

114 C.C.S.M., c. 190.

162   Survey Law in Canada

1950's, has had to meet new criteria. Money-lending practices and cautious personal investment warn against uncertain parcels of conveyance or mortgage, the uncertainty lying less with dimensions and area than with the location of structures and uses relative to the boundaries. The title to the parcel and the parcel itself may be perfectly correct for the titles system and even for the exclusions, but may be a virtual disaster as an investment for a buyer, for a lender/mortgagee and for the solicitor making the conveyance.

§4.157 Irrespective of the action decided upon by a prospective purchaser or vendor as to proceeding with a transaction, the surveyor's role is unchanged. He is to re-define the original boundaries and, for those surveys which are not solely for title records, he must provide the client with the extended range of facts about things on the land, the evident use of the land (or assumptions of use) and potential claims of which he has knowledge.

SURVEY LAW: THE FUTURE

§4.158 Issues in boundary law are continually revealing themselves. Not only are these issues emerging as a result of advances in technology, but even more as a result of the opportunities presented to the courts to adjudicate on boundary retracement and management problems. Each new decision is a further elaboration of the principles based in common law, and increases the surveyor's resource-base of guidance in the application of these principles.

§4.159 The need to be comfortable with, as well as cognizant of these principles, will remain as urgent for practitioners of boundary surveying in the future as it is today. The challenge for educators is to address this need and, at the same time, guide future practitioners in the opportunities available in new technology to manage information resulting from retracement activity. There are always academic components, but, overall, these are solid and practical matters that are day-to-day realities of professional practice and it is in the sense of realities that the topics must be taught and learned.

§4.160 Several issues which are likely topics of future research and development include the following:

(1) The relationship between standards of accuracy and precision in boundary surveys will need to be more fully developed through a comprehensive approach in regulations that reflects a more rigorous treatment of the difference between these two attributes. Not only have the standards for these attributes been confused and muddled in the

past, they have been largely an oversimplification of the real thing. The limits of precision will be re-stated as a statistical function of distance, and accuracy will be verifiable from control positions established through integrated surveys.
(2) A reconsideration of the principle of monumentation, of the notion of an iron bar as a boundary marker, and as a "permanent feature", is expected. Remoteness of the marker from the boundary being marked will become a function of such factors as risk of disturbance, property values, future jurisprudence, and reliability in terms of re-establishment of the boundary.
(3) A concerted effort to coordinate the various acts and regulations having a bearing on the surveyor's activity and profession is sure to be recognized as land information management will come to place a premium on reliability in spatial land information on a micro-scale and for legal purposes. This of course assumes the profession is prepared to recognize and come to terms with land information systems as having a direct impact on and benefit to their work.
(4) The need for statutory solutions in Canada to the grave uncertainty about most water boundaries cannot be overstated. The lack of consistent principles, and the ambiguities in administrative policy applicable to the management of water boundaries, emerges from a reading of the chapters on water boundaries in this book.

§4.161 With the further development of land information systems and the new issues that the spread of the technology in society is creating, the practitioner may find himself facing another identity crisis. The need for computer-manageable information may mean that much of the descriptive and cultural information that the practitioner has traditionally collected and represented on surveys and plans is considered secondary to the mathematical information inherent therein. Likewise, designers of information systems will need to come to terms with varying levels of legal reliability and uncertainty, both of which, while perhaps unpalatable, have always been part of the survey practitioner's reality.

Appendix A

Catalogue of Dominion Lands and Canada Lands Survey Manuals

.... (1871). Manual shewing the System of Survey adopted for the Public Lands of Canada in Manitoba and the North-West Territories with Instructions to Surveyors. (Ottawa: Queen's Printer). (Subsequently known as the 1st edition.) J.S. Dennis, Surveyor General.

.... (1881). Manual shewing the System of Survey of the Dominion Lands, with Instructions to Surveyors. (Ottawa: Dominion Lands Office, Dep't of the Interior). (2nd edition). Manual prepared by E. Deville. Lindsay Russell, Surveyor General.

.... (1883). Manual shewing the System of Survey of the Dominion Lands, with Instructions to Surveyors. (Ottawa: Department of the Interior). (This 3rd edition is in two parts: systems and procedures; instructions for township control.) Lindsay Russell, Surveyor General.

.... (1890). Manual showing the System of Survey of the Dominion Lands, with Instructions to Surveyors. (Ottawa: Queen's Printer). (Minimally revised edition of the first part of the 1883 Manual and apparently treated as also the 3rd edition; sometimes treated as the First Part (Preliminary Edition) of the 4th Edition). Dr. E. Deville, Surveyor General.

.... (1892). Manual showing the System of Survey of the Dominion Lands, with Instructions to Surveyors. (First Part (Preliminary Edition)). (Ottawa: Queen's Printer). (4th edition). Dr. E. Deville, Surveyor General.

.... (1902). Manual showing the System of Survey of the Dominion

Lands, with Instruction to Surveyors. (First Part). (Ottawa: Government Printing Bureau). (5th edition). Dr. E. Deville, Surveyor General.

.... (1903). Manual of Instructions for the Survey of Dominion Lands. (First Part). (Ottawa: Government Printing Bureau). (5th edition, reprinted) Dr. E. Deville, Surveyor General.

.... (1905). Manual of Instructions for the Survey of Dominion Lands. (First Part). (Ottawa: Government Printing Bureau). (6th edition). Dr. E. Deville, Surveyor General.

.... (1906). Amendments and Corrections to the Manual of Instructions for the Survey of Dominion Lands. (First Part). (Ottawa: Government Printing Bureau). Applied to 6th edition. Dr. E. Deville, Surveyor General.

.... (1910). Manual of Instructions for the Survey of Dominion Lands. (Ottawa: Government Printing Bureau). (7th edition). Dr. E. Deville, Surveyor General.

.... (1913). Manual of Instructions for the Survey of Dominion Lands. (Ottawa: Government Printing Bureau). (8th edition). Dr. E. Deville, Surveyor General.

.... (1918). Manual of Instructions for the Survey of Dominion Lands. (Ottawa: King's Printer). (9th edition). Dr. E. Deville, Surveyor General.

.... (1946). Manual of Instructions for the Survey of Dominion Lands. (Ottawa: King's Printer). (10th edition). F.H. Peters, Survey General.

.... (1956). (10th edition, reprinted). R. Thistlethwaite, Surveyor General.

.... (1961). Manual of Instructions for the Survey of Canada Lands. (Ottawa: Information Canada). (First Edition). 1963, French edition. R. Thistlethwaite, Surveyor General.

.... (1979). Manual of Instructions for the Survey of Canada Lands. (Ottawa: Canadian Government Publishing Centre). (Second Edition). W.V. Blackie, Surveyor General.

.... (1908, 1917, 1952). Supplements. Tables setting out the factors of the system were printed in early editions then separated as Supplements to the Manuals.

.... (circa 1900). Maps and Specimen Plans to accompany the Manual of Instructions for the Survey of Dominion Lands.

166 Survey Law in Canada

.... (1916). Manual of Instructions for the Erection of Boundary Monuments on Surveys of Dominion Lands.

.... (1917). Description of Boundary Monuments erected on Surveys of Dominion Lands 1871-1917.

.... (1918). General Instructions for Leveling on Dominion Lands Surveys, published as a separate item.

.... (1935). Manual Respecting the Survey of Quartz Mining Claims in Dominion Lands.

5

Water Boundaries — Coastal

*Susan Nichols, M. Eng., P. Eng.**

INTRODUCTION

§5.01 Canada's coastal areas have been described as a mosaic of federal, provincial, and private rights.[1] Besides rights of ownership, there are special rights of use and control over water resources and the lands covered by water. These rights have their origins in early English law, but Canadian courts and legislation have made significant modifications. Complicating Canadian water law is the division of powers among the federal and provincial governments. The extension of offshore zones and boundaries based on international law has added yet another dimension to the coastal mosaic.

§5.02 While various sources of law must be considered in specific jurisdictions, the common law provides a framework of general principles and legal presumptions. Within this framework one may determine what rights can exist, who can hold or control these rights and to what waters

---

* The author would like to thank S.T. Grant, Atlantic Regional Tidal Officer of the Canadian Hydrographic Service, Fisheries and Oceans Canada, Bedford Institute of Oceanography, Dartmouth, Nova Scotia for his comments offered on the draft on the section entitled "Tides and Tidal Reference Surfaces". The author would also like to acknowledge the helpful comments and suggestions provided by the editor, J.F. Doig.

1 *A.G. of Can. v. Higbie*, [1945] S.C.R. 385 at 431.

168 Survey Law in Canada

and lands these rights apply.[2] Exceptions to these principles, such as are provided for by provincial legislation, are critical to the surveyor's practice, but only examples can be given here. The objective of this chapter is to examine the legal framework as it affects boundary delimitation. Chapter 6 develops many of these topics for inland waters. The emphasis in the present chapter is on the delimitation of tidal boundaries — those boundaries that mark the seaward limit of private lands.[3] A brief introduction is also given to international law of the sea and jurisdictional boundaries offshore.

Water Boundaries

§5.03 A water boundary delimits property rights and jurisdictions adjacent to and within a watercourse or the ocean. A watercourse usually refers to rivers and streams in which water flows in a well-defined channel,[4] but it can also be defined in much broader terms to include lakes, marshes, and other waterbodies.[5] Unless otherwise specified, a watercourse consists of the water, the bed and the banks and must be "distinct enough to form a channel or course that can be seen as a permanent landmark on the ground."[6]

§5.04 One method for classifying water boundaries is presented in Figure 5.1, where they are distinguished by two characteristics:

---

2 See, for example, G.V.A. LaForest and Associates, *Water Law in Canada: The Atlantic Provinces* (Ottawa: Information Canada, 1973); V. Di Castri, "Water and Highway Boundaries as Affecting Title", chapter 7 in *Registration of Title to Land*, v. 1 (Toronto: Carswell, 1987); A.S. Wisdom, *The Law of Rivers and Watercourses*, 4th ed., (London: Shaw & Sons Ltd., 1979); R.G. Hildreth and R.W. Johnson, *Ocean and Coastal Law* (Englewood Cliffs, NJ: Prentice-Hall and Associates, 1983).

3 See Notes 116, 134, 158 below for Canadian references on water boundaries; others include W.A. Taylor, "Boundaries Abutting Water in British Columbia" (1956), 13(1) *The Canadian Surveyor* at 37; M. Viminitz, "Water Boundaries" (1958), 14(2) *The Canadian Surveyor* at 75; C.D. Hadfield, "The Effect of Bodies of Water in Legal Surveying" (1968), 22(5) *The Canadian Surveyor* at 451; B.N. Johnanson, "An Examination of the Law of Water Boundaries and Accretions in Manitoba" (1977), 8 *Manitoba Law Journal* 403; D.W. Lambden and I. de Rijcke, *Boundaries and Surveys* (Toronto: Carswell, 1985). A comprehensive review of coastal boundary issues, although from an American perspective, is also given by A.L. Shalowitz, *Shore and Sea Boundaries*, vol. 2 U.S. Coast and Geodetic Survey Publication (Washington: U.S. Government Printing Office, 1962).

4 Wisdom, *supra*, note 2 at 3-4; *Black's Law Dictionary*, 5th ed. (St. Paul: West Publishing Co., 1979).

5 *Ernst v. Dartmouth (City of)* (1970), 3 N.S.R. (2d) 254 (T.D.) on the interpretation of a watercourse as defined in the *Water Act*, R.S.N.S. 1967, c. 335, s. 1(k); as am. by S.N.S. 1972, c. 58, s. 1(4).

6 *Wilton v. Murray* (1897), 12 Man. R. 35 at 38 (Q.B.).

(1) the manner in which the boundaries are defined (e.g., by reference to a natural feature or as an artificial line); and
(2) the type of rights the boundaries delimit (e.g., rights of ownership or jurisdiction).

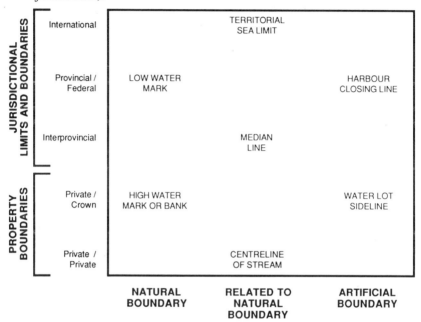

FIGURE 5.1: Classification of water boundaries with selected examples.

§5.05 *Natural and artificial boundaries* If the term water boundary is given a narrow interpretation, it is defined by the intersection of the water with the land or defined by another physical feature indicating that intersection, such as the limit of vegetation. These natural features form the water line or water mark. A slightly broader interpretation of the term water boundary would include boundaries directly referenced to or measured from the land-water interface; an example would be the centreline of a stream which can be defined as the line midway between the water marks on either side.

§5.06 As natural features, water boundaries have special characteristics that affect their status in law and their delimitation. They are natural monuments and thus have priority over measurements in the interpretation of property descriptions. This is based on the premise that a call for a natural feature best demonstrates the intention of the granting parties and

170 Survey Law in Canada

is least susceptible to error.[7] When a boundary is defined by a waterbody, its bank, shore, water line or water mark, these natural monuments will govern any incompatible survey measurements.[8]

§5.07 A natural water boundary is ambulatory; its location moves as a result of gradual, physical changes in the shore such as erosion. Any delimitation of the boundary is therefore only an indication of its position at some particular time. Since the water level in a river, lake, estuary or the open ocean can vary from hour to hour and from season to season, boundaries are often referenced to a specific water level. Defining this level and finding evidence of where it intersects the shore are the major problems in the delimitation of water boundaries along the coast.

§5.08 A tidal boundary is located by evidence of the intersection of the water with the shore when the water reaches a specific level of tide. In common law jurisdictions, the definition of this boundary can be traced to a 17th century treatise, *De Jure Maris*,[9] written by Sir Matthew Hale who was Lord Chief Justice of England under Charles II. Arguing the rights of the Crown in England's tidal waters, he defined the sovereign's coastal land as those covered by the ordinary high tides.

§5.09 To restrict water boundaries to only those defined by natural features, however, would ignore many other limits of water-related property rights and jurisdictions. If a broader interpretation is taken, then water boundaries can also include artificial lines having little or no relation to the sinuosities of the land water interface. In some cases the endpoints of the lines may be at a water-mark, but the line itself may not conform to a natural feature. Examples of such boundaries include the sidelines of a private waterlot extending from high to low water mark, a line across the mouth of a harbour marking the limit of federal jurisdiction or an international maritime boundary defined as a line drawn between two points on a nautical chart. In this more general, albeit unconventional, definition the critical test is whether the boundary delimits rights to own, use or control water resources or the lands beneath the water.

§5.10 *Property and jurisdictional boundaries* A further distinction can be made between boundaries that mark the limits of property rights and those that define jurisdictions. Of major interest in the former case are

---

7 *Fraser v. Cameron* (1854), 2 N.S.R. 189 (C.A.).
8 *Esson v. Mayberry* (1841), 1 N.S.R. 186 (C.A.); *Irving Refining Ltd. v. Eastern Trust Co.* (1967), 51 A.P.R. 155 (N.B.S.C.); *Saueracker v. Snow* (1974), 14 N.S.R. (2d) 607 (T.D.).
9 Sir Matthew Hale, *De Jure Maris* (c. 1666), as found in S.A. Moore, *A History of the Foreshore and the Law Relating Thereto*, 3d ed. (London: Stevens and Haynes, 1888) at 379.

the boundaries between lands held in private ownership and lands vested in the Crown. In coastal waters, the boundary between private lands and the Crown is *prima facie* the ordinary high water mark.[10] The boundaries of an offshore mineral lease, or of exclusive fishing rights in a nontidal stream, are examples of other property boundaries.

§5.11 Although the term jurisdiction can be found in many contexts, it will be used here mainly to refer to the territory or lands over which provinces or nations have both the power to adjudicate matters before their respective courts and the power to legislate.[11] Thus, for instance, international boundaries mark the limits of jurisdiction between two nations. Jurisdictional boundaries, particularly between the federal and provincial governments, often involve complex issues in constitutional and international law.[12] Examples of these boundaries are presented to illustrate their significance in coastal surveys, however, most of the legal issues are beyond the scope of this chapter.

§5.12 In summary, water boundaries are natural features or lines related to natural features, but there are also artificial boundaries separating property rights and jurisdictions in water bodies. Of special concern to surveyors in coastal waters is the tidal boundary: that natural boundary referenced to a specific tidal level. It is this tidal boundary — its relation to the physical regime, its significance in land ownership and its delimitation — that is considered in detail in the following sections.

TIDAL WATERS

§5.13 The law as it applies to the coast is directly related to the physical regime. Property rights and boundary definitions frequently refer to specific coastal and tidal characteristics. To understand some of the intricacies of coastal law and to interpret evidence in boundary retracement, the surveyor must first have an appreciation of both the scientific and legal aspects of tidal waters.

---

10 *A.G. v. Chambers*, (1854) 43 E.R. 486 at 490. See also *A.G. of B.C. v. Neilson*, [1956] S.C.R. 819, reversing (1955), 16 W.W.R. 625, which affirmed 13 W.W.R. 241.
11 Taken in part from J.A. Yogis, *Canadian Law Dictionary* (New York: Barron's Educational Series, Inc., 1983).
12 The constitutional issues are discussed, for example, in J. McEvoy, "Atlantic Canada: The Constitutional Offshore Regime" (1984), 8(2)*Dalhousie Law Journal* 284; and R.J. Harrison, "Jurisdiction Over the Canadian Offshore: A Sea of Confusion" (1979), 17(3) *Osgoode Hall Law Journal* 469.

Classification of Waters

§5.14 The nature and limits of jurisdictions and of property rights related to water depend on how the law classifies the waters concerned. Examples of such legal categories include: ground waters and surface waters; various types of watercourses both natural and artificial; internal waters, territorial seas and high seas in international law. Two other classifications — tidal or nontidal and navigable or nonnavigable — are considered briefly here.

§5.15 *Tidal and nontidal waters* Under the common law, tidal waters are defined as the waters of the seacoast and the arms of the sea subject to the ebb and flow of the tides. Salinity of the waters may be a factor in determining whether or not waters are tidal, but the main criterion is the ebb and flow of the daily tides.[13] As defined by Hale in *De Jure Maris*:

> [T]hat is called an arm of the sea where the sea flows and reflows, and so far only as the sea so flows and reflows; . . . although the water be fresh at high water, yet the domination of an arm of the sea continues, if it flow and reflow.[14]

§5.16 Although extraordinary tides should be disregarded when determining the tidal limit,[15] the precise extent of vertical or horizontal movement required for the waters to be tidal is not defined. In *Nash v. Newton*,[16] for example, a body of brackish water behind a seawall on Grand Manan Island was subject to a rise and fall of water level of several inches with the daily tides which percolated through the wall. The justices held differing opinions as to whether the waters were tidal or nontidal, but concurred that once the seawall was opened, the 'lake' became an arm of the sea and thus subject to the law of tidal waters.

§5.17 In English common law, the classification of a waterbody as tidal or nontidal is critical in the determination of private and public rights. Tidal waters are sometimes called public waters in which the public has the right of navigation and fishing and the soil beneath the water at ordinary high tides is *prima facie* held by the Crown.[17] Nontidal watercourses are considered private in English common law because the bed is presumed to belong to the upland proprietor *ad medium filum aquae* (to the centre line of the stream) and the exclusive right of fishing belongs to the owner of the bed. There is no public right of navigation except where acquired

---

13 Wisdom *supra*, note 2 at 259-260.
14 *Supra*, note 9 at 378.
15 *Supra*, note 4 at 58.
16 (1891), 30 N.B.R. 610 (C.A.).
17 *Supra*, note 9 at 370-413; see also *R. v. Robertson* (1882), 6 S.C.R. 53 at 88-89.

through custom or authorized by statute.[18] In Canada, this common law distinction between tidal and nontidal waters is not strictly followed and some provincial legislation has altered the circumstances quite markedly.

§5.18 *Navigable and nonnavigable waters* Canadian geography is quite different from that of England and the network of large nontidal rivers and lakes has been essential to the economic development of the nation. The law has recognized these differences and, at least in central and western provinces, the public right of navigation is based on whether the waters are navigable, without any restriction to tidal characteristics. Navigability is a matter to be established from the facts and, in waters that are *de facto* navigable, the public right of navigation exists.[19] Thus, for example, in an appeal regarding fishing rights in the Fraser River, the Privy Council held that the common law of England applied but the British Columbia Supreme Court subsequently interpreted the Privy Council decision as not limiting the public right of navigation in nontidal but navigable waters.[20]

§5.19 Canadian decisions have recognized the *prima facie* presumption that title to land bounded by nontidal waters includes title to the bed *ad medium filum aquae* and that ownership of the bed carries with it the exclusive right of fishing.[21] However, this rule has been found not to apply in the Great Lakes, in international rivers and in other large rivers and lakes that are nontidal but navigable.[22] Provincial legislation has further restricted the application of the common law rule. For instance, the bed in all navigable waters is owned by the Crown in Ontario unless explicitly granted[23] and in Nova Scotia all watercourses, including lakes and nonnavigable streams, belong to the Crown.[24]

§5.20 The situation is perhaps less clear in other Atlantic Provinces. The

---

18 *Ibid.*
19 *Simpson Sand Co. v. Black Douglas Contractors Ltd.*, [1964] S.C.R. 333. See also La Forest and Associates, *supra*, note 2 178-179; the same principles are applied in the United States and are described in F.E. Maloney, S.J. Plager, and F.N. Baldwin, *Water Law and Administration: The Florida Experience* (Gainsville: University of Florida Press, 1968) at 35-50.
20 *A.G. for B.C. v. A.G. for Can.*, [1914] A.C. 153 at 173 (P.C.); as interpreted in *Fort George Lumber Co. v. Grand Trunk Pacific Ry.* (1915), 24 D.L.R. 527 (B.C. S.C.).
21 *Clarke v. Edmonton (City of)*, [1930] S.C.R. 137, reversing [1928] 1 W.W.R. 553; *R. v. Robertson* (1882), 6 S.C.R. 52, affirming 1 Ex. C.R. 374; *Rotter v. Canadian Exploration Ltd.* (1960), 33 W.W.R. 337 (S.C.C.), reversing (1959), 30 W.W.R. 446 (B.C.C.A.).
22 *George v. Bates* (1858), 7 U.C.C.P. 116 (C.A.); *Keewatin Power Co. v. Kenora* (1906), 16 O.L.R. 184; *Ratte v. Booth* (1887), 14 O.A.R. 419, affirming 11 O.R. 491, which reversed 10 O.R. 351. See also V. Di Castri, *supra*, note 2 at 214-239 and La Forest, *supra*, note 2 at 178.
23 *Navigable Waters Act*, R.S.O 1970, c. 40, s. 1.
24 *Water Act*, *supra*, note 5 at s. 1.

174 Survey Law in Canada

courts have generally followed English common law with respect to the *ad medium filum aquae* rule in rivers, basing their decisions on whether rivers are tidal or nontidal. These include a number of decisions involving private ownership of the bed and fishing rights on the Miramichi River in New Brunswick.[25] In a Newfoundland case, *Robinson v. Murray*,[26] the majority of the court also held that English common law applied with respect to nontidal waters that are *de facto* navigable. However, the rule may not apply to lakes and private ownership of the bed and fishery in many rivers is precluded by explicit reservation or by statute.[27]

§5.21 La Forest notes that "there is a *prima facie* presumption that if the water is tidal it is navigable, a presumption that may, however, be rebutted."[28] Or, as stated in *Sim E. Bak v. Ang Yong Huat*

> The flowing of the tide is strong prima facie evidence of the existence of a public navigable river, but whether it is one or not depends upon the situation and nature of the channel . . . . The question is one of degree, and is for the jury, having regard to all the facts.[29]

Public and private rights in navigable, tidal waters are discussed further in paragraphs 5.46 to 5.59.

Tides and Tidal Reference Surfaces

§5.22 The characteristic feature of coastal boundaries is their relation to the tidal phenomena. Yet this relationship has often been ignored or misinterpreted within the law[30] and in the surveyor's evaluation of boundary evidence. Tides and tidal reference surfaces are reviewed here at a conceptual level. For more comprehensive explanations and examples, the

---

25 *Robertson v. Steadman* (1876), 16 N.B.R. 612 (C.A.); *R. v. Robertson, supra,* note 17; *Boyd v. Fudge* (1965), 46 D.L.R. (2d) 679 (N.B. C.A.); *Black Brook Salmon Club Inc. v. Seder* (1981), 85 A.P.R. 474 (N.B. C.A.); for Nova Scotia see *McNeil v. Jones* (1894), 26 N.S.R. 299 (C.A.) although this case preceded the *Water Act, supra,* note 5. See also the discussion in La Forest with regard to the right of navigation on nontidal waters, *supra,* note 2 at 178-180.

26 (1851), 3 Nfld. L.R. 184.

27 *McDonald v. Linton* (1926), 53 N.B.R. 107, (C.A.); *Nash v. Newton, supra,* note 16; *Niles v. Burke* (1873), 14 N.B.R. 237 (C.A.). See also the discussion for lakes in the Atlantic Provinces in La Forest, *supra,* note 2 at 245-246, for fishing rights at 235-236, and for the public right of navigation at 178-180.

28 La Forest and Associates, *supra,* note 2 at 180.

29 [1923] A.C. 429 at 433-434 (P.C.).

30 See Maloney, *supra,* note 19 at 75.

Canadian Tidal Manual[31] and other sources[32] should be consulted.

§5.23 *Daily tides* Tides are the periodic rise and fall of the ocean surface in response to the gravitational forces of the sun and moon on a rotating earth and the centrifugal forces caused by the revolution of these bodies around their common centres of gravity. Theoretically then, the tides are a response to the changing positions of the moon and the sun relative to the earth. However, the tides observed along the coasts are also influenced by local geography and thus there are considerable tidal variations from place to place. The predicted times and heights of the high and low tides for several hundred localities throughout Canada are published annually in the *Canadian Tide and Current Tables*.[33] These predicted tides do not take into account the effects of meteorological conditions such as strong, prolonged winds or changes in barometric pressure. In some locations the actual water levels may differ occasionally by as much as a metre from the predicted heights.

§5.24 The daily tides may be classified as semi-diurnal, diurnal and mixed tides, with mixed tides exhibiting either a mainly semi-diurnal or mainly diurnal pattern as illustrated in Figure 5.2. Where semi-diurnal tides occur, two nearly equal high waters and two nearly equal low waters are experienced approximately every 24 hours and 50 minutes. Diurnal tides consist of only one high and one low tide during the same period. Mixed tides have more significant inequalities in the heights of successive high waters and low waters. Where semi-diurnal or mixed tides occur, there is a Higher High Water (HHW) and Lower High Water (LHW) each day and a corresponding Lower Low Water (LLW) and Higher Low Water (HLW). The tidal range, which is the difference between high and low water levels, varies from day to day, from month to month and from year to year.

§5.25 *Spring, neap, and long period tides* The major monthly variations are known as spring and neap tides and are related to the orbit of the

---

31 W.D. Forrester, *Canadian Tidal Manual* (Ottawa: Canadian Hydrographic Service, Fisheries and Oceans Canada, 1983).

32 See, for example, R.E. Thompson, *Oceanography of the British Columbia Coast*, Canadian Special Publication of Fisheries and Aquatic Sciences No. 56, (Ottawa: Canadian Government Printing Centre, 1981); H.R. Hatfield, "Tides and Tidal Streams" chapt. 2, *Admiralty Manual of Hydrographic Surveying*, v. II (Taunton, Somerset, England: The Hydrographer of the Navy, 1969).

33 Fisheries and Oceans Canada, *Canadian Tide and Current Tables* (Ottawa; Tides, Currents, and Water Levels, Canadian Hydrographic Service, Fisheries and Oceans Canada).

176 *Survey Law in Canada*

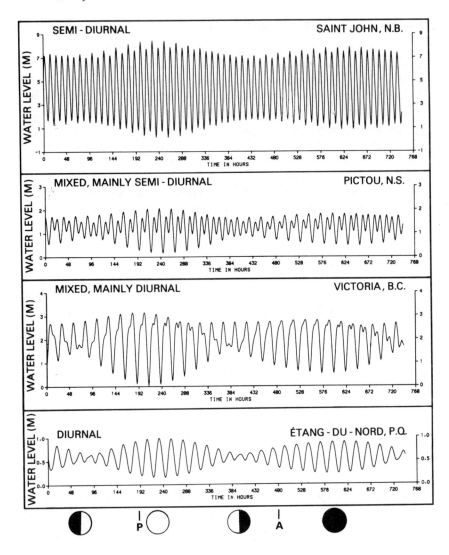

FIGURE 5.2: Predicted Tidal Levels for One Month (January, 1982) for Four Locations Showing Classes of Tide [after Forrester, 1983 Canadian Tidal Manual, p. 36][34]
Note: In the figure, the letter P stands for perigee and A indicates apogee.

---

34 Forrester, *supra*, note 31 at 36.

moon around the earth. At new moon and full moon, the gravitational forces of the sun and moon are in alignment and act together to cause spring tides that are higher than average. (See Figure 5.3). When the moon is in the first and third quarter phases, the forces are at right angles and act in opposition. This causes neap tides with ranges smaller than average. The transition between spring and neap tides during the month can be seen in Figure 5.2.

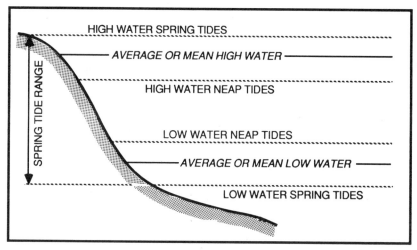

FIGURE 5.3: Relationship of Spring and Neap Tide Levels

§5.26 The tide-raising forces are greatest when the moon and the sun are closest to the earth. When the moon is at perigee (its closest position to the earth in its monthly elliptical orbit) tides with large ranges are produced. Even larger ranges occur if perigee coincides with the new or full moon. These perigean spring tides happen approximately every six to seven months and if they should also happen to coincide with onshore winds, extensive flooding can occur along the coast.[35]

§5.27 A long period of time is required to account for all possible combinations of the relative positions of the sun, moon and earth and thus all the variations in the tides. For tidal analysis and prediction, the significant changes in tidal character occur within a period of 18.61 years, known as the nodal period. The period of tidal observations used for calculating predictions and tidal datum elevations is usually rounded to a full 19 years to include complete cycles of annual sea level variations

---

35 F.J. Wood, *The Strategic Role of Perigean Spring Tides in Nautical History and North American Coastal Flooding, 1635-1976* (Washington: U.S. Government Printing Office, 1976).

178   Survey Law in Canada

caused, for example, by seasonal temperature changes and river outflow.[36]

§5.28 *Tidal reference surfaces* A vertical datum is a reference surface from which heights or depths are measured. Whereas most elevations on land are referenced to geodetic datum (an approximation of mean sea level), all depths on navigational charts and the elevations of charted rocks, wrecks and other features covered by water at high tide are referenced to chart datum. The predicted tidal heights in the *Canadian Tide and Current Tables* are also referenced to chart datum. Canadian chart datum is the Lower Low Water, Large Tides (LLWLT), defined as "the average of the lowest low waters, one from each of 19 years of predictions."[37]

§5.29 Chart datum and selected examples of other tidal reference surfaces are shown in Figure 5.4 and those defined in the *Canadian Tidal Manual*[38] are indicated. These surfaces correspond to an average height of water level at a specified stage of tide. Those defined in the *Manual* are for navigation and charting and they represent extreme rather than ordinary water levels. Other tidal surfaces are included in Figure 5.4 for reference in the discussion to follow. It should be noted that the elevation of all of these tidal reference surfaces relative to a fixed geodetic datum will vary over time as well as with location (see §§ 5.35 to 5.37).

§5.30 Although low water levels are important in surveys of some private property boundaries and in delimiting baselines for offshore jurisdictional limits, cadastral surveyors are usually concerned with high water. Corresponding to chart datum is Higher High Water, Large Tides (HHWLT) defined as the "average of the highest high waters, one from each of 19 years of predictions."[39] Theoretically this water level would only be reached about once each year. All elevations of lighthouses, other structures shown on navigational charts, and clearances under bridges and power lines are given with respect to HHWLT.

---

36 These and other influences on sea level are discussed, for example, in E. Lisitzin, *Sea-Level Changes* (New York: Elsevier Scientific Publishing Co., 1974); S.H. de Jong and M.F.W. Siebenhuener, "Seasonal and Secular Variations of Sea Level on the Pacific Coast of Canada" (1972), 26(1) *The Canadian Surveyor* 4. See also Thompson, *supra*, note 32.

37 See Forrester, *supra*, note 31 at 69; Forrester notes that in present usage, Lowest Normal Tide (LNT) is equivalent to LLWLT, but may refer to a variety of datums on older charts.

38 *Supra*, note 31.

39 *Supra*, note 31 at 69.

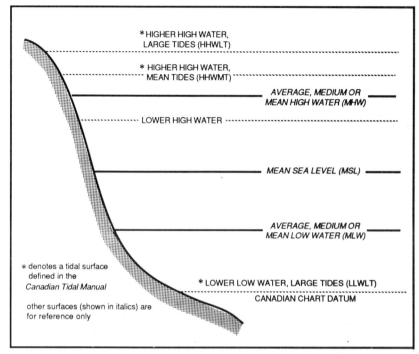

FIGURE 5.4: Selected Examples of Tidal Reference Surfaces

§5.31 A second high water datum, Higher High Water, Mean Tides (HHWMT), is defined in the *Canadian Tidal Manual* as the "average of all the higher high waters from 19 years of predictions."[40] This level is sometimes confused with average or Mean High Water (MHW). But MHW represents the average of all the high water heights and should thus be the average of both the Higher and Lower High Tides (that is, HHW and LHW) occurring each day. Only the Highest Water Level (HHW) is included in defining HHWMT elevations given in the *Canadian Tide and Current Tables*.

§5.32 Legislation[41] and property descriptions[42] sometimes refer to MHW but neither Canadian law nor the Canadian Hydrographic Service, the primary source of tidal information in Canada, provide a definition for this reference surface. In the United States, the use of tidal levels in cadastral

---

40 *Ibid.*
41 See, for example, *Beaches Preservation and Protection Act*, S.N.S. 1975, c. 6, s. 3(a).
42 See, for example, the description of the Prince Edward Island National Park in *Shaw v. R.*, [1980] 2 F.C. 608 (T.D.)

surveys has received legal recognition[43] and MHW has been defined precisely in law as a datum, being:

> The average of all the high water heights observed over the National Tidal Datum Epoch [a specific 19 year period]. For stations with shorter series, simultaneous observational comparisons are made with a control station in order to derive the equivalent of a 19-year datum.[44]

Similar definitions exist for Mean Low Water (MLW) and other tidal levels used in boundary delimitation.

§5.33 The first step in establishing elevations for these reference surfaces is to obtain a record of tidal observations at a coastal site. Through analysis of the observations, constants or constituents that describe the tidal character are determined. Once these constituents are known, they can be used to predict the tidal character over time. Tidal records greater than 19 years are usually available only at permanent tide gauge stations, which in most cases correspond to the Reference Ports (e.g., Charlottetown, Port Aux Basques, Vancouver) listed in the *Canadian Tide and Current Tables*. When only short records are available, the observations are compared with records at nearby Reference Ports that have similar tidal characteristics to correct the analysis for long term constituents. The *Canadian Tide and Current Tables* give datum elevations and predicted daily high and low water levels for Reference Ports, together with corrections to be applied for Secondary Ports. The latter typically have tidal records of one month or less.

§5.34 The tide tables can be used to establish an approximate MHW elevation at Reference Ports by averaging all of the predicted high waters for the year. An approximate correction could also be calculated for Secondary Ports. The elevation obtained, however, will differ from year to year because only a year, rather than a 19 year, record is being considered. It should also be noted that the tidal datums and daily water levels in the tide tables are referenced to chart datum and that appropriate conversions to geodetic elevations for cadastral surveys is usually necessary. The Canadian Hydrographic Service is becoming increasingly aware of the need for tidal information, including MHW elevations, in coastal surveying and mapping and is always interested in providing assistance on request.[45]

---

43 See, for example, F.E. Maloney and R.C. Ausness, "The Use and Significance of the Mean High Water Line in Coastal Boundary Mapping" (1974), 53 *North Carolina Law Review* 185.

44 F.J. Balint, "Notice of Changes in Tidal Datums Established through the National Tidal Datum Convention of 1980" (1980), 45 (207) *Federal Register* at 70296.

45 See, for example, S.T. Grant and C.T. O'Reilly, "A New Look at Tidal Datum Transfers" (1986), 40(4) *The Canadian Surveyor* 477, also published in *Papers of the XVIII International Congress of Surveyors (FIG)*, vol. 4 (Ottawa: Canadian Institute of Surveying and Mapping, 1986) at 259.

§5.35 *Spatial and temporal variations* Despite the impression created by the term water level, tidal reference surfaces are not level or flat planes. They are undulating surfaces with variations in elevation from point to point along the coast. Mean sea level (MSL), which is approximated by the National Geodetic Vertical Datum, is relatively flat. But chart datum and other reference surfaces refer to high or low water levels and these levels vary in elevation as the tidal range changes along the coast. Oceanographic and geographic influences can greatly modify the tidal characteristics and tidal range in a body of water. Such is the case in the Bay of Fundy, for example, where the tidal range changes from approximately four metres to nearly 16 metres from the mouth to the head of the bay. River discharge, predominant winds and currents, marsh vegetation and the shape of the coastline can also influence tidal elevations even over relatively short distances, as for example on two sides of a small island or in embayments and estuaries.[46]

§5.36 Consequently, even interpolation between Reference Ports or Secondary Ports with known reference surface heights only yields an approximation of the elevation at intermediate survey sites. An accurate elevation at a particular locality can only be determined by observing the local tidal levels and comparing these observations with information from control stations to correct for seasonal, annual and long term variations. The accuracies that may be achieved when local observations are made depend on the length of observations, on the method used for collecting and analyzing tidal data and on the quality of tidal datum information available for control stations.[47]

§5.37 Long term trends in regional and global sea level, as well as land subsidence and uplift, can also cause water level elevations to change over time with respect to geodetic datum.[48] In Canada, chart datum is updated occasionally to reflect these trends or to take new observations into account. Chart datum in Halifax, for example, has recently been raised by

---

46 See, for example, R.L. Swanson, *Variability of Tidal Datums and Accuracy in Determining Datums from Short Series of Observations*, NOAA Technical Report NOS 64 (Washington, D.C.: National Ocean Survey, National Oceanic and Atmospheric Administration, U.S. Department of Commerce, 1974); the variations observed at Brackley Beach, Prince Edward Island, considered in *Shaw v. R.*, *supra*, note 42, are discussed in S.E. Nichols,*Tidal Boundary Delimitation*, Technical Report 103 (Frediriction, N.B.: Department of Surveying Engineering, University of New Brunswick, 1983).
47 Methods for determining local datums are discussed in Grant, *supra*, note 45; also see, for example, J.P. Weidener, "Surveying the Tidal Boundary" (1979), 39(4) *Surveying and Mapping* 338; and Swanson, *supra*, note 46.
48 See, for example, Lisitzin, *supra*, note 36; de Jong and Siebenhuener, *supra*, note 36; and Thompson, *supra*, note 32.

approximately 0.29 metres.[49] When using water level and datum information, the surveyor must therefore take into account the date of the tidal record from which the elevation was established and any changes that have occurred since that time. This is especially critical when relating, for example, a high water elevation shown on an older plan or chart to present water levels such as those given in the tide tables.

Ordinary Tides

§5.38 In making reference to the tides and in defining tidal boundaries, Canadian law does not generally use the scientific terminology and concepts that have been presented here. The language of the law can be traced to *De Jure Maris*,[50] where Sir Mathew Hale distinguished three types of tides in defining the lands belonging to the sovereign:

> a. The high spring tides, which are the fluxes of the sea at those tides that happen at the two equinoxials; and certainly this doth not *de jure communi* [by the common law] belong to the Crown. For such spring tides many times overflow ancient meadows and salt marshes, which yet unquestionably belong to the subject...
>
> b. The spring tides, which happen twice every month at full and change of the moon; and the shore in question is by some opinion not dominated by these tides neither, but the land overflowed with these fluxes ordinarily belongs to the subject *prima facie*, unless the King hath a prescription to the contrary...
>
> c. Ordinary tides, or neap tides, which happen between the full and change of the moon; and . . . this kind of shore, viz. that which is covered by the ordinary flux of the sea, is the business of our present enquiry.[51]

§5.39 The first category of tides Hale mentions can be related to the perigean spring tides discussed in paragraph 5.26 and the second refers to the spring tides at full and new moon (§5.25). It is the last category, the "ordinary tides, or neap tides," that has led to some ambiguity because neap tides are the lowest tides occurring near the first and third quarter lunar phases. Neap tides are no more ordinary than the spring tides that Hale rejects. Furthermore, the term ordinary has no scientific meaning that would unambiguously define a tidal datum or surface for boundary delimitation. As stated in the landmark British decision *A.-G. v. Chambers*:

---

49 S.T. Grant, "The C.H.S. Permanent Tide Gauge Network"; presentation to the 80th Annual General Meeting of the Canadian Institute of Surveying and Mapping, Charlottetown, P.E.I., June 1986.
50 *Supra*, note 9.
51 *Ibid.*, at 393.

All the authorities concur in the conclusion that the right [of the sovereign] is confined to what is covered by "ordinary" tides, whatever be the right interpretation of that word.[52]

§5.40 In Canada, extraordinary or extreme tide levels have been excluded from the meaning of ordinary tides.[53] Canadian courts have followed the judgment in *A.-G. v. Chambers* in which ordinary tides were defined as

> [The medium tide between spring and neaps] . . . It is true of the limit of the shore reached by these tides that it is more frequently reached and covered by the tide than left uncovered by it. For about three days it is exceeded, and for about three days it is left short, and on one day it is reached.[54]

In 1935 the British Ordinance Survey replaced the term ordinary tides with medium tides on maps which delineate tidal boundaries and since 1965 the term Mean High Water has been used on Ordnance Survey maps for England (Mean High Water Springs for Scotland).[55] In Canada, the phrase "ordinary or neap tides" is still found in legal writings and in property descriptions in reference to the tides which define natural boundaries along the coast. This phrase is, however, generally equated with medium, average or mean tides.[56]

COASTAL LAND LAW

§5.41 The law of the coast consists of a set of special relationships among people and among governments with respect to the land and the water. Throughout the development of coastal law, these relationships have accommodated various economic and socio-political interests. At certain times, the law has favoured private interests, as during the major settlement period in Canada, but more recently the trend has been towards balancing the private development of coastal resources with environmental protection through government regulation.[57]

§5.42 To illustrate some of the limits of property rights and jurisdictions, Figure 5.5 presents the geography of the coast from a legal perspective.

---

52 (1854), 43 E.R. 486 at 490.
53 *Lee v. Arthurs* (1918), 46 N.B.R. 185 (S.C.); affirmed (1919), 46 N.B.R. 482 (C.A.); *Re McNichol* (1976), 20 N.B.R. (2d) 240 (Q.B.).
54 *Supra*, note 52 at 489.
55 Ordnance Survey, "High and Low Water Marks and Tidal Levels on Ordnance Survey Maps and Admiralty Charts", Ordnance Survey Information Leaflet No. 70 (Southampton, England: August, 1986).
56 See, for example, G.V.A. La Forest and Associates, *Water Law in Canada: The Atlantic Provinces* (Ottawa: Information Canada, 1973) at 240.
57 Anon., "The Public Trust in Tidal Areas: A Sometimes Submerged Traditional Doctrine" (1970), 79 *Yale Law Journal* 762.

184 Survey Law in Canada

Above the limit of the ordinary high water is the upland property. Ownership of the upland carries with it special rights with respect to the water called riparian rights. The land below low tide is known as the bed; the foreshore or shore is that area which is alternately covered and left dry by the ebb and flow of the ordinary tides. Along this narrow continental fringe the interests of the public, the government and the private landowner converge.

Early Development

§5.43 Under Roman law, which later influenced civil law jurisdictions, the public had unhindered use of coastal waters and of the shore as far inland as the highest wash of the winter waves.[58] In contrast, the early development of coastal rights under the common law can be seen as a struggle between the traditional rights of the riparian owner to the exclusive use of the foreshore, the private interests of the sovereign and the need to protect public navigation for trade and commerce. In feudal England, private rights to the foreshore and to the fisheries were acquired through grants from Saxon and Norman rulers or by local custom. The shore was not recorded as a parcel separate from the upland manor in the *Domesday Book*; it was commonly accepted that the manor extended to midstream in all rivers.[59] Navigation became sufficiently impeded by private fishing weirs that were prohibited in *Magna Carta* except along the open sea coast.[60]

§5.44 During the 16th and 17th centuries, Elizabeth I and subsequent rulers made unsuccessful attempts in the courts to claim sovereign ownership of the foreshore and its profits as a *prima facie* right. Professional title-hunters were sometimes employed to search private titles for flaws that would benefit the Crown.[61] Riparian owners protested this intrusion on customary rights and title to the foreshore became one of the many grievances against the monarchy in the early 1600's.[62]

§5.45 Chief Justice Hale's treatise *De Jure Maris* (c. 1666) is seen today as a turning point in the recognition of sovereign rights to the shore and

---

58 *Ibid.* at 762-763.
59 H.P. Farnham, *The Law of Waters and Water Rights*, v. I (Rochester, NY: E.R. Andrews Printing Co., 1904) at 181.
60 Sir Matthew Hale, *De Jure Maris* (C. 1666), as found in S.A. Moore, *A History of the Foreshore and the Law Relating Thereto*, 3d ed. (London: Stevens and Haynes, 1888) at 388.
61 See discussion in Moore, *supra*, note 60 at 170-171 and 212-251.
62 *Supra*, note 59 at 181.

Water Boundaries — Coastal 185

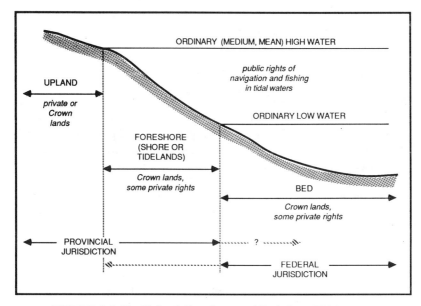

FIGURE 5.5: The Upland, Foreshore, and Bed along Tidal Waters

in the concept of *jus publicum*, or public rights of use, in coastal waters.[63] Hale wrote that the shore, those lands

> [B]etween ordinary high-water and low-water mark ... doth *prima facie* and of common right belong to the King, both in the shore of the sea and the shore of the arms of the sea.[64]

Although this was only legal theory in view of existing land holdings,

> [H]is statement has been taken in subsequent cases to have established the theory, and has been declared by judge after judge, in every case down to the present day, so that it must therefore be taken as now established law.[65]

Unlike earlier attempts to claim the foreshore for the Crown, Hale's doctrine conceded that sovereign rights could be defeated by evidence of a prior Crown grant or customary usage.[66] This concession and changes in the

---

63 For a review of the development of the *jus publicum* see, for example, F.E. Maloney and R.C. Ausness, "The Use and Significance of the Mean High Water Line in Coastal Boundary Mapping" (1974), 53 *North Carolina Law Review* 185 at 188-193. See also Note 57 above.
64 *Supra*, note 60 at 378.
65 *Supra*, note 59 at 186.
66 *Supra*, note 60 at 384.

economic and political environment in England contributed to the gradual acceptance of Hale's doctrine as law.[67]

Private and Public Rights

§5.46 If the English Crown was constrained at home in dominating the shore, it took active measures in some of the early Canadian colonies to protect these lands from private appropriation. Newfoundland represents an extreme case where settlement was discouraged until the 19th century and a wide strip along the coast, called ship's rooms, was reserved for use by the annual British fishing fleets.[68] With certain exceptions, however, Hale's synthesis of coastal law in England became the basis for coastal rights in Canada.

§5.47 *Title to the shore* In Canada, title to the foreshore is by presumption in the Crown and the upland property "extends *prima facie* only to the ordinary high water mark."[69] This presumption may be defeated by words of grant or other evidence to the contrary. Some property descriptions in the Maritime provinces, for instance, call for the limit of ordinary high water spring tides as the seaward boundary of the upland; others call for the low water mark, the line of occupation or the shore. When an upland property description calls for the shore, the parcel does not include the foreshore and the boundary is interpreted to be the limit of the ordinary high tides.[70] Similarly, if the landward boundary of a foreshore parcel is described as the shore, it is by presumption the ordinary high water mark.[71]

§5.48 Private rights to the foreshore, and in some cases the bed of tidal waters, may be acquired by grant, lease or license from the Crown. Known as water lots or foreshore lots, these parcels were frequently granted for the construction of wharves and other structures essential to the fishing and shipping industries. Some water lot grants are bounded seaward by the ordinary low water mark, but some may extend beyond this limit to

---

67 *Supra*, note 59 at 186. See also *supra*, note 57.
68 A.C. McEwen, "Land Titles in Newfoundland" (1977), 31(2) *The Canadian Surveyor* 151 at 153; this is a brief summary of A.C. McEwen, *Newfoundland Law of Real Property: The Origin and Development of Land Ownership*, unpublished Ph.D. dissertation, Faculty of Laws, University of London, 1978.
69 *Supra*, note 56 at 239.
70 *Irving Refining Ltd. v. Eastern Trust Co.* (1967), 51 A.P.R. 155 at 160; the terms shoreline and coastline are contrasted in *Chaulk v. South River (Town of)* (1971), 2 Nfld. & P.E.I.R. 1 (Nfld. S.C.).
71 *Esson v. Mayberry* (1841), 1 N.S.R. 186 (C.A.). The presumption can be defeated by words of grant as in *Doe d. Fry v. Hill* (1853), 7 N.B.R. 587 at 589 (C.A.) and *Delap v. Hayden* (1924), 57 N.S.R. 346, reversing, [1923] 4 D.L.R. 1102 (C.A.).

permit construction of wharves long enough to serve boats at low water. Legislation, such as the Nova Scotia *Beaches and Foreshores Act*,[72] also empowers government agencies to issue grants or leases for oyster cultivation or the construction of fishing weirs. Although water lots generally belong to adjacent upland owners, they may be granted to others, usually with the upland owner's consent.[73]

§5.49 Under certain conditions and unless there is legislation to the contrary,[74] private rights to the foreshore can also be acquired through possession for the statutory limitation period. Since the nature of the occupation of tidelands is generally different than that of the upland, not all of the criteria for adverse possession or colour of title may have to be met. Seasonal occupation, such as drying fish or operation of a sawmill, has been considered sufficiently continuous to establish possessory rights.[75] But structures built on the foreshore or bed that constitute a nuisance to public navigation cannot give rise to a possessory claim.[76] New legislation and federal-provincial jurisdiction in harbours have also cast doubt on the status of some water lots, including those previously granted by the Crown (see §§5.58 and 5.61).

§5.50 *Riparian rights* Riparian rights are special property rights of natural advantage that are attached to land abutting tidal or nontidal watercourses.[77] Derived from the Latin word *ripa*, or bank, the term riparian should strictly refer only to lands bordering rivers. Although the term littoral is reserved for lands bounding oceans, riparian is used in a general sense to cover all situations.

§5.51 La Forest has classified riparian rights as access to the water, drainage, flow, quality, use of the water and accretion.[78] Of these, access and accretion are the main concern in tidal waters. The right of access is fundamental because it is through access to the water that the riparian owner is able to enjoy the other rights. Along the coast, the right of access includes the right to cross the foreshore and any obstructions that bar access constitute an interference by law.[79]

---

72 R.S.N.S. 1967, c. 19.
73 See, for example, *ibid.*, s. 3.
74 Land titles legislation, for example, may limit possessory rights. See V. Di Castri, "Water and Highway Boundaries as Affecting Title", chapter 7 in *Registration of Title to Land*, v. 1 (Toronto: Carswell, 1987) at 7-202.
75 *Dawe v. Avalon Coal & Salt Ltd.* (1950), 26 M.P.R. 112 (Nfld. S.C.); *A.G. of Can. v. Acadia Forest Products Ltd.* (1985), 37 R.P.R. 184 (Fed. T.D.).
76 *Irving Refining Ltd.*, *supra*, note 70 at 162.
77 *Supra*, note 59 at 280.
78 La Forest, *supra*, note 56 at 201.
79 *R. v. Lord* (1864), 1 P.E.I. 245.

§5.52 Accretion is defined as the gradual and imperceptible increase to land bordering water through deposit of alluvium on the banks or shore or through withdrawal of the water. By the doctrine of accretion, the new lands so formed become part of the upland parcel if the parcel is truly riparian, that is, if the parcel is bounded by the watercourse and not a fixed line.[80] Erosion is the gradual and imperceptible loss of land due to the natural action of the water and title to the submerged lands belongs to the owner of the bed.[81]

§5.53 For the upland owner to gain title to accretion (or conversely to lose land due to erosion), the critical test is whether the change is gradual and imperceptible. This is a qualitative rather than quantitative criterion; it varies with individual circumstances.[82] In cases where sudden and perceptible changes to the shoreline occur, as for example due to storms or floods, the doctrine does not apply and no land is gained or lost.[83] When changes in the shoreline are the result of man's activities, title to accretion may be gained by upland owners, provided there was no intention to so benefit.[84] Dumping of miscellaneous material over an embankment in Saint John Harbour was held to be consistent with the riparian owner's right to protect the property against erosion by the forces of nature and thus title to additions to the shoreline belonged to the upland.[85] However, in *Mahon v. McCully*,[86] when a breakwater was constructed for the purpose of reclaiming land, title to the lands so formed over ten years did not pass to the riparian owner.

§5.54 A distinction has been made between vertical and lateral accretion in determining title. Vertical accretion is the accumulation of sediments on the bed that eventually rise above ordinary high water. Lateral accretion results from deposits to the upland itself. While lateral accretion belongs to the upland property, title to vertical accretion was held to remain with the Crown as owner of the bed in *A.-G. for B.C. v. Neilson*.[87] Although this distinction may be clear when islands form from the bed, in many

---

80 *Clarke v. Edmonton (City of)*, [1930] S.C.R. 137, reversing [1928] 1 W.W.R. 553; and *A.-G. of B.C. v. Neilson*, [1956] S.C.R. 819. See also, *supra*, note 74 at 7-198 to 7-214, specifically regarding the rationale for the doctrine at 7-199 to 7-200 and regarding fixed lines at 7-200.
81 *A.-G. of B.C. v. Neilson, ibid.* For further classification of accretion, erosion and avulsion, see Maloney and Ausness, *supra*, note 63 at 225.
82 *Supra*, note 56 at 226 and *Clarke v. City of Edmonton, supra*, note 80.
83 *Supra*, note 74 at 7-200 and 7-205.
84 *Supra*, note 56 at 226.
85 *Irving Refining Ltd., supra*, note 70 at 170.
86 (1868), 7 N.S.R. 323 (C.A.).
87 *Supra*, note 81.

situations determining how the accretion formed or what parts are lateral accretion may be difficult, if not impossible.[88]

§5.55 Public rights Two primary public rights in tidal waters are recognized in the common law: the right of navigation and the right of fishing. The right of fishing includes the right to harvest clams and other shellfish on the foreshore, whether this is Crown or private land.[89] The right of navigation is, however,

> [A] paramount right; whenever it conflicts with the rights of the owner of the bed or of a riparian owner it will prevail.[90]

Related to the right of navigation is the public right of floating logs and other property, a right that may apply to nonnavigable as well as navigable waters by custom or legislation. Although not recognized in English common law, this right has been essential to the forest industry in Canada and can include the right to dam waters to facilitate the transport of logs.[91]

§5.56 Public access to the shore has not received much attention in Canada but it has been protected to some degree in parts of the country by reservation of a strip of Crown land from upland grants. Similar to ship's rooms in Newfoundland, lands called fishing rooms were traditionally reserved along the open coast in Prince Edward Island to provide public areas for fishing activities.[92] Reserves, 100 feet (30.5 m) in width, are still made today from grants bordering the sea or navigable waters in the Northwest Territories and the Yukon.[93] Public access roads to waterfronts are also provided for under the British Columbia *Land Act*.[94]

§5.57 When a grant is subject to a reserve, the Crown remains the riparian owner and benefits from accretion. *Monashee Enterprises Ltd. v. Min. of Recreation and Conservation (B.C.)*[95] for example, concerned a Crown reserve along tidal waters, one chain in width, measured from the high water mark. At issue was the location of the landward boundary of the reserve after accretion had occurred along the shore. Reversing an arbitration award, the trial decision held that the reserve was ambulatory

---

88 *Shaw v. R.* [1980] 2 F.C. 608 (T.D.).
89 *Donnelly v. Vroom* (1909), 42 N.S.R. 327, affirming (1907), 40 N.S.R. 585 (C.A.).
90 La Forest, *supra*, note 56 at 185.
91 A detailed discussion of the right of floatage, with examples from provincial legislation is given in La Forest, *supra*, note 56 at 191-195, 296-297 and 310-312.
92 La Forest, *supra*, note 56 at 463; also A.C. McEwen, "Land Titles in Newfoundland" (1977), 31(2) *The Canadian Surveyor* 151 at 153.
93 *Territorial Lands Act* R.S.C. 1985, c. T-7, s. 9.
94 B.C. Reg. 556/80 *Surveyor General's Instruction Regulation*, s. 57(2), (3); pursuant to *Land Act* R.S.B.C. 1979, c. 214.
95 (1981), 28 B.C.L.R. 260, reversing (1978), 7 B.C.L.R. 388 (C.A.).

with constant width. Thus the private property adjacent to the reserve would have benefited from the accretion as an indirect result of allowing the one-chain reserve to shift seaward. On appeal the decision was reversed again and title to the accretion was awarded to the Crown. In this judgment the court reasoned that:

> It is well settled that land gained by accretion accrues to the benefit of the riparian owner . . . . It is equally well settled that to be a riparian owner, and thus to benefit from accretion, one's property must run to the shoreline . . . . In this case, the riparian owner is the Crown as the owner of the one-chain strip. The land gained by accretion is added to and becomes part of the strip.[96]

The reserve therefore varied in width and the landward boundary of the reserve between private and Crown lands was held fixed.

§5.58 *Effect of legislation* The purpose of this section is to highlight the fact that legislation in each jurisdiction must be examined to determine whether common law rules concerning public and private rights along watercourses are affected. Legislation regarding the granting of land (both current and historical) and the registration of title to land (as distinguished from the registration of deeds) are major considerations. Di Castri[97] provides a comprehensive summary of the effect of land titles legislation on coastal and inland waters, particularly with regard to the application of the *ad medium filum aquae* rule and the doctrine of accretion. In British Columbia, for example, riparian owners can benefit from accretion if the criteria of the doctrine are met[98] (see §5.85).

§5.59 An example of other legislation that affects private rights related to water is the Nova Scotia *Water Act*,[99] although the extent to which it applies to coastal waters has not been clarified. In the Act, a watercourse is defined as

> [T]he bed and shore of every river, stream, lake, creek, pond, spring, lagoon, swamp, marsh, wetland, ravine, gulch or other natural body of water, and the water within, including ground water, within the jurisdiction of the Province, whether it contains water or not.[100]

The purpose of this Act is to vest title in all watercourses in the Crown. The Act is retroactive to May 16, 1919, at which date all existing private rights were extinguished. If coastal waters are included, and it would appear

---

96 *Ibid.* at 263.
97 Di Castri, *supra*, note 74.
98 *Land Title Act*, R.S.B.C. 1979, c. 219, s. 95.
99 See also W.S. Armstrong "The British Columbia Water Act: The End of Riparian Rights" (1965), 1(5) *U.B.C. Law Review* 581.
100 R.S.N.S. 1967, c. 335, s. 1(*k*).

that some are, then the status of water lots and perhaps other rights will eventually have to be interpreted under the Act.

Federal and Provincial Jurisdiction

§5.60 A detailed discussion of the role of the Canadian Constitution in water law can be found in *Water Law in Canada: The Atlantic Provinces*.[101] In summary, under the *Constitution Act, 1867*[102] the provinces retained all property in lands, mines and minerals that belonged to those provinces at the time of union, excluding specific public works and properties granted to the Government of Canada. Among the latter were public harbours. The Government of Canada was also given exclusive legislative authority over matters of national concern, such as navigation and shipping. Since 1867 several ambiguities have called for judicial interpretation. Of special concern in coastal boundary delimitation are public harbours and the seaward limit of provincial lands.

§5.61 *Public harbours* The issues surrounding public harbours are many and complex. Only two matters are touched on here: determining whether or not a harbour is public and therefore under the jurisdiction of the federal government, and defining the jurisdictional boundaries.[103] To be under federal jurisdiction, a harbour must have been a public harbour at the time of provincial union with Canada. No one test can be applied in every case. The harbour must have been used by the public as a public harbour (for example, to anchor or dock boats), have had public money expended in harbour improvements or have been officially declared a public harbour.[104] British Columbia has had an agreement with the federal government since 1924 specifying which harbours are public, as well as the nature of the federal property.[105] In the Atlantic Provinces decisions have been made for individual harbours, but those harbours without clear status continue to present difficulties, especially in determining whether the federal or provincial Crown has the jurisdiction to grant water lots.[106]

---

101 G.V.A. La Forest and Associates, *Water Law in Canada: The Atlantic Provinces* (Ottawa: Information Canada, 1973).
102 (U.K.), 30 & 31 Vict., c. 3 (formerly *British North America Act, 1867*; now *Constitution Act, 1982*, being Schedule B of the *Canada Act 1982* (U.K.), 1982, c. 11).
103 A review of the issues can be found in G.V. La Forest, "The Meaning of 'Public Harbours' in the Third Schedule to the British North America Act, 1867" (1963), 41 *Canadian Bar Review* 519.
104 *Supra*, note 101 at 24.
105 *Supra*, note 103 at 535.
106 See, for example, C.P. Masland, "Water Lots in Nova Scotia: Their Validity and Their Usage" (1976), 31 (83) *The Nova Scotian Surveyor* 20.

192   Survey Law in Canada

§5.62 Despite some question as to the exact nature of the rights held by the federal government in harbours, there is agreement that this includes title to the bed and to at least some parts of the foreshore to the ordinary high water mark.[107] In *Re Provincial Fisheries*,[108] the Privy Council gave the opinion that only those specific areas of the foreshore that were used as a public harbour before Confederation were transferred to the federal government. As stated in a subsequent British Columbia case, this method of delimiting boundaries would result in a "patchwork of ownership both inconvenient and embarrassing."[109] The seaward extent of the federal jurisdiction in harbours also has to be established on an individual basis.[110] Determining exactly which parts of the harbour are under federal jurisdiction can therefore be a major problem unless the boundaries have been clarified by expropriation or by transfer to the federal government. Harbour improvements, for example, are often based on agreements and transfers between the provinces and the federal government to avoid title and boundary issues.

§5.63 *The seabed* Under section 109 of the *Constitution Act, 1867*, the provinces retained all lands within their territories, but the seaward limits of these territories are for the most part undefined. This fact was of minor concern until the discovery of offshore petroleum resources and the extension of national jurisdiction under international law. With vast potential revenues at stake, several attempts have been made in recent years to define the limits of provincial jurisdiction. Interprovincial boundaries defined as centrelines or median lines in the Bay of Fundy, the Northumberland Strait and Baie de Chaleur date to the creation of the provinces. It has been argued that these and other acts demonstrating provincial sovereignty offshore will possibly strengthen the positions of the Maritime Provinces should the limits of provincial jurisdiction be decided in the courts.[111] Two cases that have been given judicial consideration involve Newfoundland and British Columbia.

---

107 *Holman v. Green* (1881), 6 S.C.R. 707; for a discussion of the boundary issues see, *supra*, note 103.
108 *A.-G. of Can. v. A.-G. of Ont.; Re Provincial Fisheries*, [1898] A.C. 700 (P.C.).
109 *A.-G. of Can. v. Higbie*, [1945] S.C.R. 385 at 431.
110 *Supra*, note 103 at 532-534.
111 Nova Scotia, for example, has passed legislation that defines 'Nova Scotia Lands' as including the continental shelf, as well as the slope and seabed of the continental shelf to the limit of exploitability, in *Pipeline Act*, S.N.S. 1980, c. 13, s. 2(2). Discussion of the jurisdictional issues, including the distinction between territorial and legislative jurisdiction can be found, for example, in R.J. Harrison, "Jurisdiction Over the Canadian

§5.64 In 1984, the Supreme Court of Canada found that the Province of Newfoundland has no rights to explore or exploit the seabed resources on the continental shelf adjacent to the province. Furthermore, the province has no legislative jurisdiction regarding this area. Only a portion of the continental shelf was considered (that containing the Hibernia oil-fields) but the Court noted that it is unlikely that the legal issues would differ in any other part of the shelf.[112] The Newfoundland Court of Appeal, which addressed both the continental shelf and other issues just prior to the federal decision, found that the province had proprietary rights to a territorial sea extending three nautical miles from the coast,[113] a customary zone recognized in international law when Newfoundland joined confederation in 1949. This issue has not, however, been considered by the Supreme Court of Canada for Newfoundland or the other Atlantic Provinces.

§5.65 In a 1967 decision on the Pacific coast, the Supreme Court of Canada held that British Columbia had no jurisdiction within the territorial sea.[114] The Court had been asked to give an opinion on whether property, resource rights and legislative authority in the territorial sea and continental shelf adjacent to the British Columbia coast belonged to the Province or to Canada. In all matters, the Court found that these rights and powers belonged to Canada. The seaward limit of provincial territory was defined as the ordinary low water mark of the mainland and islands, but bays, harbours and estuaries were considered inland and therefore provincial waters. In a subsequent decision in 1984, the Court held that the Strait of Georgia and other related waters were not inland waters and therefore not included in the territory of British Columbia.[115]

TIDAL BOUNDARY DELIMITATION

§5.66 While the law defines the boundaries that delimit properties and

---

Offshore: A Sea of Confusion" (1979), 17(3) *Osgoode Hall Law Journal* 469. See also J. McEvoy, "Atlantic Canada: The Constitutional Regime" (1984), *Dalhousie Law Journal* 285; G.V. La Forest, "Canadian Inland Waters of the Atlantic Provinces and the Bay of Fundy Incident" (1963), 1 *The Canadian Yearbook of International Law* 149; E.C. Foley, "Nova Scotia's Case for Coastal and Offshore Resources" (1981), 13(2) *Ottawa Law Review* 281.

112 *Re the Seabed and Subsoil of the Continental Shelf Offshore Nfld.*, [1984] 1 S.C.R. 86; for a discussion see L.L. Herman, "The Newfoundland Offshore Mineral References: An Imperfect Mingling of International and Municipal Law" (1984), 22 *The Canadian Yearbook of International Law* 194.

113 *Re Mineral and Other Natural Resources of the Continental Shelf* (1983), 145 D.L.R. (3d) 9 (Nfld. C.A.).

114 *Re Offshore Mineral Rights* (B.C.), [1967] S.C.R. 792.

115 *A.G. Can. v. A.G. B.C.*, [1984] 1 S.C.R. 388.

jurisdictions, it is the surveyor who must find evidence of the boundary location. How the evidence is evaluated depends, to a large degree, upon the experience of the surveyor and his knowledge of local tidal conditions and accepted survey practice.[116] Although the high water tidal boundary is the focus of the following review of boundary definitions and survey practice, issues related to low water tidal boundaries are noted where appropriate. Consideration is also given to some of the problems presented by the ambulatory nature of the tidal boundary.

Boundary Definitions

§5.67 Most cadastral surveys in coastal areas entail delimitation of the boundary between private owership of the upland and Crown ownership of the foreshore. This boundary is *prima facie* the Ordinary High Water Mark (OHWM). Canadian surveyors have traditionally interpreted the OHWM in terms of physical evidence of the boundary location.[117] A more mathematical interpretation based on the establishment of the intersection of MHW with the shore, a boundary sometimes referred to as the Mean High Water Line (MHWL), has been adopted in some American[118] and Commonwealth[119] jurisdictions. Since survey methods similar to those employed in these jurisdictions have been implicitly recognized in Canada,[120] both the OHWM and the MHWL definitions are presented here.

§5.68 *Ordinary high water mark and related definitions* As discussed

---

116 See, for example, articles discussing boundary definitions and evidence in J.F. Doig, "Mean High Water — 'Nova Scotian Style'" (1979), 38 (96) *The Nova Scotian Surveyor* 3; D.K. MacDonald, "Comments — Re: J.F. Doig's Paper Entitled Mean High Water — 'Nova Scotian Style'" (1979), 38(96) *The Nova Scotian Surveyor* 8; J.F. Doig, "Mean High Water — Revisited" (1979), 39(98) *The Nova Scotian Surveyor* 14; See also S.E. Nichols, *Tidal Boundary Delimitation*, Technical Report 103 (Fredericton, N.B.: Dept. of Surveying Engineering, University of New Brunswick, 1983) at 157-171.
117 *Ibid.*; see also *Nelson v. Pac. Great Eastern Ry. Co.*, [1918] 1 W.W.R. 597 (B.C. S.C.).
118 See, for example, F.E. Maloney and R.C. Ausness, "The Use and Significance of the Mean High Water Line in Coastal Boundary Mapping" (1974), 53 *North Carolina Law Review* 185. A more up-to-date review of boundary definitions and related issues for specific American coastal states is given in a series of articles by P.H.F. Graber, "The Law of the Coast in a Clamshell" (1980-1987), 48(4) to 58(2) *Shore and Beach*.
119 For example, A.J. Baldwin, "Seaward Cadastral Boundaries" (1982), *New Zealand Surveyor*, Feb., 141; Ordnance Survey, "High and Low Water Marks and Tidal Levels on Ordnance Survey Maps and Admiralty Charts", Ordnance Survey Information Leaflet No. 70 (Southampton, England: August, 1986).
120 *Irving Refining Ltd. v. Eastern Trust Co.* (1967), 51 A.P.R. 155 (N.B.S.C.). In *Nelson v. Pac. Great Eastern Ry. Co.*, *supra*, note 117, the court recognized that tidal data should be used to determine the limit of the medium tides, but also recognized that appropriate data was not available for the lands in question (See § 5.80).

in paragraph 5.40 the term ordinary high water has not been precisely defined in Canada, but it can be equated with the average, medium or mean high tides. Survey regulations have been more concerned with specifying the evidence that can be accepted as the limit of the ordinary high tides than with defining these tides. The relation of the boundary to a visible and distinct mark left on the shore by the ordinary high tides is emphasized. In the regulations under the *Nova Scotia Land Surveyors Act*,[121] for example, the OHWM is defined as:

> [T]he limit or edge of a body of water where the land has been covered by water so long as to wrest it from vegetation, or as to mark a distinct character upon the vegetation where it extends into the water or upon the soil itself.[122]

This was derived from a similar definition given in the *Manual of Instructions for the Survey of Canada Lands*[123] under the *Canada Lands Surveys Act*.[124]

§5.69 Related definitions of the high water boundary can be found in provincial legislation such as the British Columbia *Land Act* in which a natural boundary is defined as:

> [T]he visible high water mark of a lake, river, stream or other body of water where the presence and action of the water are so common and usual, and so long continued in all ordinary years, as to mark on the soil of the bed of the body of water a character distinct from that of its banks, in vegetation, as well as in the nature of the soil itself.[125]

From the terminology used in such definitions, particularly the term bank with no mention of the shore, it would appear that they have been drafted with nontidal waters in mind. But these OHWM definitions are applied in both tidal and nontidal waters. The New Brunswick *Crown Lands and Forests Act*[126] defines the normal high water mark in a similar manner, but the definition is specifically limited to lakes and rivers.

§5.70 In the British Columbia *Land Act* provisions are also made for delimiting a conventional boundary which is there defined as:

> [A] boundary consisting of a straight line or a series of straight lines of fixed

---

121 S.N.S. 1977, c. 13.
122 N.S. Reg. 42/79, 1979, Part II, s. 11(g).
123 Energy, Mines and Resources Canada, Surveys and Mapping Branch, Legal Surveys Division, *Manual of Instructions for the Survey of Canada Lands*, 2d ed. (Ottawa: Minister of Supply and Services, 1979) at 50.
124 R.S.C. 1985, c. L-6.
125 R.S.B.C. 1979, c. 214. s. 1.
126 S.N.B. 1980, c. C-38.1, s. 1.

direction and length conforming as nearly as possible to the natural boundary, but eliminating minor sinuosities.[127]

When such a conventional boundary is established during a survey of Crown lands, the land "carries with it the rights and incidents as if it was bounded by the natural boundary."[128] This presumably includes the right to accretion but does not include any title to the bed or shore in either tidal or nontidal areas unless there is express provision.[129]

§5.71 *Mean high water line and related definitions* Although the OHWM is the most common definition of the high water boundary, reference is sometimes made to the mean high water mark or the line of mean high water. The OHWM definitions given above may be consistent with a mean high water mark because a visible feature is implied, but there are situations in which an interpretation similar to that given by the United States Federal Appeal Court in *Borax Consolidated Ltd. v. Los Angeles (City of)*[130] might be more appropriate. In *Borax* several interpretations of the OHWM boundary were discussed, including: a meander line (a surveyed line representing the natural boundary on the original patent); the limit of neap high tides; and the limit of vegetation.[131] On final appeal these interpretations were rejected and it was held that the OHWM "does not mean, as the petitioners contend, a physical mark made upon the ground by the waters."[132] Instead, the OHWM was held to be the intersection of the MHW tidal datum with the shore, that is, the line of MHW or MHWL. A precise definition of the MHW tidal datum, similar to that presented in paragraph 5.32, was recognized by the court.[133]

---

127 *Supra*, note 125 at s. 1. The conventional boundary determined here is completely unrelated to the conventional line discussed in §§9.30 to 9.45.
128 *Supra*, note 125 at s. 64(4) and also s. 76 on boundary agreement.
129 *Ibid.* at s. 52. Water boundaries and related title issues in British Columbia are further discussed by D.A. Duffy, "Accretion and Natural Boundary Adjustment" (1988), unpublished paper to be presented at a B.C.L.S. seminar, Kelowna, B.C.
130 (1935), 296 U.S. 10; for discussions of this case and its effect on American law see, for example, Maloney and Ausness, *supra*, note 118 and C.E. Corker, "Where Does the Beach Begin and to What Extent is This a Federal Question?" (1966), 42 *Washington Law Review* 33.
131 The basis for these interpretations is discussed, for example, by Corker, *ibid.*
132 *Borax*, *supra*, note 130 at 22. For a *contra* opinion see the discussion of *Dolphin Land Associates Ltd. v. Town of Southampton* (1975), 37 N.Y. (2d) 292 as discussed by J.A. Humbach and J.A. Gale, "Tidal Title and the Boundaries of the Bay: The Case of the Submerged "High Water" Mark" (1975), 4 *Fordham Urban Law Journal* 91, in which the authors, at 105, note the court held that using a line of vegetation was the customary method of establishing high water boundaries and protected the expectations of the parties.
133 *Borax*, *supra*, note 130 at 26-27. The court recognized the then U.S. Coast and Geodetic Survey MHW tidal datum definition and held that MHW was the average of all the high tides over 18.6 years or as near as possible to that length of time.

§5.72 It should be emphasized that this decision has not been expressly recognized in Canadian law and no precise definition of MHW exists for Canada. In some situations, however, boundary surveys have been based on the concepts described in the *Borax* decision.[134] *Irving Refining Ltd. v. Eastern Trust Co.*,[135] for example, considered a survey in which a water boundary was established by staking the observed water line on the shore at the precise time a calculated MHW level was predicted to occur. A 1918 decision in British Columbia is also noteworthy because the court gave the opinion that the limit of medium high tides should be established by using the elevation of these tides. The court recognized, however, the difficulties in obtaining appropriate tidal data for the survey and therefore accepted a line established by physical evidence on the shore.[136] These cases are discussed further in paragraphs 5.79 and 5.80.

Boundary Surveys

§5.73 Interpreting the OHWM as a physical mark left on the shore by the tides has certain advantages in boundary surveys. As a visible line, it generally represents the expectations of landowners and conventional survey techniques for natural boundaries can be employed. But surveyors may encounter difficulties where a distinct mark is absent or where the available evidence does not represent the limit of ordinary tides. Marshlands, harbour areas and beaches swept by winter storms can present particular problems in evaluating evidence; inconsistent interpretations among surveyors are not uncommon. Yet as the environmental and economic value of these lands increases, the precise delimitation of private and Crown boundaries will probably assume greater importance.[137]

§5.74 *The water mark as a physical feature* The OHWM is sometimes

---

134 A summary of current practices in the Maritime Provinces is given in S.E. Nichols, *Tidal Boundary Delimitation*, Technical Report 103 (Fredericton, N.B.: Dept. of Surveying Engineering, University of New Brunswick, 1983), pp. 157-171. See also J.F. Doig, "Mean High Water" (1978), 32(2) *The Canadian Surveyor* 227; S.E. Nichols and J.D. McLaughlin, "Tide Mark or Tidal Datum: The Need for An Interdisciplinary Approach to Tidal Boundary Delimitation" (1984), 38(3) *The Canadian Surveyor* 193.
135 *Supra*, note 120.
136 *Supra*, note 117 at 600-601.
137 While the boundary problem is not addressed directly, P. Harrison and J.G.M. Parkes, "Coastal Zone Management in Canada" (1983), 11(1-2) *Coastal Zone Management Journal* 1, predict a rise in land use conflicts, and thus presumably land ownership conflicts, resulting from increased intensity of competing land uses in Canada's coastal regions, both urban and rural; L. Tell, "A Tidal Wave of Claims" (1982), *The National Lawyer*, July 12, pp 1-3, describes the boundary litigation that has occurred in the United States resulting from similar conflicting interests; see also, for example, Graber, *supra*, note 118.

198 *Survey Law in Canada*

construed by surveyors to be the limit of vegetation, but in many cases the edge of vegetation does not correspond to the limit of the ordinary tides. As pointed out in *Turnbull v. Saunders*:

> [High] water mark may go clean beyond the trees along the shore. It might be 100 feet below the grass . . . . That does not affect where high water mark is . . . . To the ordinary man I do not think the question of vegetation in connection with high water mark cuts any figure at all.[138]

In this case a gravel berm formed by tidal action was found to be the boundary following an inspection of the site by the judge and jury.[139] Surveyors should therefore consider a number of factors in evaluating evidence of the OHWM. In a published discussion concerning the OHWM definition in survey regulations for Nova Scotia, one surveyor summarized the diversity of evidence:

> [T]he identification on the ground of the features as defined by these regulations, varies in complexity and exactitude depending on the nature of the geology, geography, vegetation and body of water at the particular site in question. Where you have vertical rock or earth cliffs confining the bed of the body of water, there is no great problem. On the other end of the scale, where you are confronted with the marsh lands of many parts of the coastline or inland waters, it requires a much greater understanding and examination of the subtleties of the vegetation gradients as one moves from land borne to marine borne vegetation. Where the physical geology (cliffs, precipitous banks, etc.) or the vegetation gradient (marsh lands) are not present to aid in your quest, it is necessary to examine the action of the water on the soil itself. The word soil being used of course in its broad meaning of bedrock, boulders, gravel, etc. as well as earth. Even on the most barren stretches of rocky shorelines, the continued presence and action of the water leaves its mark, subtle though it might be.[140]

§5.75 Aside from marks or discolouration on the soil, rocks, or even wharf pilings, surveyors sometimes rely on lines of driftwood and debris as evidence of the OHWM.[141] Seaweed deposited by the action of the tides indicates a water line, but there may be several lines of seaweed left, for instance, by the daily higher high water and lower high water or by the high water at spring tides. To determine whether these seaweed lines represent the medium or ordinary tides, the surveyor must know the local tidal conditions at the time of the survey. Driftwood is usually more closely related to spring or storm tides than to the limit of ordinary tides. Despite

---

138 (1921), 60 D.L.R. 666 at 670 (N.B. C.A.) where the court of appeal was citing the trial court judgment. See also Doig, *supra*, note 116; and as discussed by Doig, *supra*, note 134.
139 *Ibid.*
140 MacDonald, *supra*, note 116 at 9.
141 *Supra*, note 134.

some criticism in the British Columbia case *Nelson v. Pac. Great Eastern Ry. Co.*,[142] driftwood lines were accepted as the high water boundary in recognition of customary survey practice, and in lieu of other available evidence, to establish the boundary by reference to the level of medium tides (see also §5.80.).

§5.76 Marshlands, where vegetation extends well below the limit of the ordinary tides, present some of the greatest problems in tidal boundary surveys. In surveys made in conjunction with a dispute over lands expropriated for the Prince Edward Island National Park, a low sand ridge built up by tidal action in a marsh was accepted as the OHWM by surveyors. Evidence was presented in *Shaw v. R.*[143] to show that this ridge was frequently breached by the ordinary or mean tides and that areas landward of the surveyed OHWM were submerged at high water. Included in the evidence was an extensive study of the relationship between the occurrence of specific types of vegetation and the frequency of salt water inundation.[144] Although no decision on the boundary was reached by the court, similar cases involving such biological boundaries have arisen in the United States.

§5.77 Some recent cases in the United States have been decided in favour of surveyed mean high water lines, supported by other ground evidence, or boundaries defined using aerial photography which has been synchronized with the tides.[145] Boundaries strictly based on the interpretation of vegetation from photographs and other remote sensing imagery have been found to be less reliable in determining the limit of the ordinary or mean high tides, due to the scale of the imagery and the presence of transition

---

142 *Supra*, note 117.
143 [1980] 2 F.C. 608 (T.D.).
144 S.B. McCann, "Shore Conditions Between the Southern Gulf of St. Lawrence and Brackley Bay in the Vicinity of Brackley Beach" (1978), unpublished report prepared for Energy, Mines and Resources Canada at the Dept. of Geography, McMaster University, Hamilton, Ontario; the report findings are discussed in Nichols, *supra*, note 134 at 194-199.
145 See, for example, A.A. Porro and J.P. Weidener, "The Borough Case: A Classic Confrontation of Diverse Techniques to Locate a Mean High Water Line Boundary" (1980), 42(4) *Surveying and Mapping* 369; G.M. Cole, *Water Boundaries*, (Rancho Cordova, CA: Landmark Enterprises, 1983) at 21-30; G.M. Cole, "Where Oil, Water, Surveying and Photogrammetry Mix" (1982), *Proceedings of the Annual Meeting of the American Congress on Surveying and Mapping*, Denver, CO, March, 1982, 319-323; and P.T. O'Hargan "Demarcation of Tidal Water Boundaries" (1972), *Proceedings of the Annual Meeting of the American Congress on Surveying and Mapping*, Washington, DC, March 1972, 1-13.

200   Survey Law in Canada

zones from saltwater to land vegetation.[146] Photographs were used in *Shaw v. R.*, in part to establish past conditions in the marsh, but this evidence was also supported by tidal observations at the site in question.[147]

§5.78 Establishing the water line using tidal data   Aerial photography provides an efficient and consistent means of determining the location of visible high or low water lines for coastal mapping and charting. Aerial photography can also be used to establish coastal boundaries.[148] Where a low water line must be delimited, reliance on physical marks is not feasible and boundaries are often based on the location of the observed water line at the appropriate stage of tide. For precise delimitation, whether by field survey techniques or by aerial photography, measurements must be taken when the correct tidal level occurs at the particular site. Both the elevation of the tidal reference surface and the time at which this level occurs may differ from the elevation and time given for or prorated from Reference or Secondary Ports.

§5.79 *Irving Refining Ltd. v. Eastern Trust Co.*[149] considered a water line which had been established to delimit the high water boundary in an expropriation of virtually flat tidelands within Courtney Bay in Saint John Harbour. The elevation of MHW for the Saint John Reference Port tidal station was calculated from the tide tables. On a day and at a time when this level was predicted to occur, the observed water line was staked at the survey site. When the boundary evidence was examined by the court, there was about a 3.8 metre horizontal difference between the location of the water line shown on the survey plan and the location of the contour at the MHW elevation calculated for the Reference Port. The situation is illustrated in Figure 5.6. Although the court attributed the discrepancy to a levelling error, other factors could have contributed to the difference. For instance:

(1) the actual tide recorded at the Reference Port on that day was below the predicted level;
(2) in the time taken to stake the water line, the water level may have receded;

---

146   A.A. Porro and L.S. Telkey, "Marshland Title Dilemma: A Tidal Pheonomena" (1972), 3 *Seton Hall Law Review 323*; J.P. Weidener, "Will the Real Mean High Water Line Please Stand Up?" (1974), *Technical Papers of the American Society of Photogrammetry*, Washington, DC, September 1974, 34-43.
147   *Supra*, note 143 and 144.
148   See, for example, N.S. Reg. 42/79, s. 26.
149   *Supra*, note 120. For a further discussion of this case see S.E. Nichols, *Tidal Boundary Delimitation*, Technical Report 103 (Fredericton, N.B.: Dept. of Surveying Engineering, University of New Brunswick, 1983), pp. 172-193.

(3) due to local variations the MHW reference surface may not have the same elevation at the survey site as at the permanent tide gauge which is located in another part of the harbour;
(4) the time at which MHW occurs may also be significantly different for the two sites.

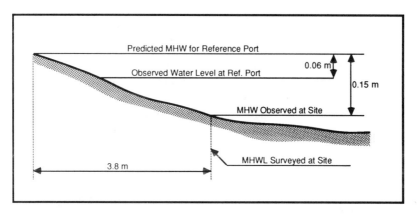

FIGURE 5.6: Mean high water levels predicted and observed in *Irving Refining et al. v. Eastern Trust*

The court recommended in its judgment that a simpler method of establishing the MHW elevation at the survey site would have been to run a line of levels at the elevation of MHW. But this method also assumes that MHW is a level surface; any local variations in the elevation of this surface would be ignored (see §§5.35 to 5.37). In resolving the issues in this case, it was held that the call for the natural monument, the MHW of Courtney Bay, governed any incompatible measurement made during the survey or shown on the survey plan.[150]

§5.80 The use of tide tables in boundary delimitation and the effect of local variations in reference surface elevations were recognized in *Nelson v. Pacific Great Eastern Railway Co.*[151] The court made the following observations:

> Plaintiffs sought to apply the English definition [*Attorney-General v. Chambers*] by adducing evidence, as to the state of tide on particular days, at the point in question [West Vancouver] and comparing it with the tides as indicated by the tide tables, at the Sand-Heads, near the mouth of the Fraser River at the same time. This would appear, upon first consideration, quite reasonable and accurate, but the evidence convinces me that it is subject to conditions which would create an important margin of error. In the first

---
150 *Irving, supra*, note 120.
151 *Supra*, note 117.

place, the tide tables are only a pre-calculation or prophesy, as to the state of the tide on certain days. While of great assistance, especially for purposes of navigation, they do not prove absolutely correct. Then again, to compare the high-water mark at West Vancouver with the Sand-Heads you would require to assume the same sea level, also that the conditions of wind and current are the same.[152]

The line of driftwood was accepted as the boundary in this case.

§5.81 Following the *Borax* decision and in response to extensive litigation in coastal areas, methods for precisely determining the MHWL have received considerable attention in some American states. To correct MHW elevations for local variations, surveyors have recommended that tidal observations should be made at the survey site and compared with simultaneous observations at nearby control stations where the MHW elevation is known. These observations, usually made with a graduated tide staff, generally span one or two consecutive high and low water cycles. Once a local MHW elevation has been calculated, either a contour can be run along the shore at this level or the water line can be staked when that elevation is observed to occur on the tide staff.[153]

§5.82 The precision of such MHWL surveys is directly related to the slope of the shore. On steep slopes rough approximations of the MHW elevation can yield acceptable results. However, in marshlands and tidal flats where there is little gradient, even small errors in determining the tidal reference surface elevation can result in relatively large discrepancies as illustrated in the *Irving* case. In many situations the methods may be no more precise than determining the OHWM by physical features. To illustrate the disparity among the various survey methods, the boundaries contested in the American case *Hughes v. Washington*,[154] as reported by Corker,[155] are shown in Figure 5.7. The U.S. Supreme Court followed the *Borax* decision in establishing the boundary as the contour of MHW elevation, but this boundary was seaward of the observed MHWL that took local effects, such as waves, into account. Although the State of Washington claimed that the upland boundary was fixed by the limit of vegetation as delineated in 1889, the court held that the boundary was ambulatory and any accretion belonged to the upland.[156]

---

152 *Ibid.* at 601.
153 Survey techniques for delimiting mean high water lines are summarized by J.P. Weidener, "Seeking Precision in the Ebb and Flow of Tides" (1982), *Professional Surveyor*, March-April, 28-33; see also, for example, Cole, *Water Boundaries, supra*, note 145.
154 (1966), 67 Wash Dec. (2d) 787, 410 P. (2d) 20.
155 After C.E. Corker, "Where Does the Beach Begin and to What Extent is This a Federal Question?" (1966), 42 *Washington Law Review* 33; Figure 5.7 is after Corker, at 46.
156 *Supra*, note 154.

Ambulatory Boundaries

§5.83 Since riparian owners have the right to accretion and conversely may lose land through erosion, the natural water boundary is an ambulatory boundary. Apart from the difficulties in determining whether or not changes in the shoreline have been gradual and imperceptible and therefore true erosion or accretion, the surveyor is frequently confronted with three related problems: conflicting measurements in title documents and plans; establishing former water boundaries; and apportioning accretion between adjacent riparian owners.

§5.84 *Effect of measurements* When a property is described as being bounded by water, this natural monument has priority over other evidence, including conflicting calls in the description. Calls for the water boundary include such terms as "to the shore" or "thence along the harbour," as well as "bounded by the OHWM." In all cases, the intention is that the body of water be the boundary and this natural feature continues to mark the boundary so long as changes in the shoreline are gradual and imperceptible and are not the result of vertical accretion or actions taken to extend the upland. Under the common law this call for the natural boundary is paramount; it governs any conflicting survey measurements depicted on a survey plan or given in a title document.[157]

§5.85 Some questions have arisen as to the effect of land titles legislation on the ambulatory boundary.[158] Where title to a riparian parcel is registered under such legislation, there is often a precise delineation of the parcel bounds, usually by reference to a survey plan. The courts have regarded the description in a Crown patent or certificate of title, including the plan as not limiting the right of the riparian owner to accretion, if that accretion is formed gradually and imperceptibly by either the deposit of alluvium to the upland or withdrawal of the water level by reliction.[159] In the British

---

157 *Fraser v. Cameron* (1854), 2 N.S.R. 189 (C.A.); *Boyd v. Fudge* (1965), 46 D.L.R. (2d) 679 (N.B. C.A.); *Saueracker v. Snow* (1974), 14 N.S.R. (2d) 607; *Clarke v. Edmonton (City of)*, [1930] S.C.R. 137, reversing [1928] 1 W.W.R. 553.
158 See, for example, M. Viminitz, "Waters and Watercourses" (1958), 29(2) *The Canadian Surveyor* 225. The issue is reviewed by V. Di Castri, "Water and Highway Boundaries as Affecting Title", chapter 7 in *Registration of Title to Land*, v. 1 (Toronto: Carswell, 1987).
159 *Clarke v. Edmonton (City of)*, *supra*, note 157; *Bruce v. Johnson*, [1953] O.W.N. 724 (Co. Ct.). In *A.-G. for B. C. v. Miller* [1975] 1 S.C.R. 556, reversing [1973] 2 W.W.R. 201 (B.C.C.A.) the water boundary on a lake as described in several plans was at issue. Although it was not shown in the case that accretion or erosion had taken place, the Supreme Court of Canada gave the opinion, at 562-563, that merely because a plan is referred to in a certificate of title, the boundaries are not unchallengeable.

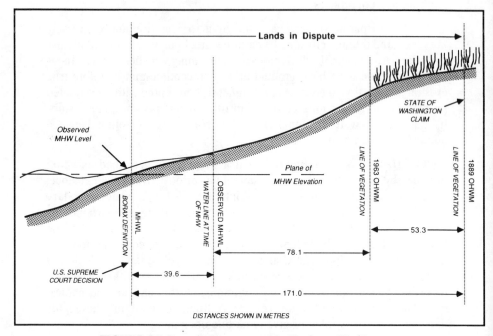

FIGURE 5.7: Boundaries contested in *Hughes v. Washington* (after Corker, 1966)

Columbia *Land Title Act*,[160] provisions are made for including accretion in the riparian owner's title by amending the plan description.

§5.86 *Establishing former boundaries* There are many situations where surveyors must establish the location of the water boundary at some former time. In some instances, the location of the former water mark or line is required to determine the nature and extent of accretion. Legislation, such as that creating reservations or expropriations, sometimes has the effect of fixing the location of the water boundary in time. Words of grant can also create a fixed boundary.

§5.87 The evidence required to establish such former boundary locations

---

160 R.S.B.C. 1979, c. 219 at s. 95. See also D.A. Duffy, "Accretion and Natural Boundary Adjustment" (1988), unpublished paper to be presented at a B.C.L.S. seminar, Kelowna, B.C.

can be extensive and diverse. In *Shaw v. R.*,[161] for example, geomorphological and vegetation analyses, as well as early aerial photographs, were used to determine both the extent of shoreline changes and how accretion was formed. Evidence of the former water line in the *Irving*[162] case included soil analysis and expert testimony as to the nature of the Saint John shipbuilding industry. Confusion over relating the various boundaries and local datums that were shown on ancient maps and charts presented as evidence considerably complicated the issues in this case. A clear indication on survey plans of the evidence found at the site and any tidal information that has been used to establish a water boundary would help to reduce such confusion in future.

§5.88 *Apportionment of accretion* The third problem related to ambulatory tidal boundaries is that of determining the sidelines between adjacent riparian properties once significant accretion has occurred — a process known as apportionment of accretion (see also §§9.55 to 9.57). Unfortunately there is little guidance in Canadian law for surveyors. *Paul v. Bates*[163] involved delimitation of the sideline between two properties along a relatively open coast in British Columbia. The court held that a line should be constructed perpendicular to a second line representing the general trend of the coastline. The situation is depicted in Figure 5.8. Despite the appropriateness of this method in some cases, it would not be suitable along deeply indented coasts where the general trend could be subject to many interpretations.

§5.89 Surveyors often apportion accretion by simply prolonging the existing sidelines. There are situations in which such a method is inequitable, as for example, where prolonging the sidelines significantly reduces or completely eliminates the riparian owner's access to water. Other methods sometimes used by surveyors are based on proportioning either the new shoreline lengths or the accreted area among the coastal properties, according to the relative lengths of the former shorelines.[164] Such methods recognize that the fundamental feature of a riparian property is the shoreline and the right of access to the water.

---

161 *Supra*, note 143.
162 (1967), 51 A.P.R. 155 (N.B.S.C.).
163 (1934), 48 B.C.R. 473.
164 Methods recognized in the United States are described in C.M. Brown, H.F. Landgraf, and F.D. Uzes, *Boundary Control and Legal Principles*, 2d ed. (Toronto: John Wiley and Sons, 1969) at 309-318; C.H. Cole, "Land Survey Law Pertaining to Accretions in Rivers and Streams", *Proceedings of the Annual Meeting of the American Congress on Surveying and Mapping*, Washington D.C., Feb. 1978, 10-27.

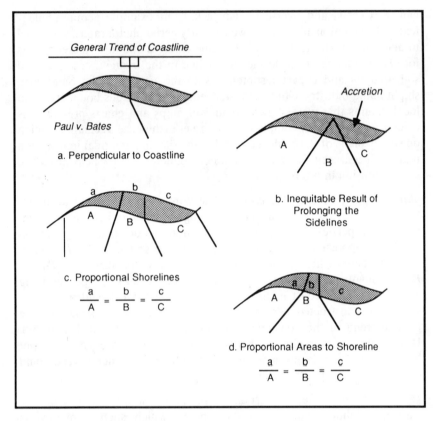

FIGURE 5.8: Examples of methods for apportioning accretion

MARITIME BOUNDARY DELIMITATION

§5.90 With the world's longest coastline (approximately 243,800 km) and the second largest continental shelf,[165] Canada has been one of the nations greatly affected by recent developments in the international law of the sea. Canada's declaration of fisheries and environmental protection zones and the development of offshore petroleum resources reflect changes in the maritime regime that have occurred world-wide over the last four decades. As coastal nations began to claim extended jurisdiction offshore, a new legal framework was required to ensure orderly development and equitable distribution of marine resources. The 1982 *United Nations*

---

165 D.M. Johnson, *Canada and the New International Law of the Sea* (Toronto: University of Toronto Press, 1985) at 1.

Convention on the Law of the Sea[166] provides such a framework.

§**5.91** Extended jurisdiction has created new national limits and new international boundaries. How these should be defined has been the subject of technical research, legal debate, and political dispute. Some of the boundary issues introduced here are treated more comprehensively by Alexander,[167] Prescott,[168] and others.[169] Kapoor and Kerr discuss each article of the *Convention* related to boundaries in *A Guide to Maritime Boundary Delimitation*, which is specifically designed "for the use of hydrographers and others who may be called upon to assist their governments in the technical aspects of maritime boundary delimitation."[170]

Development of the Law of the Sea

§**5.92** The development of the law of the sea has been a movement from customary international law towards a comprehensive international treaty to regulate ocean spaces. This codification was not easily achieved and the treaty is still not universally accepted. Over 150 nations were involved in drafting the 1982 *Convention*, each nation with its own special interests. Trying to reach a consensus on all issues was a formidable task and the *Convention* has been hailed as "a monument to international co-operation in the treaty-making process."[171] A balance was sought between a regime that would permit exclusive economic development of ocean resources by coastal nations, yet one that would protect the oceans as the common heritage of mankind, to be shared freely by all nations and protected for future generations.

§**5.93** *Emergence of a new accord* Territorial claims to extensive ocean areas are not new. As early as 1493, the Atlantic ocean was divided from

---

166 U.N. Doc A/CONF. 62/122 (October 7, 1982); reproduced in *Law of the Sea: Official Text of the United Nations Convention on the Law of the Sea with Annexes and Index* (New York: United Nations, 1983).
167 L.M. Alexander, "Baseline Delimitation and Maritime Boundaries" (1983), 23(4) *Virginia Journal of International Law* 503.
168 J.R.V. Prescott, *The Maritime Political Boundaries of the World* (London: Methuen and Co., 1985).
169 See, for example, P.B. Beazley, *Maritime Limits and Baselines: A Guide to Their Delimitation*, Special Publication No. 2, 2d rev. ed. (London: The Hydrographic Society, 1978); S.W. Boggs, "Delimitation of Seaward Areas Under National Jurisdiction" (1951), 45 *American Journal of International Law* 240; R.D. Hodgson and E.J. Cooper, "The Technical Delimitation of a Modern Equidistant Boundary" (1976), 3(4) *Ocean Development and International Law* 361.
170 D.C. Kapoor and A.J. Kerr, *A Guide to Maritime Boundary Delimitation* (Toronto: Carswell Co., 1985).
171 B. Zuleta in the introduction to *Law of the Sea, supra*, note 166 at xix.

north to south poles by Pope Alexander VI; the eastern portion, with explored Africa, was given to Portugal; the western portion, including the most of the Americas, was allotted to Spain.[172] But such exclusive claims could not be enforced and freedom of the high seas has long been a part of the customary law of nations. Excluded from the high seas was a narrow region along each nation's coast, known as the territorial sea. Although this region was considered part of the territory of the coastal nation, foreign ships had the right of innocent passage, that is, peaceful transit through these waters. The customary width of the territorial sea was three nautical miles. Some nations claimed larger zones, but these were exceptions to the generally accepted rule.[173]

§5.94 By the mid-twentieth century, the discovery of marine hydrocarbon reserves led many nations to make unilateral claims to larger areas of their coastal waters. The turning point in challenging the customary law came with President Truman's declaration in 1949 that the United States considered the continental shelf off its shores to be a prolongation of the land mass and subject to American jurisdiction and control.[174] Responding to this and many similar claims, the first United Nations Conference on the Law of the Sea (UNCLOS I) was held in Geneva in 1958; two international treaties resulted but these have proved inadequate in light of new technology and new issues. A second conference was called in 1960 to define more precisely the limits of the territorial sea, but international agreement could not be reached.[175] By the late 1960s, limits for the continental shelf, exclusive fishing zones and the territorial sea were more representative of specific national interests than of any general rules of law. There was concern for international peace as nations scrambled to claim their share of the oceans — a process likened to the division of Africa and Asia by world powers during 19th century colonialism.[176]

§5.95 UNCLOS III began in 1973 and the process of negotiating one comprehensive treaty on nearly all marine issues spanned a decade. Consensus on many issues, such as defining the seaward limit of jurisdiction over the continental shelf, followed long debate. There were also several key points

---

172 H.A. Smith, *The Law and Custom of the Sea*, 2d ed. (London: Stevens and Son Ltd., 1950) at 5.
173 *Ibid.*, at 6.
174 R.D. Eckert, *The Enclosure of Ocean Resources: Economics and the Law of the Sea* (Stanford: Hoover Institution Press, 1979) at 363-364.
175 *Ibid.* at 39. See also C. Sanger, *Ordering the Oceans: The Making of the Law of the Sea* (Toronto: University of Toronto Press, 1987) at 14-18.
176 A. Pardo, U.N. Ambassador for Malta, U.N. General Assembly Official Records, First Committee, 22nd session, Agenda Item 92, Doc. A/C.1/PV.1515 (November 1, 1967), at 12-13; as reported at *supra*, note 174 at 40.

on which the interests of the highly developed and the developing countries conflicted. On the issue of deepsea mining, for example, the United States and several other western countries were largely opposed to the international regulation and profit-sharing arrangements being proposed to ensure that all nations could eventually benefit from the exploitation of resources seaward of the continental shelf. Although consensus on all issues had not been reached, the *Convention* was voted upon in April 1982, and came open for signature in December of that year. At that time 117 nations signed the treaty, including Canada. By the closing date for signatures in December 1984, 159 countries had signed. Among the countries that have not signed are the United States, the United Kingdom and West Germany. The *Convention* will only come into force one year after it has been ratified by 60 of the signatory nations. By November 3, 1987, only 35 nations had filed instruments of ratification with the United Nations. However, most nations are complying with some if not all parts of the treaty, despite the fact that it is not yet in force. Many of its provisions may thus be considered customary international law.[177]

§5.96 *Maritime zones* The 1982 *Convention* delineates seven zones as shown in Figure 5.9; five of these can be claimed as zones of national jurisdiction. In the remaining two, the High Seas and the Area, exclusive appropriation by nations is prohibited. The zones are:

(1) Internal Waters: Those ocean areas landward of national baselines are internal waters which are part of the national territory and over which the nation has complete jurisdiction.[178] In a 1984 Supreme Court decision regarding federal and provincial jurisdiction in specific coastal waters along British Columbia, the Court held that although the Strait of Georgia and other similar areas are considered internal waters of the nation in international law, they are not necessarily inland waters subject to provincial jurisdiction.[179]

---

177 United Nations statistics from "Report of the Secretary General" in *New Directions in the Law of the Sea*, K.R. Simmonds, et al., eds. (New York: Oceana Publications, Inc., 1988). On the procedures for the treaty to come into force see, for example, Johnson, *supra*, note 165 at 2 and footnote 9 at 84. See also B.H. Oxman, D.D. Caron, and C.L.O. Buderi, eds. *Law of the Sea: U.S. Policy Dilemma* San Francisco: Institute for Contemporary Studies, 1983) for detailed discussion on the U.S. involvement in UNCLOS III, in particular, opposition to deepsea mining provisions and the potential effect of U.S. nonparticipation. The United States was one of four nations voting against opening the treaty for signature in April, 1982 (along with Venezuela, Turkey, and Israel). There were 130 nations voting for the treaty at that time and 17 abstentions, including West Germany and Great Britain.
178 *Supra*, note 166, art. 8.
179 *A.G. Can. v. A.G. B.C.*, [1984] 1 S.C.R. 388.

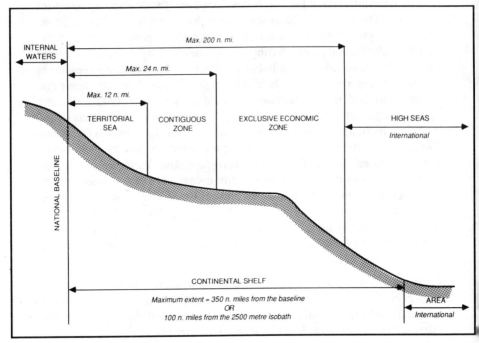

FIGURE 5.9: The Legal Regime defined in the 1982 United Nations Convention on the Law of the Sea

(2) Territorial Sea: A nation may claim a territorial sea up to 12 nautical miles from the national baselines. Within this zone the nation has complete jurisdiction over the seabed, water column and airspace, subject to the right of innocent passage.[180] Canada extended its territorial sea to 12 nautical miles in 1970.[181]

(3) Contiguous Zone: A coastal nation has the right to enforce customs, immigration, fiscal and sanitary laws within the contiguous zone which lies adjacent to the territorial sea to a maximum limit of 24 nautical miles measured from the baselines.[182] In 1970, Canada declared a special region in the north under the *Arctic Waters Pollution Prevention*

---

180 *Supra*, note 166, art. 2 and art. 17-21.
181 *Territorial Sea and Fishing Zones Act*, R.S.C. 1985, c. T-8.
182 *Supra*, note 166, art. 33.

Act,[183] measuring 100 nautical miles from the baselines of the arctic islands. This unilateral declaration was made in response to concerns over potential oil spills in the environmentally sensitive Arctic. Provisions for special national control over ice-covered areas were subsequently made in the treaty.[184]

(4) Exclusive Economic Zone (EEZ): One of the major features of UNCLOS III was the creation of a zone, with a maximum breadth of 200 nautical miles from the baselines, in which nations have exclusive rights to exploit and manage the living and non-living resources of the water and the seabed.[185] In 1977, Canada declared a 200 nautical mile fisheries zone, the seventh largest EEZ in the world, estimated to be approximately 1.29 million square nautical miles in area.[186] This declaration, as well as the potential resources of the continental shelf, has required the definition of new boundaries with Denmark, France and the United States.

(5) High Seas: The waters beyond the EEZ are known as the high seas, where all nations have freedom of navigation, fishing and scientific research as long as these activities are undertaken for peaceful purposes.[187]

(6) Continental Shelf: One of the zones that caused contention in negotiating the treaty was the continental shelf, where nations have exclusive rights to exploit, develop and manage the living and non-living natural resources of the seabed.[188] By the 1958 *Geneva Convention on the Continental Shelf*, to which Canada is a party, the continental shelf was defined as extending to a depth of 200 metres or the limit of exploitability.[189] With greater technical capabilities for extracting seabed resources further offshore, this limit became meaningless. Under the new definition, a nation may choose its limit based on geological circumstances. All nations may claim a continental shelf of 200 nautical miles measured from the territorial sea baselines. If the geological shelf extends beyond this limit, a nation may delimit its claim by straight lines, each having a maximum length of 60 nautical

---

183 R.S.C., 1985, c. A-12; the purpose and effects of the Act are discussed in D.M. McRae and D.J. Goundrey, "Environmental Jurisdiction in Arctic Waters: The Extent of Article 234" (1982), 16(2) U.B.C. *Law Review* 197.
184 *Supra*, note 166, art. 234.
185 *Ibid.*, art. 55-57.
186 *Fishing Zones of Canada (Zones 4 and 5)*, S.O.R. 77-62, January 1, 1977, by Order in Council under note 181, *supra*; see also C. Sanger, *Ordering the Oceans: The Making of the Law of the Sea*. (Toronto: University of Toronto Press, 1987) at 65.
187 *Supra*, note 166, art. 87 and 88.
188 *Ibid.*, art. 77.
189 *Convention on the Continental Shelf*, Geneva, April 29, 1958.

212   Survey Law in Canada

miles, which connect points defined by geographic co-ordinates. The location of these points may be chosen using either of the following criteria:
  (i) where the thickness of sedimentary rocks is at least one percent of the shortest distance from the points to the foot of the continental slope;
  (ii) where the points are not more than 60 nautical miles from the foot of the continental slope.

The foot of the continental slope is the point of maximum change of gradient at the base of the slope. In all cases, the maximum breadth of the continental shelf is 350 nautical miles from the baselines or 100 nautical miles from the 2500 metre isobath or depth contour.[190] Canada is one of the most favoured nations with respect to the continental shelf and played a significant role in the definition of this zone.[191]

(7) Area: Beyond the national limits of the continental shelf lies the Area, an international zone in which the exploitation of seabed resources is regulated to benefit all nations. The *Convention* calls for mining activities to be overseen by a multinational body called the International Sea-Bed Authority; provisions are made for sharing profits and transferring technology to developing nations.[192] Although most of the other provisions of the treaty had consensus, the articles creating this regime were vigorously opposed by many developed nations which have the capabilities for deep-sea mining and which wanted less interference with economic enterprises.[193]

Baselines

§5.97 The limits of each national zone are measured from the nation's baselines. As shown in Figure 5.10, baselines can also be used in constructing international boundaries. Although the baseline concepts are relatively straightforward, actual delimitation can be technically complex because each coastline presents special situations where various interpretations can be made.[194] Two types of baselines are recognized in inter-

---

190 *Supra*, note 166, art. 76(4), (5) and (7).
191 See, for example, Sanger, *supra*, note 186 at 77-81; Johnson, *supra*, note 165.
192 *Supra*, note 166, arts. 136, 141, 144.
193 See Sanger, *supra*, note 186 at 158-193. See also B.H. Oxman, D.D. Caron, and C.L.O. Buderi, eds. *Law of the Sea: U.S. Policy Dilemma* (San Francisco: Institute for Contemporary Studies, 1983).
194 Kapoor and Kerr, *supra*, note 170 at 29-56; see also L.M. Alexander, "Baseline Delimitation and Maritime Boundaries" (1983), 23(4) *Virginia Journal of International Law* 503.

national law: normal baselines and straight baselines. In addition, there are provisions for closing lines across river mouths and bays.

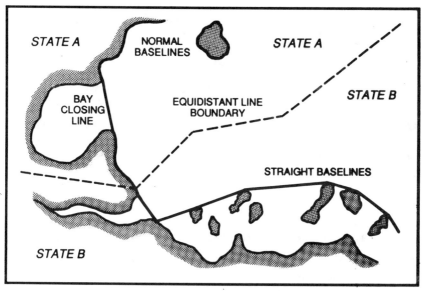

FIGURE 5.10: Baselines and an Equidistant Line Boundary

§5.98 *Normal baselines* Normal baselines follow the sinuosities of the coastline. They are defined in the 1982 *Convention* as "the low-water line along the coast as marked on large-scale charts officially recognized by the coastal State."[195] In Canada this is the line of Lower Low Water, Large Tides (LLWLT) as mentioned in paragraph 5.28. Other nations, such as the United States, use different definitions for delineating the low water line shown on charts. These differences must be taken into account in boundary delimitation in order to avoid inconsistencies in measurements which are referenced to the baselines. In the case of the boundary delimitation between Canada and the United States in the Gulf of Maine, where each nation uses different chart datums, this issue was resolved before the matter was submitted to the International Court of Justice.[196]

§5.99 *Closing Lines* The *Convention* also makes provisions for special geographic circumstances when delimiting normal baselines. Straight lines,

---
195 *Supra*, note 166, art. 5.
196 International Court of Justice, *Special Agreement Between the Government of Canada and the Government of the United States of America to Submit to a Chamber of the International Court of Justice the Delimitation of the Maritime Boundary in the Gulf of Maine Area*, Nov. 25, 1981.

known as closing lines, can be drawn across the mouths of rivers and bays, so long as certain criteria are met.[197] A bay is defined as

> [A] well-marked indentation whose penetration is in such proportion to the width of its mouth as to contain land-locked waters and constitute more than a mere curvature of the coast. An indentation shall not, however, be regarded as a bay unless its area is as large as, or larger than, that of a semi-circle whose diameter is a line drawn across the mouth of that indentation . . . . For the purpose of measurement, the area of an indentation is that lying between the low-water mark around the shore of the indentation and a line joining the low-water mark of its natural entrance points.[198]

One challenge to surveyors and hydrographers is defining precisely where the natural entrance points of a bay are located.[199] The maximum length of a bay closing line is 24 nautical miles unless there are islands at the mouth of the bay that create more than one entrance or where a nation claims an historical bay as part of its internal waters.[200] The Bay of Fundy, Placentia Bay and Hudson's Bay are Canadian examples of internal waters based on historical use and economic significance, although some of these claims may be disputed by other nations.[201]

§5.100 *Straight baselines* There are situations where normal baselines are inappropriate due to the irregularity of the coastline. Provisions are therefore made for drawing straight lines between points along the coastline "in localities where the coastline is deeply indented and cut into, or if there is a fringe of islands in its immediate vicinity."[202] These straight baselines "must not depart to any appreciable extent from the general direction of the coast."[203] Canada has drawn straight baselines between points defined by geographic co-ordinates along the Atlantic and Pacific coasts and in 1986 the Canadian Arctic archipelgo was enclosed by a series of straight baselines.[204]

---

197 *Supra*, note 166, art. 9, 10(4) and 10(5).
198 *Ibid.*, art 10(2) and 10(3).
199 See, for example, Kapoor and Kerr, *supra*, note 170 at 42-45; J.R.V. Prescott, *The Maritime Political Boundaries of the World* (London: Methuen and Co., 1985) at 50-62.
200 *Supra*, note 166, art. 10(4), 10(5) and 10(6).
201 D. Pharand, "Historical Waters in International Law with Special Reference to the Arctic" (1971), 21(1) *University of Toronto Law Journal* 599; G.V. LaForest, "Canadian Inland Waters of the Atlantic Provinces and the Bay of Fundy Incident" (1963), 1 *The Canadian Yearbook of International Law, 1963* 149.
202 *Supra*, note 166, art. 7(1).
203 *Ibid.*, art. 7(3).
204 J.B. McKinnon, "Arctic Baselines: A Litore Usque ad Litus" (1987), 66 *Canadian Bar Review* 790 at 791. See also D.C. Kapoor, "An Investigation of Canadian Baselines and the Depiction of Offshore Limits on C.H.S. Charts" (1985), unpublished study conducted for Fisheries and Oceans Canada, Ottawa.

§5.101 No limits, such as those for closing lines, are set in the *Convention* for the length of straight baselines. Baselines longer than 200 nautical miles have been claimed, for instance, by Burma and the maximum distance between a baseline and the Burmese coast (that is, the breadth of coastal waters claimed as internal waters) is 75 nautical miles.[205] In many cases nations have taken advantage of this special provision for straight baselines to enclose as internal waters greater marine areas than might be warranted by coastal configuration. Employing straight baselines can significantly extend the limits of zones that are measured from the baselines and can affect the location of equidistant lines.

International Boundaries

§5.102 When maritime boundaries are in dispute, their delimitation can be resolved by negotiation or arbitration. In some cases the delimitation is submitted, by the nations involved, to the International Court of Justice (ICJ). Unlike the common law where judicial decisions set precedents to be followed in subsequent cases when the facts are similar, Article 38 of the *Statute of the International Court of Justice*[206] gives priority in settling disputes to conventions in force, then to international custom, to general provisions of law recognized by nations and, finally, to previous judicial decisions and other sources. Thus, decisions made by the ICJ only provide general principles that may or may not be applied subsequently in a specific dispute. Two principles, equidistance and special circumstances, reoccur in recent boundary delimitations. Neither are usually strictly applied in particular cases; the objective of the Court is to determine an equitable solution given the specific situations in each case.[207]

§5.103 *Equidistance Principle* The equidistance method of delimiting lateral ocean boundaries was first proposed in 1937 and was drawn from the principle of the median line as it applied to inland waters.[208] An equidistant line is a boundary so constructed that it is an equal distance from the nearest points on the baselines of each nation.[209] Figure 5.10

---

205 See Prescott, *supra*, note 199 at 66.
206 *Statute of the International Court of Justice*, (1949) Ch. II, art. 38.
207 See, for example, M.D. Bletcher, "Equitable Delimitation of the Continental Shelf" (1978), 73 *American Journal of International Law* 60.
208 Anon., "Boundary Delimitation in the Economic Zone: The Gulf of Maine Dispute" (1979), 30 *Maine Law Review* 207 at 212.
209 See, for example, R.D. Hodgson and E.J. Cooper, "The Technical Delimitation of a Modern Equidistant Boundary" (1976), 3(4) *Ocean Development and International Law* 361; a more rigorous method is presented in G. Carrera, "A Method for the Delimitation of an Equidistant Boundary Between Coastal States on the Surface of a Geodetic Ellipsoid" (1987), 64(1) *International Hydrographic Review* 147.

shows an equidistant line between nations with different types of baselines. Where nations lie opposite each other on a body of water, the line is still sometimes referred to as a median line. Theoretically, equidistance seems to be an equitable method of dividing ocean areas but islands, promontories, and other coastal features tend to have a disproportionate effect. When in the early 1950's a panel of experts examined alternative methods, such as the prolongation of the existing boundary, those methods were also found to have drawbacks. The principle of equidistance, if not a strict equidistant line, has continued to be used in boundary delimitations and is found in the 1982 *Convention*.[210]

§5.104 Special methods have been developed to reduce the distorting effects of geographic anomalies. These have sometimes been called modified or equitable-equidistant lines.[211] One example of such a modification is that giving half-effect to a small island. Here the modified line is the line midway between an equidistant line giving full effect to the island and one which is drawn as if the island did not exist.[212] Other special circumstances may be considered to define a modified equidistant line or to define boundaries independent of the equidistance principle.

§5.105 *Special circumstances* Special circumstances gained prominence during the delimitation of the continental shelf boundaries in the North Sea where strict equidistant lines were sometimes inequitable.[213] Among the many special circumstances that might be taken into account in boundary delimitation are:[214]

(1) a configuration of the coastline that could lead to inequities;
(2) historical use or management of the area, particularly customary fishing areas;
(3) socio-economic factors such as dependence on resources or the significance of an island to a nation; and
(4) proportionality, that is, the apportionment of ocean areas according

---

210 Direct reference to an equidistant or median line is only made for the delimitation of territorial sea boundaries in Article 15 in *United Nations Convention on the Law of the Sea*, U.N. Doc. A/CONF. 62/122 (October 7, 1982). See also note 208, *supra*, at 211-212 for a review of alternative methods.
211 See, for example, *supra*, note 207; see also E. Grisel, "The Lateral Boundaries of the Continental Shelf and the Judgement of the International Court of Justice in the North Sea Continental Shelf Cases" (1970), 61 *American Journal of International Law* 562.
212 P.B. Beazley, "Half-Effect Applied to Equidistant Lines" (1979), 56(1) *International Hydrographic Review* 153.
213 See Grisel, *supra*, note 211.
214 See, for example, I. de Rijcke, "Equity and Equidistance: Continental Shelf Delimitation in the Georges Bank Area of the Gulf of Maine" (1981), 35(2) *The Canadian Surveyor* 137. See also Notes 207 and 208 above.

to the respective lengths of the nations' coasts (for an analogy with apportioning accretion, see Figure 5.8d).

§5.106 The 1982 *Convention* makes provision for delimiting lateral boundaries of the territorial sea by the equidistance principle if agreement between two nations fails and unless special historical or other circumstances exist.[215] But when boundaries are extended great distances offshore to delimit the EEZ and the continental shelf, any inequities caused by the geography of the coast can be magnified. This problem is recognized in the *Convention* for the delimitation of boundaries in these zones. For example, Article 83(1) states:

> The delimitation of the continental shelf between States shall be effected by agreement on the basis of international law, as referred to in Article 38 of the Statute of the International Court of Justice [see §5.102], in order to achieve an equitable solution.[216]

It may be easier to identify an inequitable situation than to define what is meant by an equitable solution. The ICJ sees its role "as delimitation, not apportionment"[217] and as stated in the North Sea continental shelf cases:

> [D]elimitation in an equitable manner is one thing, but not the same thing as awarding a just and equal share of a previously undelimited area.[218]

§5.107 *Canada's boundaries* With extended jurisdiction offshore, Canada has six ocean areas where international boundaries are an issue.[219] The major portion of Canada's boundary with Greenland has been agreed upon with Denmark[220] and a 1984 decision of the ICJ delimited the boundary between the United States and Canada in the Gulf of Maine except for an area surrounding Machias Seal Island.[221] Canada has three other

---

215 *Supra*, note 210, art. 15.
216 *Ibid.*, art 83. A similar definition is given for the EEZ in art. 74.
217 *Supra*, note 207 at 63.
218 *Ibid.*
219 A survey of Canadian boundary issues can be found, for example, in K. Beauchamp, M. Crommelin, and A.R. Thompson, "Jurisdictional Problems in Canada's Offshore" (1973), 11 *Alberta Law Review* 431; J.K. Renouf, *Canada's Unresolved Maritime Boundaries*, Technical Report 134 (Fredericton, N.B.: Dept. of Surveying Engineering, University of New Brunswick, 1988); and M.B. Feldman and D. Colson, "The Maritime Boundaries of the United States" (1981), 72 *American Journal of International Law* 729.
220 Canada Treaty Series, *Boundary Waters: Agreement Between Canada and the Kingdom of Denmark*, Dept of External Affairs, No. 9. (1974); see also Renouf, *supra*, note 219 at 28-33; I.T. Gault, "Jurisdiction Over Petroleum Resources of the Canadian Continental Shelf: The Emerging Picture" (1985), 23(1) *Alberta Law Review* 75.
221 A.C. McEwen, "The International Boundary Commission: Canada-United States" (1986), 40(3) *The Canadian Surveyor* 277, at 287-289.

218 Survey Law in Canada

boundary issues to be resolved with the United States: between Alaska and the Yukon in the Beaufort Sea; north of the Queen Charlotte Islands; and in the Strait of Juan de Fuca. A boundary issue that has been the cause of recent contention, particularly with regard to the fishery, has been France's claim to fishing zones and the continental shelf around its islands of St. Pierre and Miquelon. The French claim is based on the equidistance principle which extends these zones over the Grand Banks.[222] In addition to the boundary issues, there is also disagreement with the United States over Canada's enclosure of the Northwest Passage as internal waters.[223]

§5.108 In the boundary disputes between Canada and the United States, each nation has chosen to base its claims on principles from which it can derive the most benefit in the particular situation. An exception is the boundary extending from the Strait of Juan de Fuca, where both nations use equidistant or median lines that differ only slightly due to baseline delimitation.[224] In the Beaufort Sea, the United States claims an equidistant line, whereas Canada claims the 141st meridian, the prolongation of the boundary between Alaska and the Yukon.[225] The dispute over the Pacific boundary in the Dixon Entrance derives from the Alaska Boundary Tribunal decision in 1903 for the boundary of the Alaska Panhandle. The line drawn by the tribunal seaward from the Dixon Entrance is claimed by Canada while the United States maintains that the marine boundary is a separate issue that was not addressed by the 1903 tribunal.[226]

§5.109 The Gulf of Maine boundary delimitation became a landmark case in international law, as it was the first time a boundary for both the continental shelf and the EEZ was decided by the ICJ. During the dispute the boundary claims underwent modification as shown in Figure 5.11; in each instance the respective nations claimed a greater area. The American lines were primarily based on special circumstances, including the geological formation of the continental shelf and the proportionality of the

---

222 C.R. Symmons, "The Canadian 200-mile Fishery Limit and the Delimitation of Maritime Zones around St. Pierre and Miquelon" (1980), 12 *Ottawa Law Review* 145. See also Beauchamp *et al.*, *supra*, note 219 at 443-447.
223 D. Pharand, "The Northwest Passage in International Law" (1979) 17 *Canadian Yearbook of International Law*, 1979-99. See also J.B. McKinnon, "Arctic Baselines: A Litore Usque ad Litus" (1987), 66 *Canadian Bar Review* 790.
224 See Renouf and Feldman *et al.*, *supra*, note 219.
225 See Beauchamp *et al.* and Renouf, *supra*, note 219; an American perspective is given by K.L. Lawson, "Delimiting Continental Shelf Boundaries in the Arctic: The United States-Canada Beaufort Sea Boundary" (1981), 16(2) *Virginia Journal of International Law*, 1976 at 221.
226 C.B. Bourne and D.M. McRae, "Maritime Jurisdiction in the Dixon Entrance: The Alaska Boundary Re-examined" (1976), 14 *Canadian Yearbook of International Law, 1976* 175.

coastlines.[227] First claiming an equidistant line, Canada later modified this boundary by excluding Cape Cod and surrounding islands as baseline points, claiming that these areas had too great an influence on the equidistant line given their relatively small socio-economic significance. In the final judgment, the ICJ constructed a line based partly on the

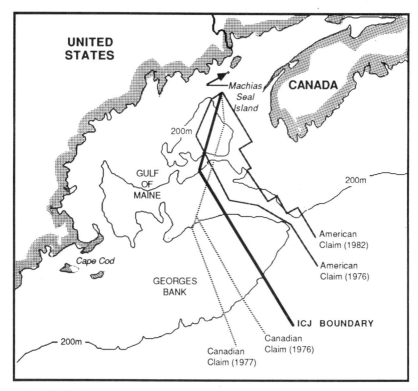

FIGURE 5.11: Portion of the Canada - U.S. Boundary in the Gulf of Maine delimited by the ICJ and previous Canadian and American boundary claims.

---

227 See, for example, I. de Rijcke, "Equity and Equidistance: Continental Shelf Delimitation in the Georges Bank Area of the Gulf of Maine" (1981), 35(2) *The Canadian Surveyor* 137; J. Cooper, "Delimitation of the Maritime Boundary in the Gulf of Maine Area" (1986), 16(1) *Ocean Development and International Law* 59; and N.T. Terres, "The United States/Canada Gulf of Maine Maritime Boundary Delimitation" (1985), 9 *Maryland Journal of International Law and Trade* 135.

equidistance principle, partly on the configuration of the coastline and to much a lesser extent, on other special circumstances.[228]

The Surveyor's Role

§5.110 Of particular significance to surveyors in the Gulf of Maine case is the *Special Agreement*[229] made between the two nations when the boundary delimitation was submitted to the ICJ. This *Special Agreement* made specific provisions for:

(1) resolving the differences between the baseline definitions and chart datums of the two nations which could affect the boundary delimitation;
(2) having the final boundary defined in terms of geographic co-ordinates rather than as a line drawn on a chart;
(3) providing technical expertise in interpreting charts and other evidence during the course of the proceedings.

The inclusion of these provisions recognized technical difficulties that have been encountered in previous boundary delimitations and ICJ decisions.[230]

§5.111 As the expert in boundary delimitation, the surveyor has responsibility for ensuring that boundaries defined in law are unambiguous and technically correct. This includes precise datum and baseline definition. Positioning systems are also required to provide consistent relocation of jurisdictional boundaries at sea for resource development, for environmental monitoring and for national security. Among the major tasks in Canadian waters will be charting the complex limit of our continental shelf. Along the coast, boundary delimitation has often tested the surveyor's knowledge and skills, but in the offshore the challenge has just begun.

---

228 See Cooper and Terres, *ibid.*
229 International Court of Justice. *Special Agreement Between the Government of Canada and the Government of the United States of America to Submit to a Chamber of the International Court of Justice the Delimitation of the Maritime Boundary in the Gulf of Maine Area*, Nov. 25, 1981.
230 D.A. Colson, "The United Kingdom-France Continental Shelf Arbitration" (1978), 72 *American Journal of International Law* 95 points out the problems which arose in the delimitation of the continental shelf boundary between France and Great Britain by the International Court of Justice, where the boundary was delimited by a line drawn on a Mercator chart. The difference between this line and computed geodetic lines was significant but the use of co-ordinates and geodesics was not an issue raised until after the decision. A method to avoid such ambiguities is proposed by G. Carrera, *supra*, note 209.

6

Water Boundaries — Inland

David W. Lambden*
B. Sc. F., F.R.I.C.S., F.I.S. Aust., M.N.Z.I.S., C.L.S., O.L.S.

RIPARIAN RIGHTS AND THEIR DEVELOPMENT

§6.01 In the common law definition of 'land', the several natural rights incident to the ownership of the soil include, if the property is riparian, certain rights to water and the right of title to accretions attached slowly and imperceptibly by the natural action of water along the watercourse and, equally, the loss sustained from erosion. This last right is known to the law as the 'doctrine of accretion'. These riparian rights arise where the land adjoins water. Direct right of access to the water at all points along the water frontage is inherent in the riparian right and there is a property right to the water irrespective of the ownership of the land under the water. A riparian proprietor of the upland enjoys rights of use of the water as to quantity and quality, as to flow and as to drainage. The rights are accordingly matched by the duties.

§6.02 This common law developed at a time and in a land where there was never a problem of supply, but rather, a need for drainage to improve

---

* Appreciation must be recorded to many people for thought-provoking comments. Towards the end of writing the manuscript, critical reviews were given by Max Viminitz of Nanaimo, British Columbia, formerly Examiner of Surveys of Saskatchewan, W.E. Shepherd of Winnipeg, James N. Gardiner and R.A. Stocker of Toronto. We know there are many contentious aspects; the views expressed are my own responsibility.

agricultural potential. Considered as part of the 'land', the right to the use of water was by the established ancient right; the right to drainage needed some statutory development and local government involvement and this was achieved by statutory provisions.[1] The present common law rules respecting waters were established by the 19th century.

§6.03 More or less concurrently, the common law recognized navigation as a *prima facie* fact on all waters subject to the ebb and flow of the tide, to which was added that the beds of such waters were the property of the Crown. For non-tidal watercourses, the common law put title to the bed with the adjoining owner of the upland; if navigation in fact existed, that right was of the nature of a public highway arising from grant, user or by statute. By contrast, in France, the property in the bed of all waters navigable in fact belonged to the Crown. It is this law that held in New France and, by one line of reasoning, became the inheritance of the law of Ontario, at least for that portion south of the Hudson Bay watershed.

§6.04 The rise of modern cities and industrial complexes brought new and greater demands on water resources far beyond those recognized by the common law doctrines of a simpler time. The law has developed in all jurisdictions by enacting statutory provisions covering matters that range from simple use to hydroelectric power generation, from drainage and land development to retention and environmental concerns, from watercourse management to surface and percolating waters.

CONSTITUTIONAL BASIS

Summary of Basic Constitutional Provisions

§6.05 A statement for Ontario and some comparative comments on the situation in other provinces is the most that can be attempted here.

> Before their union under the *B.N.A. Act* [now *Constitution Act*, enacted by the *Canada Act*, 1982 (U.K.), c. 11], the Colonies of the Province of Canada, New Brunswick and Nova Scotia were each governed separately under the Crown. In each colony Her Majesty owned property. Her Majesty could also in each colony claim properties as royalties under prerogative rules. The B.N.A. Act allocates these properties and royalties between the newly created governments of Her Majesty in right of Canada and of Her Majesty in right of each of the four provincial governments.
>
> The allocation of Her Majesty's property and royalties between Her Majesty in right of Canada and Her Majesty in right of the provinces is to be determined under the provisions of the B.N.A. Act relating to such

---

1 *Commissioner of Sewers Act*, 1531, 23 Hen. 8, c. 5.

allocation. The bestowal of legislative power on Parliament or on the Legislatures in relation to a class of subjects under ss. 91 or 92 of the B.N.A. Act does not carry with it any right of Her Majesty to property falling within that class of subject, e.g., the authority conferred on Parliament to make laws in relation to "fisheries" did not carry with it Her Majesty's property rights in fisheries.

Wherever in the B.N.A. Act public land or property are described as "the property of" or as "belonging to" Canada or a province, these expressions merely import that the right to its beneficial use, or to its proceeds, has been appropriated to Her Majesty in right of Canada or the province, as the case may be, and is subject to being dealt with by Her Majesty subject to the control of Parliament or the Legislature; the land itself being vested in Her Majesty. The rights of management or disposal of the land or property can only be exercised by Her Majesty under the advice of the Ministers of Canada or the province to which the beneficial use of the land or its proceeds have been appropriated, and by an instrument under the seal of Canada or the province or as otherwise provided by Parliament or the Legislature of the province.

By s. 108 of the B.N.A. Act the public works and property of each province, i.e., each pre-existing colony, enumerated in the Third Schedule to the B.N.A. Act were declared to be the property of Canada. These comprise the following classes of property of interest at the present time.

(a) Canals with lands and water power connected with them
(b) Lighthouses and piers and Sable Island
(c) Public harbours
(d) Rivers and lake improvements
(e) Lands reserved for military purposes

The residue of the properties of Her Majesty, both real and personal, owned in right of the colonies that existed prior to their union, that are not specifically allocated to Her Majesty in right of Canada, are allocated by the B.N.A. Act to the Crown in right of the province, subject to a reservation in favour of Canada to resume any land or public property required for fortifications or for the defence of the country.

The beds [and the foreshores] of all navigable rivers where the tide flows and re-flows and of all estuaries or arms of the sea are by law vested in Her Majesty and allocated to her in right of the province. [The beds of inland non-tidal navigable rivers and lakes is vested in Her Majesty and allocated to her in right of the province.] Her ownership is subject to the public right of navigation and neither she nor a grantee from her can interfere with this public right. No grant of lands bordering on a navigable body of water shall be deemed to have [conveyed any portion of] the bed of the water in the absence of a further express grant. (Revised statement.)[2]

---

2 D.W. Mundell and J.T.S. McCabe (1975). *Crown*. (Canadian Encyclopedic Digest (Ont. 3rd), Title 40), §§86-93. §95 is here restated after review of the earlier editions of the C.E.D. A distinction must be drawn between tidal waters and inland waters. It would

Jurisdiction and Reception of Laws in Ontario

§6.06 By the Treaty of Paris of 1763, French control in the lands of New France ended and the newly acquired territory became 'property' of the English Crown. Blackstone, writing in the mid-1700's, divided the English colonies into those claimed by occupancy and settlement and those claimed by conquest, stating that there followed a difference in the laws that applied.[3]

§6.07 Between 1763 and 1791, the laws of Quebec were more or less resolved. In 1764 Murray's Ordinance[4] had stated English law to prevail, but in 1774 the *Quebec Act*[5] restored the old laws of Canada for civil actions. This was not a pattern that appealed to the English settlers and United Empire Loyalists who had located far to the west of Montreal. After 1774, they petitioned for the creation of a separate province which was effected by an order-in-council under the *Constitutional Act* of 1791.[6] The first statute[7] of the new Province of Upper Canada provided that:

> In all matters of controversy relative to property and civil rights, resort shall continue to be had to the laws of England, as they stood on the 15th of October, 1792, as the rule for the decision of the same, except insofar as the said laws may have been repealed by Imperial legislation having force in the Province, or by Acts of the provincial legislature.

The basis here is "the laws of England"; the earlier *Quebec Act* of 1774 had adopted "the laws of Canada".

§6.08 George III in 1763 acceded to all rights of the French king, and the continuance — as per Blackstone's views — of the French laws until these might be altered by specific legislation. By that French law — most nearly, it appears, the *coutume de Paris* as a species of the Roman or Civil law (and this predates the Napleonic Code[8]) — the beds of navigable non-tidal rivers remained vested in the Crown, not as a Royal or national domain but as lands of which the public have the use, hence not to be the subject

---

appear that the bed of tidal waters was generally considered Crown land under federal control but see the definition of Nova Scotia lands in S.N.S. 1980, c. 13. As inland waters are deemed non-tidal, there is not in proper usage a foreshore on such waters. A legally acceptable meaning of the colloquial usage may be emerging.

3 Sir William Blackstone *Commentaries on the Laws of England*, 4th ed. (Dublin: John Exshaw, 1765-70), Vol. 1 "Of the Countries subject to the Laws of England" at 106.
4 Governor Murray's Ordinance of 17 September, 1764.
5 1774, 14 Geo. 3, c. 83.
6 1791, 31 Geo. 3, c. 31.
7 *Law of England* S.U.C. 1792, 32 Geo. 3, c. 1, which is here quoted is now included in section 1 of the *Property and Civil Rights Act*, R.S.O. 1980, c. 395.
8 J.-G. Castel, *The Civil Law System of the Province of Quebec* (Toronto: Butterworths, 1962), at 14, 53.

of grant; that is, the title of riparian owners extended to the water and not to the middle thread as is the rule in the common law.

§6.09 These principles were considered in *Dixson v. Snetsinger*[9] where Gwynne J., stated:

> [T]here is no doubt that the river Saint Lawrence was a river the bed and waters of which were vested in the Crown, for the benefit of the public, according to the law of Canada: that, in effect, the rule of the civil law, and not that of the common law of England, which is limited to the extent of the flux and reflux of the tide, prevailed.

THE ONTARIO DEVELOPMENTS

Navigablity, Water Power and the Bed of Navigable Waters Act (1911)

§6.10 From 1872, the date of the *Dixson* case, to 1908 a number of decisions were given in Ontario on the matter of title to land under water, navigability and fisheries. These were reviewed by Mr. Justice Anglin in *Keewatin Power Co. v. Kenora; Hudson's Bay Co. v. Kenora*,[10] an issue arising from a power development proposal on the outlet river of the Lake of the Woods. Reference was made[11] to the opinion of Sir Henry Strong, Chief Justice of Canada, in *Re Provincial Fisheries*:[12]

> Assuming that the Upper Canada cases ... [here cited] ... were well decided, as I hold they were, the soil of all non-tidal navigable rivers, so far as it has not been expressly granted by the Crown, was, at the date of confederation, vested in the provinces, and was reserved to them by section 109 of the Confederation Act.

Mr. Justice Anglin[13] summed up that:

> [The weight of judicial opinion of authority in this Province distinctly supports the view that the soil of our rivers navigable in fact is] presumed to remain in the Crown unless expressly granted. It follows that a Crown grant of lands bordering upon such rivers gives title to the grantee only to the water's edge.

§6.11 Mr. Justice Anglin chose in his decision to recite and analyse the alternative points of view about the law applicable in the province of Ontario. It was his view that the common law had been altered and it was immaterial whether this was seen as arising from the English Crown's assumption of the rights of the French Crown or from colonial variation

---

9 (1872), 23 U.C.C.P. 235 at 243 (Ont. C.A.).
10 (1906), 13 O.L.R. 237 (T.D.).
11 *Ibid.* at 250.
12 (1895), 26 S.C.R. 444 at 527, and see [1898] A.C. 700 (P.C.) on certain matters appealed.
13 *Supra*, note 10 at 263.

of the laws adopted by the first statute of Upper Canada in 1792:[14] title in the non-tidal navigable waters of the province did not go *ad medium filum aquae*. His decision was appealed and in 1908 the decision of the Court of Appeal was given with judgments by Moss C.J.O. and Meredith JJ.A., to which concurrence was given by Osler, Garrow and MacLaren JJ.A. Moss and Meredith JJ.A. saw the 1792 enactment of the legislature of Upper Canada adopting the laws of England as providing the rule that the common law, as set forth, must control the decision. Moss C.J.O. stated:[15]

> In my opinion, the rule of the common law as to the presumption of title in the beds of the streams, whether navigable or non-navigable, still prevails in this Province, and is to be applied in the first instance. Whether there exist circumstances or conditions sufficient to repel the presumption is a question to be dealt with in the particular case.
> I have already said that in the case of the Great Lakes and some of the rivers rebutting circumstances and conditions would not be far to seek.

§6.12 No further appeal was made, but the decision now rested contrary to government policy. With new power development proposals envisaged, the legislature passed the *Bed of Navigable Waters Act*[16] in 1911 to reverse the result of the judgment of the Ontario Court of Appeal. A few years later, Meredith C.J.O. gave his judgment in the Court of Appeal decision of *Haggerty v. Latreille*.[17] He stated that the Act had overridden the prior decisions of the courts respecting the *ad medium filum* rule applying to waters determined to be navigable and in the absence of a grant to the bed.

A Further Ontario Provision Excluding Land Under Certain Waters

§6.13 In Ontario, there is a further exception of land under water stated in section 11 of the *Surveys Act*.[18] This might ordinarily be seen as a strange place to have this provision — the *Public Lands Act*[19] would appear to be the more appropriate Act — were it not for the fact that the section has such a significant effect on the subsequent division of larger parcels by the rules that pertain to aliquot parts.

> 11. (1) Where a lake or river is shown on an original plan of Crown lands

---

14 *Supra*, note 7.
15 *Supra*, note 10 varied on appeal (1908), 16 O.L.R. 184 at 192 (C.A.).
16 S.O. 1911, c. 6, now R.S.O. 1980, c. 40.
17 (1913), 14 D.L.R. 532 at 540-541 (Ont. C.A.).
18 *Surveys Act*, R.S.O. 1980, c. 493, s. 11 (originally S.O. 1913, c. 33, s. 2 enacted following *Williams v. Salter* (1912), 23 O.W.R. 34 (Ont. C.A.)).
19 R.S.O. 1980, c. 413.

and a parcel of land shown thereon is given an acreage covering the land area only, such parcel of land does not include any land covered by the water of the lake or river.

(2) Subsection (1) does not affect the rights of any person where such rights were determined by a court before the 8th day of July, 1913.

§6.14 It is important to note that in these enactments and decisions about property in the bed of watercourses, none has stated that the upland proprietor is *not* riparian. No matter how the natural water boundary may be described, or whether title to the bed is held by the Crown or by a party other than the upland parcel owner, the terms of the description will not detract from the riparian entitlement in the absence of specific exclusionary clauses.

Current Ontario Law Respecting Riparian Entitlement

§6.15   The common law distinguishes between tidal waters and non-tidal waters in so far as the rights of riparian proprietors are concerned. Tidal waters are deemed prima facie to be navigable, and the boundary of an upland proprietor prima facie is at the line of the mean high water mark. On inland non-tidal waters, whether rivers or lakes and whether navigable or not, the ordinary presumption or rule of construction at common law is ownership of the bed by the adjoining riparian proprietors to the middle thread of the river or lake whatever the size of the water space may be. The presumption is rebuttable, as where in the grant the title to the bed is expressly reserved to the Crown.

In Ontario the common law has been varied by statute and by decisions on grounds of the fact of navigability and the interpretation of Crown grants, and title to the bed of waters is not determined on the distinction of tidal or non-tidal character.

The determination of whether the natural boundary of an upland riparian property falls at the bank or at the middle thread of a waterway rests upon the title to the bed and therefore in first instance upon the original patent by the Crown. The bed of a body of water may be reserved to the Crown by specific words of description; or by statute; or it may be retained by a common phrasing of Crown patents that reserves to the Crown the beds of navigable waters. The latter requires first the determination of navigability in law as at the date of the grant for it is at that date that the title to the bed passed to the grantee or was reserved to the Crown as the case may be.[20]

The Ontario courts have laid down the following well-established rules: (1) riparian owner's rights are not founded on the ownership of the bed of the lake or river, but on right of access to the water; (2) a grant of land to the water carries with it to the grantees the right of access to and from the water from any spot on their own lands; (3) any grantee of the Crown

---

20 D.W. Lambden and I. de Rijcke (1985) *Boundaries and Surveys* (Canadian Encyclopedic Digest (Ont. 3rd), Title 19), §§14-16.

228 Survey Law in Canada

must take subject to the right of navigation; (4) the riparian owner has the right to the natural flow and quality of the water, subject to the same rights as his riparian neighbours; (5) the riparian owner is entitled to accretions; (6) the riparian owner and the public have the right of navigation in navigable waters; (7) the right of navigation is a public right, but it may be connected with a right to the exclusive access to particular land on the bank. This latter right is a private one, invasion of which may form a ground of action for damages. By the law of England, when the lands of two coterminous proprietors are separated from each other by a running stream of water, each proprietor is prima facie owner of the soil of the alveus or bed of the river ad medium filum aquae.[21]

§6.16 The last statement was specifically altered by the *Beds of Navigable Waters Act*[22] of Ontario. Since 1911 the common law presumption in respect of ownership of the bed of waters deemed navigable by a court has not applied unless there has been an express grant of the bed or the matter of private ownership has been decided by a court prior to the enactment in 1911.

THE LEGAL SITUATION IN THE PRAIRIE PROVINCES

Statutory Change of Common Law Rules

§6.17 In the prairie provinces a different situation appears to exist. Following the 1869 surrender of the lands of the Hudson's Bay Company to the federal Crown, the *Dominion Lands Act* was enacted in 1872 as the first legislation containing provisions for survey and for disposition of the Crown lands.[23] The Act and the Manual set out a strict survey system of townships, sections and quarter sections, defined the meaning of boundary,[24] and provided that the parcel was to be within these bounds and no more or less.[25] The Act provided generally that no sale to one person would exceed a section, or 640 acres[26] — that is, the section, or lesser unit, within the boundaries fixed by the scheme of the survey was to prevail as to the area granted.

§6.18 The question of entitlement to the bed of non-navigable waters in areas surveyed under the Act was considered in the Saskatchewan case

---

21 H.P. Eccles, (1981). *Waters and Watercourses.* (Canadian Encyclopedic Digest (Ont. 3rd), Title 148), §33.
22 R.S.O. 1980, c. 40, s. 1.
23 *Dominion Lands Act*, S.C. 1872, 35 Vic., c. 23.
24 *Ibid.* s. 129.
25 *Ibid.* s. 130.
26 *Ibid.* s. 29.

of *R. v. Fares*, where, in the unanimous decision that reversed the lower court judgment, Mr. Justice Anglin, Chief Justice of Canada, stated:[27]

> [T]he *ad medium filum* rule of English law . . . ordinarily applicable in Saskatchewan to non-navigable waters . . . is, at the highest, a rule of interpretation, and the rebuttable presumption thereby created yields readily to proof either of circumstances inconsistent with its application, or of the expressed intention of a competent Legislature so to exclude its application. . . . the intention of the Dominion Parliament — an authority competent so to provide — to exclude the application of the rule to Dominion lands in the North West Territories, was sufficiently manifested by the provisions of the *Dominion Lands Act* [R.S.C. 1886, c. 54].

§6.19 A decade earlier, the inapplicability of the rule to non-tidal waters whether navigable or not, had been decided for Alberta in *Flewelling v. Johnston*[28] and nearly simultaneously for navigable rivers in Manitoba in *Re Iverson and Greater Winnipeg Water District.*[29]

§6.20 At the root of these decisions lies the constitutional principle of adoption of the English law in the prairie provinces. By the *North-West Territories Act* of 1886:[30]

> Subject to the provisions [of this Act] the laws of England relating to civil and criminal matters, as the same existed on the . . . [15th July, 1870] . . . shall be in force in the Territories, *in so far as the same are applicable to the Territories*, and . . . [as changed by statute, etc.]

It was the absence of the *underlined* phrase from the Ontario Act that led to the rejection by the Ontario Court of Appeal of the decision given at trial in the *Kenora* case.[31]

§6.21 The decision in *R. v. Fares*[32] was that the *ad medium filum* rule of the common law was not applicable to lands surveyed and granted under the *Dominion Lands Act*. However, in the British Columbia case of *Rotter v. Canadian Explorations Ltd.*,[33] the Supreme Court of Canada held the rule applied to non-tidal, non-navigable waters and distinguished *R. v. Fares* as having been decided on the applicability of the *Dominion Lands Act* and the interpretation of the grants.

---

27 [1932] S.C.R. 78 at 80, reversing [1929] Ex. C.R. 144.
28 (1921), 16 Alta. L.R. 409 (C.A.); and see also *R. v. Miller*, [1941] 1 W.W.R. 483 (Alta. Dist. Ct.).
29 [1921] 1 W.W.R. 621 (Man. C.A.).
30 S.C. 1886, c. 25, s. 3.
31 *Supra*, note 15.
32 *Supra*, note 27.
33 [1961] S.C.R. 15, reversing (1959), 30 W.W.R. 446.

230   Survey Law in Canada

§6.22 In Alberta the question of ownership of the bed is effectually resolved by the *Public Lands Act*;[34] navigability of the waters is not a criterion.

> 3. (1) Subject to subsection (2), the title to the beds and shores of all rivers, streams, watercourses, lakes and other bodies of water is hereby declared to be vested in the Crown in right of Alberta and no grant or certificate of title made or issued before or after the commencement of this Act shall be construed to convey title to those beds or shores. [Subsection 2 provides for certain exceptions.]

These statutory provisions, with the date of June 18, 1931, for Alberta and July 15, 1930 and April 1, 1931, for the similar provisions for Manitoba and Saskatchewan respectively,[35] provide continuity with the earlier federal enactment, the *North-west Irrigation Act*, originally passed in 1894;[36] where the Act applied the bed of the waters would not pass with the grant.

§6.23 The prairie provinces laws are statutory and generally exclude the continued operation of the common law principle of entitlement to the land under any water, except where a judicial interpretation of a grant might rule otherwise. In Ontario the law on this matter remains the common law except for the statutory provision that waterways that are navigable, in fact, are excluded from the title and except for the special exclusion in certain grants that arises from section 11 of the *Surveys Act*.[37] The western statutes[38] go much farther than this and provide that the property in any water to which the legislative authority extends is vested in the province. By contrast, the common law, as pertains in Ontario, sees no property in water until it is possessed in an entirely personal manner.

§6.24 The laws about the beds of waters are distinctly different in the Canadian provinces, but four common factors remain: first, the original patent from the Crown must be considered and assessed in light of the law in force when issued; second, later legislation may have changed the

---

34   R.S.A. 1980, c. P-30, s. 3 and 4 (originally S.A. 1931, c. 43, s. 8).

35   For Manitoba, see the *Crown Lands Act*, R.S.M. 1970, c. C-340, s. 5. [Per W.E. Shepherd, M.L.S., letter of 5 February, 1988. Section 5 (1) . . . allows for certain reservations with respect to water boundaries (i.e., a reservation of a strip of land one and one-half chains wide above the ordinary high water mark of navigable waters, and the bed of a body of water below the ordinary high water mark). It has always been assumed that the Act dealt only with grants made after 1930 and was not retroactive as in the case of the Alberta Act which is quoted. Section 38 of the Manitoba Act allows the application of the *Dominion Lands Act* as it existed in 1930 on such matters that are not inconsistent with or varied by provincial legislation.] For Saskatchewan, see the *Provincial Lands Act*, R.S.S. 1978, c. P-31, s. 12.

36   *North-west Irrigation Act*, S.C. 1894, c. 30, s. 5; amended by S.C. 1920, c. 55.

37   *Supra*, note 18 at s. 11.

38   Alberta *Water Resources Act*, R.S.A. 1980, c. W-5, s. 3; Manitoba *Water Rights Act*, S.M. 1982-83-84, c. 25, s. 2; Saskatchewan *Water Rights Act*, R.S.S. 1978, c. W-8, s. 7.

situation with retroactive provisions; third, title subsequent to grant must be searched as granted parcels in the bed of waters may have reverted to the Crown by conveyance or by dedication; fourth, the decisions of the courts on matters of title must be followed.

Boundaries: Provisions for the Survey of Dominion Lands

§6.25 In each of the western cases discussed above, reference was made to the 1918 *Manual of Instructions for the Survey of Dominion Lands*[39] which, at first glance, seems specific in defining the natural boundaries. Clause 193 reads:

> Land abutting on tidal waters is bounded by the line of ordinary high water. In the case of an inland lake or stream, the boundary, if the parcel does not include the bed, is the edge of the bed of the lake or stream, which edge is called the bank.

Previous manuals had used the word 'bank' without definition; the 1892 edition[40] added in clause 100 "[t]he plans must show the actual water line at the time of the survey and the deductions from areas are to be calculated accordingly."

§6.26 Shorelines were usually traversed to locate the extent and area of water bodies for their exclusion from the granted parcels. The traverse line was always known to be a feature of surveying methods and not a boundary feature; the 1903 *Manual* stated this principle in clause 134: "[t]he courses of a traverse are not boundaries of the parcels fronting on bodies of water" and this rule was restated in the subsequent editions.

§6.27 Following on several court decisions (which were not cited, but see, for example, *Hindson v. Ashby*[41]), clause 134 of the 1903 Manual contained the following:

> In the case of a lake or navigable stream, the boundary is the edge of the bed of the lake or stream, which edge is called the *bank*. The bed of a body of water has been defined as the land covered so long by water as to prevent vegetation, and the bank is the line where vegetation ceases. The *shore* is the space between the bank and the water.

The changes in wording for more explicit meaning appear to have followed

---

39 Dr. E. Deville, Surveyor General, *Manual of Instructions for the Survey of Dominion Lands*, 9th ed. (Ottawa: King's Printer, 1918).

40 Dr. E. Deville, Surveyor General, *Manual Showing the System of Survey of the Dominion Lands, with Instructions to Surveyors*, 4th ed. (Ottawa, Queen's Printer, 1892).

41 [1896] 1 Ch. 78, reversed on the facts by [1896] 2 Ch. 1 (C.A.), but the law stated by Romer, J., was endorsed. See also the cases therein cited.

on the 1894 *Irrigation Act* which used the words 'bed' and 'shore'. This was repeated in the 1905 and 1910 editions. In 1918, the provisions were restated as follows:

> 194. The *bed* of a body of water has been defined as the land covered so long by water as to wrest it from vegetation, or as to mark a distinct character upon the vegetation where it extends into the water or upon the soil itself. According to this definition, the limit of the bank is the line where vegetation ceases, or where the character of the vegetation and soil changes.
>
> 195. The *foreshore* or *shore* is the strip of land lying along tidal waters, over which the daily tide ebbs and flows; it is the space between high and low water marks at ordinary tides.

§6.28 'Bank' was, therefore, specifically a term for use on inland waters and it was given a linear and not an areal meaning. Clause 194 of the Manual was not itself a definition and was not in fact a particularly good exposition of what it was attempting to define. The case law is of much more value in provoking serious thought as to what does constitute the line of the natural boundary.

THE AD MEDIUM FILUM RULE

§6.29 By the common law, ownership of upland property adjoining a watercourse gives rise to the presumption that title extends to the middle thread. It is a rebuttable presumption — that is, evidence to rebut the presumption is always admissible.[42]

§6.30 In what terms must a grant be couched to make the upland property riparian? It must be such as to imply contact with the water.

> In Ontario, it has been well settled by the Courts that when applied to navigable, non-tidal bodies of water the terms 'bank', 'line of the bank', 'shore', 'line of the shore', 'margin of the water', and 'water's edge' are synonymous as lines of demarcation. And so also 'to the bank of lake', 'to the lake' and 'to the shore' mean to the edge of the water in its natural condition at low water.
>
> No description in words or by plan or by estimation of area is sufficient to rebut the presumption that land abutting on a stream carries with it the land ad medium filum merely because the verbal or graphic description describes only the land that abuts on the stream without indicating in any way that it includes land underneath that stream. This doctrine is not a mere inference or dedication; it is not a mere convenience in conveyancing; but it is, and is nothing less than, a presumption of, and applicable to, ownership itself.

---

42 Atkinson, P. (1976). *Evidence*. (Canadian Encyclopedic Digest (Ont. 3rd), Title 57), §116.

This is too deeply embedded in the law to be disturbed or doubted, but the rule of construction is a rebuttable presumption.[43]

§6.31 In Ontario, therefore, the common law rule of title extending *ad medium filum* is applicable now only to non-navigable waters. The rule of the statute law, the *Beds of Navigable Waters Act*,[44] is that title adjoining navigable waters is limited to the natural water boundary, properly called the 'bank' of rivers and the 'shore' of lakes. In either case, it would appear to be the same natural physical limit that must be located by survey as the limit of the upland riparian proprietor abutting navigable waters, or as the land/water interfaces on both sides of non-navigable waters, the distance between which is then split across the watercourse to locate the middle thread.[45]

§6.32 In 1940 a significant, but short lived, amendment[46] was made to the Ontario *Beds of Navigable Waters Act* stating the synonymity of the various terms and equating all to the single term 'high water mark'. In 1951, the legislature repealed[47] the amending provisions of 1940, restoring the common law rule that determines the line of land/water interface for the natural boundary. This, of course, applies in the absence of a specific reservation to the contrary, such as a shore road allowance or other reservation of the upland which would make it non-riparian.

---

43 *Supra*, note 41 at §§26 and 32. The mode of description seems to be at the root of several contests in the courts and of many actions on policy by the Crown contending that by not wording grants with specific mention of a water boundary (e.g., "water's edge") leads to the upland parcel not being riparian. The argument is contrary to the decisions of the highest courts. See, for example, *Bruce v. Johnson*, [1953] O.W.N. 724, and commentary thereon by Professor C.A.G. Palmer in 32 *Canadian Bar Review* (1954), at 328. See also the comments of Professor Palmer in 31 *Canadian Bar Review* (1953), at 713. Also note the views expressed in *Chuckry v. R.*, [1973] S.C.R. 694, reversing (1972), 27 D.L.R. (3d) 164 per Guy, J.A., at 172 and Dickson, J.A., at 180 (Man. C.A.).

44 *Supra*, note 22.

45 However, in the *Manual of Instructions for the Survey of Dominion Lands* (Dr. E. Deville, Surveyor General, 5th ed., (Ottawa: Government Printing Bureau, 1903)), the thalweg, the middle thread of the main channel ("middle being in the sense not of midway between the banks, but of the line following the centre of the main body of water flowing in the channel") was stated to be the line of boundary ad filum aquae for lands bordering upon unnavigable waters. In the 8th edition of 1910, clauses 136 and 137 were amended, the specific words omitted and the meaning to be derived for "middle of the stream" left, perhaps, uncertain. Presumably the titles that followed on the basis of surveys and plans made under these instructions created the thalweg as the boundary. This is not the common law meaning as applied in Ontario: *R. v. Carleton* (1882), 1 O.R. 277 at 282 (C.A.); *Bartlet v. Delaney* (1913), 29 O.L.R. 426 at 433 (C.A.).

46 *Statute Law Amendment Act*, S.O. 1940, c. 28, s. 3.

47 *Beds of Navigable Waters Amendment Act*, S.O. 1951, c. 5.

THE SPECIFIC PROBLEM OF INLAND WATER BOUNDARIES

Terminology

§6.33 In the absence of legislation on water boundaries, recourse must be made to the decided cases for answers. Over time these have been more explicit as the need for more strict resolutions arose. Thus we read in Mr. Justice Romer's trial decision in *Hindson v. Ashby*:[48]

> I know of no authority in English law which has expressly decided this [how the bed of a river is to be ascertained], but on principle I think it ought to be so decided. In the United States the question has been judicially considered, and a view expressed in accordance substantially with that formed by me.

On appeal the decision of Romer J. was reversed on the facts but the criteria for determining the extent of the waterway were unanimously approved. It was these criteria that were followed for their good reasoning in *Clarke v. Edmonton*.[49]

§6.34 The present problem of water boundaries has become the issue of 'high water mark' and 'low water mark' as terms to define the limit. Mr. Justice Romer rejected the analogy with a tidal water regime. Nor did he accept arguments for the lowest or for the highest lines of flow during an average or ordinary year.[50] In *Parker v. Elliott*,[51] Chief Justice Macaulay argued for high water mark as the line of demarcation on the Great Lakes but the majority opinion of the Appeal Court was against any legal meaning attaching to the term on inland non-tidal waters.

§6.35 In the *A.G. for Ont. v. Walker*[52] decision of the Supreme Court of Canada, affirming the Ontario Court of Appeal which had affirmed the trial court decision, the matter may be seen to have been put to rest. On inland waters, the 'bed' is that land covered by water below the low water level, and the 'bank' of river, like the 'shore' of a lake (though the terms have been used interchangably), is the space from the low water level up to that line that has received the equally uncertain designation of 'high water mark.'

§6.36 However, the term 'high water mark' cannot be avoided because it is in many acts, regulations, instructions and documents of conveyance.

---

48 *Supra*, note 39 at 84; reversed on the facts [1896] 2 Ch. 1. (C.A.).
49 [1930] 1 S.C.R. 137.
50 *Supra*, note 39 at 84.
51 (1852), 1 U.C.C.P. 470, 49ln at 481 (Ont. C.A.).
52 [1975] 1 S.C.R. 78, affirming [1972] 2 O.R. 558 (C.A.), which affirmed [1971] 1 O.R. 151 (H.C.).

Its meaning in legal terms appears to have come from the general descriptions for finding the 'banks' as the line of boundary, such descriptions as are endorsed in *Hindson v. Ashby*[53] More recently these new terms themselves have been argued to control the limit, that is, the 'bank' or the 'shore', thus converting the tests into the object of the test which is to find the riparian boundary. This progression is illustrated in the 1961 *Manual of Instructions for the Survey of Canada Lands*[54] where clause 19 of Definitions (chapter 2, part B) introduces the term 'high water mark' as the limit on inland waters under Canada lands administration:

> The 'ordinary high water mark' of a body of water is the limit or edge of the bed of the body of water and in the case of non-tidal waters may be called 'the bank'.

This definition equates 'ordinary high water mark' and 'bank' in the sense of line of boundary. The definition clause is restated with the same substance in the second edition of 1979 (chapter B2, clause 20). In fact, this new terminology appears to have originated with a memorandum to Dominion Land Surveyors on December 18, 1952, stating that notwithstanding clause 193 of the 1918 and 1946 editions the term 'ordinary high water mark' was to be used in order to concur with the *Territorial Lands Act* when referring to a boundary between a water body and land in the Territories. Despite changing terminology, the manuals maintained the statements that all upland parcels so variously bounded held the riparian rights.

§6.37 If one allows that the amendment to the Ontario *Beds of Navigable Waters Act* as it existed from 1940 to 1951 sought a normal, or ordinary or average or general pattern of high water mark, and not the highest water mark, the Ontario approach was identical to the continuing federal instruction. Where the provincial effort failed was in definition and in the understanding of that definition by administrators and property owners and by the surveying profession caught in the middle in the execution of their duties. To argue otherwise is to impute to the Crown an intention of entirely improper expropriation by statutory provisions without compensation.

Water Levels in Rivers and Lakes

6.38 The differences in the water level patterns of rivers and lakes is recognized. In rivers there are some years of fairly steady higher levels

---

53 *Supra*, note 39.
54 R. Thistlethwaite, Surveyor General, *Manual of Instructions for the Survey of Canada Lands* (Ottawa: Information Canada, 1961).

236 Survey Law in Canada

and others of lower levels, a pattern that is repetitive but not regular, as the after-effects of short-term climatic conditions. There are also the extremes of floods and droughts as abrupt changes of short-term weather patterns. The smaller inland lakes also reflect much the same patterns and at much the same times.

§**6.39** The Great Lakes have 'high water levels' and 'low water levels' each year with a range of some 12 to 15 inches; there are longer term extreme high and low levels of irregular occurrence ranging as much as six feet in Lake Huron and Lake Ontario. This latter variation of advance and recession of waters is the result of longer term changes such as periods of continuing drought reflected in the low levels of the early 1930's and again in the mid 1960's. The tidal analogy is not applicable in the legal sense and designating an ordinary or normal or average high water mark or low water mark, or an ordinary or normal or average water's edge, is substantially a fiction. A natural water boundary is a very real thing — it can be seen and it can be felt. On the other hand, numerical location by measurements in the field requires a feature more or less strictly definable to which measurements can be made or a repeat performance will not give the same position. More to the point, the later survey will almost assuredly give a different location simply because the feature has moved.

§**6.40** The high and low water levels of rivers are well recognized in law. So also are the resultant difficulties in defining the bed. Romer J. in *Hindson v. Ashby*[55] noted that

> [I]f a river preserves a tolerably even flow, and does not fluctuate in volume much, except on extraordinary occasions, there is no difficulty in determining its bed. But when the river is one ... which, apart from extraordinary floods or droughts, changes considerably in volume in the course of each year, being, as a rule, much higher in some months of the year than in others, then the question as to how its bed is to be determined is not so easy.

§**6.41** In *Flewelling v Johnston*,[56] evidence was received of survey investigation about the water boundary.

> [T]he evidence as to low and high water marks is I think of value in deciding the question of navigability ... The river is stated ... to be about 300 feet wide at the point in question. Low water seems to be at the place in question about three feet deep. With a river of such a size and depth and an ordinary yearly rise of 9.5 feet lasting about two months of about seven months while it remains unfrozen, and with [the] admission that ... boats could navigate it ... and with one's general knowledge of the use to which all such rivers

---

55 *Supra*, note 39 at 83.
56 *Supra*, note 28 at 417-421 per Beck J.

are put . . . for commercial purposes, it seems to me that it is evident that the Pembina River is as a matter of fact and law a navigable river.

. . . .

In taking these findings of fact I would hold that there is no rule of law in Canada carrying *prima facie* the title of a bordering owner to the middle thread of the water in the case of waters navigable in fact; that the boundary in such cases is ordinary high water mark.

§6.42 In this Alberta decision we have the specific statement that 'high water mark' is the boundary (but no guidance as to how to find it, as that matter was not in issue), whereas until 1962, as earlier noted, the *Manual of Instructions for Dominion Land Surveyors* had used only the term 'bank'; the means to find its location were descriptive explanations without use of the expression 'high water mark'.

§6.43 Faced with the indisputable fact that there are high and low water levels — long term, annual, and momentary — the proposition in the legal aspect is to determine the rules that are currently seen to apply and to find out what to do about the boundary to rationalize the interpretation. There are the constructions to be put upon the grants, on the survey records and on any statutory provisions, regulations, guidelines and instructions.

The Interpretation of Water Boundary Terms

§6.44 The rule for the interpretation of grants is well settled and has been often repeated in Canadian decisions. As stated by Maclaren J.A. in *Williams v. Pickard*,[57] it is:

In construing the language of the patent the proper rule . . . is that, 'upon a question of the meaning of words the same rules of common sense and justice must apply whether the subject matter of construction be a grant from the Crown or from a subject; it is always a question of intention, to be collected from the language used with reference to the surrounding circumstances.'

There is also the presumption that deeds are made with caution and forethought and on good advice. The interpretation of a deed is a serious matter for the courts.

§6.45 Legal decisions provide an answer to the question of what is the natural boundary of the land/water interface and how it is to be found. In *Hindson v. Ashby*,[58] there are the often quoted words of Mr. Justice Romer.

---

57 (1908), 17 O.L.R. 547 at 549 (C.A.) citing *Lord v. The Commissioners of Sydney* (1859), 14 E.R. 991 (P.C.).
58 *Supra*, note 39 at 84; reversed on the facts [1896] 2 Ch. 1 (C.A.), but the principles stated by Romer J. were endorsed.

I think that the question whether any particular piece of land is or is not to be held part of the bed of the river at any particular spot, at any particular time, is one of fact, often of considerable difficulty, to be determined, not by any hard and fast rule, but by regarding all the material circumstances of the case, including the fluctuations to which the river has been and is subject, the nature of the land, and its growth and its user.

§6.46 The English and American decisions reviewed and endorsed in *Clarke v. Edmonton*[59] provide a means of finding the bed of a river. They have also been seen as giving a definition of 'high water mark' or 'ordinary high water line'. For example, the definition given in the *Manual Relating to Surveys and Surveyors*[60] is interpretive, but not a ruling —

> 'High Water Mark' means the mark made by the action of water under natural conditions on the shore or bank of a body of water, which action has been so common and usual and so long continued that it has created a difference between the character of the vegetation or soil on one side of the mark and the character of the vegetation or soil on the other side of the mark.

This is effectively a precis of the words of Mr. Justice Curtis of the United States Supreme Court in *Howard v. Ingersoll*.[61] Curtis J. went on to say that "neither the line of ordinary high-water mark, nor of ordinary low-water mark, nor of a middle stage of waters can be assumed as [the] line." What is given is the definition of 'bank' in its sense as the line of boundary.

§6.47 In the amendment of 1940 to the *Beds of Navigable Waters Act*,[62] which was in effect for 11 years until repealed in 1951,[63] the following definition was given:

> 'high water mark' shall mean the level at which the water in a navigable body of water has been held for a period sufficient to leave a watermark along the bank of such navigable body of water.

There are inevitably a series of watermarks and it is undefined which one is to be taken.

§6.48 In these definitions, resort to natural objects is recommended. The definition of the statute, and to a lesser degree the definition in the Ontario

---

59 *Supra*, note 49.
60 (Toronto: Assoc. of Ontario Land Surveyors, 1959, 1973). See also *Crown Lands and Forests Act* S.N.B. 1980, c. C-38.1, s. 1, where 'normal high water mark' is defined to mean
> [T]he visible high water mark of a lake or river where the presence and action of water are so usual and so long continued in ordinary years as to mark upon the bed of the lake or river a character distinct from that of the bank thereof with respect to vegetation and the nature of the soil itself.
61 (1851), 54 U.S. 381 (corrected headnote at 54 U.S. xii) (U.S. Sup. Ct.).
62 *Statute Law Amendment Act*, S.O. 1940, c. 28, s. 3.
63 *Beds of Navigable Waters Amendment Act*, S.O. 1951, c. 5.

Manual, fails to provide the specifics which are needed. These can be substantially obtained in the full statements of Curtis J. and Romer J. and, as these have been adopted and approved in other decisions, it is inevitably a matter of opinion.

§6.49 Behind all these decisions stands the matter of riparian status and one clear fact emerges — the name attached to the limit is secondary to the riparian status arising from access to and contact with the water. In light of the decisions, it may be argued that the term 'high water mark' is actually nonsensical, if not invalid. In effect, this takes us back to the majority decision in *Parker v. Elliott*[64] where it was determined that a distinction between high and low water levels cannot be drawn on inland non-tidal waters for the purposes of defining boundaries.

§6.50 The Court of Appeal judgment of *Attersley v. Blakely*[65] recites in full the decision of Lane Co. Ct. J., which it affirmed, and which includes the following:

> It would seem, therefore, that the old common law rule as to the boundary between land and water placing it at the water's lowest mark is the law as it stands at the moment.

The reason behind specifying the water's lowest mark is not evident from individual cases, but collectively it may, perhaps, be logically deduced.

§6.51 In *Stover v. Lavoia*,[66] Chancellor Boyd had pointed out that this interpretation was well understood to be the case in both this country and the United States. While there is no doubt that this is the old common law interpretation, reference to Farnham[67] shows that there is considerable division of opinion in the United States. It would appear that the high water mark as a boundary was the later and much contested development in the common law, arising from the Crown's effort to gain, or regain, control of the waterways, non-tidal as well as tidal. There seems little doubt that in Canada the use of the term 'high water mark' had its origin with the Crown lands administration of Upper Canada. A meaning has to be ascribed to the term. At the moment it can be fairly summed up that the image conveyed by the term, and consequently the positioning of the mark as a limit on the ground, is not consistent between Crown officers, riparian proprietors, surveyors and the public. By contrast, where

---

64 *Supra*, note 51.
65 [1970] 3 O.R. 303 at 308 (Co. Ct.); affirmed [1970] 3 O.R. 313 (C.A.).
66 (1906), 8 O.W.R. 398 (H.C.); affirmed 9 O.W.R. 117 (C.A.).
67 H.P. Farnham, (Rochester: Lawyers' Co-op Publishing Co., 1904), chapter iv, The Law of Waters and Water Rights.

title extends to the 'water's edge', the boundary feature is a wholly tangible thing and it is immediately known as ambulatory.

§6.52 Uncertainty also arises in attempting to isolate precise meanings for 'bank' and 'shore' along a watercourse, that is, whether they are used in a popular or in a strictly legal sense; whether they are areas above or below the line of water; when they are to be treated as areas and when as lines; and, if it is an area above the water line, when to recognize a 'top of bank' and a 'bottom of bank' and the circumstances of the particular case which will determine which or neither is the boundary.[68]

§6.53 A definition of the term 'shore' has not been found in Ontario survey literature or Crown instructions for surveys. While the word appeared in early editions of the *Manual for the Survey of Dominion Lands*, it was not until the fifth edition of 1903[69] that a definition was given in clause 134: "The shore is the space between the bank [here a line] and the water."

§6.54 In the 1910 Manual[70] there was a reference to section 7 of the *Northwest Irrigation Act*[71] of 1894. Clause 136 of the Manual stated that

> [N]o grant shall be made by the Crown of any exclusive property or right in the land forming *the bed or shore* of any lake, river, stream or other body of water. *The word 'shore' in this section is presumed to be intended to designate that part of the bed which is uncovered when the water is low.* [emphasis added]

§6.55 In the trial decision of *A.G. for Ont. v. Walker*[72] Mr. Justice Stark asserted that

> [A]ny Crown patent which indicates that one of the boundaries of the lands granted is to be a boundary of water . . . establishes that boundary as at the water's edge and not upon any bank or high water mark unless . . . the grant clearly reserves by description or otherwise a space between the lands granted and the water boundary or unless the boundaries of the lot can be so clearly delineated by reference to an original plan of survey as to clearly except or reserve to the Crown a space between the lands granted and the water's edge.

---

68 The need to recognize a 'bank' above the waters is evident in words of grants such as "reserving the right of access to the shores of all rivers, streams or lakes for all vessels, boats and persons, together with the right to use so much of the banks thereof, not exceeding one chain in depth from the waters edge, as may be necessary for fishery purposes." And see, for example, *Williams v. Pickard, supra*, note 57; *Robertson v. Watson* (1874), 27 U.C.C.P. 579 (C.A.). Also, see note 43.
69 *Supra*, note 45.
70 Dr. E. Deville, Surveyor General, *Manual of Instructions for the Survey of Dominion Lands*, 7th ed. (Ottawa: Government Printing Bureau, 1910).
71 S.C. 1894, c. 30.
72 *Supra*, note 52 at 181.

The term 'bank' is here used in its general sense as an area of land lying above the water's edge. It was the latter line which was determined on the precedents to be the line of boundary.

§6.56 In *Georgian Cottagers' Assn. v. Flos (Twp.)*,[73] Mr. Justice Gale deduced a meaning for 'banks' in the sense of area rather than line. By the legal decisions, 'shore' is strictly applicable only to the space between ordinary high and low water marks of waters that are subject to the ebb and flow of the tides. It is, however, applied to inland non-tidal rivers and lakes, either as synonymous with 'bank' as a line or as denoting that portion of the bank as an area which touches along its lower side the margin of the waters of the lake or stream at low water. The terms are synonymous in both line of boundary sense and area sense.

§6.57 Along a watercourse, the term 'bank' has both a wide general meaning of 'any steep acclivity' or 'bordering slope' and a narrower legal meaning that defines the limits of adjoining upland property and determines whether or not it is riparian.

> [W]hat are usually termed riparian rights, or rights of proprietors of land on the banks of streams, [arise], strictly speaking, not from the ownership of the bed over which the water flows, but from the right of access which such proprietors have to the water. In the case of non-tidal waters, where the owner of land on the banks is *prima facie* owner of half the bed, this may appear a fine-drawn distinction; but on the banks of tidal waters, where the ownership of the bed is *prima facie* in the Crown, the distinction will be manifest — as the origin of such rights cannot be referred to ownership of the bed.[74]

§6.58 The 1848 edition of Barclay's *A Complete and Universal Dictionary of the English Language*[75] defines 'bank' as "a rising ground on each side of a river, *washed by its waters*, which it hinders from overflowing. [emphasis added]"

§6.59 At the very root of the definition of 'banks' is the definition of 'watercourse'. It is always recognized to consist of the bed, the banks, and the water; that is, a watercourse is the flow of water in a defined channel. Without banks there can be no channel and hence riparian rights do not arise; the waters then are surface waters. A lake is a body of water but may not be a watercourse; for example, a channel of flow is absent in the Great Lakes. There is, however, containment or pondage that distin-

---

73 [1962] O.R. 429 at 441 (H.C.).
74 S.R. Hobday, ed., *Coulson and Forbes on the Law of Waters, Sea, Tidal and Inland and Land Drainage*, 6th ed. (London: Sweet & Maxwell, 1952).
75 Rev. James Barclay, *A Complete and Universal Dictionary of the English Language*, ed. by B.B. Woodward (London: James S. Virtue, 1848).

guishes such features from superficial waters and riparian rights attach to the abutting uplands and follow the advance and recession of the waters from natural causes.

§6.60 The key to the existence of riparian rights is in the access to the water. The expressions 'to the bank' and 'to the shore' have consistently been determined by the courts as meaning contact with the water; one cannot go 'to' the feature unless that feature lies beyond and the riparian entitlement is fulfilled by the contact with the water. Angell[76] says: "A bank of a river is that space of rising ground above low-water mark which is usually covered by high water, and the term, when used to designate a precise line, may be somewhat vague and indefinite." The 'banks' in this legal sense are the holding sides of the river or the lake — more properly called the 'shore' of a lake — short of overflowing under conditions of flooding. Angell is here recognizing 'bank' in both the linear and the areal sense.

§6.61 In the New Zealand case of *Kingdon v. Hutt River Bd.*,[77] it was determined that in law the 'bed' of the river extends from bank to bank. It is not confined to a narrowed channel in which the water may for a time flow in dry weather, nor does it extend beyond the banks to land over which the water flows in time of flood. The case was decided without use of the term 'high water mark' which is, as earlier noted, an indicator or test and not the object of the test. Other cases have established that a watercourse is still at law a watercourse even if its channel is occasionally dry;[78] the abutting upland owners remain riparian proprietors.

§6.62 Mr. Justice Stark observed in the trial decision of *A.G. for Ont. v. Walker*:[79]

> [I]t was pointed out in evidence that in the granting of water lots, greater accuracy would be obtained by tying the dimensions of the water lot to the northerly corners of the [township] lot, which were in fact fixed, rather than tying the dimensions to the high water mark, which though not as variable as the water's edge, is still recognized as being a line which may vary from year to year.

Any shoreline feature is variable. The evidence in *Servos v. Stewart*[80] showed erosion of some 790 feet in some parts of the shoreline in a period of 108 years.

---

76 J.K. Angell, *A Treatise on the Law of Watercourses*, 5th ed. (Boston: Little Brown and Co., 1854) see notes at 2.
77 (1905), 25 N.Z.L.R. 145 (S.C.).
78 See, for example, *Stollmeyer v. Trinidad Lake Petroleum Co.*, [1918] A.C. 485 (P.C.).
79 *Supra*, note 52 at 173.
80 (1907), 15 O.LR. 216 (H.C.).

§6.63 While 'bank' and 'shore' are defined terms within clear legal concepts that recognize the ambulatory nature of natural boundaries, there is still the irreconcilable fact that there is no conclusive definition of the term 'high water mark' or 'low water mark'. The very word 'mark' is suggestive of fixing and this notion does appear in the policy of Crown officials, despite the decisions of the courts to the contrary. The attempts at definitions are interesting and sometimes useful, but they are not conclusive. They are attempts at specifics in things that are of natural origins, which things rather defy specifics, except that all the terms relative to the natural water boundary import an ambulatory line of boundary.

§6.64 In the British Columbia case of *Dunstan v. Hell's Gate Enterprises Ltd.*,[81] one of the issues was the possible occurrence of accretion and the extension of a road over the accreted land from an old to a new high water mark on the inland non-tidal, navigable Fraser River. Mr. Justice Cumming, at trial, considered the instructions issued to land surveyors under federal and provincial authority to "simply reflect rules of the common law and standard surveying practices which have existed for many, many years." It is important to note, however, that in the early Dominion Land Survey instructions for surveys, the word used was 'bank' not 'high water mark'. Mr. Justice Cumming looked then to *Clarke v. Edmonton*[82] and cited the comments of Lamont J. for the legal specification of the natural boundary terms.

Summary: the Natural Boundary of Lakes and Rivers

§6.65 It is inevitable that some marks are left by the high levels of waters on inland lakes and rivers. These marks may be seen as the line of pieces of wood and other debris left by receding waters, by some broken line of the general edge of terrestrial vegetation, by changes in soil characteristics or by the edge of some embankment particularly scored by the action of water. It is a variable line in characteristic indicators and distinctiveness, but for any lake it will be a level line to all intents and purposes whereas on a river it will be a line of variable gradient. It is found by the considerations of all visible evidence, not alone by one indicator.

§6.66 Along a river, the variations of very low (or drought) and very high (or flood) conditions are, by all the decisions, excepted from the extent of the river which flows between the lines of the banks. Thus, a river is

---

81 (1985), 22 D.L.R. (4th) 568 at 581 (B.C. S.C.); reversed (1987), 45 D.L.R. (4th) 677 (C.A.). The reversal was on matters other than the survey aspects here discussed.
82 *Supra*, note 49.

said to be out of its banks when it is in flood and the law finds the bed to extend from bank to bank when the river is a a normal high flow. There is an essential channel of flow in the case of rivers — that is, a channel which cannot be blocked or narrowed without creating an undue hazard and an actionable wrong. The banks of a river contain the flow of the river at its fullest normal natural state, not when temporarily overflowed by flood waters, and the term 'high water mark' is the indicator of this line of the normal natural higher level of the water.[83]

§6.67 The majority opinion expressed in *Parker v. Elliott*,[84] and consistently confirmed, is that a legal distinction cannot be drawn in respect of the high and low levels on the inland non-tidal waters. The effect of this opinion is that a grantee of an upland parcel adjoining waters is not to be restricted in so far as enjoyment of the riparian rights is concerned; that is, the proprietor of a title to the upland adjoining water has a boundary with the water whether that level is seasonally high or low or advances or recedes in the long term irregular patterns of the Great Lakes.

§6.68 The issue in *A.G. for Ont. v. Walker*[85] was, simply stated, whether the original grant was subject to a newly invoked terminology — 'high water mark' — and was therefore to terminate short of the water's immediate edge. The court would not so hold. It is interesting to note that in the cases where 'water's edge at low water' has been ruled by the court to be the boundary, it has been in the face of argument for 'high water mark'.[86] What the courts have consistently done is determine the riparian right as extending to the water's edge — a limit variable in position — and, often, emphasize this feature by taking the title to the low level to ensure that it would never be restricted but always touch the water. This addition is tautology, but it appears to be unavoidable.

§6.69 In *Throop v. Cobourg and Peterboro Ry. Co.*,[87] a parcel was described as bounded by the 'high water mark' of Lake Ontario. The Court of Appeal

---

83 *New Hamburg v. Waterloo* (1892), 20 O.A.R. 1 at 8 (C.A.); reversed on other grounds (1893), 22 S.C.R. 296; *Plumb v. McGannon* (1871), 32 U.C.Q.B. 8 (Ont. C.A.); *Kingdon v. Hutt River Bd. supra*, note 77 (the 'essential' channel of the river is defined without recourse to the term 'high water mark' as a river is contained within its 'banks'); *Clarke v. Edmonton, supra*, note 49; *Hindson v. Ashby*, [1896] 1 Ch. 78, reversed on the facts [1896] 2 Ch. 1 (C.A.); *Dunstan v. Hell's Gate Enterprises Ltd., supra*, note 81.
84 *Supra*, note 51.
85 *Supra*, note 52.
86 See, for example, *Carroll v. Empire Limestone Co.* (1919), 45 O.L.R. 121 (C.A.), affirmed (1919) 17 O.W.N. 295 (S.C.C.); *Servos v. Stewart, supra*, note 80; *Stover v. Lavoia, supra*, note 66.
87 (1856), 5 U.C.C.P. 509, affirmed (1859), 2 O.A.R. 212n. (C.A.).

attached the accretion to the parcel thus leaving no uncertainty that a parcel described to the 'high water mark' is a riparian upland parcel.

§6.70 The terms 'shore' or 'bank' must be distinguished, in the case of lakes and rivers, for popular and for legal meanings. In the legal sense of determining boundary for the extent of an upland riparian parcel, the 'shore' or 'bank' is the line of level of a lake in its holding capacity at the fullest normal natural state or it is the line of variable gradient in the case of a river in its fullest normal natural state of flow. This is the 'line of the shore' or 'line of the bank'. It is identified in survey by those indicators previously mentioned that are collectively and imprecisely called the 'normal ordinary water's edge' or 'normal high water mark'. These are terms that, at best, are the collective identifiers or tests of the riparian limit.

§6.71 In the areal sense, as arises from some reservations and exceptions in grants, the 'shore' or 'bank' is that part of the 'bed' that lies between the normal high and low levels of a lake or high and low gradient lines of rivers. Or, otherwise stated, the 'bed' includes the 'bank' or 'shore', a conclusion from those cases where the specifics were not involved; otherwise, the 'bed' lies below the low water line.

§6.72 In the popular sense, 'bank' is any embankment sloping to water, but no Canadian case has been found by this writer in which this expression so used has been the subject of legal decision.

SURVEY SYSTEMS AND THE SURVEY OF WATER BOUNDARIES

The Concept of True and Unalterable Boundaries

§6.73 The cultural landscape of Ontario and of the provinces to the west reflect statutory prescriptions for the division of Crown lands rather than a random system as generally covers the Maritime Provinces. There is a strict pattern of lines on the ground forming regular blocks of farm lots regardless of topography. These systems were paramount to the concept of land division. So also was the inviolable nature of the original lines run by the Crown surveyors. In this respect, the statutes for surveys restate the common law rule that lines of boundary, as first established and sanctioned, are the true boundaries.

§6.74 The original plans of the Crown's original surveys also show the natural boundaries of lakes and rivers. An erroneous notion arose that the natural boundaries of land/water by the original survey were also 'true and unalterable' in position like the land lines of the system.[88] (Unless varied by some legitimate means, natural boundaries are, of course, true and unalterable in their character.)

§**6.75** The argument is taken out of context of the various Surveys Acts and loses track of the essence of the Acts. Natural boundaries are not established by survey; they are objects of representation. They are monuments in their own right, as is emphasized by the priorities of evidence that govern in the determination of boundaries. Nevertheless, the argument is often raised and consequently has been considered by the courts which have, with a consistency to be expected, ruled that natural boundaries are paramount as monuments and that the water boundary is ambulatory in nature and not fixed by the surveys. The Survey Acts do not override the common law rules about natural boundaries.[89]

Lot and Parcel Lines run to the Water Boundary

§**6.76** It follows, then, that a post planted by a surveyor along the side line of an upland riparian property is not a property corner but is a post-on-line, an indicator of the line direction that continues to the water boundary. The key is, of course, that the upland parcel is riparian, bounded by water as a natural boundary and the natural boundary does not need, in fact cannot take, a surveyor's mark to designate a fixed limit; for a water boundary is ambulatory under recognized conditions of legitimate accretion and erosion, recession and advancement of the waters.

§**6.77** This is not to discount the significance of a surveyor's post-on-line and a recorded distance to water, or of shore traverses and numerous ties to the water feature, for these may be the evidence that settles a contest about the existence of accretion or erosion, or whether the change was by way of filling or extraction of sand and gravel.

---

88 See, for example, the Ontario *Surveys Act*, R.S.O. 1980, c. 493, s. 9; and consider, D.H.L. Lamont, *Real Estate Conveyancing*. (Toronto: Law Society of Upper Canada, 1976) which, at 278, states the opposing view.

89 See, for example, *A.-G. for B.C. v. Neilson* (1954), 13 W.W.R. (N.S.) 241 per Wilson J. at 250-252 (sub nom. *Re Quieting Titles Act and Neilson*) (B.C.S.C.), affirmed (1955), 16 W.W.R. 625, reversed [1956] S.C.R. 819 at 822 per Kerwin C.J.C.: "It is unnecessary to consider if s. 2 of the *Official Surveys Act*, R.S.B.C. 1948, c. 321, prevents the operation in British Columbia of the English common law in regard to accretion as that point which was taken before Wilson, J. was abandoned by counsel for the appellant." See also trial decision of Stark J. in *A.G. for Ont. v. Walker supra*, note 52 at 172-178, commented upon and followed in *Merriman v. New Brunswick* (1974), 7 N.B.R. (2d) 612, per Dickson J. at 628-631 (C.A.).

Shore Road Allowances and Shore Reservations

§6.78 A road allowance, or other exception in fee simple, along the natural boundary creates no new situation in respect of the upland riparian parcel, for here it is the space, usually 66 feet in Ontario, 99 feet in Manitoba, that is retained in fee simple by the Crown or may be vested in an organized municipality. It is, like any other fee simple parcel adjoining waters, subject to loss by erosion or gain by accretion. The character or quality of use as a road is added, so that an addition by accretion means an extended road area and erosion means a limited road area.

§6.79 Road allowances were rarely reserved in early 19th century surveys of townships in Ontario. With the adoption by the Crown in the 1850's of an amended scheme of the 1000-acre section system, the practice of defining and excluding a shore road allowance was standardized for this system and the similar 1800-acre section system later adopted for part of northern Ontario.

§6.80 In New Brunswick, section 4 of the statute of 1884[90] provided for a reserve, one chain in width, from the banks of specified inland waterways and that the riparian ownership of the streams remained with the Crown. The width of the reserved strip was later increased to three chains, then provision made for disposal of the strip,[91] reserving a public right of passage over a ten metre width and retaining the bed below the normal high water mark.

§6.81 The inner limit of a shore road allowance, or any other shore reservation where the fee simple is retained by the Crown, is a different matter from a natural boundary, although it is dependent on the natural boundary at a particular moment in time, the time of the grant, with the exception noted below which is a feature of Ontario law. Generally, the fee simple owner of an upland parcel receives a parcel of fixed limits and the parcel is not riparian where there is a saving and excepting clause excluding the strip along the shore. The inner limit of such a shore reservation, reserved in fee simple by the grantor out of the whole parcel described, is fixed in position by the original survey marks, if found as evidence, and as at the date of the grant. The trick is for the surveyor to locate that original natural boundary where the grant may have been given 100 or more years earlier.[92] Where, in Ontario, road allowances were

---

90 *An Act to provide for the survey, reservation and protection of lumber lands*, 1884 47 & 48 Vic., c. 7, s. 4.
91 *Crown Lands and Forests Act*, S.N.B. 1980, c. C-38.1, ss. 15-18.
92 The notion of an ambulatory strip has some interesting history. The New Zealand case of *Pipi Te Ngahuru v. The Mercer Road Bd.* (1887), 6 N.Z.L.R. 19, held on the facts

248 Survey Law in Canada

defined by the Crown surveys and never formed part of the township lots, they were shown on the township survey plans as extended parcels without a break. They were therefore bounded by an irregular line, distant the specified distance inland from the natural boundary as it existed at the time of the original Crown survey. The Ontario *Surveys Act* provides (at least since 1849, 12 Vict., c. 35, s. 32) that all side lines and limits of lots surveyed are true and unalterable and the courts have consistently held that the inland fee simple parcel is definite as to extent. This inner limit of the road allowance is not ambulatory as is the water boundary from which it was first defined by dimension. If this was not the case, there would be non-alignment of the road allowance lying in front of the lots granted. The *Surveys Act* speaks of lines and limits surveyed; regardless of the earlier comment that natural boundaries are not established by survey but represented as monuments in their own right, it would be trite to argue that the inner limit is of the same nature and therefore that the date of grant would prevail. On the other hand, in the general case, a grant of a parcel to the water boundary saving and excepting a road allowance or other strip must place the inner limit with relation to the natural boundary at the date of the grant and non-alignment may occur.

Significance of the Date of Grant

§6.82 Reference must be made to the patent as the basic title document to know what constitutes the parcel and for the first evidence needed to

---

of erosion into a road reservation along a river that the public were entitled to a road of full width from the new bank. This stood until *A.-G. and Southland County Council v. Miller* (1906), 26 N.Z.L.R. 348 where with the same facts and argument it was decided that the owner of the inland parcel had no liability to provide new land for the public road and that the Crown, if it wished to have full width, must acquire the land and pay compensation. See also *Smith v. Renwick* (1882), 3 L.R. (N.S.W.) 398 (Sup. Ct. F.C.); *A.-G. for N.S.W. v. Dickson*. [1904] A.C. 273 (P.C.); and *McGrath v. Williams* (1912), 12 S.R. (N.S.W.) 477 for similar Australian considerations on these shore reservation. The Ontario case of *Herriman v. Pulling & Co.* (1906), 8 O.W.R. 149 (T.D.), broached the question ("It may be upon a matter of nice law that the road reserved . . . would shift . . . over the accretion . . . but this was not argued . . . .") and no answer was forthcoming. However, the earlier decision of *McCormick v. Pelee (Township)* (1890), 20 O.R. 288 (Ch. D.), rejected the notion of an ambulatory strip for a shore road allowance. The most recent Canadian decision on this subject is *Re Monashee Enterprises Ltd. and Minister of Recreation & Conservation for B.C.* (1981), 124 D.L.R. (3d) 372 (B.C.C.A.) where the concepts and the law are neatly presented. Per Seaton, J.A., delivering the judgment of the Court (at 374-375): "It seems to me that the inconvenience of a mobile boundary is such that it should only be found to exist where it is unavoidable. It is unavoidable at the shoreline, but it is not unavoidable at the upland side of the one chain strip." And: "The land gained by accretion is added to and becomes part of the strip."

establish the bounds. Where there is a plan of survey that appears to show the bounds of the parcel with all necessary clarity, it must still be remembered that it is the date of the grant that controls, or where the specific features of the Ontario systems pertain, the date of the survey. Setting up a paper parcel that does not exist on the ground is not realistic conveyancing; specifically, if a grant is larger at the time of patent or conveyance as a result of the addition by accretion or smaller because of loss by erosion, it is the existing upland parcel at the time of grant this is transferred and not the diagrammed parcel of some survey that was prepared distant in time, either before or after the grant. The survey plan may be, however, the only evidence available, and hence may become the best evidence; and it is on the objector that the onus lies to dispute the natural boundary. It would appear that it is for just this reason that a watercourse has traditionally been "considered the safest boundary of real estate, as it is a natural boundary."[93] This is still true despite a tendency to confuse the importance of the measurement as against the thing that is measured.

SYNONYMITY OF TERMS AND ACCURACY OF SURVEYS

§6.83 While the instructions for the surveys of Dominion lands in the western provinces stated some reasonable specifics for natural boundaries from the very beginning and the *Irrigation Act* of 1894 added acceptable terms that applied in respect of patents, it is evident that in Ontario there was no settled policy of intent about natural boundaries for the surveys and grants. Mr. Justice Dickson, in his trial decision in the *Merriman* case,[94] provides a fair analysis of this use of words, speaking in this case of the parallel usage of terms in New Brunswick:

> The various terms employed in the early grants to designate the waterside boundary of lots, i.e., 'to the bank or edge', 'to the bank or shore', 'to the border of the lake', etc., do not in my view reflect any precise effort to indicate fine distinctions in the exact location of the boundary, but rather indicate merely that lack of preciseness and employment of nonstandardized terminology which could only naturally be expected to have marked the usage of the early surveyors. Those terms must therefore be deemed, as has been variously held, to be synonymous.

§6.84 There is little indication in the Crown records that it was a matter

---

93 J.K. Angell, *A Treatise on the Law of Watercourses*, 5th ed. (Boston: Little Brown and Co., 1854), at 10 (1878 edition at 15).
94 *Supra*, note 89 at 631.

250   Survey Law in Canada

of concern at an early time.⁹⁵ Except where there was a road allowance or other shore reservation, no restriction was imposed on the fulfilment of the riparian interests of an upland grantee. Farnham in *The Law of Waters and Water Rights* stated:⁹⁶

> The policy of the common law is to assign to everything capable of ownership a certain and determinate owner, and for the preservation of peace and the security of society, to mark by certain indicia, not only the boundaries of such separate ownership, but the line of demarcation between rights which are held by the public in common, and private rights. If capable of occupancy and susceptible of private ownership and enjoyment the common law makes it exclusively the subject of private ownership; but if such private ownership and enjoyment are inconsistent with the nature of the property the title is in the sovereign as trustee of the public, holding it for common use and benefit.

§6.85 In *Badgely v. Bender*,⁹⁷ Chief Justice Robinson had this to say about the origins and meanings to be given to patents:

> [K]nowing, as we do, that the grants were framed from this plan [in the records of the Surveyor General's office], and meant to carry into effect the scheme which it exhibits, it becomes the duty of courts and juries to give to the words of the patents such a construction (if they are capable of it), as will be most consistent with the known and evident intention. It is only by doing this that public convenience and private interests can be secured.
>
> I am well aware, and so is every one at all conversant in such things, and I have heard it several times proved in other trials, by surveyors on their oaths, that unless surveyors are directed by the government for some special purpose to lay down accurately the river, lakes, streams which border upon or intersect the townships which they are surveying, they never do so in fact, but content themselves with giving a general idea of them from the eye, or by noticing the points where their lines actually intersect them, not pretending to delineate minutely their several indentations. To do so, would require a multitude of offsets, and make the survey much more tedious and expensive. It is not necessary either for the object of the survey, because the grants of land are

---

95 In 1821, an enquiry was addressed to the Lieutenant Governor in Council of Upper Canada in respect of retaining road allowances on the top of the banks of navigable waters. It was so ordered: "that on Navigable Waters the Rule should be general to reserve for the Public the Beach and one chain for a Road." However, a week later on the 9th May: "On revisal of the order ... it is respectfully submitted that the objects of the Council may be fully attained by including in the Patents a proviso that free access to the Beach may be had by all Vessels, Boats and Persons. Ordered accordingly." The Surveyor General was so advised. In the particular case of the inquiry, the patent shortly issued with both reservations.
96 Farnham, *supra*, note 67 at 217.
97 (1834), 3 O.S. (C.A.) 221 at 227-228., Some understanding of the standards of accuracy for survey and of the details which might be recorded in the course of original surveys may be gained by referring to the Dr. E. Deville, Surveyor Genral, 5th ed. *Manual of Instructions for the Survey of Dominion Lands* (Ottawa: Government Printing Bureau, 1903), clause 145.

made with the qualifying words, more or less, as to quantity of land, and length of side lines, so as to extend to the front of the water where that is meant, whether it be more or less remote than is supposed.

§6.86 Quite apart from the inaccuracy of the surveying approaches and techniques used to locate the land/water boundary there is the equally significant question as to precisely which points the measurements were being taken.

THE DOCTRINE OF ACCRETION

Principles

§6.87 A Canadian case, *Chuckry v. R.*,[98] considered the implications of the doctrine of accretion. The Supreme Court of Canada endorsed the views of Dickson J.A. in the first appeal to the Manitoba Court of Appeal wherein his dissenting views about accretion were set forth in words that are best quoted directly for their succinct expressions of the principles.

> It would seem evident that if the boundary line of a parcel of land is related to a particular river bank the boundary line, and acreage, of the land will in fact vary as the bank of the river advances or recedes through the effect of the forces of nature....
>
> [T]he doctrine of accretion [was] born of Roman law and developed as an integral part of the English common law ... The doctrine has been often defined, perhaps nowhere better than [by Blackstone][99]
>
> And as to the lands gained from the sea, either by alluvion, by the washing up of sand and earth, so as in time to make terra firma; or by dereliction as when the sea shrinks back below the usual water-mark; in these cases the law is held to be, that if this gain be by little and little, by small and imperceptible degrees it shall go to the owner of the land adjoining ... In the same manner, if a river, running between two lordships, by degrees gains upon the one, and thereby leaves the other dry; the owner who loses his ground thus imperceptibly has no remedy.
>
> The river and the doctrine can give, and they can also take away.
>
> If silt carried by the waters of a river gradually and imperceptibly washes up on shore to form firm ground, the doctrine holds that this accumulation of land belongs to the owner of the land immediately adjacent to it. The principle is said to be founded upon "the general security of landholders and upon the general advantage" (per Lord Shaw of Dunfermline in *Attorney-General of Southern Nigeria v. Holt & Co.*,[100]) and on "the necessity which exists for some such rule of law, for the permanent protection and adjustment

---

98 [1973] S.C.R. 694 reversing [1972] 3 W.W.R. 561 at 570-576.
99 Sir William Blackstone, *Commentaries on the Laws of England*, 4th ed. (Dublin: John Exshaw, 1765-1770), Vol. 2 at 261-262.
100 [1915] A.C. 599 at 612, (P.C.).

of property" (per Lord Abinger, C.B., in *Re Hull and Selby Railway*[101]).

An example will make this clear. A farmer owns land which runs along the edge of a river. He uses the waters of the river for domestic purposes, for watering his livestock and for irrigating his fields. Imperceptibly, over the years, alluvial deposits create a narrow strip of land between the farm land and the river. If the doctrine applies this strip of land belongs to the farmer. If it does not apply it belongs to the Crown. If it belongs to the Crown, the Crown can sell it. In any event a barrier is created which, whether inches wide or yards wide, effectively denies the farmer access to the river. This could create a situation of manifest injustice; a landowner would be required to suffer land loss caused by erosion but be denied land gain resulting from accretion. It would seem to me that in fairness the person who suffers the losses should also enjoy the gains.

The Mechanics of Accretion

§6.88 Dickson J.A. continues his dissenting viewpoint:

For accretion to have occurred it must be the result of a well-understood mechanism operating in a definite way. This mechanism can take one of two forms or be a combination of both. These are: (1) the retreating of the high-water line away from its former position, thus exposing land that was until then submerged; or (2) the build-up of land through the process of alluvium, thus pushing back the high-water line: *Attorney-General of British Columbia v. Neilson*;[102] *Re Bulman*.[103]

The second requirement is that the process operate slowly and gradually so that the land growth is imperceptible. If any additions are made as a result of flooding they must be those that would occur in the natural course of events: *Clarke v. Edmonton*;[104] *Re Bulman*;[105] *Bruce v. Johnson*.[106]

There was further clarification of the concept by the Supreme Court of Canada in the *Nielson* case. Their Lordships were of the opinion that the drying-up of the river bed due to the formation of silt dykes[107] or the formation of islands unconnected to anyone's land[108] did not constitute accretion and that the land thus formed belonged to the Crown. This statement of the law was accepted by Ruttan J. in the *Bulman* case, when,[109] he said that the vertical

---

101 (1839), 5 M. & W. 327 at 332, 151 E.R. 139.
102 [1956] S.C.R. 819 at 826 per Rand J.
103 (1966), 56 W.W.R. 225 at 230 per Ruttan J. See also *Dunphy v. Williams* (1874), 15 N.B.R. 350 (C.A.).
104 [1930] S.C.R. 137 at 141.
105 *Supra*, note 103.
106 [1953] O.W.N. 724 at 727 (Co. Ct.).
107 *Supra*, note 102, per Rand J. at 827.
108 *Supra*, note 102, per Locke J. at 840.
109 *Supra*, note 102 at 229.

development through sedimentation which was not a gradual extension of existing upland did not establish any accretion.

...

During argument a question was raised whether the doctrine of accretion could apply in circumstances such as are here present, namely, the former boundaries of the land are related to a defined line on a plan, i.e., the plan of 1875, and the lot acreages are stated on the plan. At one time it was doubtful whether the doctrine of accretion applied when the former boundaries of the land concerned were defined or ascertainable. The law now seems clear, however, that so long as the change is gradual and imperceptible the doctrine applies.[110] [Canadian authority is found in *Clarke v. Edmonton*.[111]] In that case appellant's title gave him "All that portion of River Lot Twenty-one (21) . . . lying north of the north boundary of the Dowler Hill Road". The patent conveyed "Lot numbered twenty-one, in Edmonton Settlement aforesaid . . . containing by admeasurement, one hundred and sixty-three acres, more or less". Lamont J., said,[112]

> A plan of land abutting on a river which shows the east and west boundary lines of a lot as running northerly to the river line and having no defined northern boundary, is, in my opinion, to be construed as having the river (i.e., the edge of the river bed) for the northern boundary of such lot.

The learned Judge held that the northern boundary of lot 21 was the edge of the river bed and that the accreted land was a true accretion.

More recently, in *Government of State of Penang v. Beng Hong Oon*,[113] Lord Cross of Chelsea, speaking for the majority of the Privy Council, had occasion to say:

> It is, of course, well settled that if the boundary of land conveyed is the line of medium high tide the mere fact that the acreage of the land conveyed is given and that the position of the line of the medium high tide at the date of the conveyance can be established — whether or not it is delineated on a plan — will not prevent land which subsequently becomes dry land through the gradual and imperceptible recession of the sea from being added to the land conveyed.[114]

The test is whether the land in fact comes to the water's edge under the grant and not upon the manner of land description within the grant.

...

In the leading case of *Clarke v. Edmonton*[115] Lamont J., for the Court, held that the English law as to accretions became the law of the Territories; that even accepting the Crown as owner of the bed of the Saskatchewan River,

---

110 "The following cases are cited in 39 Hals., 3d ed., p. 561, in support of that proposition: *Gifford v. Lord Yarborough* (1828), 5 Bing. 163, 130 E.R. 1023 (H.L.); *A.G. v. McCarthy*, [1911] 2 I. R. 260; *Brighton & Hove General Gas Co. v. Hove Bungalows Ltd.*, [1924] 1 Ch. 372; *Secretary of State for India v. Foucar & Co.* (1933), 50 T.L.R. 241 (P.C.)." (49 Hals., 4th ed. at 163).
111 *Supra*, note 101.
112 *Ibid.* at 152.
113 [1972] A.C. 425 at 436 (P.C.).
114 Citing *A.G. of Southern Nigeria v. Holt & Co.*, [1915] A.C. 599 (P.C.).
115 *Supra*, note 111.

all accretions became the property of the riparian owner to whose land they attached, and the ownership of accretions was binding on the Crown. Accretion is no way dependent on the riparian owner's ownership of the bed of the body of water; it is one of the riparian rights incident to all lands bordering on rivers.

RIPARIAN RIGHTS AND TITLE RECORDS

§6.89 That the title to a parcel of land is in documents of record in a deeds registry or is the subject of certification under a land titles system has no bearing on the interests attached to riparian properties unless statutory provisions have imposed restrictions on water privileges or restricted rights of ownership of the bed; and in particular the law applicable to accretion and erosion is not varied whatever the nature of the titles system. This applies even though the subject appears to gain at the expense of the Crown, a matter of no small importance, as by the *Beds of Navigable Waters Act*[116] of Ontario, and similar legislation of other provinces, the bed of navigable waters is Crown provincial land except for insignificant portions vested in the Crown federal, harbour commissions and private ownership. These statutes state a position deemed essential for the Crown.

§6.90 Ownership of land is a matter of property law. Demonstration of that ownership by deed or title records is a matter of administrative policy and the bookkeeping of records systems. It was perhaps inevitable that the entitlement of upland riparian owners to the beds of waters in front, and to natural accretions, became a problem in need of an effective and realistic means of resolution. In Manitoba, the problem arose after the decision in *Chuckry v. R.*,[117] and in British Columbia, after an opinion on a specific reference was given by the Inspector of Legal Offices in 1954. Solutions were provided by new legislation and by administrative procedures.

§6.91 The conceptual problems seem to emerge most frequently in registration systems based on the Australian Torrens title model, although throughout Australasia it was settled at an early date after the system was introduced. The New Zealand case of *Auty v. Thompson*[118] adjudged that the doctrine of accretion is applicable to land under the system and in Western Australia the Commissioner of Titles similarly advised in 1898.[119] These approaches fit the common law principles enunciated in the English

---

116 *Beds of Navigable Waters Act*, R.S.O. 1980, c. 40.
117 *Supra*, note 98.
118 (1903), 5 Gaz. L.R. 541 (N.Z.S.C.).
119 Noted in D. Kerr, *The Principles of the Australian Land Titles (Torrens) System* (Sydney: Law Book Co. of Australasia, 1927) at 28.

case of *Mercer v. Denne*[120] that "where imperceptible accretion has occurred, the piece of land is to be treated as having been as it is from the commencement of legal memory." The fact of accretion having been settled, there is not a new dry land entity or parcel; the new land is part and parcel of the upland.

§6.92 The real problem of accretion is not one of registering the existence of new legal rights, but restatement of the limits. How this is achieved appears to be a matter of choice in each jurisdiction, but it is not improper or unfair to suggest that some of the solutions are based on faulty concepts and ponderous administrative precepts.

§6.93 The administrative uncertainty of the effect of accretion and the consequent ambulatory nature of the natural boundary of a parcel of guaranteed title is paralleled by the problem of acceptance of the *ad medium filum* rule as an operative but rebuttable presumption. The Australian writer, John Baalman, called it "the most clumsy doctrine of a legal system notorious for its obscurity, the English common law," and said further "that in any jurisdiction which aims to have an efficient conveyancing system the rule, (with perhaps others) should be sharing a common grave." There were, and are, opposing views.[121] The rule is not inconsistent with the principles of a system of title by registration.

§6.94 The problem is not really so serious; the resolution is one that calls for simple common sense in recognizing the continuing enjoyment of riparian rights as incidents of the upland riparian ownership of parcels abutting watercourses, and that survey does not, and should not, fix natural ambulatory boundaries.

THE SURVEY OF UPLAND RIPARIAN PARCELS

§6.95 Regulations for surveys generally require that parcels be fully surveyed and that on the plans the boundaries be shown by lines of some specified character. There are also requirements by title and deed registration systems, and by planning legislation, that the whole of parent

---

120 [1905] 2 Ch. 538, affirming [1904] 2 Ch. 534.
121 J. Baalman (1951), 25 *Australian Law Journal* 449-453, 538-541, cited in D.J. Thom, *Thom's Canadian Torrens System*, 2nd ed., by V. Di Castri (Calgary: Burroughs & Co., 1962). The views of P.M. Fox (25 *Australian Law Journal* 678-680) are also noted. "Mr. Fox . . . while not suggesting that the rule is either 'a good thing or in the spirit of the Torrens system' proceeds on the basis that the rule is included in the existing body of law and advances the proposition that the rule may not after all be necessarily inconsistent with the Torrens system."

parcels be represented in some fashion and accounted for in registered transactions.

§6.96 In jurisdictions where the common law only applies, the surveyor locates the extent of title of the uplands adjoining inland non-tidal waters at the middle thread on the presumption of ownership (the presumption being rebuttable); and, if the waters are in fact navigable, the existence of the public right of navigation may be indicated over the space between the line of bank or shore and the line of the middle thread. Where the waterway is, by statute or otherwise, excluded from the private ownership of the upland riparian parcel (that is, where the presumption is rebutted), the line of boundary to be shown is the line of the bank or shore.

§6.97 Under current Ontario law, the first matter to be resolved is, however, the navigability of the watercourse from which then follows the ownership of the bed. A decision on navigability is, therefore, effectively one of title. The upland owner remains riparian, but if the waters are navigable, his title does not include the land under water (unless it has been decided by a court prior to March 24, 1911, or a Crown grant of land under water had been given), whereas if non-navigable, the rebuttable presumption applies that extent of title is to the middle thread.

§6.98 A court decision (before March 24, 1911) that the bed is in the ownership of the upland riparian proprietor is a document that may not necessarily be of record in a land titles or deed registry record, but if it exists it must be followed. A decision made subsequent to that date is equally necessary to determine the extent of title; here it is the interpretation of the grant — whether the wording is to be construed as an express grant of the bed. Present day navigation-in-fact on the waters above such granted land, as over any grant of a water lot, remains as a public right under federal control.[122]

§6.99 The crux of these problems of surveys, in Ontario certainly, is that the surveyor cannot determine navigability; that is a matter solely for decision by the courts. If he limits the upland parcel at the bank or shore, the surveyor has failed to show the extent of parcel if the water is non-navigable. The regulations for surveys, planning and registration of deeds or of title require the surveyor to demonstrate the whole of the title and he must elect to show some one kind of natural boundary; he is thus placed in the invidious position of a marked risk that he will be found wrong if the matter comes before a court for the determination of navigability.

---

122 *Re Coleman and A.G. for Ont.* (1983), 27 R.P.R. 107 (Ont. H.C.).

§6.100 The surveyor, however, serves his client and is paid for his opinion. In light of the current issues as to both the title to the bed of watercourses and the line that should be represented as the interface of land and water, it would appear advisable for the surveyor to report in writing with some words of qualification to alert the client and lawyer on these matters which are essentially issues of title.

REFERENCES

Gisvold, P. *A Survey of the Law of Water in Alberta, Saskatchewan and Manitoba*. Ottawa: Canada Dept. of Agriculture, Prairie Farm Rehabilitation Administration and Economics Division, 1959.

LaForest, G.V. and Associates. *Water Law in Canada — the Altantic Provinces*. Ottawa: Information Canada, 1973.

Roberts, A.C. *Riparian Rights: a compilation of various papers, judgments, historical records and references*. Winnipeg: Department of Renewable Resources & Transportation Services, Surveys and Mapping Branch, 1976.

7

Evidence

Izaak de Rijcke
B.Sc., LL.B. O.L.S.

INTRODUCTION

§7.01 Perhaps no other single field or area of law is more significant for the land surveyor than the law of evidence. This may seem trite and simplistic, but many surveyors fail to appreciate that the core activity of their role that is highest in profile lies in being providers of professional opinions on matters of retracement. A retracement survey is nothing more than and nothing less than an exercise in collection and assessment of evidence and formulation of opinion based on that evidence. The matter of collecting evidence and the resources that might be available to the land surveyor in order to obtain all the facts and information relating to a boundary are not the main thrust of material dealt with in this chapter. Instead, the assessment of the information and facts collected by land surveyors and the law applicable to the use of that information in the formulation of an opinion by him is focal to this chapter.

§7.02 Before one delves into a treatment of the substantive law of evidence, it may be useful to clear away a misconception of the role of the land surveyor. The term "quasi-judicial' is a descriptor that has been used in the past in reference to the land surveyor's role.[1] Quasi-judicial can mean and does mean different things to different people. To land surveyors, the key word in the term seems to be 'judicial'. Land surveyors have viewed themselves as having a role that is not unlike that of a judge, that setting forth an opinion as to the location of a boundary based on an assessment and collection of facts and evidence relating to that very question is analogous to the role of a judge.

§7.03 But the true mandate and extent of the role of a land surveyor does not extend so far as to actually *empower* him to make a decision which would be binding on any party who is an owner of an interest in land that abuts the boundary in question. Instead, the whole notion of *opinion* seems to have become eclipsed by the concept of a *decision*. It is the interface between evidence collection, with the assembly of all of the information pertinent to a boundary's location and the formulation of a decision as to where that boundary in fact is that is the basis for this misconception.

§7.04 The law of evidence is, therefore, fundamental to the land surveyor whenever he engages in any opinion or activity which falls within this interface. The single guiding light which perhaps must and always will remain relevant for the land surveyor is to anticipate the treatment which a court of law, or other decision-maker having binding jurisdiction, will accord to the evidence that he has collected. The land surveyor, in forming his professional opinion, must therefore assess the evidence in the same manner as a court of law or other decision-maker would consider that same evidence if presented to it. The opinion must be determined on the same principles as a decision might be made.

§7.05 One may appreciate that the underlying drift of the introduction of this chapter is none other than negligence and loss prevention. Whenever a land surveyor forms an opinion which could never have reasonably been formed on the basis of facts which should have been available to him and are ignored or not considered or, on the basis of principles of evidence which should have been applied by him in assessing that evidence but are not, a question of liability arises.

§7.06 The reasons for even making reference to negligence at this point and in this chapter are twofold. Firstly, there seems to be an increasing sensitivity on the part of clientele and members of the legal profession today as to the liability of professionals in general. Whenever professional

---

1 W. Marsh Magwood, Q.C. "The Law and the Surveyor" in *Legal Principles and Practice of Land Surveying* (Ottawa: Surveys and Mapping Branch, 1961), reprinted from *The Canadian Suveyor*, Vol. XV, No. 5, 1960.

opinion forms part of the confidence and decision to proceed with legal action and it later turns out that the professional opinion was incorrect, the professional is considered as one source of compensation. This general perception definitely extends to land surveyors. Secondly, no professionals today, including land surveyors, can afford to practise with a blind eye to the potential for liability on their part. Insurance and the protection that insurance may afford has nothing to do with these concerns. One may expect to experience in the future increasing demands on the part of the public and the legal profession for the formulation of correct opinions. These are to be distinguished from opinions which are merely plausible. That an opinion could be plausible, but turns out to be incorrect, because a decision-maker such as a tribunal or a court of law decides differently will in the next years not be readily forgiven insofar as a losing party to a proceeding may be concerned. Therefore the need remains, and may increase, for opinions and judgments to be as correct as possible insofar as the proper application of evidentiary principles to the available facts are concerned in formulating such opinions and judgments.

§7.07 This chapter purports to be a review of the law of evidence. It should be stated at the outset that this may well be an impossible task. There are whole textbooks which provide general overviews of the law of evidence.[2] For this chapter to represent everything that may have a bearing on the law of evidence for the land surveyor, or that should be taken into consideration by land surveyors, is unrealistic. At best, it may provide an appreciation and guide to further resources.

§7.08 The law of evidence draws from both the fields of what may be characterized as procedural law and substantive law. The rules of evidence are intended to control the presentation of facts before a court. The purpose of the rules is to facilitate the introduction of all relevant facts without doing violence to any fundamental principle or policy of law which may be more important than the ascertainment of the full truth. That anything may be more important than ascertaining the full truth may be surprising, but the question of whether the truth should be disclosed or not is not always answered with a unanimous 'yes', especially in the field of criminal law. There may be situations where the law considers it more expedient and more desirable for the truth not to be ascertained than to permit the admission of certain types of evidence. An example of such a situation in law is in the general principle that prevents a spouse from testifying against an accused. The spouse

---

2 For example, see G.D. Nokes, *An Introduction to Evidence* (London: Sweet & Maxwell, 1967) and J. Sopinka and S.N. Lederman, *The Law of Evidence in Civil Cases* (Toronto: Butterworths, 1974).

may be the only person who can provide oral evidence or testimony as to an accused's guilt on a charge of murder. However, the spouse may not be compelled as a witness for the Crown which is doing the prosecution since such a measure would fundamentally jeopardize the trust and confidentiality which exists between husbands and wives. This is also an example of an evidentiary principle which creates an exclusionary rule which in some cases could operate to prevent a total ascertainment of the truth.

§7.09 Pleadings, discoveries and other measures designed to facilitate agreement between parties in civil actions as to relevant facts assist in identifying the issues and matters which are in dispute and any fact that relates to these issues is, therefore, relevant. These facts may be proved by the collection, introduction and presentation to a court or a decision-maker of the evidence which tends to establish these facts.

§7.10 While the procedural/substantive distinction may not be formally recognized by all writers, the substantive law does not deal with practice and procedure. Instead, substantive law sets out what must be proved in a particular case in order to obtain a particular outcome at law. Evidence could, therefore, be summarized as 'that which tends to establish proof.' The establishment of proof and the manner in which it is done is a matter for the procedural law. That which has to be proven for purposes of obtaining a certain outcome is a matter for substantive law.

ADMISSIBILITY

§7.11 Many approaches to the law of evidence are based on a classification of different types of evidence. Classification of evidence can be formulated on, again, a number of different criteria. Classification systems could be based on whether or not the evidence is admissible, on the nature or character of the evidence or whether the evidence in itself is the result of common law rules or the result of legislation. Perhaps no better explanation exists than that of Mr. Justice Rose in the decision of *Scott v. Crerar* wherein classification of evidence was dealt with as follows:

> One often becomes confused in an endeavour to formulate a rule under which evidence may be received, and to classify it under headings that have been adopted for convenience. It must be remembered that the classification could only be after it had been determined in the interests of justice that certain kinds of evidence were admissible and certain kinds inadmissible: that arbitrary headings were not first decided upon under one of which all evidence must be arranged ere it could be admitted. The varying circumstances and

needs of mankind must be met by an equal varying of rules to meet the exigencies, else the substance will be strangled by the form.[3]

The approach in this chapter will, therefore, be based on the more popular and recognized approach to evidence classification. With a view to convenience, it is basically organized on the principles of admissibility, types of evidence, proof or the formalities of proof, documentary evidence and the role of witnesses.

§7.12 In entering into a discussion of the grounds for the admissibility of evidence, it is useful to briefly review the forums in which evidence may need to be presented. One must never lose sight of the ultimate decision making forum that the evidence may be presented in or to. The rules of evidence, the admissibility of a particular item of evidence and its treatment may vary accordingly. The most obvious example of variation arises from the nature of the tribunal hearing the evidence. The strict and formal requirements of the proof of documentation before a court of law is often different from that of a tribunal. A court of law will probably tend to insist on certified copies with the author or party to the document testifying orally as to its validity. However, an administrative law tribunal might have more flexibility and tolerance in terms of admitting and allowing for consideration of evidence which could simply be in the form of photostatic copies of true documents. The evidentiary rules and criteria to permit the admission of some forms of evidence have been relaxed somewhat by recent procedural statutes governing practice before administrative law tribunals. One reason for this may be that tribunals tend to be composed of decision-makers who are better trained in or more familiar with the matter that is going to be the basis of applications made to them and determinations made by them. One would, of course, then ask why the same amount of discretion would not be available to a judge in a court of law. But under no circumstances would a judge necessarily exclude a photostatic copy of a document simply because no one is available to testify as to its validity or originality. The tendency of courts today appears to be leaning more towards an attitude of let's admit the copy of the document into evidence and, if we find some other evidence that contradicts it we will attach less weight to this evidence or else dismiss it altogether.' In other words, the exclusionary rules seem to be losing ground to the task of the decision maker to weigh the evidence and attach the appropriate amount of weight to each item of evidence in formulating a decision.

§7.13 For evidence to be admissible it must be relevant. Again, this may seem like a very trite issue and one may tend to think that anything which

---

[3] (1887), 14 O.A.R. 152; reversing (1886), 11 O.R. 541 at 562-563.

has a bearing on the very question before the court is relevant. But the courts have used the criterion of reasonably relevant in order to determine whether a particular fact in evidence is admissible. What does reasonably relevant mean? One writer has defined relevance as:

> Any two facts to which [relevance] is applied are so related to each other that according to the common course of events, one either taken by itself or in connection with other facts, proves or renders probable the past, present, or future existence or non-existence of the other.[4]

If a judge or a decision-maker finds a particular item of evidence not relevant, it may be ordered excluded.

§7.14 From a land surveyor's point of view, relevance does not usually create a problem. Instead, the determination of whether a particular item is a fact or not poses greater difficulty. The distinction here is more than just an academic one. If, for example, one were to say, as a witness in a court, that one saw a brown car pass through an intersection through a red light, one would be making a statement of recollection of what had been perceived while standing as a pedestrian at the intersection. However, if one also stated that one thought the car was travelling 40 miles per hour as it entered the intersection, it would be an expression of an opinion on a matter which could not definitely be established as a fact. One may have to qualify the opinion by indicating that one was in the habit of observing vehicles at different speeds and that one was, perhaps, knowledgeable as a witness in terms of being able to assess speed of vehicles as a pedestrian or witness. But even if this obstacle could be overcome, the statement as to the speed of the vehicle would be inadmissible as an opinion. However, it could be admissible if it were a matter which the court decided was the kind of thing a person of ordinary intelligence would be able to speak about. This is not to be confused with opinion evidence in the form of an expert's opinion which would be admissible as an exception to the hearsay rule. This will be dealt with later in this chapter.[5]

§7.15 To take the above example further: a witness who had observed a vehicle entering an intersection at a high speed and through a red light might think that the person who was driving the car was intoxicated. If the witness were to form and articulate this opinion, the statement by itself would not be admissible in evidence. The witness would have no personal knowledge of the driver's sobriety and the conclusion which was drawn

---

4 Sir H.L. Stephen and L.F. Sturge, eds. *Stephen's Digest of the Law of Evidence* 12th ed. (London: MacMillan & Co., 1948).

5 See §7.54 to 7.59, *infra*.

from the conduct of the driver would not be sufficient grounds for the witness to express the opinion that the driver was inebriated.

§7.16 Another basis for the admissibility of evidence, which may otherwise not have a direct bearing on the issue to be decided by a court, can be characterized as similar fact evidence. For example, the survey which one prepares may deal with a property that fronts on a municipal road allowance which is, at the time of the survey, fenced. However, the fences are only 50 feet apart and the question that may arise in one's mind is how wide is the road allowance? Suppose the original survey for the municipality designates a width of 66 feet but the old fencing would appear to be only 50 feet wide at the widest part in the road allowance. What is the significance of the fencing as evidence of the original boundary of the road allowance? If it could be established that, as a matter of practice, the municipality's residents and first settlers really fenced the road allowances at 50 feet wide rather than 66 feet, such evidence might be admissible in order to establish a pattern of usual conduct or habitual practice. This would be the case despite the fact that the evidence brought before the court would having nothing to do at all with the specific parcel which is the subject of survey.

§7.17 Most Canadian provinces have enacted statutes for the purpose of amending the common law of evidence.[6] The *Canada Evidence Act*[7] accomplishes the same effect at the federal level. The object of most of the legislation is the simplification and codification of the common law rules, in addition to certain changes to the common law. The statutory amendments in the Evidence Acts have, as a general object, the relaxing of the rules of admissibility. Therefore, matters which have any bearing at all and which would appear, on the surface, to be relevant to the determination of the truth of a matter in issue, would be admissible. This, however, is subject to certain significant qualifiers. As was indicated earlier, there exist overriding policy considerations and strong exclusionary rules whereby certain evidence might never be admissible in a court. However, despite these rules, the general statement could be made that most of the evidence that could be characterized as relevant to the determination of an issue would be admissible. The thrust of the evidence legislation is to change questions of admissibility into questions of weight. Once a document or item has been admitted into a proceeding as evidence, it is then weighted as to the appropriate reliance that should be placed on it in making the final decision or determination.

---

6 For example, *Evidence Act*, R.S.O. 1980, c. 145.
7 R.S.C. 1985, c. C-5, as amended.

EXCLUSIONARY RULES

§7.18 Exclusionary rules cannot be overlooked by a land surveyor in his own assessment and sifting through of information relating to a lost boundary. Take, as an example, a boundary survey that is being performed by a land surveyor who takes into consideration all the results of his field work and his research of other surveyors' field notes, land registry office records and other sources. Suppose that when the deed and title information are compared to the occupation, he concludes that there is an apparent five-foot discrepancy; the fence which is the only evidence of occupation is a very old fence, but, no connection is found between the circumstances that surrounded the construction of the fence long ago with the original position of the monuments which were planted in the original survey. Leave aside the question of burden of proof and whether one is entitled to presume that the fence is the best evidence of the original monuments unless such presumption has been rebutted. On the basis of the information, one might conclude that the old fence is the boundary line and that there is no discrepancy between occupation and title to the property. The next issue is then the extent to which one is entitled to rely on evidence and information of the boundary's location in any document, either creating a boundary or dealing with a parcel that has boundaries, which does not make reference to any survey or to any other documents or to any occupation on the ground.

§7.19 If one answers this question differently from what may later be determined or found to be correct, one would have arrived at a conclusion based on parameters different from those that a court of law would apply. What has been arrived at by the surveyor is merely an opinion based only on information that seemed correct and appropriate at the time. However, when faced with the same array of information, a court may very well say that the evidence of some person is not admissible in the proceedings before the court. The end result is that the surveyor's opinion was based on certain evidence that he took into consideration but that resulting opinion was different from a court's decision. Exclusionary rules are one explanation for a court not making the same decision or arriving at the same conclusion that a land surveyor might. Quite apart from the liability considerations that are present, one should always bear in mind the need to adopt the same rules of evidence and evidence consideration as are used by a court.

ORIGINS OF THE HEARSAY RULE

§7.20 The rules of evidence place certain restrictions on the admissibility of material in evidence. One such restriction, known as the hearsay rule, is the result of a natural suspicion of evidence which is seen as having an inherent untrustworthy quality. There remain doubts and suspicions as to the accuracy of such evidence and the rule against hearsay evidence has survived for centuries for this reason. Stated very simply, the rule suggests that written or oral statements or communicative conduct made by persons who are not testifying as witnesses are not admissible if such statements or conduct are tendered either as proof of their truth or proof of assertions that are implicit in the communications themselves.

§7.21 The law assumes that all natural human testimony is unreliable. Accordingly, certain safeguards have been built in such as the oath, the opportunity to cross-examine a person and the existence of the crime known as perjury. Special attention has also been given to hearsay or second or third party evidence as being inherently untrustworthy. Some of the reasons that have been given for assigning such a considerable amount of suspicion to hearsay evidence are:

(1) the author of the statement is not under oath and is not subject to cross-examination;
(2) accuracy tends to deteriorate with each repetition of the statement;
(3) the admission of such evidence lends itself to the perpetration of fraud;
(4) hearsay evidence results in a decision based upon secondary and, therefore, weaker evidence rather than the best evidence available; and
(5) the introduction of such evidence will lengthen trials.

§7.22 The general proposition, therefore, is that any evidence which is hearsay is excluded and is simply not admitted in any judicial or administrative law proceeding. This initial position has been softened through exceptions to the hearsay rule; it is the exceptions which now receive more attention than the rule itself.

EXCEPTIONS TO THE HEARSAY RULE

Declarations Against a Party's Interest

§7.23 A declaration that is made by a person which is essentially and fundamentally contrary to his own interest constitutes an exception to the hearsay rule. The law presumes that nobody will end up saying, for example, he owes money to John Doe unless, in fact, that is true. Such an acknowledgement is seen as being such a strong indication of the existence of a debt that the law creates a special exception to the hearsay rule for those types of statements or declarations that are made by parties which are fundamentally against their own interests. However, the exception to the hearsay rule in the form of a declaration against a party's interest also has certain ingredients, all of which are essential and all of which must be present as a condition to make the declaration admissible. These elements can be summarized.

(1) The declarant must be deceased. The reason for this requirement is that, if the declarant were alive, he would be brought under subpoena or under some other method of examination to testify as to the truth or validity of matters which are in issue before the court.
(2) The declaration must be against the pecuniary or proprietary interest of the declarant at the time it was made. An example of the type of admission that is not, therefore, within the exception to the hearsay rule is a statement by a person that he had killed someone; that statement is admissible as an 'admission'. A 'declaration against interest' must be limited to property or monetary interest; any statement beyond this narrow limitation would not be a declaration against interest within the meaning of this exception to the hearsay rule.

Declarations Made in the Course of a Business Duty

§7.24 Another exception to the hearsay rule is declarations which are made in the course of a business duty. This former common law exception was justified simply on the basis of necessity. The declarant was no longer available to give evidence and the statement could be said to have possessed a certain guarantee of truth arising from the declarant's fear of discipline or dismissal if an employer should discover inaccuracies in his statement or find out that he was in the habit of making entries which proved to be untrue.

§7.25 The traditional common law rule could be summarized as follows:

> Statements made by a deceased declarant under a duty to another person to do an act and record it in the ordinary practice of the declarant's business or calling are admissible in evidence, provided they were made contemporaneously with the fact stated and without motive or interest to misrepresent the facts.[8]

§7.26 This common law exception to the hearsay rule has been superseded in the Province of Ontario and a number of other provinces by legislation which allows for the admissibility of business documents if certain criteria are met. These criteria today include requirements that the declarant must be dead, the duty must be one to do a specific act and then to record it when it is done, and the act must have been completed. The act must also have been performed by the declarant himself and the statement must have been made contemporaneously with the act. There also must not

---

8 Sopinka and Lederman, *supra*, note 2 at 66.

be any motive or possible reason for the declarant to misrepresent the facts recorded by him in the business records; collateral matters relating to the making of the business records would not, in themselves, be admissible.

§7.27 This brings one to the latest developments in the law with respect to declarations in the course of business duty. Judicial reforms of the common law exceptions make sense in certain circumstances for several reasons. First, there is the consideration of simple commercial convenience. To call all the persons responsible for keeping a record in a large institution is too inconvenient and may, in fact, serve to disrupt the business of that institution by taking a multitude of witnesses away from their work. Second, one must consider the expense to the litigants in seeking out and subpoenaing all relevant witnesses. Third, the cost to the public and the great length of time that could be consumed at trial by the testimony of witnesses called merely to prove a business record should be considered. All of these factors give rise to a circumstantial guarantee of trustworthiness insofar as business records are concerned which would justify the further modification of the law to allow the receipt of documents without the necessity of calling the authors who made these notes or documents as witnesses.

§7.28 If this trend in the modification or reform of the law is to be extended to land surveyors, one can draw certain conclusions. For example, most land surveyors do not see themselves, their employees or any of their associates, as having a material or other interest in the determination of a boundary. It is for this reason, as well, that under no circumstances a land surveyor or his employee survey any property in which any form of conflict of interest might subsequently lead to a challenge of either the validity or the admissibility of the field notes.

§7.29 The direction of judicial reform would seem to suggest that an oral explanation of what was done, based on a surveyor's field notes, may no longer be permitted, Surely one must question the ability of a land surveyor to remember the specific circumstances of a survey performed 20 years ago when, 20 years after the fact, he is presented with the field notes and is asked to explain exactly what he did. The field notes must be able to stand on their own. They must be prepared by him and his staff today with a view to being independent business documents that can stand at a later date as a complete explanation of what was done today. An opportunity for the surveyor to explain, by way of oral testimony, the conduct of either himself or his field crew may not be possible at a later date.

Declarations as to Reputation

§7.30 A third significant exception to the hearsay rule relates to declarations as to reputation. Statements that are made by persons as to reputations of public or general rights, marital relationships and ancient historical matters are admissible under this common law exception. The rationale for this exception, like the others, turns on the elements of necessity and circumstances which affect reliability. It is necessary because the subject matter of the declaration is so ancient in time that no first hand evidence to substantiate the fact exists. It also carries with it a certain degree of reliability on the ground that, because the reputation affects the community as a whole or a family, it is probably trustworthy since the reputation is assumed to not have developed otherwise.

§7.31 The right or the interest which is in question must also be a right or interest of a public or general nature as opposed to a private right. The rights are public if they affect the interest of the community as a whole; such matters as right of highway, the right to use a beach, the width of a road, or the public nature of a square within a municipality have been recognized as public rights within the meaning of this exception. Reputation as to the private rights of an individual, a specific boundary or matters which do not affect the community as a whole do not have the enjoyment of this exception rule since they cannot be said to be public in nature.

§7.32 Consider, as an example, the court action brought in the City of Toronto regarding the public nature or character of a square which was shown on a map of the City of Toronto and marked 'Bellevue Square'.[9] Maps which were prepared in 1857 and 1858 by surveyors for the City showed the square clearly marked as Bellevue Square. Other witnesses were called to testify that the Bellevue Square site within the City was a public square. In attempting to establish the boundaries of the square, the maps were also relied on in argument, but were *excluded* as evidence for the purpose of showing boundaries and were admissible only for purposes of determining that a square known as Bellevue Square was public and existed in the City of Toronto. One must watch for these fine distinctions and always appreciate that such fine narrow interpretations can always come to light at a later time. The risk, therefore, exists that the full evidentiary framework for the surveyor's retracement work is replaced by a different framework, based on what is admissible.

---

9 *Van Koughnet v. Dension* (1885), 11 O.A.R. 699.

Statements in Ancient Documents

§7.33 A further exception to the hearsay rule is found in statements contained in ancient documents. Ancient documents such as deeds or leases which affect or create an interest in property have been admitted by the courts as evidence of possession of the property. Most authors do not treat the admissibility of such evidence as an exception to the hearsay rule. They are inclined to treat such evidence as being presumptive of possession and thus as original evidence in its own right. Other authors and writers, however, feel that the documents are tendered not only to establish the inference of possession from the mere existence of the documents themselves, but are submitted as proof of the truth of the statements contained therein and can only be accepted as an exception to the hearsay rule. Most of the authorities that this writer has reviewed seem to indicate that the latter view is the more popular one. Once admitted as authentic a document is seen as reliable and it would not be logical to simply limit its use to that of serving as evidence for the drawing of inferences of possession.

Statements in Public Documents

§7.34 Yet a further exception to the hearsay rule can be found in statements contained in public documents. Founded upon the belief that public officers will perform their tasks properly, carefully and honestly, an exception to the hearsay rule was created for written statements prepared by public officials in the exercise of their duty. When it is part of the function of a public officer to make a statement as to a fact coming within his knowledge, it is assumed that, in all likelihood, he will do his duty and make a correct statement. There are a number of preconditions which must be established for this exception to the hearsay rule to be invoked:

(1) the subject matter of the statement must be of a public nature;
(2) the statement must have been prepared with a view to being retained and kept as a public record;
(3) it must have been made for a public purpose and be available at all times to the public for inspection; and
(4) it must have been prepared by a public officer in pursuance of his duty.

Admissions of a Party

§7.35 The final exception to the hearsay rule relates to the admissions of a party; these are to be distinguished from declarations against interest. A declaration against one's own interest is clearly a positive statement that a certain state of affairs exists which conflicts with that of, or derogates from, one's own proprietary or financial interest. An admission of a party can, however, be viewed as a specific subcategory of declarations against interest. Admissions of a party are receivable only as evidence *against* the party and, normally, a party who has made a self-serving or favourable admission cannot take advantage of the admission for his personal gain. One can summarize by saying that if such statements were admissible, then there would exist the danger that every person would seek to improve his own position in pending or anticipated litigation by making statements in his own favour. For example, the admission by a person, other than the accused in a murder trial, that he committed the crime must be looked at very carefully and should not, of itself, be used as a basis for exonerating the accused.

§7.36 An admission may consist of an oral or written statement made directly by or on behalf of a party. It may take many forms. As in the case of other kinds of admissions in criminal law, the party who made the admission may later lead evidence at trial to reveal the circumstances under which the admission was made in order to reduce its negative effect. It should be noted that before a plea of guilty is admissible in a subsequent civil action, the latter proceeding must have arisen out of the same or similar circumstances which forms the basis of the criminal charge.

§7.37 Statements which are made by a representative of a party and in that legal capacity may be binding as an admission against the party himself. Usually the relationship between the agent and the principal against whom the admission is made has been found to lie in employment situations. However, admissions made by predecessors in title or other persons having a legal or contractual connection with a party are also found frequently admissible in evidence against the party. The same kind of reasoning is behind the use of admission by one partner in a business enterprise made in the course of business affairs of the partnership as evidence against all the partners. Admissions are binding only as against the party who made them. Accordingly, an admission by one party is not evidence against a co-defendant.

PRIVILEGE

§7.38 Hearsay, opinion and character evidence, as a general rule, are excluded because of their inherent unreliability, lack of probative worth and susceptibility to fabrication. These are all dangers related to the adversary method of ascertaining the truth. Moreover, in order to minimize the risk to the trier of fact, in his search for truth, relying on untested and untrustworthy proof, such evidence is excluded from the fact-finding process. The exclusionary rule of privilege, however, rests upon a different foundation. It is based upon social values, external to the trial process. Although such evidence is relevant, probative and trustworthy and would advance a just resolution of disputes, it is excluded because of overriding social interests. Thus, in recognizing the exclusionary rule of privilege the courts have been prepared to sacrifice, in some measure, their ability to inquire into all material facts in order to preserve a societal interest in non-disclosure. The protection of privilege may be sought by a litigant or a witness either at trial or at an earlier stage. The witness may decline to reveal certain information or to produce a document that is requested at trial. Similarly, at an examination for discovery, a party may refuse to answer questions on the grounds of privilege and, for the same reason, decline to disclose pertinent documents.

§7.39 The primary kinds of communication which enjoy the benefit of the exclusionary rule of privilege are those that take place between a solicitor and his client. The communications that pass between a solicitor and his client are, and at law must remain, immune from disclosure. The extent of the rule and its application has been considered by the courts since as early as the 16th century and a statement of the modern law could be given as follows:

> That rule as to the non-production of communications between solicitor and client says that where . . . there has been no waiver by the client and no suggestion is made of fraud, crime, evasion or civil wrong on his part, the client cannot be compelled and the lawyer will not be allowed without the consent of the client to disclose oral or documentary communications passing between them in professional confidence, whether or not litigation is pending.[10]

§7.40 The justification for this exclusionary rule is based on the premise that full and frank confidence between a client and his legal advisor is necessary for proper consultation and the rendering of effective legal assistance. To preserve the basic right of individuals to prosecute actions and to prepare defences, there must be no impediment to their seeking legal advice.

§7.41 A common misconception that remains is that the relationship between a priest and a person who engages in confession is immune from an obligation to disclose. This, however, is not a privilege that is recognized at law and there is no protection at law from a compulsion to have this type of communication disclosed. However, a judge will, in many cases, exercise his discretion and rule that such communications should not be

---

10 *Re Dir. of Investigation and Research and Can. Safeway Ltd.* (1972), 26 D.L.R. (3d) 745 at 746 (B.C.S.C.).

disclosed, despite the absence of a legal rule giving no protection to this type of disclosure.

§7.42 If a land surveyor is, therefore, going to rely on certain statements which a client makes to his lawyer and the statements are made in correspondence which is furnished by the lawyer to another party including the land surveyor, he may think that this type of information could be referred to and included in the information that is available to him for the making of a determination or the formation of an opinion. It may, however, turn out that an attempt to rely on such information in a court proceeding would result in the information in question being barred from admission as testimony. The grounds for the exclusion would, of course, be the privilege that exists between a solicitor and his client. Accordingly, if a land surveyor is to obtain information of this nature, he should ensure that the information is obtained directly from the client.

PAROL EVIDENCE

§7.43 Some writers have reviewed the parol evidence rule as not a rule of evidence at all but, instead, a matter of substantive law that establishes which facts are legally effective when there is difficulty in interpreting a written document. Regardless of how the parol evidence rule is classified, it is a doctrine of law which can result in the exclusion of certain facts and it is therefore a matter which is ordinarily treated as a topic under the law of evidence.

§7.44 The parol evidence rule has been articulated in the classic statement of Denman C.J.:

> By the general rules of the common law, if there be a contract which has been reduced into writing, verbal evidence is not allowed to be given of what passed between the parties, either before the written instrument was made, or during the time that it was in a state of preparation, so as to add to or subtract from, or in any manner to vary or qualify the written contract.[11]

Although this rule is known as the parol evidence rule, it encompasses all prior or contemporaneous transactions between the parties whether oral or written. It is well to bear in mind the reason for the rule: when parties deliberately put their engagements in writing in such language as to import a legal obligation, it is only reasonable to presume that they have intended it to be a full and complete statement by introducing into it every material term and circumstance; consequently, all parol testimony of conversations or declarations made by either of them, whether before

---

11 *Goss v. Nugent* (1833), 110 E.R. 713 at 716.

or after or at the completion of the contract, will be rejected because such evidence, while deserving far less credit than the writing itself, would inevitably tend, in many instances, to substitute a new and different contract for the one really agreed upon.[12]

§7.45 No rule in law is more simply stated but more difficult in practice for the land surveyor than the parol evidence rule. How many times, as a land surveyor, does one formulate an opinion on the location of a boundary when indulging in the liberty of taking into consideration matters and evidence which lie outside the deed itself? How often does one interpret a metes and bounds description to mean something other than that which the deed specifically calls for? On the other hand, to rely on the simple specifications of a metes and bounds description in all instances would result in absurdity. The hierarchy of evidence, which has so often been referred to in earlier literature prepared for land surveyors about the law, always calls for the most reliance to be placed on those things of which men are least likely to be in error. The result of this hierarchy of evidentiary authority and its application in boundary surveying has invariably led to primary reliance being placed on monumentation or on occupational evidence of the specific items called for in metes and bounds descriptions and, only after that, on the specific distances and bearings called for in the courses. The conflict between this practice and the parol evidence rule is a difficult one to resolve. There is no simple answer as to the actual treatment that this type of a situation would receive in a court of law in any particular instance. One suspects that the parol evidence rule would be relaxed somewhat from a strict application of the requirements of the rule, but this is never a certain outcome. There would appear to exist sufficient grounds to form the basis of a strong argument for the exclusion of evidence which is not either referred to in the document itself or which is not necessary as an aid to the interpretation of that document.

§7.46 A written contract, and also a metes and bounds description in a deed, may be written in a language which is not entirely clear and definite in meaning and may be subject to various interpretations. In such a case, parol evidence is permitted to assist the court in determining the intentions of the parties and to resolve ambiguities in the construction of the document. A court, however, cannot consider evidence outside the written document unless some ambiguity arises. The reader would do well to refer to authoritative materials on the principles and rules of construction and give serious consideration to the maxim, *falsa demonstratio non nocet*.[13]

---

12 *Eillis v. Abell* (1884), 10 O.A.R. 226 at 247.
13 For a good treatment and discussion of the maxim, see *Eastwood v. Ashton*, [1915] A.C. 900 at 914 ff (H.L.).

Although the parol evidence rule is based on sound reasoning and, in most cases, would lead to a fair outcome, the rule sometimes leads to harsh and arbitrary results. However, most of the current proposals for reform of the parol evidence rule lie in the field of consumer transactions. One can readily appreciate the harsh results that flow from a decision on the part of a court excluding all oral representations by a retail sales person to a consumer in a transaction of which only a written contract remains.

BEST EVIDENCE RULE AS YET ANOTHER EXCLUSIONARY PRINCIPLE

§7.47 The best evidence rule emerged in the common law and was first articulated as "the best proof that the nature of the thing will afford is only required."[14] The rule rapidly became regarded as the single most important rule in the law of evidence. If better evidence was not available, then the best that was at hand could be allowed and admitted into evidence. Therefore, the rule operated both on an exclusionary basis and an inclusionary basis. That which was not the best evidence of a particular fact or issue would be excluded. Only that evidence which was the best, the most reliable, the most trustworthy and had the most relevance would be included.

§7.48 The original rule and its application led to absurd results. It has long fallen into disuse and today considerations such as the prevention of fraud and perjury, which motivated the early framers of the rule, affect the weight of the evidence rather than its admissibility. This, along with an obvious advantage to a party to adduce the best evidence, has contributed to the decline of the rule itself. Although the rule is still referred to with respect to matters other than the admission of secondary evidence of documents, it is rather by way of a rationale for assigning less weight to the evidence tendered than as a reason for its exclusion.

§7.49 The rule, as it applies to admission of secondary evidence of documents, is not necessarily very narrow. The rule as it has evolved to the present, is quite limited in that a court will generally not limit itself to the best evidence. All relevant evidence is admitted. The worth of it goes only to weight and not to admissibility.[15]

§7.50 There also exists a general distinction between what is known as primary and secondary evidence and this distinction might be summarized best by the following test:

---
14 *Ford v. Hopkins* (1701), 91 E.R. 250.
15 *Garton v. Hunter* [1969], All E.R. 451 at 453.

The law does not permit a man to give evidence which from its very nature shows that there is better evidence within his reach, which he does not produce.[16]

When parol evidence is, therefore, tendered of a written instrument, it is apparent that there exists or existed better evidence. The absence of the better evidence must be accounted for before the secondary evidence can be entered and admitted to the court. There are a number of circumstances in which secondary evidence would be admitted. Such circumstances could include a court being satisfied that the original document existed and has been lost or destroyed; if the documents are in the possession of another party who is not prepared to produce the document or to respond to a subpoena; or, if the document is lodged in a public or official document registration system.

OPINION EVIDENCE AS AN EXCLUSIONARY RULE

§7.51 Generally speaking, opinion evidence is not admissible in a court of law. The reasons for such a rule are obvious and include the following policy considerations:[17]

(1) if the matter in question is non-technical, a court can just as readily as the witness draw the appropriate inferences from the facts and thus the witness' opinion is not relevant;
(2) if a witness is permitted to give his opinion on a matter which a jury is capable of determining by itself without assistance, there is the danger that numerous witnesses will be called to give their opinion resulting in a waste of a court's time and confusion of the issues;
(3) there is the concern that a jury may too readily accept the opinion of influential witnesses without exercising their own independent judgment; and
(4) the admissibility of opinions would permit a witness to testify without any fear of prosecution for perjury.

§7.52 Having stated the general exclusionary rule against the admissibility of opinions, it must be emphasised that the implementation of this principle in practice is not workable. The exclusion of opinion evidence is based upon the assumption that there is a clear demarcation between inferences and the facts which give rise to them. Witnesses can testify only to the latter. They are not entitled to speculate on what conclusions follow from the facts. However, between the two extremes, there exist situations where the distinction between fact and inference becomes blurred. The distinction

---

16 *Doe & Gilbert v. Ross* (1840), 7 M. & W. 102 at 106, 151 E.R. 696.
17 Sopinka and Lederman, *supra*, note 2 at 297-8.

between opinion and fact is, therefore, only one of degree and the courts have been prepared to give a wide latitude to statements of inference where practical considerations such as the saving of time justify their reception.

§7.53 A lay witness will, for example, be allowed to give identification evidence of a person's handwriting if he can first establish his competence to do so on the basis of actually having watched the person write or being able to recognize the person's handwriting by reason of a regular exchange of correspondence with him.

EXPERT OPINION

§7.54 This exception to the exclusionary rule of opinion evidence relates, of course, to the opinions of experts. The testimony of experts is commonplace in civil litigation today. If the opinion of the expert is merely conjecture and not on a subject requiring any special study or experience, it will be disallowed. The word expert refers to the qualifications of the witness in a field of specialization. It does not imply that the witness is necessarily a proficient or good witness.

§7.55 An expert witness is usually called for one of two reasons. He provides basic information to the court necessary for its understanding of the scientific or technical issues involved in a case. In addition, because the court alone may be incapable of drawing the necessary inferences from the technical facts presented, an expert is allowed to state his opinion and conclusions. His usefulness in this respect is circumscribed by the limits of his own knowledge. Before a court will receive his testimony on matters of substance, it must be demonstrated that the witness possesses sufficient background in the area so as to be able to appreciably assist the court. This process is known as qualifying the witness, that is, having the witness recognized as an expert before he actually begins his testimony. As long as the court is satisfied that the witness is sufficiently experienced or educated (or both) in the subject matter at issue, the court will not be concerned whether his skill was derived by specific studies or by practical training, although the source of the skill may have a bearing on the weight which the court will give to his evidence.[18]

§7.56 The significance of the surveyor's report and the thoroughness of information shown on the plan cannot be overstated.

> In virtually every case, the written [survey] report is the cornerstone of the evidence which will be presented by the [surveyor]. In *viva voce* evidence, the [surveyor] may highlight and elaborate upon the written report and will

---

18 Sopinka and Lederman, *supra*, note 2 at 310.

be called upon to substantiate and defend its data and conclusions. The report itself, however, remains the cornerstone. Consequently, great care and attention must be devoted to drafting a comprehensive report which leads to sound and logical conclusions.[19]

§7.57 There has been much written on the subject of expert witnesses, their qualifications and their role within a court of law. However, this brief overview must limit the matter of expert witnesses to a very narrow explanation or analysis of their role. For the land surveyor, it is often said that his role is essentially one of being an expert in the retracement of boundaries. This is a confusing statement when the word expert is also used in the phrase 'expert witness' itself. The person acting as a land surveyor is an expert because of his skill or expertise in collecting or assimilating evidence. His opinion, which would be the very thing that a court of law would be asked to decide on, is not necessarily the most significant part of his testimony. The significance of the surveyor's testimony lies in the evidence he has found on the ground and gathered from elsewhere. Arguably, if his opinion were that important, that reliable and that trustworthy, then why is the very subject matter of the surveyor's opinion the issue of the litigation?

§7.58 In most cases, a land surveyor's opinion will be solicited as a witness, but once this opinion begins to overlap with the very issue that the court has to decide, caution should be exercised. A court of law will not necessarily appreciate a land surveyor, or any expert for that matter, giving an opinion of the outcome of the very issues which the court itself must decide.[20] This would be encroaching on the jurisdiction of the court and some judges have been known to become downright annoyed when a witness is prepared to volunteer an opinion on that which the judge himself must make a decision.

PRESUMPTIONS

§7.59 A presumption *of law* is a consequence annexed by the law to particular facts. A presumption of law is an inference which the law imputes, given certain facts, irrespective of their legal effect. Presumptions

---

19 K J. Boyd, "The Appraiser as Expert Witness", *The Canadian Appraiser*, Vol. 29, Book 3, Fall 1985 at 26.
20 See, for example, *Gorgichuk v. American Home Assurance Co.* (1985), 5 C.P.C. (2d) 166 (Ont. H.C.) where the plaintiff tried to admit expert evidence as to the rules of grammar and syntax applicable to the interpretation of a contractual document. Interpretation of the contract was seen as the core judicial function and therefore general opinion testimony as to grammar and syntax were held inadmissible.

of law can be either conclusive and irrebuttable or rebuttable. Presumptions of fact are always rebuttable.

§7.60 From the surveyor's point of view, presumptions of fact and not presumptions of law are the primary interest and concern in his activity. When facts raise a presumption of fact, they give rise to a permissive inference which the trier of fact may, but need not, draw. There are certain circumstances or combinations of fact which occur frequently. It has been found that whenever these circumstances or facts are present, it can safely be assumed that a certain other fact exists. This latter fact is the inference to be drawn from the circumstances or collection of facts. In order to assist the trier of fact, these recurring circumstances or combination of facts are said to give rise to a presumption of fact. There will, of course, be situations where there will arise conflicting presumptions. The general rule is that when two presumptions conflict, they cancel each other out, leaving the matter to be decided in accordance with the legal burden.

BURDEN OF PROOF

§7.61 Proof is a word that is frequently used in the English language, but few people have a clear understanding of the exact meaning of the concept. A standard of proof determines the degree of probability that must be established by the evidence to entitle a party having the burden of proof to succeed in proving either the case or an issue in the case. Generally, there are two levels of probability, commonly referred to as the civil onus and the criminal onus. The civil onus requires the party having it to establish the case or an issue therein on a balance of probabilities. Within this broad category, there are degrees of probability which may vary according to the subject matter. This does not alter the basic standards of proof but casts on the party having the burden of proof the obligation of collecting and presenting more evidence or evidence of greater reliability and, on the part of the decision-maker, the obligation of subjecting the evidence to closer scrutiny.

§7.62 The ultimate or final burden of proof is the burden that is assigned by the substantive law to the parties in a case with respect to the various issues disclosed by the pleadings. The allocation of the ultimate burden of proof is determined by the substantive law. It is sometimes said that the ultimate burden does not shift throughout a trial. Notwithstanding this general rule, however, judges do from time to time, when speaking of ultimate onus, say that it shifts. For example, in a breach of contract situation that is the subject of litigation, the burden of proof rests with the plaintiff throughout the trial to demonstrate the facts that constitute the cause of action. Once having discharged this burden in chief, the onus falls on the defendant to raise a sufficient defence to bar or otherwise negate the claim. Ultimately, the onus and burden lie with the plaintiff.

DOCUMENTS

§7.63 The general rule of law is that the substance of a written document or plan must be produced as the original writing itself and not by a copy, or *viva voce* testimony. The general rule outlined here is a derivative of the best evidence rule.[21]

§7.64 Naturally, where the original of a document has been lost or where the author of it is no longer alive, secondary evidence of its authenticity and origin may be admitted. Some forms of documents require no proof as to their authenticity; these include public documents made for public use and reference and certified copies of court documents such as judgments, orders and decrees.

§7.65 In contrast to public documents and court documents, private documents may also be tendered in evidence, but their validation or proof of authenticity must also be undertaken. This process usually entails the comparison of the writing or document with a proven original or the calling of a witness to the execution of the document to attest to this issue.

§7.66 The significance of field notes cannot be overstated. Their admissibility is rarely an issue today (provided the notice provisions under the provincial Evidence Acts are complied with) and their significance in validating the information shown on a plan of survey is substantial. It is perhaps the background of field notes to a plan that distinguishes a plan from a mere map as an exercise in geography.

> Maps, when they have no conventional or statutory significance, should be regarded merely as representing the opinions of the persons who constructed them, they furnish at best no adequate proof, and none when it appears that they are founded upon misleading or unreliable information or upon reasons which do not go to establish the theory or opinion represented, and when they have not the qualifications requisite to found proof of reputation. Some of the later printed or coloured maps issued by the department of Colonization or of Crown Lands represent the seignory in accordance with the respondent's contention, others adopt that of the Crown. These maps embrace large districts, if not the whole province; they are issued for departmental use. One realizes that publications, documents and information not infrequently find their way into the Crown Lands and other departments of the Government from which inferences may be drawn adverse to the public right. Claimants are vigilant to avail themselves of any consent which may be afforded to introduce to the records information which may serve their interests. Territorial limits and

---

21 See §7.47, *supra*.

the boundaries of wilderness grants are, perhaps more frequently than not, lacking in definition or precision of statement, and when a general map of a province or district is in course of preparation, the attention of the departmental draftsman is not apt to be specially directed to careful consideration of the particular features or details upon which claims may depend, and sometimes, not unnaturally, particulars creep into the draft without due consideration of their use or trustworthiness. They are matters of detail, perhaps proper to be shown if verified, but not contributing to the main purpose of the work, which is not essentially concerned to verify them. These maps are prepared and issued not for the purpose of establishing facts or as admissions; they merely illustrate, and the proof must come from sources outside the maps.[22]

STANDARDS OF PROOF

§7.67 The standard of proof is established by the degree of probability that the law requires be demonstrated for proof to be said to have been accomplished. Generally speaking, there is no difference as to the rules of evidence between civil and criminal cases. A fact must be established in accordance with the rules, irrespective of whether civil or criminal consequences follow.

§7.68 However, the onus that must be discharged in criminal cases is higher than that in civil cases. While in criminal cases the decision-maker must be satisfied that proof has been effected beyond all reasonable doubt, the civil standard requires that proof be established merely on the balance of probabilities.

> [I]n every civil action before the tribunal can safely find the affirmative of an issue of fact required to be proved it must be reasonably satisfied and whether or not it will be so satisfied must depend upon the totality of the circumstances on which its judgment is formed including the gravity of the consequences of the finding.[23]

REAL EVIDENCE

§7.69 Real evidence, which has also been called demonstrative evidence, consists of physical objects made the subject of inspection by a court. In surveying matters, real evidence most commonly involves a decision-maker attending at the scene of the property or boundary in dispute.

§7.70 Attending at the scene of a boundary problem must be for the purpose of better enabling the decision-maker to understand and make sense out of the evidence. No new evidence can be presented during the course of

---

22 *R. v. Price Bros. & Co.*, [1926] S.C.R. 28 at 45-46.
23 *Smith v. Smith*, [1952] 2 S.C.R. 312.

a viewing; its purpose cannot be to enable any fresh evidence to be collected. Evidence is properly presented during the court or hearing process; evidence cannot be independently assimilated by the court. A recent case involved a judge attending at the scene of an accident to use his own observations, as evidence on which to base his own findings of fact. But the use of a view to acquire additional evidence on which to base the judgment was held to be improper.[24]

---

24 *Swadron v. North York* (1985), 8 O.A.C. 204 (Div. Ct.).

8

Settlement of Boundary Uncertainties

James F. Doig*
C.D., B.Sc., B.Ed., N.S.L.S., C.L.S.

There must have been a recognizable boundary to the Garden of Eden or Adam and Eve could not have been evicted.[1]

INTRODUCTION

Aim

§8.01 The aim of this chapter is to indicate the source and the nature of uncertainties and the means available to resolve them. Since the means used to resolve uncertainties are not uniform throughout the common law provinces and since the means depend upon the statute law of the jurisdiction or upon case law, two cautions are in order. First, statute law is referred to in this chapter for the purposes of illustration and by way of stating general principles; details must be taken from the statutes themselves. Second, and with respect to case law:

---

* Two associates — C.W. MacIntosh Q.C., Land Registration and Information Service, Halifax, N.S. and R.G. McBurney PEng., College of Geographic Sciences, Lawrencetown, N.S. — were kind enough to review my work when it was just about finished. Their suggestions and comments were very helpful and much appreciated. The opinions expressed are, of course, my responsibility.

1 R. Rowton Simpson, *Land, Law and Registration* (Cambridge: Cambridge University Press, 1976) at 126.

286   Survey Law in Canada

A case is only an authority for what it actually decided. [A] case can [not] be quoted for a proposition that may seem to logically follow from it. Such a mode of reasoning assumes that the law is necessarily a logical code, whereas every lawyer must acknowledge that the law is not always logical at all.[2]

Therefore the case reports here cited must themselves be read in order to become aware not only of the principal points at issue, but of the often unique statutory provisions which surround them.

Settlement

§8.02 The pattern and purpose of the original land grants in any area are often primary factors in subsequent uncertainties. The prairie provinces saw the establishment of an orderly mathematical system of survey in advance of homesteading.[3] Exceptions were made to this system for the Metis settlements where the Quebec pattern of long, narrow river front lots had been adopted and arranged without benefit of formal survey. In eastern Canada, things had been rather different. With the exception of Prince Edward Island, none of the Atlantic Provinces had been surveyed under a single connected system. Early settlement in Nova Scotia, for example, was most often by a collective land grant. There were no roads; water was the means of communication and transportation. Along the coast, or along a river of some consequence, an area of perhaps one hundred thousand acres (about 12 miles square) would be set aside as a township; from this lots would be created and granted to individuals through the township proprietors.[4] Most of the early surveyors were conscientious and capable, but there were never enough of them; and some were neither overly skilled nor attentive to their instructions. In one of the more flagrant cases, what a Crown grant intended as 100 acres became 772 acres on the ground.[5] Even under the best of circumstances a number of questionable practices were bound to occur. One of the more remarkable was in New Brunswick where a 1500-acre sale of Crown lands was divided into 13 tracts, in different parts of the province, "consisting of narrow strips with extensive fronts on the rivers, often embracing both sides of the stream, containing three islands and several mill privileges and other desirable sites."[6]

---

2 *Quinn v. Leathem*, [1901] A.C. 495 at 506.
3 D.W. Thomson, *Men and Meridians*, Vol. 2 (Ottawa: Queen's Printer, 1967) gives a very thorough account of things in each province.
4 *Boehner v. Hirtle* (1912), 46 N.S.R. 231; reversing 9 E.L.R. 258 (C.A.); reversed without written reasons, 50 S.C.R. 264. The property was in Lunenburg Township, which had not all been granted.
5 *Davison v. Benjamin* (1874), 9 N.S.R. 474 (C.A.).
6 G.A. Rawlyk, *Historical Essays on the Atlantic Provinces* (Toronto: MacLellan & Stewart, 1967) at 131.

§8.03 In Newfoundland for many years it was forbidden to settle on shore.

> All owners of [English] ships trading to Newfoundland [are] forbidden to carry any persons not of the ships Company or such as are to plant or do intend to settle there . . . speedy punishment may be inflicted on the offenders (Confirmation of the rules of 1663).[7]

Naval parties were regularly sent ashore to burn or tear down such shelters as had been put up since the fishing fleet had departed the previous autumn. Land grants eventually were given out, but not until the early years of the 19th century. The records of many of these were lost when the registry housing them was destroyed by fire. Documents attesting to original grants are still coming to light from safekeeping in private homes; these parchments are the only existing written records of severances long ago made by the Crown.[8] Parts of Ontario were surveyed hurriedly under the township system to accommodate the influx of Loyalists. Baseline frontages were laid out bordering lakes or rivers; waterfront parcels were surveyed for immediate occupancy; work on the rows of lots behind would be done later as the need arose.[9] Though the Dominion Lands Survey system extends into British Columbia from the prairies, it was not there to begin with. Early surveys were unconnected to each other. They were made as required for settlement without much stress on regularity or ties between one group of lots and another.[10]

Transfers in Writing

§8.04 Just as boundaries are made by people, so they are tended or neglected by them. Much depends upon circumstances. Some uncertainty in the boundaries of land used as orchard or hayfield is quite tolerable

---

[7] A.B. Perlin, "An Outline of Newfoundland History", in *The Book of Newfoundland*, Vol. I (St. John's: 1937) at 162-194.

[8] A.C. McEwen, "Land Titles in Newfoundland", *The Canadian Surveyor*, Vol. 31, No. 2, June 1977 describes the evolution of land titles. Possessory claims still remain the basis of a substantial portion of all land titles in the province. See also F. Shortall, "A History of Survey and Land Title for the Island of Newfoundland", unpublished Technical Report, University of New Brunswick, Department of Surveying Engineering, December 1987.

[9] B.J. Reiter, R.C.B. Risk and B.N. MacLellan, *Real Estate Law*, 2d ed. (Toronto: Edmond-Montgomery Ltd., 1982). Chapter 14 contains R.L. Gentilcore's "Lines on the Land; Crown Surveys and Settlement in Upper Canada". W. Brown and H. Senior, *Victorious in Defeat: The Loyalists in Canada* (Agincourt: Methuen, 1984) gives an excellent account of the frustrations, the complaints, the political difficulties, some of the technical difficulties and the accomplishments attendant upon Loyalist land grants in New Brunswick, Nova Scotia, Prince Edward Island and Upper Canada.

[10] F.C. Green, "Settlement and Surveys in British Columbia", *The Canadian Surveyor*, Proceedings of Annual Meeting, 1940.

and of no practical significance, while the land is being used for such purposes. Rural land values themselves may make it uneconomical to resolve uncertainties after lines have become blurred. The settlement of uncertainties is a continuing process. Sometimes it occurs through the adjudication of individual cases; sometimes more general provisions are made. Such was the case when the *Statute of Frauds* was enacted by the Parliament of England in 1677.[11] While it was an unsatisfactory piece of legislation in many respects and subsequently caused great difficulty in matters of contract, it was a milestone in the development of the law of real property. Title to land could no longer pass orally; agreements for the transfer of interests in land henceforth had to be in writing. As it pertains to real property transfers, the statute is found in each of the common law jurisdictions of Canada either as an act of the legislature or through the Imperial statute being in force as received English law.[12] The British Columbia statute, for example, stipulates that "an agreement concerning an interest in land is not enforceable by action unless evidenced in writing signed by the party to be charged or his agent."[13]

Exceptions

§8.05 No rule can be final;[14] there must always be exceptions. Consequently, ever since 1677 there have been exceptions to the requirement that title pass only by written contract. The doctrine of adverse possession is probably the most notable of these; so are some of the means employed to resolve boundary uncertainties. Nevertheless, the fundamental requirement of the *Statute of Frauds* as it pertains to the transfer of land should always be borne in mind. Any agreement relating to property lines should be reduced to writing whether required to be done by one of the parties or not. Indeed, present experience seems to suggest a tightening up of things as far as permissible exceptions are concerned. This is illustrated

---

11 1677, 29 Cha. 2, c. 3.
12 Alberta, Northwest Territories, Manitoba, Newfoundland, Saskatchewan, and the Yukon have no direct legislation. P.E.I. has the *Statute of Frauds*, R.S.P.E.I. 1974, c. S-6, but it does not deal with the matter of land transfer being in writing; *Delima v. Paton* (1971), 1 Nfld. & P.E.I.R. 317 (P.E.I.S.C.) ruled that the English statute had been in force since the Province was settled in 1773, having been brought to the Island as part of the law necessary to make transactions effective in the colony. See also A.H. Oosterhoff and W.B. Rayner, *Anger and Honsberger Law of Real Property*, Vol. 2, 2d ed. (Aurora: Canada Law Book Inc., 1985) at 1263-1264.
13 *Statute of Frauds*, R.S.B.C. 1979, c. 393, s. 1.
14 Except perhaps this one, which calls to mind the paradox about the village barber who shaved everyone who didn't shave himself.

by cases involving conventional lines. Some jurisdictions appear far less prone than previously to accept lines as lost or uncertain.

Causes of Uncertainty

§8.06 The written record goes a long way toward diminishing uncertainty — at least one has no doubt about what was said — though doubt may remain about what was meant. Imperfections are inevitable. They arise from imprecise phraseology, unclear statements of intent, inattention to detail, carelessness in safeguarding one's interests, the use of different words for the same thing,[15] alterations in attitudes and changes in adjacent owners. The initial purchase of a property, for example, is often primarily to satisfy that most pressing need of a place to live. With increasing disposable income, thoughts may later turn to developing the property for leisure, recreation or just with an eye to resale. This can lead to a desire or a need for more precision and greater certainty about boundary lines. On the other hand, there may be no concern about dividing lines until a new neighbour arrives who has a more enquiring mind or who holds a more jealous attitude toward his possessions than had his predecessor.

After Retracement

§8.07 The retracement of boundaries, the interpretation of deed descriptions and the doctrine of adverse possession are not the principal concerns of this chapter. But since boundary uncertainties usually involve one or more of these topics one will find frequent reference to them. Uncertainties arise from descriptions found in deeds, from the usage to which property has been put and from obliterated boundary lines; many of these are resolved by retracement. Evidence is found on the ground; information comes from neighbouring deeds; the application of well-established principles of law makes clear that which had been in doubt. The concern of this chapter is to deal with the resolution of those uncertainties which remain when the surveyor's work of retracement does not provide solutions.

SOURCES AND NATURE OF UNCERTAINTIES

§8.08 There would be no uncertainty at all if each parcel had good documentary title, if its boundaries were well defined on the ground and were kept that way, if encroachments were stopped before they were started

---

15 *Spearwater v. Seaboyer* (1984), 65 N.S.R. (2d) 280 (T.D.). A property was derived from four separate lots. Each of the descriptions used a different term to describe the water boundary, though this was not the point at issue.

and if there were no overriding interests. Title means the right of ownership and the evidence of that right;[16] boundaries are the physical limits of ownership. Strictly speaking, matters relating to title and to some overriding interests need not directly affect parcel boundaries. Similarly, boundary problems can present themselves where there is no question of title. In practice, however, a question about boundaries often raises the matter of title and *vice versa*.

Title

§8.09 Questions about the chain of title usually relate to one of two topics: Is there good documentary title? Are there any overriding interests? While both matters are properly the province of the lawyer, the surveyor needs some understanding of each.

§8.10 In a deed registry system good documentary title is a clear unbroken title chain going back either to a Crown grant or as far or farther than the period of time required for title searches.[17] For a number of reasons documentary title may be lacking. Earlier instruments may never have been placed on the register;[18] land may have passed within families by mutual agreement and without any written transfer; property may have been acquired by adverse possession. In some instances there may even be doubt whether the land in question was ever granted by the Crown.[19] Perfectly good possessory title, on the other hand, may be in the owner from long occupation instead of the owner shown on the record; in cases of this kind statutes like the *Land Titles Clarification Act* or the *Land Title Inquiry Act*[20] may solve the problem. The need to demonstrate documentary

---

16 *Re Vancouver Improvement Co.* (1893), 3 B.C.R. 601 (C.A.).
17 Forty years is prescribed in Newfoundland (*Limitations of Actions (Realty) Act*, R.S.N. 1970, c. 207, s. 16), Prince Edward Island (*Investigation of Titles Act*, R.S.P.E.I. 1974, c. I-7, s. 2(1)) and Ontario (*Registry Act*, R.S.O. 1980, c. 445, s. 105(1) amended 1981 c. 17). No specific period is prescribed in New Brunswick and Nova Scotia; in these provinces it is a matter of common law rather than one of statute law; the period of search may be 20, 40, 60 or more years until a good root of title is found. Much depends upon the state of title in a given area and the assessment of the relevant documents by the individual who is responsible for certifying the title. Any search period is, of course, abbreviated by the existence of a document creating a new good root of title such as a deed from the Director, the *Veterans Land Act* (R.S.C. 1970, c. V-4), C.W. MacIntosh, "How Far Back Do You Have To Search", *Nova Scotia Law News*, Vol. 14, No. 3, December 1987 traces the origin and development of the period necessary for a title search, notes Canadian practice, and summarizes some recent cases in New Brunswick and Nova Scotia.
18 There is no statutory requirement, at least in the Atlantic Provinces, to register deeds.
19 *Boehner v. Hirtle*, supra, note 4; *Scott v. Smith* (1979), 36 N.S.R. (2d) 541 (C.A.).
20 *Land Titles Clarification Act*, R.S.N.S. 1967, c. 162; *Land Title Inquiry Act*, R.S.B.C. 1979, c. 220.

title usually arises when land is being sold out of the family, or mortgaged, or a house is being built upon a portion of it.

§8.11 *Overriding interests* A variety of restrictions exist on the use of real property; particular ones may range from loss of title through adverse possession, to the minor inconvenience of providing access over unimproved lands for someone to go fishing.[21] These interests are not the general province of the surveyor at all, though he would be expected to be acquainted with expropriation, dower, watercourses and the like. The matter of overriding interests is raised, not to suggest that the surveyor become expert in them, but rather to show the range of circumstances that can affect title. The cure for existing or potential problems rests with the solicitor, assisted on appropriate occasions by the surveyor. But the circumstances must first be recognized. Many overriding interests are enforceable without registration in the registry of deeds. The detail in paragraph 8.12 relates to the Province of Nova Scotia; it is not meant to suggest that these circumstances are precisely mirrored elsewhere.[22]

§8.12 During research connected with development of draft legislation, all incidents were listed which had been created in Nova Scotia by statute and which affect title to real property. The result: 32 in 1976 and 36 in 1984:[23]

> Civic Tax Lien
> Civic Betterment Charge
> Estate Tax Lien re deaths within a particular time frame
> Land Tax Lien
> Non-resident Income Tax Lien
> N.S. Power Corporation Bill Lien
> Lien under the Worker's Compensation Act
> Liens under the Labour Standards Code
> Mechanics' Liens during unexpired registration period

---

21 *Angling Act*, R.S.N.S. 1967, c. 9. A very different arrangement prevails in New Brunswick where fishing leases are auctioned by the Crown. This is a practice unknown in Nova Scotia. The neighbouring provinces are similar in geographic, economic and other circumstances; and both, of course, function under the English common law. But this is a good example of how differently the same topic can be treated by the statute law of similar jurisdictions. Pronounced differences are found also, between the two provinces, in the matter of title to lands used as public highways.
22 Reiter, Risk and McLellan, *supra*, note 9. Chapter 13 gives a glimpse of things from an Ontario perspective. The expressive American phrase "Off-the-Record Claims" is used as the chapter title.
23 C.W. MacIntosh and M.L. Gavin, "Overriding Interests in Land", Nova Scotia Law News, Vol. 2, No. 3, January 1976; and C.W. MacIntosh and R.C. Penfound, "More Overriding Interests in Land", *Nova Scotia Law News*, Vol. 11, No. 3, December 1984.

Unpaid Vendor's Lien
Dower Rights re property conveyed prior to the coming into force of the Matrimonial Property Act
Public Highways
Abandoned Highways
Restrictions under the Public Highways Act
Necessary Rights of Way — The Private Ways Act
Crown Rights of Expropriation, Access and User
Utility Corporation Easements to Maintain Overhead Wires
Watercourses — The Water Act
Marshland Reclamation
Rights of Access to Fish — The Angling Act
Flooding Rights
Beaches Preservation and Protection
Treasure Trove
Mining Leases and Licenses — The Mineral Resources Act
Prospecting and Drilling Licenses — The Petroleum and Natural Gases Act
Salvage Yards Act
Title by improvements by adjoining owner[24]
Adverse Possession and Prescription
Condemnation under the Public Health Act
Street Tree Lien
Quieting Titles and Land Titles Clarification Application[25]
Airport Zoning Regulations
Restrictions on Rental Property
Interests under the Matrimonial Property Act
Heritage Property Act restrictions
Approvals required by the Shopping Center Development Act

A lawyer can always endeavour to protect himself against liability by giving his client a certificate of title sufficiently restrictive in nature as to relieve himself from such liability with respect to the existence of most overriding interests. However, the more professional approach to the problem is seen as a duty to become knowledgeable of these interests so that they can be detected and their nature and effect explained to the client before a purchase is finalized.[26]

This is good advice for the surveyor too, appropriate changes being made to suit the circumstances.

---

24 See §§8.50, 8.58-8.60.
25 See §§8.77-8.79.
26 *Supra*, note 23.

Settlement of Boundary Uncertainties

§8.13 Overriding interests are not relevant to irregularities under provincial planning acts.

Boundaries

§8.14 The boundary uncertainties which are considered in this chapter are those which arise because:

(1) boundaries have been lost; that is, the boundaries exist but their location is not known to the adjoiners;
(2) there is lack of coincidence between resurveys[27] of the same line by different surveyors;[28]
(3) a dispute occurs over the validity of the resurvey of a property line by a single surveyor;[29]
(4) improvements have been made or encroachments have occurred;
(5) lands newly acquired, and contiguous to a parcel already held, must be apportioned.

Responsibilities of Surveyors

§8.15 The surveyor, like anyone else, has many responsibilities. Three of special importance arise when resurveys are made: the duty to be impartial, the duty to search for the best evidence, and the duty to consult with other surveyors.

---

27 *Manual of Instructions for Canada Lands Surveyors*, 2d ed. (Ottawa: Canadian Government Publishing Centre, 1979) Chapter B7. A distinction is made between resurveys, restoration surveys and retracement surveys. "Resurveys" is here used in the sense of all three.

28 For some accounts see: *Re Janes* (1977), 17 N.B.R. (2d) 600 (Q.B.) (retracement of an old woods line); *Drysdale v. Drysdale* (1977), 18 N.B.R. (2d) 429 (Q.B.) (one surveyor's plan was based solely upon hearsay); *Steeves v. Downey* (1972), 4 N.B.R. (2d) 646 (S.C.) (one surveyor did no research); *Pitre v. Robinson* (1978), 15 Nfld. & P.E.I.R. 63; reversing 19 Nfld. & P.E.I.R. 475 (P.E.I.S.C.) (two surveyors with different approaches arrived at different conclusions); *Re Risser's Beach* (1975), 20 N.S.R. (2d) 479 (T.D.) (two surveyors drew different inferences from the same situation); *Holden v. Strickland* (1976), 13 Nfld. & P.E.I.R. 439 (Nfld. Dist. Ct.) (two approaches to the same problem by two surveyors); *Boyd v. Luscombe* (1986), 57 Nfld. & P.E.I.R. 242 (Nfld. Dist. Ct.) (errors in laying out Crown grants apparently never were pinned down despite the efforts of four surveyors; one surveyor was agreed upon by counsel to review the surveys of two others).

29 For some accounts see: *Thompson & Purcell Surveying Ltd. v. Burke* (1977), 39 N.S.R. (2d) 181 (Co. Ct.) (surveyor and client disagree about line location); *Benoit v. Benoit* (1976), 15 N.B.R. (2d) 233 (S.C.), affirming 15 N.B.R. 59 (C.A.) (surveyor resolved all problems in favour of his client); *Finley v. Sutherland* (1969), 2 N.S.R. 1965-69 197 at 203 (C.A.) (surveyor acting under instructions of client to give neighbour benefit of any doubt); *MacDougall v. Layes* (1969), 2 N.S.R. 1965-69 96 (S.C.) (acceptance of surveyor's work and extent of correlation between different types of evidence on the ground).

294   Survey Law in Canada

§8.16 *Impartiality* No better advice on this topic can be found than that given by the Surveyor General of Canada:

> A boundary when surveyed marks not simply the limits of one property but the division between two or more properties. The surveyor must, therefore, consider the rights of all adjacent owners and must include in his search all pertinent evidence created in surveys of their properties.
>
> In resurveying boundaries of occupied properties the surveyor should be very cautious about doing anything that would upset the established limits of occupation. Settled possession, where it can reasonably be related back to the time of the original survey, may provide the courts with satisfactory evidence of the original boundary. There will be cases where, if the surveyor replaced monuments in what he believed to be their original positions, the results would lead inevitably to a dispute. In this event the surveyor can do no more than report the circumstances and his opinion to the parties concerned.
>
> In any case of a disputed boundary, the surveyor can only advise the disputants and give his opinion as to the correct or most equitable position of the boundary. In addition to this, he should take care not to perform any act that might have the effect of prejudicing the case of either party. So long as the dispute continues, no surveyor can lay down the boundary since its determination is of necessity a judicial act, and must be judged in court according to law after hearing of evidence.[30]

§8.17 *Best evidence* By statute,[31] the Ontario Land Surveyor is enjoined to "obtain the best evidence available respecting the . . . boundary . . . but if the . . . boundary . . . cannot be re-established in its original position from such evidence, he shall proceed as follows . . ." (and then are listed explicit procedures, depending upon the township within which the lands are found). Note that the primary requirement is for the best evidence available, so that the line can be re-established in its original position. Only when the re-establishment of the line in its original position is impossible may the surveyor have recourse to the mechanical procedures set out in the applicable part of the statute. Surveyors in other jurisdictions can not go far wrong in adopting the principle of striving to obtain the best evidence.

§8.18 *Consultation* Surveyors have a duty to communicate and consult with one another, for the benefit of their clients, particularly when they are engaged upon surveys of the same line or of adjoining properties.

---

30 *Supra*, note 27, Chapter B7 at 58.
31 *Surveys Act*, R.S.O. 1980, c. 493, ss. 13(2), 17(2), 24(2), 31(2), 37(2), 44(1). The *Land Survey Act*, R.S.B.C. 1979, c. 216, s. 5(1) also calls for the best evidence in like situations, as do the *Surveys Act*, R.S.A. 1980, c. S-29, ss. 31(1) and 42(1), and the *Land Surveys Act*, R.S.S. 1978, c. L-4, s. 24(1). See also A.C. McEwen, "The Meaning of Best Evidence," *The Canadian Surveyor*, Vol. 35, No. 1, March 1981.

Although the judicial process in Canada is based upon the adversarial system, this does not, or should not, affect the capacity and willingness of surveyors to make known to each other the evidence they have found about boundaries. Cooperation and the free exchange of information is but an extension of the principle that when adjoining properties have been surveyed and the surveyors are not in agreement, matters should be discussed between them before any public pronouncements are made about mistakes or disagreements. One mark of a competent person, secure in his capacity to conduct affairs creditably, is his willingness to cooperate with others in matters of mutual concern.

§8.19 *Clients' affairs* Disclosure of evidence relating to boundaries is one thing; a client's business arrangements are another. Unless required by law, the surveyor must not disclose the business affairs of his client. The surveyor has no right to privileged communications.

RESOLUTION OF UNCERTAINTIES

Courts

§8.20 When real property boundaries are uncertain or come into dispute there are two fundamental means of resolving the issue. The owners of the adjoining lands may come to an agreement between themselves, though it is best if they are assisted by others, or the matter can be placed before the courts.

§8.21 *Surveyor's authority* The surveyor has no power to decide the location of any boundary on re-survey; he can only express an opinion as to where the boundary is to be found based upon a search for and study of evidence he has found in the field, in the testimony of others and in searches within the appropriate registry offices, Crown records and other offices of record.[32] It should surprise no one, however, to learn that the surveyor can, and often does, bring different levels and sorts of pressures to bear upon disputants in order to encourage them to settle their differences without going to court. A careful survey and an objective written report are indispensable to the surveyor's efforts in this direction.

§8.22 *Quality and extent of title* In the broadest sense a property dispute may involve questions relating both to the validity of title (ownership) and to the extent of title (boundaries), though both elements need not necessarily be at issue in any one situation. There may be lands of perfectly good

---

32 See also W.D. McLellan, "Notes of Interest to Surveyors Pertaining to Various Aspects of the Canadian Law of Real Property", *The Nova Scotian Surveyor*, Vol. 45, Nos. 123 and 124, 1986, and Vol. 46, No. 125, 1987.

title with uncertain limits, just as one may find very clear and uncontested lines of demarcation which enclose a parcel of very dubious title. In order to distinguish broadly between a variety of circumstances, it is useful to consider the situation under land titles systems and the situation under registry systems.

§8.23 Land title systems When land is brought under a land title system, the title of the parcel is examined closely. In effect, a title search is done one last time. The title and its encumbrances, if any, go on the record, and title as subsequently reflected by the record is guaranteed. The boundaries of the parcel are examined at the same time but the rigour of the examination varies with the jurisdiction and the enabling legislation. All land title systems in Canada are limited or qualified systems in that boundaries are not guaranteed under them.[33] Once land is within a land titles system, boundary uncertainties may be addressed through resurveys by private land surveyors, or through special surveys ordered by the governing authority, which surveys may be appealed to the courts.[34]

§8.24 Registry systems A registry system is a depository for documents relating to land. There is no examination of title or of boundaries when a document is presented for registration be it a deed, will, plan or other instrument. If the document is in proper form and has the required endorsements (such as planning board approval), it is accepted and filed. All registry system jurisdictions provide, by statute, a means of quieting or certifying title.[35] These arrangements require the judicial investigation of the title of the particular parcel and an examination of its boundaries. The boundaries are confirmed when the title is certified.[36] Some, but not all of these jurisdictions, make statutory provision for the resolution of boundary uncertainties whether or not the lands have had titles certified and whether or not the title of the lands is at issue.[37] In all jurisdictions,

---

33 The *Land Titles Act*, R.S.O. 1980, c. 230, s. 141(2), for example, stipulates that the "description of registered land is not conclusive as to the boundaries or extent of the land." Within Ontario, of course, one may combine a first registration under land titles with an application to confirm the parcel boundaries under the *Boundaries Act*, R.S.O. 1980, c. 47.
34 See §§8.74-8.76 for further details about special surveys.
35 *Land Title Inquiry Act*, supra, note 20; *Quieting Titles Act*, R.S.O. 1980, c. 427 (which was repealed in 1984) and *Certification of Titles Act*, R.S.O. 1980, c. 61; *Quieting of Titles Act*, R.S.N.B. 1970, c. Q-4; *Quieting of Titles Act*, R.S.N. 1970, c. 324; *Quieting Titles Act*, R.S.N.S. 1967, c. 259; *Quieting Titles Act*, R.S.P.E.I. 1974, c. Q-2.
36 See §§8.77-8.80 for further details. Also New Brunswick is an exception; a plan of survey is not required by the statute, nor are the boundaries confirmed at the time title is certified. The abstract of title is validated.
37 *Boundaries Act* supra, note 33; *Boundary Lines and Line Fences Act*, R.S.M. 1970, c. B-70, s. 2.

except perhaps within land title systems, boundary agreements can be made by adjoining owners providing proper conditions are met.[38]

§8.25 The Atlantic Provinces — New Brunswick, Newfoundland, Nova Scotia and Prince Edward Island — make no provision by statute for the resolution of boundary uncertainties unconnected to a question of title. Neither are there statutory obligations for the sharing of any costs of survey work needed to help resolve boundary disputes. There is no obligation (other than a moral one) upon any adjoiner to defray the cost of the re-survey which retraces or recovers his uncertain or lost line. If he likes the result he may accept it; if he doesn't, he can challenge it. In neither case does he have to bear any part of the expense of it unless he has earlier and voluntarily agreed to do so.

§8.26 Under such circumstances, if agreement cannot be reached about the location of a disputed or lost boundary, the matter can only be resolved through court action. One may proceed by an action to quiet title or by an action in trespass. The choice between the two approaches may lie principally in the fact that an action to quiet title requires a survey of the perimeter of the parcel (except in New Brunswick), while that in trespass can proceed with the survey of only the uncertain line and carries prospects of an award in damages. Damages can be a major consideration when logging operations or the destruction of fences or ornamental trees and shrubs are at issue.

§8.27 One may encounter the expression "dummy trespass". This denotes circumstances where there has been no actual trespass other than perhaps the use of land up to a disputed boundary line and where an approach to the courts is the only way to resolve a problem not amenable to solution by mutual agreement.[39] This situation may call to mind the imaginary lease to the imaginary John Doe who was being ejected by the equally imaginary Richard Roe in proceedings, now long ago, for the recovery of land.[40] At any rate the court's final decision will normally settle the question of boundary location and of title (if title is at issue) for valid title must be in an owner for his boundary claim to succeed.[41]

---

38 See §8.30 *et seq.*
39 *Dixon v. Crowhurst* (1976), 14 N.B.R. (2d) 401 (C.A.) is perhaps an example. A father's city lot was divided between brother and sister. After some dozen years living side by side a dispute arose over a latent ambiguity in the description which had divided the original lot.
40 J. Burke, *Osborn's Concise Law Dictionary*, 6th ed., (London: Sweet and Maxwell, 1976).
41 *Rollings v. Smith* (1977), 15 Nfld. & P.E.I.R. 128 (P.E.I.S.C.) is probably an exception in that boundary details were left to the disputants. Describing the parcel confirmed to the defendant, the court ordered: "If its exact location cannot be agreed upon . . .

§8.28 Placing a question before the courts is an expensive process, as everyone knows. But there is rather more to the matter than that. By and large, the individual's opportunity for compromise is gone; negotiations are over; one no longer knows what one will be giving up for what one will be getting. Moreover, there are very few boundary problems in which all the good arguments are on one side. Once in court things often appear very different from earlier assessments of how things will go. Certainties may no longer appear quite so certain. Virtually every case has its strengths and weaknesses; the courts will examine each in detail as the appropriate principles of law are applied. Precedents from similar cases will be seen — more clearly than ever before — to be applicable in those particular situations only. "In all actions brought to determine the true boundary line between properties, the burden of proof lies upon the plaintiff who seeks to change the possession."[42] Courts do not exist to split differences.

§8.29 Some cases seem to be forlorn hopes from the very beginning and one wonders how and why they got into court at all.[43] In contrast, there appear from time to time points of considerable importance and of fundamental principle. It is a matter of some regret that our system of law on occasion requires the expenditure of so much of a private citizen's time, money and persistence to resolve an important point ultimately for the benefit of many others.[44]

Conventional Lines

§8.30 *Doctrine* The conventional line, like so many other things, is a simple concept fraught with many difficulties in execution. Broadly speaking, if a boundary line has been lost and is not just unknown because sufficient enquiry has not been made nor surveys performed, then adjoining owners may, between themselves, agree upon a boundary line dividing their lands. This concept is a long-standing and well established practice, particularly in those parts of Canada where no survey systems were initially emplaced, though it is by no means confined to those or any other particular areas. It is easy to see its relevance to rural properties such as pasture land, wood lots and timber holdings though it has never been restricted

---

a qualified land surveyor [shall] be engaged . . . and the cost of such a survey [will] be borne in equal shares."

42 *Palmer v. Thornbeck* (1876), 27 U.C.C.P. 291 at 302 (Ont. C.A.).
43 *Ewing v. Publicover* (1975), 13 N.S.R. (2d) 346, affirmed 14 N.S.R. (2d) 159 (C.A.). Plaintiff made a claim, through adverse possession, to title to lands in a public highway.
44 *Merriman v. New Brunswick* (1974), 7 N.B.R. (2d) 612 (C.A.). The ownership of the beach on a lake was at issue. The judgment fully explores the law of New Brunswick in the matter of water boundaries.

to such. The doctrine has found acceptance in urban as well as rural districts. Indeed, two rather prestigious cases involving conventional lines between states went to the Privy Council.[45] Unalterable boundaries can be created by parties who are not land surveyors.[46]

§8.31 *Application* Perhaps only within recent years has there been the awareness that the doctrine of the conventional line had at least the potential to be in conflict with subdivision regulations and with planning acts. Then, too, there may have been instances when the conventional line remedy has been wrongly applied. Lines may have been regarded as lost and an agreement made on that basis, rather than incur the expense of a survey to determine where the line really was. It is not hard to imagine this happening. Were this done, the circumstances could be obscured, fully or partially, by long acceptance of the line, and by the principle of equity which prevents parties from going back on their agreements. Whatever the situation, the decision handed down in *Bea v. Robinson*,[47] caused quite a stir: the doctrine of the conventional line is indeed a valid one, but conditions applicable to its use must be fully adhered to in order for the line to be binding upon both parties.

§8.32 *Future* The decisions handed down in *Sullivan v. Lawlor*[48] and *Bea v. Robinson* may have put an end to the doctrine of conventional lines where written agreements are not involved. Certainly *Crossland v. Dorey*[49] demonstrates the usefulness of having reduced the agreement of the adjacent proprietors to writing; otherwise there is no written record and at a later date reliance must be placed on oral evidence to establish the conventional line claimed.[50] With these circumstances before him, the surveyor might be acting in the best interests of both parties to provide documentary evidence in every instance. A plan of survey, signed by both proprietors, ought to give conclusive evidence of the agreement. Fuller assurance might be given by the exchange of quit claim deeds after fixing

---

45 *Re Labrador Boundary* (1927), 43 T.L.R. 289; *South Australia v. Victoria*, [1914] A.C. 283 (P.C.).
46 *Dell v. Howe* (1857), 6 U.C.C.P. 292 (C.A.). See R.J. Stewart, "Conventional Lines: Estoppel and Boundaries", *The Ontario Land Surveyor*, Spring 1983. It might be noted that the *Land Act*, R.S.B.C. 1979, c. 214, s. 1 defines a conventional boundary as one consisting of straight lines conforming to the natural boundary of a body of water. The *Boundary Surveys Act*, R.S.A. 1980, c. B-10 defines a conventional boundary-line as a portion of the provincial boundary. These definitions have absolutely nothing to do with the kind of conventional line which is considered in this chapter.
47 (1978), 18 O.R. (2d) 12 (H.C.).
48 (1981), 45 N.S.R. (2d) 325 (C.A.).
49 (1977), 27 N.S.R. (2d) 139, affirmed 28 N.S.R. (2d) 91 (C.A.).
50 *O'Melia v. Himmelman* (1985), 69 N.S.R. (2d) 271 (T.D.).

the location of the mutual boundary, though legal opinion should first be secured to ensure a planning act is not violated by the exchange of deeds.

§8.33 No alienation Settlement of boundaries is not an alienation because, if fairly made without collusion, the boundaries so settled are presumed to be true and ancient limits.[51] The establishment of a conventional line is not a breach of the *Statute of Frauds*[52] or of the Ontario *Planning Act*.[53]

§8.34 Crown cannot be party Except through special statutory authority, the Crown cannot convey land otherwise than by its grant under the Great Seal. The Crown cannot acquiesce in a conventional line.[54] Thus an owner of lands adjoining those of the Crown cannot ordinarily claim title to any lands of the Crown by reason of acquiescence or recognition of a line to which he claims. Statutory provision is probably made in each province, however, for settlements with respect to highway boundaries. In Nova Scotia, for example, the *Public Highways Act* gives the Minister authority, "in the event of a dispute", to fix the boundaries of any highway; since the Minister's decision can be appealed to the courts, there is a measure of agreement implicit in the process.[55] Of course title to lands in a public highway cannot be gained through occupation or possession.[56]

§8.35 Owners only can agree The doctrine of a conventional line has no application where one of the parties to an agreement does not have title.[57] An agreement between an intending purchaser of lands and the owner of the adjoining lands, as to their dividing line, did not create a binding boundary between them; such an agreement must be between the actual owners of the properties.[58] However, an agreement was upheld which had

---

51 *Grasett v. Carter* (1884), 10 S.C.R. 105 citing *Penn v. Lord Baltimore*.
52 *Davison v. Kinsman* (1853), 2 N.S.R. 1 and 69 (C.A.).
53 R.S.O. 1980, c. 379. See *MacMain v. Hurontario Mgmt. Services Ltd.* (1980), 14 R.P.R. 158 (Ont. Co. Ct.).
54 *Mersereau v. Swim* (1914), 42 N.B.R. 497 (C.A.). See, for example, the *Crown Land Act*, S.N.S. 1987, c. 5, s. 15. But see *A.-G. for Ont. v. Booth* (1923), 53 O.L.R. 374 (C.A.) re conventional lines for purposes of cutting timber under licence. The *Land Act*, R.S.B.C. 1979, c. 214, s. 78(1) provides for the establishment of a boundary by agreement.
55 R.S.N.S. 1967, c. 248, s. 14(2), (3). The *Towns Act*, R.S.N.S. 1967, c. 309, s. 147 gives a committee of a town council authority to fix a street line which a "person dissatisfied with the order" may appeal. Thirty years ago in the older parts of the City of Halifax, and before redevelopment, it was not unusual to find a block or a portion thereof for which a street line was not recorded or for which the record was suspect. A new line would be recorded which gave the street its accepted width, satisfied the deed descriptions of the adjoiners, cleared all existing structures, and joined smoothly with existing lines in adjacent blocks.
56 In Nova Scotia, for example, by virtue of the *Public Highways Act*, s. 16.
57 *Mooney v. McIntosh* (1887), 14 S.C.R. 740, affirming 19 N.S.R. 419.
58 *Smith v. Anderson* (1942), 16 M.P.R. 287 (N.S.C.A.).

been made by an occupier who had acted as agent for the owner.[59] A boundary agreement made between tenants of adjoining properties was not binding even when they later became owners of the previously leased land.[60] A wife was not bound by her husband assenting to a division fence altering the line by which she was entitled to hold.[61]

§8.36 *Boundary discoverable* Under these conditions, for whatever reason, a conventional line agreement is not valid. This has always been the rule, though the rule has not always been respected. Arguing 100 years ago on behalf of a client, a barrister said: "There is no conventional boundary to bind us, because the true boundary was not uncertain or unascertainable, and the Statute of Frauds applies."[62] His view of the law was confirmed in 1978 by Madame Justice Boland:

> When parties do not know the location of the line because they have made no inquiries or other attempts to discover it, that is not an uncertain boundary that can be varied by agreement.
>
> . . .
>
> [A] boundary agreed upon by adjoining landowners can only be presumed to be the true and ancient limit of the property when there is no registered instrument to contradict the agreement. That is to say when a parol agreement as to a boundary is at variance with the boundary that may be determined by reference to a deed or plan, then the agreement is an unenforceable attempt to convey land without the formal requirements of writing and registration.[63]

The conventional line has the obvious attraction of being a simple solution to a problem as opposed, otherwise, to spending time, effort, and money on the complexities of occupation, estoppel and the interpretation of deed descriptions. The temptation can be very strong. The surveyor should only counsel the conventional line after he has satisfied himself that such action is proper, having made diligent search for evidence of the line both on the ground and in the records.

§8.37 *Essentials* Above all else, the boundary must be lost; that is, its location must be unknown to the title holders on either side of the line. Given this condition, the next concern in establishing a conventional line is that the parties should have agreed on a boundary line between their adjoining lands. It is not necessary that there should have been a dispute;

---

59 *Jollymore v. Acker* (1915), 49 N.S.R. 148 (C.A.).
60 *Fraser v. Kirk* (1858), 3 N.S.R. 290 (C.A.).
61 *Dill v. Wilkins* (1853), 2 N.S.R. 113 (C.A.).
62 *Mooney v. McIntosh, supra*, note 57.
63 *Bea v. Robinson, supra*, note 47 at 18-19. But see *Woodberry v. Gates* (1846), 3 N.S.R. 255 at 257 (C.A.). Note that while one may discover a lost boundary through an analysis of deed descriptions or other evidence, one cannot *change* a boundary. This point was made in *Buchanan v. Saulnier* (1984) (N.S.S.C.).

it is not necessary that such boundary be marked by a fence, so long as it is clearly defined by blazing or spotting or by monuments or otherwise; it is not necessary that this conventional line should have been acquiesced in for any special period after the agreement. The essential matters are the making of the agreement and afterwards such an alteration of one party's position as would estop the other from disputing the conventional line. Thus if one erects a building, relying on the conventional line, the other party is estopped from denying it. The erection of a fence or any expenditure of money or labour might also be sufficient; certainly cultivating and cropping fields was judged so.[64]

§8.38 Evidence of agreement In order to establish a conventional line dividing two lots of land, it must be shown that there was an agreement, not necessarily in writing, between the respective owners of the lots to change the former dividing line which was afterwards adopted and lived up to by them.[65] Adjoining owners asked a surveyor to establish a line between their woodland properties; he did so in their presence; both signed an agreement and sketch in the surveyor's notebook; the signatures were witnessed by the surveyor and another person; when later the validity of the line was questioned it was held to be correct.[66] A boundary line in dispute was fixed or agreed upon by an exchange of deeds by the predecessors in title.[67] In another situation a conventional line had not been established because there was not the clear and cogent evidence required to show that there was ever any agreement between adjoining owners; the court quoted from the American *Corpus Juris Secundum*: "Inasmuch as the ownership of land is affected so far as the agreed line varies from the true line the proof of the agreed location should be clear."[68] It is exceedingly dangerous, when one of the parties is dead, to accept implicitly the testimony of the other as to a conventional line which supercedes the boundaries of documentary title deeds.[69] Uncontradicted evidence of the defendant is not sufficient in law to establish a conventional line.[70] While evidence of general reputation of boundaries is admissible

---

64 *MacMillan v. Campbell* (1951), 28 M.P.R. 112 (N.B.C.A.).
65 *Piers v. Whiting* (1923), 50 N.B.R. 363 (C.A.).
66 *Crossland v. Dorey, supra,* note 49.
67 *O'Neil v. Steele* (1979), 33 N.S.R. (2d) 514 (T.D.).
68 *Sullivan v. Lawlor* (1981), 45 N.S.R. (2d) 325, reversing (1980), 38 N.S.R. (2d) 332 (C.A.), from *Corpus Juris Secundum* (1916), Vol. 9 at 232. *Corpus Juris Secundum* is an encyclopaedic statement of the principles of American laws (St. Paul: West Publishing Co.).
69 *McGregor v. Webber* (1917), 51 N.S.R. 226 (C.A.).
70 *McDonald v. Mahoney* (1895), 31 N.S.R. 523 (C.A.).

to prove public rights, it is not admissible to prove private boundaries.[71]

§8.39 Establishing line If the respective owners of adjoining lands are in dispute as to the location of the boundary between them and they meet and agree upon a boundary line or have a boundary line located upon the ground and marked and both parties acquiesce in that agreement, they have by thus doing established a conventional line between their lands. The line so established becomes the actual and fixed boundary between their properties; no length of time is necessary after an agreement is reached. The erection of a fence on the agreed line is not necessary. Delay in objecting may and frequently does establish acquiescence. Such an agreement does not require a conveyance of any land from one party to the other. It is simply an agreement acknowledging the correct location of the boundaries and settling a dispute;[72] it does not involve a breach of the *Statute of Frauds.*[73]

§8.40 Agreement binds successors When owners of adjoining lands, fully cognizant of the dispute as to the location of the line dividing their properties, jointly agree upon a certain line as a division between them, jointly put up or continue a fence along such chosen line as the common boundary of their respective holdings and for years limit their respective occupation and cultivation of said properties by such fence, in the absence of fraud, each successor in title is bound by the line so agreed upon.[74] After the agreement upon a line, the fact that one of the parties thereafter objects to the line and takes steps to obtain another survey, even though such survey is correct and the conventional survey line incorrect would not interfere with the validity of the agreement once made.[75] The mutual agreement of the parties to abide by such a line constitutes a valuable consideration and specific performance could be ordered.[76] Once the agreement is made unconditionally, it is effective immediately and in the absence of fraud cannot be cancelled or repudiated at the will of one of the parties.[77] Previous owners had agreed on a line and had erected a

---

71 *Dunphy v. Phillips* (1929), 1 M.P.R. 227 (N.B.S.C.) which also pointed out that where the title deeds of parcels in dispute are put in evidence, conversations between previous owners about the boundaries are not admissible.
72 *Wilbur v. Tingley* (1949), 24 M.P.R. 175 (N.B.C.A.). The fact that agreement was reached because of threats of legal action did not invalidate the agreement.
73 *Davison v. Kinsman* (1853), 2 N.S.R. 1, 69 (C.A.); *Lawrence v. McDowall* (1838), 2 N.B.R. 442 (C.A.); *McLean v. Jacobs* (1834), 1 N.S.R. 9 (C.A.).
74 *Phillips v. Montgomery* (1915), 43 N.B.R. 229 (C.A.); *Kaneen v. Mellish* (1922), 70 D.L.R. 327 (P.E.I.C.A.); *Steeper v. Harding* (1884), 24 N.B.R. 143 (C.A.); *Lawrence v. McDowall, supra,* note 73; *Ross v. McKenzie* (1872), 9 N.S.R. 69 (S.C.).
75 *Inch v. Flewelling* (1890), 30 N.B.R. 19 (C.A.).
76 *Grasett v. Carter, supra,* note 51; *Joyce v. Smith* (1984), 66 N.S.R. (2d) 406 (T.D.).
77 *Supra,* note 70.

board fence along it; a subsequent purchaser was deemed to have known the fence was the boundary.[78]

§8.41 *Line run by surveyor* Several owners agreed upon a particular surveyor to determine the boundary between their land and lands adjoining; their agreement was held to have established the boundary.[79] Adjoining owners agreed upon a surveyor to run a line between their lands and abide by the results; they helped him in the survey; subsequent allegations of trespass resulted in confirmation that a conventional line had been established.[80] When an agreement is made and the line run by the surveyor, it must be adhered to even though the results are not those expected, unless there be evidence of mistake or deception.[81] A boundary line run between adjoining properties and continued by the surveyor, so as to indicate the line between other properties, was found to have been accepted as a conventional line.[82] The boundary line between the lands of a plaintiff and the lands of a defendant had for some time been unclear; subsequently the line was established by a surveyor hired by the plaintiff; the defendant knew of the survey and accepted the new line; subsequently he repudiated the acceptance of the new boundary and trespassed on lands in possession of the plaintiff; the boundary line was the line established by the surveyor.[83]

§8.42 *Survey line wrong* Plaintiff employed a surveyor to run the boundary lines of his land two years before the defendant went into possession of his land. The latter, for some years, both by words and acts, recognized the line between them, as claimed by the plaintiff, as being the true line. It appeared subsequently, however, that this line was not in fact the true line. It was held that since the defendant had acted under the misapprehension of the facts, not being acquainted at the time with the real boundary

---

78 *Mooney v. MacIntosh, supra,* note 57; *Jones v. Morgan* (1882), 22 N.B.R. 338 (C.A.).
79 *Merritt v. Toronto* (1913), 48 S.C.R. 1 affirming 27 O.L.R. 1, which affirmed 23 O.L.R. 365.
80 *Carrigan v. Lawrie* (1909), 7 E.L.R. 108 (N.S.). But taking part in a survey does not of itself imply agreement. See *Forrest v. Turnbull* (1909), 14 O.W.R. 478, affirmed 1 O.W.N. 150 (Div. Ct.).
81 *Steeper v. Harding, supra,* note 74. See also *Snowball v. Ritchie* (1888), 14 S.C.R. 741, reversing 26 N.B.R. 258. Surveyors chose a line run earlier by one of them instead of making the new survey by which the adjoining owners had agreed to be bound; the line agreed upon by the surveyors was not binding. In *Boyd v. Luscombe* (1986), 57 Nfld. & P.E.I.R. 242 (Dist. Ct.) at the conclusion of the presentation of evidence in the case, at the suggestion of the court and with the agreement of counsel, a third surveyor was engaged to review the work of two others.
82 *Crowley v. Feeney* (1932), 5 M.P.R. 248 (N.B.C.A.).
83 *Penney v. Gosse* (1974), 6 Nfld. & P.E.I.R. 344 (Nfld. S.C.). The court also granted an injunction restraining the defendant from further trespass and awarded plaintiff general damages for trespass.

of his lot, there was nothing in the acts or declarations so made which established a conventional line independently of right.[84] A surveyor ran a line which varied from the proper line; the surveyed line was consented to in ignorance of the fact that a mistake had been made. The assent given was not sufficient to establish a conventional line for there was no uncertainty as to where the proper boundary was, since it appeared on the face of the deed.[85] Owners of adjoining properties agreed in the erection of a fence between them to control cattle. There were no representations made and no dispute as to boundaries was intended to be settled by a conventional line. Neither party altered position by making improvements beyond the true line. The line so adopted is not binding if found to be wrongly run by an unskilful surveyor.[86]

§8.43 *Survey line correct or not* Where a division line is agreed to between the respective owners of adjoining property and it is understood that it is to govern whether correct or not, a conventional line is established which is binding on the parties, whether in fact correct or not.[87] However, an agreement may properly be conditional upon a line being correct.[88]

§8.44 *Lines at variance with descriptions* Where the owners of adjoining lands run a line between their properties, agree that it shall be the boundary line between them, erect a fence upon it, mark its course throughout its length and occupy their lands according to it for a number of years, the division line so agreed upon will be considered to be the true line even though it may not agree with the descriptions in the respective deeds.[89]

§8.45 *Lines never run* There are many instances where the dividing lines between lots have never been ascertained by the respective owners, owing to the land being in a state of wilderness and of low value; in some cases, the side lines were traced by the surveyors on the original surveys no farther than was sufficient to mark their respective courses. Public policy, as well

---

84 *McDonald v. McDonald* (1867), 7 N.S.R. 42 (C.A.).
85 *Roach v. Ware* (1886), 19 N.S.R. 330 (C.A.).
86 *Les Soeurs de Misericorde v. Tellier*, [1932] 2 W.W.R. 357 (Man. C.A.). Some jurisdictions provide by statute that so-called cattle fences have no effect on title. See, for example, *Fences and Impounding of Animals Act*, R.S.N.S. 1967, c. 104, s. 12.
87 *Martin v. Weld* (1860), 19 U.C.Q.B. 631 (C.A.); *Woodberry v. Gates, supra*, note 63.
88 *Reddy v. Strople* (1910), 44 N.S.R. 332, reversed on other grounds 44 S.C.R. 246.
89 *Kingston v. Highland* (1919), 47 N.B.R. 324 (K.B.). The decision rendered by Barry J. bears a noteworthy resemblance to Cooley's paper, "The Judicial Functions of Surveyors" c.1890, without making any reference to it. The full paper may be found as an appendix to *Theory and Practice of Surveying* by J.B. Johnson and L.S. Smith, 17th ed. (New York, 1914). Its broad principles are as relevant today as when the article was written.

306   Survey Law in Canada

as private convenience, require that every facility should be given to the settlement and adjustment of such boundaries.[90]

Estoppel

§8.46 *General* Closely associated with the doctrine of the conventional line is that of estoppel. This is the rule which "precludes a person from denying the truth of some statement formerly made by him, or the existence of facts which he has by words or conduct led others to believe in."[91] The rule has many applications within the common law; its connection with the conventional line arises because quite frequently one party makes improvements to land, based upon the words or actions of another. Often, rather lengthy periods of time are associated with conventional lines and long periods of time can be connected with estoppel. But an estoppel (or a conventional line) can be created in a very short time, indeed within moments. "As to deeds it has long sagely been said: 'Tell me what you have done under such a deed and I will tell you what that deed means'."[92]

§8.47 *Defence* Estoppel provides a shield, not a sword. It cannot provide the basis for an action; it is, rather, a defence against one. A plaintiff was estopped from denying the existence of a defendant's right of way by reason of the plaintiff's own deed.[93] An owner was estopped from denying to a bank that he held title to a property, having held out to the bank in mortgaging the property that he had title.[94]

§8.48 *Concept* When one person by his conduct leads another to believe that certain facts are true and these are acted upon, then in any subsequent proceedings this person cannot deny the truth of such facts.[95] If a man, either by words or by conduct has intimated that he consents to an act which has been done and that he will offer no opposition to it, although it could not have been lawfully done without his consent, and he thereby induces others to do that from which they otherwise might have abstained, he can not question the legality of the act he had so sanctioned, to the prejudice of those who have so given faith to his words or to the fair

---

90 *Lawrence v. McDowall, supra*, note 73.
91 J. Burke, *Osborn's Concise Law Dictionary*, 6th ed. (London: Sweet and Maxwell, 1976).
92 *Re Luness* (1919), 51 D.L.R. 114 at 131 (C.A.) which refers to *Watcham v. A.-G. of the East African Protectorate*, [1919] A.C. 533. See also *Marr v. Davidson* (1867), 26 U.C.Q.B. 641 (Ont. C.A.).
93 *Myers v. Brennan* (1973), 37 D.L.R. (3d) 79, affirmed 10 N.S.R. (2d) 391 (C.A.).
94 *Hanley and Fall v. Bank of N.S.* (1976), 13 Nfld. & P.E.I.R. 261 (P.E.I.S.C.). See also *Central Trust Co. v. Thistle* (1986), 58 Nfld. & P.E.I.R. 1 (Nfld. Dist. Ct.).
95 J.A. Yogis, *Canadian Law Dictionary* (Woodbury, N.Y.: Barron's, 1983) referring to *Capital Trust Corp. v. Gordon*, [1945] O.R. 277 (S.C.); *Joyce v. Smith, supra*, note 76.

inference to be drawn from his conduct.[96] If the owner of land gives permission for the doing of an act on his land and that act is completed, then, generally speaking, he will be too late to complain of it and the owner's proprietary right will to that extent be extinguished.[97]

§8.49 *Elements* Four essential elements must be present in order for the words or conduct of an owner to estop him in equity from complaining of the violation of his rights: (1) the owner must know of his legal right to the land; (2) the occupier must be unaware of the owner's legal right to the land; (3) the occupier must spend money or do some act to his prejudice in relation to the land; and (4) the owner must know of the occupier's mistaken belief as to his legal right to the land.[98]

§8.50 *Improvements* There being no indication of the line between two lots owned respectively by the parties, the plaintiff's surveyor asked the defendant where the boundary line lay. The defendant pointed out the remains of a fence and the surveyor marked it. A builder employed by the plaintiff marked it a few days later; on neither occasion did the defendant object. When the plaintiff completed the walls of a house, the defendant interfered with them, claiming they encroached four inches on his land. The defendant was estopped from disputing the plaintiff's line.[99] Whatever may be the effect of a recognition by the owners of an adjoining lot, of a boundary line, by maintaining a fence thereon, it must, like any other admission, be perfectly free from doubt and ambiguity and must also be made with a full knowledge of the facts. If made under error and mistake, the party making the admission is not bound thereby, but may show that it was so made, unless, indeed, the admission has induced another to alter his position and has thus affected his rights in which latter case it will have the effect of an estoppel.[100]

§8.51 *No improvements* In a boundary dispute the parties claiming according to the original grants are entitled to succeed unless they have lost their rights by reason of a conventional line having been established. Where a line between two adjacent lots has been set out in a wrong place

---

96 *McGugan v. Turner*, [1948] O.R. 216 (H.C.) citing *Cairncross v. Lorimer* (1860), 3 Macq. 827 (H.L.); *Brown v. Norbury*, [1931] 2 W.W.R. 863 (Alta. C.A.).
97 *Quartermain v. Stevens* (1972), 4 N.B.R. (2d) 266 (S.C.) citing 38 Halsbury 3d ed. at 751.
98 *Rollings v. Smith* (1977), 15 Nfld. & P.E.I.R.128 (P.E.I.S.C.). See also A.C. McEwen, "The Effect of Estoppel Upon Title to Land", *The Canadian Surveyor*, Vol. 34, No. 3, September, 1980 which is based upon this case.
99 *Grasett v. Carter* (1884), 10 S.C.R. 105. This is a classic case. An especially important point is that of elapsed time; an estoppel was created over a period of a few days.
100 *Dill v. Wilkins* (1853), 2 N.S.R. 113 (C.A.); *Benoit v. Benoit* (1976), 15 N.B.R. (2d) 233 (S.C.).

and there is no dispute, the mere acquiescence in its location and the occasional cutting of trees up to such line, does not furnish evidence of estoppel and either owner may assert his right to have the line correctly run.[101] A line was blazed by one of two adjoining owners. The other party did nothing beyond refraining, when cutting trees, from taking any beyond that line. There was nothing in what happened to estop him from exercising his rights up to his true boundary when he saw fit to do so. When it is asserted that a line between the lands of two persons has become a conventional line superceding the true line, some situation making it inequitable and improper that the true line be the measure of the right of the so-called trespasser must be shown. This may be an agreement for consideration or a standing-by while the other party changes his position.[102]

§8.52 *Occupation only* Where an owner of land without any knowledge of the true boundary line between his and his neighbour's land occupies up to a line which he believes to be the correct one, his occupation will not conclusively bind him unless he has led his neighbour to do something to his detriment on the strength of the former's occupation.[103] Where two persons take possession of adjoining properties under a common error as to the true line of division so that one of them is in fact in possession of a portion belonging to the other, the doctrine of estoppel may be raised against the latter.[104]

§8.53 *Admission* A survey was made and the surveyor's plan was shown and explained to the defendant. He thereupon signed a letter, addressed to the plaintiff, admitting that the line established by the surveyor was the proper one and agreeing to accept it as the line between the properties. While the letter created no estoppel, nothing having been done or omitted to be done by the plaintiff in consequence of it, yet it might be treated by the court as an admission by the defendant that the line as laid down

---

101 *Murray v. McNairn* (1952), 30 M.P.R. 200 (N.B.C.A.); *Re Hunter* (1978), 23 N.B.R. (2d) 130, reversing 19 N.B.R. (2d) 710 (C.A.).
102 *Sutherland v. Campbell* (1923), 25 O.W.N. 409 (C.A.).
103 *Byram v. Violette* (1893), 32 N.B.R. 68 (C.A.). There was no proper evidence presented as to the bounds of the plaintiff's grant; "it [was] exceedingly difficult, perhaps impossible, to follow [the surveyor's] evidence . . . a good deal of time and labour might have been saved . . . if the plaintiff's lot . . . had been properly run out." Some discussion occurred about the surveyor having been asked the very question the jury was to decide. Sir John Allen C.J. commented:
> The position of the corner of lot No. 405 is not a questiuon of science, but a fact dependent upon the construction of the grants under which the parties respectively claimed title, and the accuracy of the survey of the plaintiff's grant, and ascertaining where its rear or eastern line was.
104 *McGugan v. Turner*, [1948] 2 D.L.R. 338 (Ont. H.C.).

by the surveyor was the correct one.[105] During the course of a retracement, a surveyor showed his plan to the owner of adjoining lands and discussed his placement of the common boundaries with him; the owner subsequently engaged another surveyor to determine the boundaries of his lot; the second surveyor referred to the first surveyor's plan, discussed the placement of the common boundaries with his client and then showed the same boundaries on his plan as were on the plan made by the first surveyor; the owner subsequently testified that he had not agreed to the location of the boundary lines and stopped complaining only because he could not afford legal action; the owner was bound by the location of the boundaries shown on the plans.[106]

Apportionment of Accretion

§8.54 *General* Accretion refers to a gradual addition of sediment to the shore by the action of water.[107]

> The term "accretion" denotes the increase which land bordering on a river or on the sea undergoes through the silting up of soil, sand or other substance, or the permanent retiral of the waters. This increase must be formed by a process so slow and gradual as to be, in a practical sense, imperceptible, by which is meant that the addition cannot be observed in its actual progress from moment to moment or from hour to hour, although, after a certain period, it can be observed that there has been a fresh addition to the shoreline. The increase must also result from the action of the water in the ordinary course of the operations of nature and not from some unusual or unnatural action by which a considerable quantity of soil is suddenly swept from the land of one man and deposited on, or annexed to, the land of another.
>
> The fact that the increase is brought about in whole or in part by the water, as the result of the employment of artificial means, does not prevent it from being a true accretion, provided the artificial means are employed lawfully and not with the intention of producing an accretion, for the doctrine of accretion applies to the result and not to the manner of its production.[108]

Accretion is gradual and imperceptible as opposed to avulsion which is the sudden change in the course of a river or stream.[109] Alluvion is the

---

105 *McIntyre v. White* (1911), 10 E.L.R. 248, affirming 40 N.B.R. 591 (C.A.).
106 *Spearwater v. Seaboyer* (1984), 65 N.S.R. (2d) 280 (T.D.).
107 *Canadian Law Dictionary, supra* note 95. See also A.H. Oosterhoff and W.B. Rayner, *Anger and Honsberger Law of Real Property*, 2d ed., vol. 2, (Aurora: Canada Law Book Inc., 1985) at s. 3004.
108 *Clarke v. Edmonton (City of)*, [1930] S.C.R. 137.
109 *Shey v. McHeffey* (1868), 7 N.S.R. 350 (S.C.). In 1837 the Avon River at Windsor breached an oxbow (avulsion) and a new marsh subsequently formed (accretion). All proprietors except plaintiff and defendant appeared "content to hold according to lines and boundaries made and agreed upon between themselves."

material deposited through accretion. There is ample precedent in Canadian law, and in the common law generally, to show that any land formed by accretion belongs to the owner of the lands adjoining the accretion.[110]

§8.55 Methods of Division Broadly speaking there are four general approaches to the division between affected landowners of land gained through accretion: the prolongation of sidelines; the determination of proportional areas; the representational baseline; and the determination of proportional shorelines. Neither the prolongation of sidelines nor proportional areas have been sanctioned by Canadian courts. Since prolonging sidelines is the most obvious, it may appear the simplest solution. It is equitable, though, only when sidelines are generally perpendicular to the shore or when their prolongations meet in a point. Proportional areas require that adjoining owners receive accreted areas proportional to the length of shoreline originally held; this has an obvious attraction where the value of the new land is higher than the value of access to the water which it borders.[111] Division by proportional shorelines is a more general method of division than that of the representative baseline.

§8.56 Representative Baseline This solution was applied on the first occasion when the division of accretion was decided by a Canadian court.[112] The precedent case had arisen in Wales.

> Take a line representing the line of the shore drawn at such distance seawards as to clear the sinuosities of the coast, and let fall a perpendicular from the end of the land boundary. Of course a line representing the whole coast of the bay [is not meant] but a line fairly representing the average line of the shore, extending on either side of the land boundary.[113]

§8.57 Proportional Shorelines Here, adjoining owners receive new shorelines proportional in length to their former shorelines. No regard is paid to the direction of the landward sidelines of the several parcels. This approach recognizes the generally accepted principle that water frontage is rather more important than the quantity of land immediately behind that frontage. The right to be secured and protected is not so much the right to the accreted land itself, but rather the right to retain access to

---

110 *A.G. of B.C. v. Neilson*, [1956] S.C.R. 819, wherein the law of accretion is discussed and the situation before the court was regarded as a widespread emergence of land owned by the Crown, not accretion; *Chuckry v. R.*, [1972] 3 W.W.R. 561 (Man. C.A.).
111 C.M. Brown, W.G. Robillard and D.L. Wilson, *Evidence and Procedures for Boundary Location*, 2d ed. (New York: John Wiley & Sons, 1981) at 252-255 refers to a number of American situations and their solutions.
112 *Paul v. Bates* (1934), 48 B.C.R. 473 (Co. Ct.).
113 *M'Taggert v. M'Douall* (1867), 5 M. 534.

the river or sea since such is regarded as more valuable than the mere land. Thus the division of land accreted to contiguous properties will generally not have reference to the land mass attaching to a particular parcel but rather to the river or sea frontage lost or gained.[114] The British Columbia Supreme Court quoted from two American cases:

> Each proprietor will take a larger or smaller proportion of the alluvial formation and of the newly formed river or shore line, according to the extent of his original line on the shore of the river.[115]

> When general course of shore approximates a straight line, division of accretion is made among adjoining riparian owners by lines perpendicular to general course of original bank, or original mark of shore; but when it curves or bends, general rule is to divide newly formed shore line by giving to each owner a share of new shore line in proportion to what he held in old shore line, completing division by running a line from boundary between parties' lands on old river bank to point thus determined on newly formed shore.[116]

The Canadian court then said that "the rules applicable in the United States, founded as they are on the same principles as the rules in English law . . . must be held to be properly applicable here."[117]

Improvements or Encroachments

§8.58 Improvements to adjoining properties and encroachments thereon can give rise to boundary uncertainties. Statutory provisions address these situations in some jurisdictions. As a class, the statutes relating to improvements follow the Saskatchewan model: where a person has made lasting improvements on land under the belief that the land is his own, he will be entitled to a lien on the land to the extent that its value has been enhanced by the improvements; or he may even be entitled to retain the land on making compensation as directed by a court.[118] The working

---

114 *Re Brew Island*, [1977] 3 W.W.R. 80 (B.C.S.C.).
115 *Deerfield v. Arms* (1835), 34 Mass. (17 Pick.) 41, 28 Am. Dec. 276 (Supreme Court of Massachusetts).
116 *New Hampshire v. 6.0 Acres of Land* (1958), 139 A. (2d) 75 (Supreme Court of New Hampshire).
117 See also J.A. MacRae, *The Surveyor and The Law*, vol. 2 (Auckland: N.Z. Institute of Surveyors, 1984), at s. 5.11.6. This general rule in New Zealand is based upon *Riddiford v. Feist* (1902), 22 N.Z.L.R. 105, 5 G.L.R. 43.
118 *Improvements Under Mistake of Title Act*, R.S.S. 1978, c. I-1, s. 2. See also: *Law of Property Act*, R.S.M. 1970, c. L 90, ss. 27, 28; *Land Titles Act*, R.S.A. 1980, c. L-5, s. 183; *Conveyancing and Law of Property Act*, R.S.O. 1980, c. 90, s. 37(1). *Re Robertson and Saunders* (1977), 75 D.L.R. (3d) 507 (Man. Q.B.) dealt with the Manitoba Act in some detail and made the point that possessory right was still possible even after land had been brought under the *Real Property Act*, R.S.M. 1970, c. R-30. See also *Lutz v. Kawa* (1980), 15 R.P.R. 40 (Alta. C.A.).

of land and the erection of a fence were found not to be lasting improvements.[119]

§8.59 A few years ago in Nova Scotia a house was built on the wrong parcel of land and was lived in for almost a year before the mistake was discovered. A realtor had pointed out and sold the lot adjacent to the one actually intended. A surveyor had subsequently certified as correct the building foundation which had been placed upon the wrong lot. The later civil action[120] resulted in a court order to remove the offending structure. It seems reasonable to suggest that had provincial legislation been in effect about improvements under mistake of title, the outcome would have been different.

§8.60 As well as improvements, Manitoba treats encroachments through statutes. Where it is found that a building encroaches upon adjoining land, the court may declare that the owner of the building has an easement during the life of the building upon making compensation or the court may vest title to the land so encroached in the owner upon payment of compensation, or it may order the owner to remove the encroachment.[121] *Re Robertson and Saunders*[122] clearly distinguished between an encroachment resulting from a building which was partly on one parcel and partly on another and an improvement which was entirely within a lot of land. Comparable legislation in British Columbia addresses improperly located fences as well as buildings.[123]

Fences

§8.61 *General* The concern here is for fences as they relate to the solution of boundary uncertainies. This is not an exposition on their construction, maintenance or the adjudication of those disputes about them which are the province of fence viewers. Fences, of themselves, are not conclusive

---

119 *Tuckwell v. Guay* (1917), 34 D.L.R. 106, affirmed 37 D.L.R. 805 (C.A.). W.H. Hurlburt, "Improvements Under Mistake of Ownership: Section 183 of The Land Title Act", Alberta Law Review, Vol. XVI, No. 1, 1978 briefly reviews a number of cases; in one situation, an owner in 1976 relied, to his detriment, on a survey certificate prepared in 1959. See *Anger and Honsberger Law of Real Property, supra,* note 107 at 1125-1126 regarding up-to-date surveys; see also N.L. Petzold, "Re-use of Old Survey Documents", *Terravue,* Summer, 1986.
120 *Brean v. Thorne* (1982), 52 N.S.R. (2d) 241 (T.D.). See J.F. Doig, "The Ultimate Trespass", *The Canadian Surveyor,* Vol. 37, No. 1, 1983 for a summary and comment.
121 R.S.M. 1970, L 90, s. 28.
122 *Supra,* note 118.
123 *Property Law Act,* R.S.B.C. 1979, c. 340.

as to boundary location. Nor do fence viewers have power to determine boundaries.[124]

§8.62 *Purpose* If at times fences seem to have caused more problems than they have solved, it is not the fault of the structures themselves. Rather, the villain of the piece is the imperfect understanding of their origin and purpose on the part of those who would use them as evidence or ignore them altogether. Robert Frost's observation that "[g]ood fences make good neighbours" had its roots in far more than the fence's physical condition or its continued maintenance. His remark presupposed a full understanding on the part of the adjoiners as to the fence's origin and purpose. Thus a fence's evidentiary value in the resolution of boundary uncertainties depends upon the extent to which its origin and purpose can be discovered.

§8.63 Nevertheless, "as between old boundary fences and any survey made after the monuments have disappeared, the fences are by far the best evidence of what the lines of a lot actually are."[125] The reasons why adjoining owners of rural lands would normally place a fence or stone wall along their common boundary rather than elsewhere are plain enough. As a general rule old fences and stone walls should be regarded as valuable evidence of boundary location, only to be set aside when conclusively proved to be unreliable or irrelevant for such purposes. Of course not all old fences are line fences. Often the irregularity of a cattle fence, or its direction not matching the general run of the lines in the district, combined with the topography at the site, will attest to its purpose. But other fences may be found which are quite regular with an orientation matching that to be expected within the locality, without being boundary fences. Ideally, the evidentiary value of a fence will rest upon agreement with evidence from other sources, be it conformity with deed calls or the testimony of older residents. There may be a temptation to abandon an old fence or stone wall when it fails to agree with other evidence found on the ground. The case for one defendant failed because of this; his surveyor "was inclined, throughout his work to accept survey markers or pins placed by

---

124 *Delamatter v. Brown* (1908), 13 O.W.R. 58 varied on other grounds (1909), 13 O.W.R. 862 (Div. Ct.); *Griffin v. Catfish Creek Conservation Authority*, [1968] 1 O.R. 574 (Co. Ct.). See also, however, *Doherty v. McDevitt* (1892), 31 N.B.R. 526 at 530-531 (C.A.):

> The duty of a fence viewer . . . is to decide which part and how much of the fence between the two occupiers is to be made by each. He cannot decide, in case of dispute, where the line is, but it is the duty of each occupier, after such decision, to make his part of the fence somewhere, so that there may be a fence between the two; and he must at his peril put it on the true line.

125 *Naugle v. Naugle* (1970), 2 N.S.R. (2d) 309 (C.A.), quoting *McIsaac v. McKay* (1916), 49 N.S.R. 476 (C.A.) which in turn referred to *Diehl v. Zanger*, 39 Mich. 60.

another surveyor, at an earlier time, in preference to stone walls and fences marking the lines."[126]

§8.64 *Limits* An owner of land may build and maintain a fence. The common law rule is that no man making a ditch can cut into his neighbour's soil, but he usually cuts it to the very extremity of his own land. He is of course bound to throw the soil which he digs out upon his own land and, often if he likes it, plants a hedge on the top of it; therefore if he afterwards cuts beyond the edge of the ditch which is the extremity of his land he cuts into his neighbour's land and is a trespasser. The same principle must of course apply to the making of a partition fence, which a party is not at liberty to make beyond the extremity or boundary of his own land.[127]

§8.65 *Convenience* Owners may arrange fences solely for the control of cattle or other livestock, making no representations as to boundaries; such fences have no relevance to boundary determination or to title.[128] The mere enclosure of another's land by the adjoining proprietor, with a fence put up with the consent of and by arrangement with the owner, for the purpose of protecting the lands of both, does not dispossess the owner. Nor does it prevent him maintaining an action against anyone using his land for purposes other than that for which it was enclosed.[129]

§8.66 *Long acceptance* The question at issue in one case was whether a fence or the middle line of a stream was the legal boundary of the plaintiff's land. It was found for all practical purposes that the fence had been treated as the boundary line, but the court said this fact appeared of no great importance because the plaintiff's predecessors hardly ever wished to use the stream. The question was concluded by the specific admissions of one of the plaintiff's predecessors in title that his line went to the fence. Therefore, whatever was the original purpose in erecting it, the fence had become the true boundary.[130] In another instance, the plaintiffs could not contend that the fence line, however dilapidated or in disrepair or even crooked, was not now the boundary line when it was recognized as such since the plaintiffs came to reside upon the property shortly after 1945 and were there until a survey was made in 1972.[131]

§8.67 *Conventional line* Adjacent owners putting up a fence, and abiding

---

126 *Naugle v. Naugle* (1969), 1 N.S.R. (2d) 554 at 560 (S.C.).
127 *Hunter v. Ronne* (1870), 8 N.S.R. 113 (C.A.).
128 *Les Soeurs de Misericorde v. Tellier*, [1932] 2 W.W.R. 357 (Man. C.A.).
129 *Brookman v. Conway* (1903), 35 N.S.R. 462, affirmed 35 S.C.R. 185; *Carson v. Musialo*, [1940] O.W.N. 398 (C.A.).
130 *Nourse v. Clark*, [1936] O.W.N. 563 (C.A.).
131 *Nelson v. Varner* (1977), 20 N.S.R. (2d) 181 (S.C.).

by it, may create a conventional line,[132] though a conventional line may be established without the erection of a fence being necessary.[133] A fence erected by one party along a boundary recently surveyed, and meeting no objections from the adjoiner, was sufficient to confirm the line as a conventional one.[134]

§8.68 *Protraction* A fence had been erected between two properties on a line agreed upon between the owners for a part of the length of the respective lots, but there was no implied agreement resulting therefrom that the direction of the fence should form the line throughout and there was accordingly no right to have the line of the fence protracted to form the boundary between the lots. Where the line formed by the protraction of a fence has been held to constitute the boundary line there was always a line run, put, or marked out, and observed and acted upon by each party exercising acts of ownership according to it; and it was this observing of the line that was decisive and not the line claimed being the continuation of the fence.[135] However, in another instance, plaintiff and defendant, who were adjoining owners, had, over a period of 20 years, occupied their lands in the front part of their properties according to a certain fence that was never extended to the rear. The front part was cleared land while the rear was uncleared. In the absence of actual possession up to any line in the uncleared portion it might be considered that the parties intended to hold according to the projection of the line so fenced.[136]

§8.69 *Parallelism* In an action for trespass the properties owned by the parties adjoined each other. Two surveys, made on the same basis in 1902 and 1910 respectively to replace lost marks, conflicted with each other; it was held that the earlier survey should be peferred as it was in accord with the old division fence and was parallel to all the adjacent lines in the vicinity.[137] By law, regard must be had for fences and as nearly as possible parallel lines must be established.[138] The use of other fences in the subdivision on other lots to determine boundaries in question, where those fences do not conform to the location of the sidelines as determined

---

132 *Perry v. Patterson* (1874), 15 N.B.R. 367 (C.A.); *Stevens v. Stevens*, [1930] 3 D.L.R. 762 (N.S.C.A.).
133 *Wilbur v. Tingley*, [1949] 4 D.L.R. 113 (N.B.C.A.).
134 *Steeper v. Harding* (1884), 24 N.B.R. 143 (C.A.).
135 *Charbonneau v. McCusker* (1910), 22 O.L.R. 46 (C.A.).
136 *Belyea v. Belyea* (1857), 8 N.B.R. 588 (C.A.).
137 *McIsaac v. MacKay, supra*, note 125; *Richards v. Gaklis* (1984), 63 N.S.R. (2d) 230, (T.D.) contains a lengthy assessment of the location of an old stone fence. Choosing this line lent sense to the language and interestingly enough the rest of the description fell into place.
138 *K. & W. Enterprises Ltd. v. Smith* (1974), 7 N.S.R. (2d) 411 at 423 (S.C.).

by survey based on the description in the deeds, was an improper approach to establishing a disputed boundary line. It was the fences on the boundary lines in dispute that were relevant. Acts of possession over many years on lots down the street were not to be accorded any great weight.[139]

§8.70 *Conflict with monuments* The original survey of an Indian reserve took place before the surveys preliminary to settlement were made. Owing to the proximity of the reserve, the sections in the area, when it was finally surveyed, were smaller than the usual size. The defendant took title to the fractional quarter section. He was shown by a government surveyor a plan giving his acreage, but did not check the monuments on the ground. He erected a fence which included 40 acres of the reserve. In an action to recover the 40 acres, it was held the Crown should recover. By section 62 of the *Dominion Lands Surveys Act* the monuments governed.[140]

§8.71 *Confirmation by plan* A plaintiff claimed title to a strip of land which a defendant had enclosed by a fence. A surveyor was unable to find any original stakes or monuments at any point, but it was proved that, many years before, fences were built dividing the lots in question and the lots in the rear of them and that the owners of these recognized them as being and treated them as marking the boundary line between the lots; and there was evidence that the fence ran between the lots in a straight line; and it was also proved that the line between the lots of plaintiff and defendant was a straight line, according to the registry office plan. The line claimed by the defendant departed from the straight line to the extent of several feet. It was held, upon the findings of fact as to the old fence, that a proper inference might be drawn that the old fence was built when the original monuments were in existence and on the true boundary line. The line contended for by the plaintiff, being in accord with these findings, was therefore declared to be the true line.[141]

§8.72 *Onus of proof* A plaintiff sued in an action of trespass alleging that the defendant, the owner of adjoining lands, had reconstructed an old fence which was on the plaintiff's property. The plaintiff himself had erected the fence about 25 years previously on the line pointed out to him by his predecessor in title and never at any time had he repudiated the reconstructed portion of the fence or suggested that it was not the true

---

139 *O'Melia v. Himmelman* (1985), 69 N.S.R. (2d) 271 (T.D.). The boundary between two house lots was in dispute; the area was very small. Two surveyors were engaged; both testified and both were cross-examined extensively. The court agreed with O'Melia's surveyor; $1.00 damages were ordered for trespass and Himmelman was to remove the fence. See also *Weston v. Blackman* (1917), 12 O.W.N. 96 (C.A.).

140 *R. v. Weremy*, [1943] Ex. C.R. 44 (Ex. Ct.).

141 *Weston v. Blackman, supra,* note 139.

dividing line of the lots. From the evidence of surveyors and the judge's own inspection of the ground, it appeared that there were substantial facts indicating that the old fence line was the true line. The plaintiff had failed to discharge the onus which lay on him of showing that he was entitled to the land in dispute.[142]

Statutes

§8.73 Throughout Canada there is a good deal of legislation directed toward the resolution of boundary uncertainties. For the most part, the remedies established by this legislation seem to fall into one of three categories: special surveys, quieting of titles or resurveys. It is, however, one thing generally to classify statutory provisions and to direct attention to them. It is another to get an appreciation of how they are employed and to make an assessment of their usefulness. All that can be done here is to give a very *general* idea of the broad concepts through particular examples or illustrations.

§8.74 *Special surveys* Other than in the Maritime Provinces, jurisdictions with land titles systems provide for special surveys in one way or another.[143] Special surveys are established either by a particular Act of the legislature or by the appropriate provisions being incorporated within some other Act.[144] Special surveys appear to be intended to address boundary uncertainties within a community, or within a portion of it, rather than to resolve isolated problems which arise between adjoining property owners. Costs, which can be considerable, are sometimes (and perhaps

---

142 *Fitzpatrick v. McSorley* (1920), 48 N.B.R. 162 (C.A.); *Lake Erie Excursion Co. v. Bertie* (1912), 4 O.W.N. 111, affirming 3 O.W.N. 1191 (C.A.); *Bragg v. Rogers* (1875), 25 U.C.C.P. 156 (C.A.); *Palmer v. Thornbeck* (1877), 27 U.C.C.P. 291 (Ont. C.A.).
143 Land titles legislation exists in each of the Maritime Provinces. Nova Scotia and Prince Edward Island adopted it in 1904 and in 1971 respectively. Nova Scotia's enactment (*Land Titles Act*, S.N.S. 1903-4, c. 47) was modelled on the English legislation of 1875. It was proclaimed in three counties: Annapolis, Colchester and Halifax. At last count one parcel in Annapolis County was in the system. The Prince Edward Island legislation (*Land Titles Act*, R.S.P.E.I. 1974, c. L-6), which bears a resemblance to the Acts of the western provinces, has never been proclaimed. The New Brunswick statute of 1914 (under which no properties were ever registered as far as anyone has been able to discover) has been replaced by the *Land Titles Act*, S.N.B. 1981, c. L-1.1. Albert County was designated a district under this Act on 9 July 1984.
144 *Land Title Act*, R.S.B.C. 1979, c. 219, ss. 322-365; *Surveys Act*, R.S.A. 1980, c. S-29, ss. 22-25, 62-68; *Land Surveys Act*, R.S.S. 1978, c. L-4, ss. 60-94; *Special Surveys Act*, R.S.M. 1970, c. S. 190; *Canada Lands Surveys Act*, R.S.C. 1985, c. L-6 ss. 48-60 (for lands in the Northwest Territories or in the Yukon Territory). The *Boundaries Act*, R.S.O. 1980, c. 47 applies to lands in the land titles system, to lands in the registry system, and to Crown lands.

quite often) absorbed by the governing authority which has authorized the survey. But costs may be assessed to the municipal units involved or they may be shared among the owners of the lands involved. The *Land Titles Clarification Act*[145] of Nova Scotia falls in this category of legislation; the Act addresses boundary and title problems within particular communities.

§8.75 The legislation supporting special surveys of lands situated in the Northwest Territories or in the Yukon Territory is a good representation of legislation of this kind: special surveys of territorial lands may be made for the correction of errors or supposed errors in existing survey plans; for the subdivision of lands not previously subdivided or for the showing of divisions of land not previously or correctly shown on an existing plan of subdivision; for fixing the location or width of roads or highways; for establishing any boundary lines, the positions of which, due to incorrect placing or loss or obliteration of monuments defining the same on the ground, have become doubtful or difficult of being ascertained and for any other purpose deemed necessary by the Minister administering the territorial lands or by the respective Commissioners for the respective lands.[146] The Surveyor General has the management of special surveys; no person other than a Canada Lands Surveyor can make them.[147] On completion of a survey the Surveyor General directs the plotting of the plan, signs it and forwards the plan and such supporting documents as he thinks necessary to the person ordering the survey, who appoints a Hearing Officer to enquire into and report upon any complaints that may be made against it.[148] The Hearing Officer holds hearings in, or as near as practicable to, the locality in which the survey has been made after public notice has been given of the scope and purposes of the survey and the lands affected thereby.[149] Where no complaints are received, the Surveyor General confirms the plan; where written complaints are received, the Hearing Officer hears them, receiving any evidence he thinks proper, calling any witnesses and exercising any of the powers of a commissioner under the *Inquiries Act*.[150] On receiving the Hearing Officer's report, the person ordering the survey decides whether the plan should be approved or amended as a result of the complaints; a notice of his decision and

---

145 R.S.N.S. 1967, c. 162.
146 *Canada Lands Surveys Act supra*, note 144 at s. 49. Special surveys do not apply to Canada Lands which are within the provinces; they can apply, however, to private lands within the Northwest Territories and the Yukon.
147 *Ibid.*, s. 50(2), (3).
148 *Ibid.*, s. 51, 52(1).
149 *Ibid.*, s. 52(2), (3).
150 *Ibid.*, ss. 53, 54; R.S.C. 1985, c. I-11.

the right of appeal to the Supreme Court in the Northwest Territories or in the Yukon Territory is sent to persons whose complaints have been heard and to persons whose interests are affected by his decision.[151] If there are no appeals or if appeals are subsequently withdrawn, the Surveyor General confirms or amends the plan as directed by the person ordering the survey.[152] When an appeal is taken to the Supreme Court, the court hearing it may confirm or amend the decision of the person ordering the survey and the Surveyor General accordingly confirms or amends the plan.[153] The plan, as confirmed or amended by the appropriate authority, is filed in the appropriate land title office. It is then substituted for all, or corresponding portions of all, former surveys or plans of the lands thereby affected.[154]

§8.76 A glance at special surveys within the Province of Manitoba will provide some idea of how they are arranged in that jurisdiction and under what circumstances. Special surveys are normally initiated by the Land Titles Office, Winnipeg though they may be undertaken by the Surveys and Mapping Branch of the Department of Natural Resources, if problems have arisen with respect to Crown lands. Municipalities occasionally request special surveys where problems are discovered, usually related to the location of boundaries for municipal works. A private land owner may be able to arrange a special survey by bringing a boundary problem to a municipal council or to a Crown agency where the council or agency has *other* sufficient reason to pursue one. The criteria used to initiate special surveys relate to ambiguous descriptions, overlaps or boundaries that do not coincide. There may be areas in which monumentation has deteriorated over the years to the extent that land surveyors are having great difficulty in carrying out surveys for clients. In such instances, the special survey will re-establish block and street limits, but not the individual lot boundaries. All special surveys are processed and examined through the office of the Examiner of Surveys, Winnipeg. About 100 special surveys were made in the province in the seven years between 1979 and 1986.[155]

§8.77 *Quieting Titles* Every jurisdiction which operates a registry system provides the owner of land, embraced by the system, a means of obtaining a certificate of title.[156] Though this legislation is primarily intended to

---

151 *Ibid.*, s. 55(1), (2), (3).
152 *Ibid.*, s. 56, 57.
153 *Ibid.*, s. 58.
154 *Ibid.*, s. 59.
155 Correspondence from L.E. Boutilier M.L.S., A.C. Roberts M.L.S., and E.C. Tacium O.L.S., M.L.S., S.L.S., March, 1986.
156 *Land Title Inquiry Act*, R.S.B.C. 1979, c. 220; *Certification of Titles Act*, R.S.O. 1980, c. 61. See Reiter, at note 22, *supra*, at 463-466:

320   Survey Law in Canada

establish good title and not to resolve boundary uncertainties, it can be used to achieve the latter end; an action brought to quiet title will also settle boundary questions. A surveyor's plan of the property is required, except in New Brunswick, to support the action. The confirmation of title and the concomitant approval of the plan, as presented or amended, fixes the boundaries. While uncertainty in the matter of boundary location does not of itself imply doubtful title, sometimes doubts arise on both counts when the validity and extent of land ownership are investigated closely.

§8.78 The Newfoundland statute[157] is representative of those in other jurisdictions: Any person claiming ownership of land may have title judicially investigated; the Attorney General may apply for the investigation of the title of the Crown to any land; an application is made to the Supreme Court or a judge thereof; the application must be supported by the title deeds and evidence of title in the possession of the applicant, certified copies of all registered instruments affecting land, an abstract of title, a concise statement of facts which do not appear in produced documents, proof of any fact required to be proved, an affidavit by the person whose title is to be investigated, a land surveyor's diagram of lands claimed showing metes and bounds thereof; the affidavit must state whether anyone is in possession of the land and under what claim; the judge may receive and act upon any evidence that is received by the Supreme Court on a question of title whether the same is or is not receivable or sufficient in point of strict law if the same satisfies the judge of the truth of the fact

---

    It is a kind of bastard land titles system grafted onto the registry system ... [S]ince the procedures to obtain a certificate are virtually the same as the procedures on an application for the first registration under the *Land Titles Act* ... the choice of the land titles application will almost always be preferred;

    Quieting of Titles Act, R.S.N.B. 1973, c. Q-4; *Quieting of Titles Act*, R.S.N. 1970, c. 324; *Quieting Titles Act*, R.S.N.S., c. 259; *Investigation of Titles Act*, R.S.P.E.I. 1974, c. I-7. See *Legge v. Scott Paper Co.* (1970), 3 N.S.R. (2d) 206 (T.D.) which dealt with a claim for a certificate of title: the certificate went to the defendant.

157 *Quieting of Titles Act*, ibid. *Re Lundrigans Ltd. and Prosper* (1982), 38 Nfld. & P.E.I.R. 10 (Nfld. Dist. Ct.) considered the application of a legal owner, and possessor of good documentary title, for a certificate of title. An occupier had laid a possessory claim to a portion of the lands. The claim failed because the occupation had not been open and notorious: a cabin had been used for more than 20 years but was not readily visible from the air or from the ground, nor was it shown that others were aware of its existence. Surveyors discovered the cabin while working on a proposed real estate development. The certificate was granted. But see *Afton Band of Indians v. A.G. of Nova Scotia* (1978), 29 N.S.R. (2d) 226 (T.D.). The band was not a body corporate and hence not entitled to bring an action; the individuals who made up the band could not gain possessory title as a band.

intended to be established thereby;[158] it is not necessary to produce or account for the originals of any registered documents; if the judge is not satisfied with any initial evidence he must give reasonable opportunity for further evidence to be produced; before a certificate is granted public notice of the application is given; any person having an adverse claim may file a statement of claim; in case of a contest the judge may either decide the question of title on the evidence before him or he may refer the same to the Supreme Court or direct another mode of investigation; every claim is subject to exception by reasons of reservations in the original grant, charges, title or lien by possession, lease or agreement, or any public highway; a certificate of title shall be conclusive and the title thereon shall be absolute and indefeasible; the judge's decision may be appealed to the Supreme Court.[159]

§8.79 Among the statutes of Nova Scotia, in addition to the regular provisions for the quieting of title, is a piece of legislation of a kind not found elsewhere: the *Land Titles Clarification Act*. Within the province, title is doubtful throughout a number of small communities which grew from squatter settlements many years ago.[160] For the most part these have been out-of-the-way hamlets to which industry has never been attracted and whose residents have travelled to work elsewhere. Legislation was enacted in 1964 to provide the means of acquiring good title to lands within these communities without judicial investigation. Where the residents of an area of a municipality are in necessitous circumstances as a result of lack of property development in the area and where there appears to be confusion as to the ownership of land, the Governor in Council may designate the area as a Land Titles Clarification Area.[161] Where an area is designated, the Minister shall file a plan of the area in the registry of deeds.[162] Any person *resident* in the province and claiming ownership of land in the area may apply for a certificate of claim. The application must contain a description of the land sufficient to identify and distinguish it from all other lands, a statement of facts on which the claim is based,

---

158 *Quieting of Titles Act*, R.S.N. 1970, c. 324, ss. 2(1), 3-5, 6(2) and 8(1). A summary of the evidence of boundaries (which differs in kind and in degree) is to be found in *Halsbury's Laws of England*, 4th ed., Vol. IV (London: Butterworths, 1973) at 370-381.
159 *Ibid.*, ss. 8(2), 10, 11(1), 16(1), 17, 22(1), 26 and 34.
160 See P. Burroughts, "The Administration of Crown Lands in Nova Scotia, 1827-1848", Nova Scotia Historical Society, Vol. 35, 1966.
161 R.S.N.S. 1967, c. 162, s. 2(1).
162 *Ibid.*, s. 2(3). The statute does not require a plan of survey to be filed with an application for a certificate. In practice, detailed surveys would have been made and corresponding plans produced by the Surveys Division, Department of Lands and Forests before applications were made. Twelve areas have been designated for purposes of land title clarification since this statute was adopted.

322 Survey Law in Canada

the names of those who have occupied the land or who have claimed ownership or an interest in it.[163] The application must be accompanied by an abstract of title showing all records in the registry of deeds that affect or may affect title, a statutory declaration of the history of the occupation of the lot as far as it be known and a statement of the names of any person who holds any charge on the lot.[164] When it appears from the application that the applicant is entitled to the lot of land, the Minister may issue a certificate of claim. Otherwise he may appoint a barrister of the Supreme Court as a commissioner to examine the claim; on examination of the claim, the commissioner must either recommend issuance of a certificate of claim or report his reasons for not making this recommendation.[165] The interests of others are protected; no certificate of claim can be issued unless encumbrances have been satisfied or unless the holder of same agrees in writing.[166] A certificate of claim is filed in the registry of deeds and public notice is given.[167] Any person claiming an interest in the lot may file notice with the Minister; when the Minister receives a notice within 60 days he shall revoke the certificate of claim; no notice having been received within 60 days, the Minister may grant a certificate of title.[168] A certificate of title vests title in fee simple, absolute and indefeasible, but subject to any liens mentioned in the certificate.[169]

§8.80 Resurveys Much of the legislation relating to resurveys follows a pattern similar to the provisions of the *Canada Lands Surveys Act*,[170] though it may go into greater detail. The federal statute provides that these lands:

---

163 *Supra*, note 161, s. 3(1), (2).
164 *Ibid.*, s. 3(3).
165 *Ibid.*, 4(1), (2), (3), (4).
166 *Ibid.*, s. 4(6).
167 *Ibid.*, s. 4(7).
168 *Ibid.*, s. 6(1), (2), (4).
169 *Ibid.*, s. 6(5).
170 R.S.C. 1985, c. L-6:

    24(1) In this Part, "Canada Lands" means
    (a) any lands belonging to Her Majesty in right of Canada or of which the Government of Canada has power to dispose that are situated in the Yukon Territory, the Northwest Territories or in any National Park of Canada and any lands that are

        (i) surrendered lands or reserves as defined in the *Indian Act*; or

        . . .

    (b) any lands under water belonging to Her Majesty in right of Canada or in respect of any rights in which the Government of Canada has power to dispose.

        . . . .

33(1) Canada Lands may be resurveyed under this Part
(a) for the purposes of correcting errors or supposed errors or re-establishing lost monuments; or
(b) at the request of the member of Her Majesty's Privy Council for Canada or the Commissioner charged with administering the Canada Lands in respect of which the resurvey is to be made.[171]

(2) Plans of Canada Lands that are resurveyed shall be dealt with in accordance with this Part and shall, after confirmation thereof by the Surveyor General, be deemed to be the official plans under this Act of the lands thereby affected and shall be substituted for all, or corresponding portions of all, former official plans of the lands thereby affected.[172]

Matters of detail relevant to resurveys can and do, of course, crop up in statutes other than those which deal with the broader principles. An example is found in the *Highways Act* of New Brunswick: "Where any doubt or dispute as to the boundaries of highway arise a line drawn along the centre line of the travelled portion of such highway shall be deemed *prima facie* to be the centre of such highway."[173] It would be neither practical nor useful to attempt to summarize all such legislation here. One provincial statute, however, and parts of two others are of sufficient importance and interest to warrant individual consideration.

§8.81 The *Boundaries Act*[174] Where doubt exists as to the true location on the ground of any boundary, including the boundary of a public highway, an application may be made to the Director of Titles to confirm the true location of the boundary on the ground.[175] Applications may be made by the owner, the council of a municipality, a Minister of the Crown, the Surveyor General of Ontario, the Surveyor General of Canada or, with the consent of the owner of the parcel, a surveyor.[176] The Director himself, on his own initiative, may start proceedings and may engage a surveyor.[177] Broadly speaking, the costs are levied against all parcels included in the

---

26(2) Where surveys of those Canada Lands affect or are likely to affect the rights of landowners of adjoining lands that are not Canada Lands, the surveys shall be made by a surveyor of the province in which those surveys are made.

In practice this means a Canada Lands Surveyor who is also commissioned in that province, or a surveyor of that province engaged especially for that purpose and who must work both to the provincial standards and those specified by the Surveyor General of Canada.

171 *Ibid.*, s. 33(1).
172 *Ibid.*, s. 33(2).
173 R.S.N.B. 1973, c. H-5, s. 30(4). See *Barnes v. Belyea* (1880), 19 N.B.R. 541 (S.C.) about the centre line of a railway.
174 R.S.O. 1980, c. 47.
175 *Ibid.*, ss. 2, 3(1).
176 *Ibid.*, s. 3(2).
177 *Ibid.*, s. 5(1).

324   Survey Law in Canada

application.[178] Notice of application is given to appropriate persons; objections to the location of the boundary must be delivered to the Director of Titles within the time fixed by the notice of application, stating the grounds of the objection; where objections are received a hearing will be held to determine the boundary's validity.[179] The Director of Titles may confirm the location of the boundary as shown on the plan of survey or he may order that the survey and plan be amended to confirm the boundary in that location.[180] He may also order the removal of any monument that conflicts with any confirmed boundary.[181] Any party aggrieved by an order of the Director may appeal to the Divisional Court.[182] Such court may decide the matter, order trial of the issue, dismiss the appeal or order the survey and plan be amended and confirm the boundary or boundaries.[183] When no appeal has been made, or an appeal has been disposed of, the Director certifies the confirmation of the boundary. Notwithstanding any other Act, the confirmed and certified boundary is deemed to be the true boundary of the parcel; a copy of the plan is placed on the register.[184] Approval of the plan under the *Planning Act* is not required.[185]

§8.82 The *Boundaries Act* (which is, in fact, a special surveys Act) thus applies throughout the whole of Ontario. It is as applicable to lands within the land title system as to lands within the registry system. The statute provides for the resolution of boundary problems through an adjudication process which resembles that of a land court in some other jurisdictions;[186] specialists are applying the law in specialized situations; an owner can bring an action with some reasonable prospect of his adjoiner bearing a part of the cost. Reports of hearings under the *Boundaries Act* are published regularly in the quarterly of the Association of Ontario Land Surveyors. These reports will be found helpful by surveyors in other provinces. The evidence given at the hearing is summarized, the applicable statute law is quoted, relevant case reports are identified and commented upon and a sketch of the locus is provided.[187] The rules which the surveyor

---

178 *Ibid.*, ss. 5(2), 6.
179 *Ibid.*, ss 7, 8(1), 8(2).
180 *Ibid.*, s. 9(1).
181 *Ibid.*, s. 10.
182 *Ibid.*, s. 12(1).
183 *Ibid.*, s. 12(2).
184 *Ibid.*, ss. 13, 15, 16(1).
185 *Ibid.*, s. 17.
186 See R.L. Woodbury and G.R. Perry, "Evidence and The Massachusetts Land Court", *The Canadian Surveyor*, Vol. 29, No. 1, 1975 for a brief outline of the functions of this court.
187 See, for example, *The Ontario Land Surveyor*, Winter, 1983 at 7-12.

in Ontario must follow when making resurveys are set out, of course, under the *Surveys Act*.[188]

§8.83 *Replotting* Some jurisdictions have legislation which goes far beyond the resurvey of an individual lot boundary, the resurvey of a whole property or even that resurvey which would confirm the perimeter of a block of properties. Manitoba[189] and British Columbia[190] provide a resurvey and a regranting arrangement which has obvious benefits with respect to the enhancement and further development of real property, whether the lands be corporate or individual, public or private, large or small. The British Columbia statute provides that: a municipal council may authorize a scheme for the replotting of a district to indicate the proposed relocation and exchange of parcels of real property; for the purpose of replotting, all the parcels and highways and all the other real property in the district is thrown together and forms one common mass of real property; from the common mass is to be taken the real property necessary for highways, parks or public squares; the remainder of the common mass is divided into parcels for allotment to the owners in a fair and equitable manner, so that as far as possible the value of the new parcels allotted to them shall be equal to the value of their former parcels; endeavour is made to allot to owners new parcels in approximately the same location as their former parcels; before the scheme is initiated notice must be published; the owners of seventy percent of the total of the assessed value of all the land in the district must consent in writing; on completion of the scheme the allotments are binding on all the owners; each owner who does not consent has the right to compensation; an appeal from the decision of the commissioner, appointed to hear complaints as to compensation, may be made to the Supreme Court.[191]

§8.84 The *Boundary Lines and Line Fences Act* of Manitoba.[192] In the matter of resurveys, this legislation is unique within Canada:

> Where a question arises as to the boundary line between adjoining parcels of land, if an owner of a parcel gives one month's notice of his intention to do so to each party interested, he may employ a Manitoba Land Surveyor to survey the line; and each party shall pay his proportional share of the expense of the survey.[193]

---

188 R.S.O. 1980, c. 493.
189 *Municipal Board Act*, R.S.M. 1970, c. M 240, Part IV. The procedures for replotting are not as comprehensively set out as in the B.C. statute.
190 *Municipal Act*, R.S.B.C. 1979, c. 290. Equivalent arrangements in the United Kingdom are known as rural re-allotments; the French term is *remembrement* (regrouping of land).
191 *Ibid.*, ss. 887-890, 893, 899, 901 and 908.
192 R.S.M. 1970, c. B-70.
193 *Ibid.*, s. 2.

The legislation only applies within rural areas or unorganized municipalities.[194] Nevertheless one wonders how it was ever accepted by a legislature and why it has remained on the books for many years. There is an interesting story here which waits to be told one day in another forum. The statutory right to oblige a neighbour to share costs has, apparently, not been invoked within recent memory.

> I can recall property owners coming into my office with legal boundary problems, usually after they had obtained an estimate of cost for a retracement survey from a private surveyor. They would ask for help from government but of course funding and staff could not be made available for such private surveys. I would usually refer them to the Act as a *last resort* to solve their problem, but always emphasized that forcing an adjoining owner to participate would most likely seriously injure any neighbourly relations. They always acknowledged that aspect, and to our knowledge did not pursue that road. This lack of personal knowledge does not say that such owners didn't refer the purport of the Act to their adjoining property owner, thereby persuading him or her to cooperate in the survey of their boundary.[195]

Miscellaneous

§8.85 *Quit claim deeds* The grantor of a quit claim deed does not represent that he holds any interest in the property for which the deed is given. The grantor simply conveys to the grantee whatever interest he may have. Since conventional line agreements (paragraphs 8.30 to 8.45) may subsequently be called in question, quit claim deeds given by all parties concerned and placed in the registry, will provide later confirmation of mutual understandings.

§8.86 *Statutory declarations* Surveyors have been known to hold statutory declarations in little esteem, as well as to place too much confidence in their efficacy. If the declarant is not available as a witness and thus not subject to cross-examination, little attention may be paid to his sworn statement. However, courts *may* accept a statutory declaration in a title investigation[196] or when the declaration has acquired some standing by reason of its age.[197] Important testimony of older residents about boundaries should be preserved through a statutory declaration placed in the registry.

---

194 *Ibid.*, s. 12.
195 Correspondence from A.C. Roberts M.L.S., formerly Director, Surveys and Mapping Branch, Department of Natural Resources, Province of Manitoba, March 1986.
196 *Quieting of Titles Act*, R.S.N. 1970, c. 324, s. 8(1).
197 *Vendors and Purchasers Act*, R.S.O. 1980, c. 520, s. 1; see also A.H. Oosterhoff and W.B. Rayner, *Anger and Honsberger Law of Real Property*, 2d ed., vol. 2 (Aurora: Canada Law Book Inc., 1985) at s. 1904.10.

§8.87 Tax deeds While repair of title defects through the medium of tax deeds is not fully germane to this chapter, the matter deserves some mention. The acquisition of good title has been known to put a stop to boundary disagreements of at least a minor nature. But in recent years tax deeds have not always met the standard of that sufficient and thus satisfactory root of good title they had previously enjoyed. Tax sales came well recommended.[198] and the process was cheap. Payment of taxes was withheld and after three years or so the property would be auctioned off by the municipality. Bidding by the owner would be prefaced by an announcement that it was a tax sale. This action usually sufficed and the subsequent deed from the sheriff would give good title — absolute and indefeasible. Even if someone was unsporting enough to top the owner's bid, the latter had a year to redeem the property. The only real penalty was that one had then to start the play all over again. A number of cases make it now quite clear that "proof is required of strict compliance with all the requirements . . . respecting the assessment because on such assessment all sale proceedings are founded",[199] that tax deeds are not cure-alls[200] and that they cannot convey lands on which taxes are not in arrears.[201]

Bornage (*See also* Boundary Determination §§10.185 to 10.218)

§8.88 General The civil code of the Province of Quebec makes provision for the resolution of boundary uncertainties through a process different from those in other provinces. The difference is so pronounced that it seems appropriate to include a short account of things here for purposes of comparison. The process has much to commend it. Some of the Atlantic Provinces — where there are no statutory provisions for resurveys — might do well to consider the advantages it has to offer.

§8.89 Development *Bornage* (literally "monumenting") developed from early times as a logical, practical means of keeping up boundaries or of settling disputes about them. The enabling article of today's civil code is essentially the same as that in the Napoleonic code. This, in its turn,

---

198 D.H. Lamont, *Real Estate Conveyancing*, (Toronto: The Law Society of Upper Canada, 1976) at 350; Oosterhoff and Rayner *ibid.*, at s. 2303.
199 *Hebb v. Hebb* (1944), 17 M.P.R. 276 (N.S.C.A.); *Re Kirton and Frolak*, [1973] 2 O.R. 185 (H.C.); *Carnegy v. Godin* (1982), 52 N.S.R. (2d) 697 (T.D.).
200 *Devereaux v. Saunders* (1977) 26 N.S.R. (2d) 301 (T.D.).
201 *Scott v. Smith* (1979), 7 R.P.R. 10 (N.S.T.D.), affirmed on appeal 36 N.S.R. (2d) 541. For specifics relating to tax deeds and other problems, see C.W. MacIntosh, "How Perfect is A Tax Deed Title?", N.S. Law News, Vol. 8, No. 5, April 1982, and "Decisions to Terrify Property Practitioners", N.S. Law News, vol. 13, No. 6, June 1987.

formalized an earlier doctrine or custom brought forward from Roman law. Until a dozen years ago, *bornage* could be pursued in one of two ways: either strictly between neighbours, or through a judicial process. Now both approaches begin in the same fashion. The judicial route is available at a later stage if agreement between contiguous owners cannot be obtained earlier or if the initial agreement between them breaks down.

§8.90 This presentation is not a definitive statement of the law with respect to *bornage*. It is simply a brief introduction to it through a description of the principal parts of the process. *Le bornage* by Paul Lachance[202] has been the indispensable reference for this purpose.

§8.91 *Definition* The civil code stipulates that every proprietor may oblige his neighbour to settle the boundaries between their continguous lands:[203] "[b]oundaries may be determined either by mutual consent between neighbours, and by their mere act, or with the intervention of judicial authority."[204] The civil code thus establishes a fundamental principle which is left to be more fully detailed through certain articles found within *le code de procedure civile* — the equivalent of regulations under an Act in other provinces. But neither the code nor its regulations define *bornage* directly.

> Bornage is the process which aims to provide by means of visible and durable markers, called monuments, the common boundary line between two contiguous properties, such as had been determined through the earlier process of delimitation.[205]

Bornage therefore may be viewed as both a right and as a servitude. Every proprietor has the right to bring his adjoiner to *bornage*; equally, every proprietor has the obligation so to respond to his adjoiner. *Bornage* is made up of two distinct stages: (1) the delimitation of the dividing line between two properties; and (2) the demarcation of that line by visible and lasting monuments.

---

202 P. Lachance, *Le Bornage*, 2d ed. (Quebec: l'Université Laval, 1974). Additional references are: M-L. Beaulieu, *Le bornage, l'instance et l'expertise, la possession et les actions possessoires*, (Quebec, 1961) and W. deM. Marler, *The Law of Real Property — Quebec* (Toronto: Burroughs, 1932).

203 Metis of the Saskatchewan area in 1874 adapted the Quebec landholding customs to their special circumstances. There was no surveyor within the community. One of the Metis regulations provided for a three-man commission to resolve disputes over boundaries — perhaps performing the essentials of fence viewing as well as those of *bornage*. G. Woodcock, *Gabriel Dumont, The Metis Chief and his World* (Edmonton: Hurtig Publishers, 1975) at 100.

204 L. Saintonge-Poitevin, *Les Codes Civils* (Montreal: Wilson & Lafleur Ltée, 1986) at Art. 504, le code civil du Bas Canada.

205 Lachance, *supra*, note 202 at 2, footnote 1.

Settlement of Boundary Uncertainties

§8.92 *Necessary conditions* In order that *bornage* be exercised, the following conditions must be satisfied: (1) the properties be adjoining; (2) they belong to different owners; and (3) the work be capable of being done in a useful and profitable way. The requirement about usefulness or profitability calls for some explanation. For the most part, this provision serves to except from the process of *bornage* the middle thread of waterways, party walls and boundaries of lands in the public domain. Other procedures apply when the limits of public lands come into question. The surveyor is an essential part of *bornage*. The adjacent owners must agree upon a surveyor. He then proceeds to investigate the locus. He then produces a written report in which are addressed all matters taken into consideration. Should the report be accepted by both proprietors, the joint undertaking is at an end except for marking the line. Should the surveyor's report be unacceptable to one of the parties, the judicial portion of the process comes into play and carries on to a conclusion.

§8.93 *Circumstances appropriate* *Bornage* is open to a proprietor when:

(1) two adjoining properties have never been subjected to the process before;
(2) the monuments marking the line have disappeared;
(3) fences or other line markings have been wrongly placed; or
(4) adverse possession has carried ownership beyond a formerly monumented line.

§8.94 *Two stages* The work of *bornage* is carried out in two stages: the first is the investigative work leading up to the delimitation of the boundary; the second is the demarcation of the common line. But before he proceeds with his investigation, the surveyor must determine that he himself is not subject to challenge. The same reasons for challenging judges apply to surveyors when *bornage* is involved. The regulations under the civil code stipulate several bases for challenge. One such involves a family relationship to either party; another concerns any previous connection to the matter at issue. Should the surveyor find himself liable to challenge, he must apprise both proprietors of the circumstances. They, in turn, may challenge or they may set any objection aside. Similarly, if a surveyor finds himself nominated by a court to proceed with a *bornage*, he must notify the court of any grounds for challenge. The question of whether the owners themselves are entitled to *bornage* must also be addressed by the surveyor. Clearly, the owner in fee simple of lands is so entitled. Also entitled, for example, under Quebec practice, is one who has the right to cut wood on a parcel or one who has the right of passage over an easement or right of way.

§8.95 *Delimitation* The investigative work of the surveyor calls for a survey of the locus and an examination of title documents. The survey commitment requires the surveyor to search out the essential circumstances and the incidental details that support or relate to the primary evidence. There are no clear dividing lines between the two. In the first category, one will recognize such things as existing monumentation, hedges and fences. The second category might well be concerned with the monumentation of other parcels in the neighbourhood. Each *bornage* is different and varies according to circumstances. The examination of title involves the study of deed descriptions, the inspection of plans, the questioning of witnesses (on oath or by solemn declaration) and the receipt of admissions or avowals of either proprietor against their own interests, if such be offered. The acquisition of title by adverse possession is also a matter to be dealt with. It might be noted that in Quebec possession may ripen into title under certain conditions in a ten-year period; title may be acquired unconditionally after 30 years. The written records of earlier *bornage* in the vicinity of the locus may be relevant and may be referred to.

§8.96 *Report of survey* Following his consideration of all the evidence and upon deciding the location of the boundary in question, the surveyor is required to give the parties a report of his work. He must indicate the division line which he believes to be correct. The report is the key document at this first stage of *bornage*; the surveyor reports to the parties the task given him; he identifies the problem, he sifts out from the examination of title and of possession, the factors which will serve as premises for the establishment of property rights; he highlights facts of singular importance; he must marshall logically the reasons underlying his choice of the place where the line must be monumented. In essence the report must present the complete results of true expertise, that is, its terms must present the justification and the facts so as to enable the parties to appreciate the opinion expressed by the surveyor, indicating the position of the division line.

§8.97 *Demarcation* The surveyor's report having been received, neither party may wish to proceed. If so, they can leave things as they are. But the originator of the *bornage* has the capacity to carry the matter into court and there obtain a decision. When both parties are in agreement, the work of monumenting their mutual line can begin right away. The monuments are placed before witnesses. The best witnesses are the proprietors themselves in the company of others. But whatever the desires or the capacities of the owners, witnesses must be present. The surveyor must give formal notice to the proprietors that he will be setting monuments on a specified date. Naturally, previous consultation on the matter of suitable dates will smooth the path for such arrangements. The surveyor must make an official report (*un proces-verbal*) on his work of marking the line. A plan showing the

Settlement of Boundary Uncertainties   331

position and kind of monuments placed, their relation to the lot lines and relevant topography is an indispensable portion of this report. Formerly there was no obligation for such a report to be placed in the land registry. This is now a requirement by virtue of the *Land Surveyors Act*.[206] The surveyor is responsible to see that this is done. The registration of the surveyor's official report completes the *bornage*. It confirms the proper dividing line between proprietors. The work is done.

FUTURE

§8.98 It will be a long time before the resolution of boundary uncertainties becomes a minor part of the work of surveyors, solicitors and the courts. Meanwhile, property mapping programs and surveys carried out within integrated or coordinated survey districts are helping to eliminate the difficulties which arise when individual parcels have been surveyed in virtual isolation, unconnected to other parcels within a community. Property mapping is placing cadastral map sheets in the hands of the public for the very first time. Coordinated surveys make for easier and more consistent retracements.[207] Within the Atlantic Provinces there is a need for legislation which will address boundary retracement and improvements made under mistake of title. These matters ought to be dealt with directly, rather than approached through quieting of title procedures or through actions in trespass or just ignored. Periodically the topic of guaranteed boundaries arises; usually this is when land titles systems are discussed and an analogy is being made with the guaranteed title which these systems provide. The first jurisdiction which guarantees boundaries will have interested audiences in many other jurisdictions.

---

206 R.S.Q. 1977, c. A-23, s. 53(4).
207 Property mapping and coordinated surveys can function as separate entities, but in most instances they are manifestations of a land information system of which the land surveyor, by the very nature of his work, is a part: P.F. Dale and J.D. McLaughlin, *Land Information Management: An Introduction with Special Reference to Cadastral Problems in Third World Countries* (Oxford: Clarendon Press, 1988) is a practical handbook on the fundamentals of these systems.

9

Liability in Negligence and Contracts

*Izaak de Rijcke**
B.Sc., LL.B., O.L.S.

INTRODUCTION

§9.01 The practice of land surveying in Canada, whether as a private practitioner, a government employee or the employee of a corporation, is performed in a social, legal and economic context that imposes potential liability for each practitioner and places certain consequences upon any misconduct. This liability and the consequences of misconduct are the subject of this chapter.

§9.02 The treatment of liability in negligence and liability arising out of contract in one chapter is not an easy task. Their relatedness is primarily historical and, while today liability in both contract and negligence for the practitioner of a profession can be concurrent, their respective foundations are rooted in separate legal doctrines.

§9.03 Another significant difference between liability in negligence and contract is the source of the legal framework that imposes the liability itself. On the one hand, contractual liability may be seen as arising out of the breach of contractual duties or obligations or the failure to deliver

---

* The author wishes to acknowledge the contributions to this Chapter made by Brian Campbell and David N. Jardine, and thanks the contributors for their kind permission to include excerpts from their previous writings.

or perform such duties or obligations. The nature of a duty or obligation is, however, usually the product of a voluntary process of negotiation by the parties to the contract themselves. Voluntariness is itself a hallmark of the freedom to contract or refrain from contracting. The duties and obligations are therefore largely the result of a process whereby the professional himself has direct say in defining the parameters of the contract.

§9.04 On the other hand, liability in negligence may be seen as arising out of a breach in performing certain duties or obligations that are the result of either statute law or common law or both. Negligence in failing to perform to a certain standard or level of care can therefore arise out of the professional's relationship to a client, whether that certain standard or level of care was pre-defined by contract or not. The levels of care or duty and their nature may only be defined, in the final analysis, by looking to statute law and case law and are therefore requirements that are imposed on the practitioner, whether he likes it or not. They are largely involuntary and their existence in the relationship between the professional and his client is largely imputed by law.

§9.05 A certain standard of conduct may therefore constitute a breach of contract or negligence, or both. Other differences and similarities between contract and negligence arise. For example, the general principle in contract law is that the injured party is entitled to such damages as would reasonably place him in the same position as if the contract had been properly performed. In negligence, the general measure of damages are those damages which are a reasonably foreseeable consequence of the actions causing the injury. Depending upon which of these principles is applied to the facts of a particular case, the damages can vary significantly. The courts have often chosen to accept that basis for liability which brought the greatest recovery to the client.

§9.06 A second example is in the area of limitation of actions. Under the limitation statutes in Canadian common law provinces, the general limitation period for an action in negligence is six years from the date of injury. An action in contract must normally be brought within six years from the date of the breach of contract. The commencement dates for the two periods are therefore not always identical. Where concurrent liability exists, the courts will usually apply the rules most generous to the plaintiff.

§9.07 One area in which the distinction can still be significant is in the area of standards of performance. As discussed above, the general standard of care expected of a surveyor would be that of a reasonable surveyor performing a similar survey. The accepted standard would be that of careful, generally accepted and competent practice, and if a surveyor conformed to that standard of practice, he would not be liable.

§9.08 In an action in negligence, the mere fact that there has been a mistake does not mean that the surveyor is liable in negligence. A surveyor is not a guarantor and, if the mistake or error in judgment occurs despite the surveyor having conformed to proper and prudent practice in accordance with the standards of the profession, there may be no liability.

§9.09 The situation is different in contract. If a contract contains specifications and standards of performance, the surveyor will generally be held to those standards even if they exceed the standards normally required by the surveying profession. By executing a contract which imposes such standards, the surveyor binds himself to a higher standard of care and level of performance than would be normally expected in the profession. In effect, the surveyor will have guaranteed a certain level of performance and may be held liable if that standard of performance is not met whether or not the surveyor was negligent or even, in fact, if the standards of performance were impossible to attain. A surveyor must, therefore, exercise considerable caution in agreeing to specifications and standards of performance set out in a contract.

§9.10 In contract, a surveyor will be held liable only to the client with whom he is contracting. This is in accordance with the general principle of contract law that the only persons entitled to enforce a contract are the actual parties to the contract. However, in negligence the scope of liability is considerably wider. A surveyor may be held liable in negligence to parties other than the client if the court considers it reasonable in the circumstances that the surveyor owes a duty to these other parties.

§9.11 For example, a surveyor may contract with a client to perform a construction survey. If, as a result of the surveyor's negligence, the survey is performed incorrectly, there may be other parties besides the client who reasonably relied upon the survey and suffered damage. Given the nature of a legal plan of survey, the potential number of persons who may rely upon the survey over a period of years is very large as is the potential for liability. While the courts have still not clearly defined the precise scope of this liability, it remains a significant risk faced by a surveyor.[1]

§9.12 The professional land surveyor in Canada is, by and large, familiar with the legal requirements of how to conduct a survey and what must be considered in order to address the substantive law on this topic. It is the social, legal and economic contexts in which lie his relationship to

---

1 D.N. Jardine, "The Liability Potential." *ALS News* (Winter, 1987) at 22 and 23.

client, government or employer that have raised both interest and concern in recent years. These contexts are also shared with others in almost all professional groups who find themselves practising in, and meeting the demands of, a society that has become both more demanding and more conscious of its entitlements as a consumer. Being "almost right" is no longer good enough. Blunders are simply not tolerated. It may well be that changes in the law of negligence and contract, together with the advent of liability insurance, have had a greater effect on the professions than any other single development.

§9.13 While this chapter is meant to be only an overview, it borrows heavily from other well-written articles and materials on the topic and the reader is well advised to refer to the sources cited. Liability in negligence will be addressed first, followed by liability in contract and then concluding observations are made on liability insurance.

LIABILITY IN NEGLIGENCE

§9.14 The courts are fond of making general statements regarding what amounts to negligence. For example, in *Stafford v. Bell*,[2] the court held that "there must be evidence of a want of reasonable skill and knowledge." In the 1886 case of *Badgely v. Dickson*[3] it was stated that "a professional person is responsible if he fails to do his work with an ordinary and reasonable degree of care and skill." Similarly, the finding that "a professional person is bound to exercise reasonable care, skill and diligence in preparing plans and supervising work" is contained in the 1905 case, *Russell v. McKercher*.[4]

§9.15 These definitions are of little assistance in providing a surveyor with practical guidelines as to what specific actions will amount to negligence. The courts also frequently add to such general statements the comment that whether a professional person has been negligent is a question of fact which depends on all of the circumstances of the case. This is similarly unhelpful in determining in any specific way what is improper or negligent behaviour.

§9.16 As can be seen from these legal definitions of negligence, the concept is not one which can be pinned down or strictly defined because it is a concept which depends on all the facts in a particular situation. However, there have been, as expected, various decisions which have concluded that

---

2 (1881), 6 O.A.R. 273 (C.A.).
3 (1886), 13 O.A.R. 494 (C.A.).
4 [1905] 1 W.L.R. 138 (Man. C.A.).

a surveyor was or was not negligent and it may be useful to briefly review some of these to determine what specific behaviour has constituted negligence in the court's eyes.

§9.17 One of the earliest cases is the 1824 English decision in *Moneypenny v. Harland*.[5] There, a surveyor who had been retained to complete the survey necessary for the construction of a bridge foundation relied on information provided by a third party to the effect that the soil was suitable and adequate for the purpose. It turned out that the soil was poor and could not support the foundation which resulted in an additional expense to the plaintiff of more than £1,600 sterling. The court concluded that the surveyor was negligent for failing to inform himself of the nature of the soil and that he should not, without careful examination of the ground, have relied on information supplied from other sources. In concluding its judgment, the court said:[6]

> If a surveyor, who makes an estimate, sues those who employ him for the value of his services, and it appears that he was negligent, that he did not inform himself, by boring or otherwise, of the nature of the soil of his foundation, and it turned out to be bad, this goes to his right of action; and if he went upon the information of others, which now turns out to be false, or insufficient, he must take the consequences; for every person, employed as a surveyor, must use due diligence. Whether the plaintiff has used due diligence or not, is a question for the jury; and if the plaintiff went on the statements of others, that is no excuse, as it was his duty to ascertain how the fact was, or to report to his employers that he only went on the information of others, or that the fact was uncertain.

§9.18 *Harries Hall & Kruse v. South Sarnia Properties Ltd.*, a 1928 Ontario Court of Appeal decision,[7] concerned the laying out of a subdivision in Sarnia. As is often the case, there was great urgency and the surveyor, in defence of certain errors which subsequently became apparent, argued that the area surveyed was large, covered with brush and other obstructions and that the owners were in a hurry to place the lots on the market.

§9.19 In dealing with this defence, the court said:[8]

> One of the objections raised by counsel for the surveyor was that the lands were covered with brush and other obstructions. If so, and the surveyor could not proceed with safety and obtain accurate information, it was his duty to have notified his employer or to have stated in his report to the architects that there might be discrepancies. It may be that such obstructions would make more difficult the staking out of the lots upon the ground, but with

---

5 (1824), 171 E.R. 1227.
6 *Ibid.* at 1228.
7 (1928), 63 O.L.R. 597 (S.C.).
8 *Ibid.* at 603.

an outline survey completed, so far as the plan and its accuracy is concerned, the fact that the land was covered with brush or other obstructions would, in my mind, present no difficulty in the way of an experienced skillful surveyor.

§9.20 The facts in this case were that the owners retained the services of certain architects to prepare a general scheme of subdivision and a street and block plan. As the architects required a survey, surveyors were retained to determine, among other things, the location of the south side of St. Clair Avenue in Sarnia.

§9.21 The surveyor prepared the required information and forwarded it to the architects. Relying on this information, the architects prepared a street and block plan and instructed the surveyor to stake out the property and to prepare for registration a street and block plan showing the blocks that were subdivided into lots.

§9.22 Before the staking was completed, the architects, at the request of the owners, prepared a further plan and also a sales plan so that the owners could commence selling lots based upon these plans. The surveyor was provided with copies of these plans.

§9.23 In staking out the property, the surveyor then discovered two discrepancies between the architects' plans and the conditions upon the ground. One error resulted in St. Clair Avenue appearing to be further south upon the plans than it actually was upon the ground. The surveyor admitted that he had given an incorrect measurement in this regard. His mistake was caused by having accepted as accurate a row of stakes planted by another surveyor in staking out an adjoining property. As a result, the owners incurred additional costs by way of architects' and surveyor's fees in having the errors corrected and also had to return many deposits to purchasers who had entered into agreements of purchase and sale based on the erroneous plans.

§9.24 The court concluded that the incorrect location of the south side of St. Clair Avenue was caused by the surveyor.[9]

> The incorrect location of the south side of St. Clair Avenue, I find, was caused by the surveyor, instead of examining the descriptions, the plans of the adjoining subdivisions, and locating the true monuments, accepting as accurate a stake planted on the bank as the correct angle stake of the intersection of the south side of St. Clair Avenue with the west side of Scott Road, and accepting another stake which apparently was some seventeen feet south of the true line. This error or mistake and the erroneous angles given, for which I can find no excuse, resulted in making it impossible for

---

9 *Ibid.* at 604.

the architects in the preparation of their plan to locate with accuracy the intersection of Tashmoo Avenue with Vidal Street.

§9.25 Although the surveyor was clearly responsible for this error, the court concluded that he had no liability to the owners for the damages which flowed from the early sale of the lots because it was clear that the owners knew that there were certain problems and inconsistencies in the survey. The owners well knew of the difficulties encountered by the surveyor and the architects, yet proceeded to sell the lots. The surveyor's liability was, therefore, limited to the extra time and cost spent on correcting the errors.

§9.26 The case illustrates a fundamental difference between negligence (or even incompetence as defined in the Regulations passed pursuant to the Ontario *Surveyors Act*[10]) and the legal concept of negligence. If the Discipline Committee were reviewing the surveyor's actions as set out above, it seems clear that the surveyor would be found to be negligent (or incompetent) for failing to exercise the ordinary skills of a member. However, if a court were considering the same case on the same facts with the same errors, it is conceivable that negligence would not be established. *A finding of negligence in law depends not only on the errors themselves, but also on damages resulting directly from those errors.* That is, in law, negligence cannot be established without an injury resulting from the errors committed. If there is no damage, there is no finding of negligence at common law.

§9.27 The factor of "being rushed" or a sense of urgency on the part of the owner in requiring the results of the survey in a very short time also played a role in *MacLaren-Elgin Corp. v. Gooch*.[11] The plaintiffs brought suit against the surveyor for his alleged negligence and incompetence in the preparation of a survey and in breach of his contract of employment. In 1959, the plaintiffs ordered a survey of a parcel of land that they had recently acquired for the construction of an apartment building. Although the exact purpose of the survey does not appear to have been specified at that time, the court considered that it must have been obvious to the defendant that it was required for architectural design.

§9.28 During the design of the building, following completion of the survey, it was discovered by the plaintiffs that a sewer pipe shown on the defendant's plan was in fact non-existent, and also that the proposed building, for which structural steel had already been ordered, would not fit the actual dimensions on the ground. The plaintiffs further complained that the survey

---

10 S.O. 1987, c. 6.
11 [1972] 1 O.R. 474 (H.C.).

measurements and other data were incorrect, for a second survey made by the defendant revealed that the parcel was not exactly rectangular, as appeared to be indicated by the first survey plan, and that an encroachment of overhanging eaves from an adjoining house had been omitted.

§9.29 The court dealt in turn with each of the alleged defects of the first survey plan. First, the court found that since the plaintiffs had acquired their land subject to an existing encroachment of eaves, which would be required to be removed in any event, the encroachment could not result in any damage to them. Thus, the failure of the defendant to show the overhanging eaves on his plan had not placed the plaintiffs in any worse position than they were in before the survey was made.

§9.30 Second, the defendant's first survey was clearly wrong in showing a non-existent 12-inch sewer and this was corrected on his second plan which indicated a 9-inch pipe in a different location. The architect required accurate information for the plumbing design for the proposed building and, even if this purpose was not disclosed to the defendant, he, as a careful and skillful surveyor, should not have supplied erroneous data which could be relied upon to the user's detriment. The learned judge held that this error, which by itself necessitated an alteration of the architect's plans and accounted for considerable delay, represented a failure by the defendant to exercise reasonable care and skill in the circumstances, resulting in damages to the plaintiffs.

§9.31 Third, the lengths of the front and rear boundaries of the parcel were each shown on the first survey plan as 68 feet, whereas the correct dimensions were later found to be 67.90 feet and 67.93 feet respectively. Also, since the two side lines had a depth of 110 feet, it had appeared from the first survey plan, on which no angular measurements were given, that the parcel was exactly rectangular, whereas the second survey showed the southeast angle to be 88°58′30″. The variations in boundary measurements and the discovery that the property corners were not right angles required a redesign of the building and alterations in the specifications for steel. The absence of any angles or bearings on the first survey plan, together with the identical measurements given for the front and rear and the two side dimensions of the parcel, had led the architect to the erroneous presumption that all four corners were right angles. In fact, the surveyor's field notes contained at least one angular measurement and the defendant stated that angles would have been shown on the plan had it been drawn for architectural purposes. In finding that the architect was not justified in making this presumption and acknowledging that the purpose of the survey should have been obvious to the surveyor, the court concluded that the defendant was under no contractual duty to include angles and bearings on his plan and that the discrepancies in the front and rear measurements of the parcel were "within tolerable limits, for ordinary purposes of a survey." It therefore found the defendant to be not accountable or responsible for damages under this part of the complaint.

§9.32 Accordingly, it was held that the plaintiffs were entitled to recover damages arising only from the incorrect information regarding the sewer pipes. The court appears to have been at least partly influenced in its decision by the urgency of the plaintiffs' request for a survey and it accepted the defendant's argument that he would have supplied additional information had he known the purpose of the survey, for in the court's view the surveyor would not otherwise "have allowed himself to be rushed, but would have taken extra time-consuming precautions."

§9.33 These legal principles were confirmed by the Supreme Court of Canada in the 1984 decision of *Parrot v. Thompson & Monty*.[12] In that case, the defendant surveyor accurately prepared a survey in 1956, but erroneously calculated the square footage of the property as 80,340 feet instead of 52,111 square feet. It appeared that the error occurred when the surveyor made a miscalculation when converting acres into square feet. The plaintiffs purchased the property in 1973 assuming that they were acquiring approximately 80,000 square feet. However, when they came to sell, the error was discovered and the plaintiffs ultimately sold at a reduced price. An action was commenced against the surveyor for the reduction in the purchase price.

§9.34 The Supreme Court noted that to succeed in establishing negligence, a plaintiff must show a "positive damaging"; in other words there must be damages which result from a nonjustifiable act or an omission to perform a duty to the injured party.

§9.35 In that case, the court concluded that the plaintiffs had failed to establish that whatever damage had been suffered was a direct consequence of the surveyor's error. As a result, the surveyor was not liable to the plaintiff.

§9.36 In coming to this conclusion, the Supreme Court said the following:[13]

> It remains to be seen whether, as between respondents and appellant, these damages are a direct consequence of the wrongful error of the latter, which he has always denied.
>
> To decide this point, the situation must be seen with the error excluded, as if it had not occurred. Knowing that the area was 52,111 square feet [instead of the 80,340 square feet indicated on the survey], respondents could only

---

12 (1984), 51 N.R. 161 (S.C.C.).
13 *Ibid.* at 179-180.

have undertaken to sell that area to Dr. Lasalle. Would they in those circumstances have obtained a higher price per square foot from Dr. Lasalle than that which they finally obtained for the same area? If so, the difference in price would presumably represent direct damage caused by the error contained in the cadastre. However, there was certainly no evidence of this and the burden of presenting it lay with respondents, the plaintiffs.

§9.37 The plaintiffs/respondents also alleged that, in order to mitigate their damages, they had no alternative but to bow to the leverage position of the purchaser and sell at a reduced price per square foot.

§9.38 The court said:[14]

> If this is the case, and Dr. Lasalle [the purchaser] took advantage of the situation at the expense of respondents . . . the damage would be his doing, not a direct consequence of appellant's error.

In other words, the Supreme Court failed to find a direct causal link between the damages suffered and the error made by the surveyor.

§9.39 The Supreme Court hypothesized that there were situations where the surveyor might have been liable.[15]

> It does not follow that a surveyor could not be liable to a third party as the result of an error of the type committed in the case at bar. There might be a case where respondents, for example, relying on the technical description, themselves had plans prepared for the building occupying the entire area. In view of the error, the fees spent would have been a total loss. That conceivably could be damage resulting directly from the error, though of course there is no need to decide the point.

§9.40 Negligence at law, therefore, requires something beyond improper actions. There must be damages suffered as a direct consequence of those improper actions. Because of this additional requirement, it is conceivable that a Discipline Committee could find a surveyor negligent or incompetent while a court, on the same facts, would not make a finding of negligence. Although the two situations appear to be very similar, different results may occur.

§9.41 Other factors relevant to a legal determination of negligence are knowledge by the surveyor of the purpose of the survey and knowledge that the person who ultimately suffered as a result of any errors in the survey would rely on the surveyor's professional advice. A plaintiff need not prove actual knowledge; it is sufficient if a reasonable person in the same circumstances would have realized the purpose of the survey and that reliance would be placed on it.

---

14 *Ibid.* at 180-181.
15 *Ibid.* at 181.

Liability in Negligence and Contracts 343

§9.42 For example, in the 1977 British Columbia case, *Beebe v. Robb*,[16] the plaintiff purchased a boat from the defendant. Prior to the purchase the defendant asked that a marine survey be completed and such a survey was carried out by a marine surveyor on behalf of the vendor. The surveyor failed to discover extensive dry-rot in the hull of the boat and was subsequently sued by the purchaser for negligence.

§9.43 The court concluded that there was no liability against the marine surveyor because a professional person giving advice is liable for economic loss suffered by another person who he knows or should know will place reliance on his advice. In this case the surveyor had been retained by the vendor of the boat and not by the purchaser/plaintiff. The surveyor thought he had been retained by the vendor to complete a survey for purposes of a bank loan. There was no direct relationship between the purchaser and the surveyor. The surveyor had no way of knowing that anyone other than the vendor and possibly his banker would rely on the survey. Therefore, even though his survey was undoubtedly negligent, or even incompetent, the plaintiff could not succeed in negligence against the defendant surveyor. The court stated:[17]

> I suspect that Davis [the surveyor] did not do a very thorough job. . . . If he were negligent in his survey his obligation was confined to where his duty lay, i.e., to the defendant Robb who hired him and who told him he required the survey solely for the purpose of a bank loan. I fail to see any duty owing by Davis to Beebe and therefore any resulting liability for Beebe's resulting loss. There was no evidence that Davis knew a sale was in process, nor that half his fee came from Beebe, or that the sale was contingent upon the survey being satisfactory.

§9.44 It may be that a surveyor is better protected against claims in negligence if he knows as little as possible about the purpose of the survey and who will ultimately rely on it. The relationship, or lack of it, which exists between the surveyor and the "ultimate victim" is obviously simply another way of characterizing the reliance principle. That is, if it can be established that there was a sufficient relationship between the parties, the result of that relationship will be that a reasonable person would know, or in fact did know, that his carelessness would result in damage to the other party. For example, in the 1980 Alberta case of *Canadian Western Natural Gas Co. v. Pathfinder Surveys Ltd.*,[18] surveyors were retained to plot the location of a gas pipeline from A to B. The right of way required that the pipeline curve in the middle. The curve was to have been staked by the surveyor. This required him to determine the point of intersection

---

16 (1977), 81 D.L.R. (3d) 349 (B.C.S.C.).
17 *Ibid.* at 351.
18 (1980), 21 A.R. 459 (C.A.).

of the tangents, then to measure back to the point of commencement of the curve and then to stake the actual curve itself. The surveyor failed to complete the last step of measuring back and staking the curve.

§9.45 When the contractor was excavating, it was realized that the angle at the junction of the tangent lines exceeded angles permitted for transmission lines. The contractor therefore put the curve where the angle was acceptable. Unfortunately, the place where the contractor put the curve was not on the right of way. The gas company was successful against the surveyor in negligence.

§9.46 In coming to this conclusion, the court discussed when a duty of care arises. Such duty must necessarily be established before a plaintiff can succeed in negligence.

> In reaching this conclusion it [the court] found support in the decision of the House of Lords in *Anns v. London Borough of Merton*,[19] and in particular in the passage of Lord Wilberforce's judgment found at page 498:
>
> First one has to ask whether, as between the alleged wrongdoer and the person who has suffered damage there is a sufficient relationship or proximity of neighbourhood such that, in the reasonable contemplation of the former, carelessness on his part may be likely to cause damage to the latter, in which case a *prima facie* duty of care arises. Secondly, if the first question is answered affirmatively, it is necessary to consider whether there are any considerations which ought to negative, or to reduce or limit the scope of the duty or the class of person to whom it is owed or the damages to which a breach of it may give rise.[20]

§9.47 In other words, the principle of reliance or foreseeability must first be established. Once this has been done, there may be other factors which reduce the liability of the defendant, such as actions on the part of the owner or victim which have contributed to the extent of the damages suffered.

§9.48 There are a variety of situations in which a surveyor may find himself liable to persons who are not immediate parties to a contract for services with that surveyor. There have been, in recent years, major changes in this area of law. There now exists in Canada a solid legal basis upon which to find third party liability on the part of a surveyor. The traditional view of the law was to disallow recovery in tort for negligent misrepresentation unless there was a contractual or fiduciary relation between the parties or unless the misrepresentation was fraudulent. The courts have been

---

19 [1977] 2 All E.R. 492.
20 *Ibid.* at 470.

clearly apprehensive about creating, "liability in an indeterminate amount for an indeterminate time to an indeterminate class."

§9.49 Until 1964, the closest any common law court came to supporting liability in tort for negligent misrepresentation was a dissenting opinion by Lord Denning in *Candler v. Crane, Christmas & Co.*[21] In that case, the plaintiff had subscribed to shares in a company on the strength of certain accounting documents. The accountant's report, however, had been carelessly prepared and gave a wholly misleading picture of the financial state of the company. As a result, the plaintiff lost the whole of his investment and an action was taken against the company and the chief accountant in his personal capacity. The majority of the Court of Appeal dismissed the action on the ground that the defendants owed no duty of care to the plaintiff as there was no contractual relation between them. Lord Denning, however, in an innovative dissenting opinion, made it clear that he would not be bound by such traditional, theoretical constraints. In Lord Denning's view, the professional person's duty of care in preparing reports and accounts was not limited simply to the contracting party, but extended as well to any third party acting in *reasonable reliance* thereon.

§9.50 The real turning point in the law of negligent misrepresentation came later, however, in *Hedley, Byrne v. Heller*.[22] In that case the plaintiffs, who were advertising agents, sought financial information from their bankers as to the financial stability of a company with which they wished to place substantial advertising orders. The bankers, in turn, made inquiries from Heller and Partners, the defendants, who were bankers of the company with whom the orders were to be placed. The bankers' reply was favourable but they cautioned that their advice was without responsibility as to its accuracy. In reliance upon these references, the advertising agents placed orders which resulted in substantial financial loss. As a result, the plaintiffs brought an action against Heller and Partners for damages for negligence.

§9.51 The issue of liability in *Hedley, Byrne* was resolved ultimately in favour of the defendant. The defendant had given the financial information subject to the *caveat* that it would not be responsible for the accuracy of that information. But the House of Lords took advantage of the opportunity presented in *Hedley, Byrne* to break new ground. Following the dissenting opinion of Lord Denning in *Candler v. Crane*, the Law Lords stated that a negligent, though honest, misrepresentation spoken or written gave rise to a cause of action for damages for financial loss caused thereby, apart

---

21 [1951] 2 K.B. 164 (C.A.).
22 [1964] A.C. 465 (H.L.).

from any contractual or fiduciary relationship. As Lord Morris stated:[23]

> [I]f someone possessed of a *special skill* undertakes, quite irrespective of contract, to apply that skill for the assistance of another person who relies on such skill, a duty of care will arise. The fact that the service is to be given by means of or by the instrumentality of words can make no difference. Furthermore, if in a sphere in which a person is so placed that others could reasonably rely on his judgment or his skill or on his ability to make careful inquiry, a person takes it on himself to give information or advice to, or allows his information or advice to be passed on to, another person who, as he knows or should know will place *reliance* upon it, then a *duty of care* will arise.

§9.52 While the outer limits of the *Hedley, Byrne* principle have not yet been conclusively determined, the Canadian decision in *Haig v. Bamford*,[24] seems to have pushed the extent of liability originally contemplated one step further. In *Haig v. Bamford*, a firm of chartered accountants had been retained to prepare financial statements on a company which was indebted to the Saskatchewan Economic Development Corporation. The defendant chartered accountants had been advised by the company that the financial statements would be used to apply for further loans from the Development Corporation and to induce other persons to invest in the company. The plaintiff, Mr. Haig, who was completely unknown to the defendants, invested in the company on the basis of the financial statements which had been inaccurately prepared. Mr. Haig suffered financial loss as a result thereof and brought an action against the auditing firm. Damages were awarded at trial, but on appeal to the Saskatchewan Court of Appeal this decision was reversed. On further appeal, the Supreme Court of Canada upheld the decision of the trial judge and imposed liability for the amount advanced in reliance on the inaccurate financial statements. Mr. Justice Dickson summarized his reasons:[25]

> In summary, Haig [the plaintiff] placed justifiable reliance upon a financial statement which the accountants stated presented fairly the financial position of the Company as at March 31, 1965. The accountants prepared such statements for reward in the course of their professional duties. The statements were for benefit and guidance in a business transaction, the nature of which was known to the accountants. *The accountants were aware that the Company intended to supply the statements to members of a very limited class. Haig was a member of the class.* It is true the accountants did not know his name but, as I have indicated earlier, I do not think that is of importance. I can see no good reason for distinguishing between the case in which a defendant accountant delivers information directly to the plaintiff . . . and the case in which the information is handed to the employer who, to the knowledge of the accountant, passes it to members of a limited class (whose identity is

---

23 *Ibid.* at 502 and 503.
24 (1976), 72 D.L.R. (3d) 68 (S.C.C.).
25 *Ibid.* at 80.

unknown to the accountant) in furtherance of a transaction the nature of which is *known* to the accountant. [emphasis added]

§9.53 It is suggested that the decision of the Supreme Court of Canada in *Haig v. Bamford* represents a significant extension of the limits of liability as stated in *Hedley, Byrne*. The range of possible plaintiffs is clearly expanding. The step which the court took in *Haig v. Bamford* was from liability in cases where an auditor actually knows the identity of a specific plaintiff relying on his financial report to situations where the auditor has mere knowledge of the existence of a limited class of people who will rely on such reports. And it is further submitted that there is nothing peculiar in the nature of the accounting profession which would allow one to distinguish *Haig v. Bamford* on its facts and claim that it has no application to a case of surveyor's negligence.

§9.54 The practice of providing copies of old plans of survey to surveying colleagues and members of the public should be critically examined in light of the recent developments in the law regarding third party liability. While the sharing and exchange of information and old plans with colleagues may very well be a professional duty, the practice of releasing old prints of a survey plan, even for a nominal fee or charge, should be discontinued altogethr. The risk of the outdated plan being treated at law as a negligent misrepresentation is very real and one must question whether liability in such circumstances can ever be effectively disclaimed.

LIABILITY IN CONTRACT

§9.55 A contract is an agreement between two or more persons which creates legally binding obligations enforceable by the courts. Breach of the terms of the contract by one of the parties may give rise to various remedies including a claim for damages. The essential elements of a contract include the intention to create legal relations, an offer and an acceptance, and sufficient certainty of terms to identify what was agreed to between the parties. Contracts may be either written or verbal, but certain contracts must be in writing if they are to be enforceable because of the provisions of acts such as the *Statute of Frauds*, the *Sale of Goods Act* and the *Guarantees Acknowledgement Act*.[26]

§9.56 To establish an action for breach of contract, the plaintiff must establish that a contract exists between the parties and that the defendant

---

26 *Statute of Frauds*, R.S.O. 1980, c. 481; *Sale of Goods Act*, R.S.O. 1980, c. 462; *Guarantees Acknowledgment Act*, R.S.A. 1980, c. G-12.

has breached the contract. In order to recover damages, the plaintiff must also establish what injury has been caused by the breach of contract.

§9.57 While it is relatively easy, as above, to define the elements of a contract, complex issues can arise in an action for breach of the contract. It may be difficult for the plaintiff to establish with certainty the existence of a contract and the exact nature of its terms. As well, it may be difficult in particular circumstances to determine whether the standard of performance by the defendant constitutes a breach of contract. Difficult issues can also arise as to what constitutes damages arising out of a breach of contract and whether those damages will be compensated by the courts. In general terms, the object of awarding damages for breach of contract is to place the plaintiff as far as money can do so in the same situation as if the contract had been performed. In practice, this is often difficult to assess and there are also difficult issues when the courts are called upon to consider whether certain consequential damages flow from the breach of contract or are too remote to be compensated.

§9.58 In most instances where a surveyor is being sued by a client, there will have been some form of contract between the surveyor and the client whereby the surveyor undertook to perform certain work in return for payment of a professional fee. In such an action, the court will examine the dealings between the parties in an attempt to determine what services were to be performed and the standard of performance. To the extent that these matters are set out in writing, the court's task will be made easier, but where these issues are not dealt with in a written agreement, the court will have regard to the dealings between the parties, the general standards of the profession and the nature of the work to be done.

LIABILITY INSURANCE

§9.59 Some surveying associations, as other professional organizations, have recently adopted measures which require individual members to carry liability insurance for errors and omissions. The advent of these measures has been partly a response to the need for self-preservation and partly a response to the need, on the part of the public, for a source of compensation when a professional person errs. Various plans and types of insurance have been tried and considered. On a theoretical level, there has always been a balance between the consequences of totally eliminating the threat of tort liability and the consequences of having no ability to compensate at all due to an inevitable bankruptcy.

§9.60 Whether a particular insurance plan will serve as an effective mechanism to minimize civil liabilities will depend upon a number of factors: Is it a mandatory insurance? Are the premiums based upon experience? Is there a substantial, uninsurable deductible requirement? The examination of the insurance packages available will emphasize the importance of these variables. To the extent that a particular insurance plan employs a premium rating not based upon experience or permits a nominal or non-existent deductible, the incentive rationale of the mechanism as a means of encouraging or continuing competence is virtually eliminated.

§9.61 To achieve optimal utility, the civil liability mechanism must trigger a response not only at the individual level but at the level of the profession as a whole. Continuing professional competence is unattainable unless the incidents of individual professional malpractice can be collected, identified and evaluated in a systematic manner in order to provide the necessary direction for the implementation of loss control seminars or continuing education programs. Several factors are at play here. First, there is the profession's attitude towards compulsory insurance. With a universal mandatory insurance scheme, the details of claims and data on loss-by-origin are not only more easily obtainable but reflect more accurately the total incidence of malpractice. Second, it is relevant to ask whether the profession's insurance arrangements include a substantial self-insurance component. To the extent that the profession as a whole undertakes a significant self-insurance arrangement, the incentive for stricter scrutiny and for appropriate corrective efforts is increased. Third, and of paramount importance, is the nature and the extent of the inter-relationship between the profession and its insurance carriers. Does the insurance carrier provide the profession with all the data that is necessary for effective loss control? Are the files on claims systematically analyzed with a view toward professional education? Has a loss control program been instituted? Are loss control bulletins distributed in order to advise the profession of recent developments or particular problems? The fourth and final factor that can limit the effectiveness of the civil liability mechanism as an incentive for the competence of the profession as a whole is the nature and extent of the profession's disciplinary proceedings. The effectiveness of the mechanism is increased if the disciplinary body views professional incompetence as a matter properly falling within its jurisdiction and acts accordingly.[27]

§9.62 While professional liability insurance is virtually essential for surveyors practicing in today's environment, the fact that a surveyor has professional liability insurance does not mean that he can stop being concerned about the possibility of an action against him for negligence.

---

27 E.P. Belobaba, "Civil Liability as a Professional Competence Incentive" Professional Organizations Committee (Toronto: 1978), at 11 and 12.

Professional liability insurance policies provide a great deal of protection, but they do not completely insulate a practitioner from the consequences of an action by a client.

§9.63 The need for professional liability coverage is obvious. No one can guarantee that they will not be sued. A practitioner requires protection not only in respect of the damages which might be awarded against him, but also for the costs involved in defending an action. It is often assumed by the public that a successful defendant can recover all his legal fees and other expenses (such as fees for expert witnesses) from the plaintiff if the plaintiff's claim is dismissed. This is not necessarily correct because the costs which can be recovered are based on standard amounts. These amounts are usually substantially less than the actual fees incurred; but in a long or complex trial, the costs may become prohibitive. Even those costs which are awarded to a successful defendant may be of little use if the plaintiff does not have enough assets from which recovery can be made. A proper insurance policy provides protection against both an adverse judgment and the costs of defending actions, either of which might otherwise bankrupt a surveyor or his firm.

§9.64 However, having an adequate policy of insurance does not mean that a surveyor, through prudent practice, no longer has to be concerned about avoiding claims. One obvious reason is that most policies carry a deductible amount which must be paid by the insured. In many professional liability policies, the deductible amount can be substantial and it is usually applied to each claim made during the policy period. A series of successful and unsuccessful claims may therefore result in a considerable expense to the surveyor involved.

§9.65 A second factor which is often overlooked is that the insured is required under the terms of the policy to assist the insurer in defending the action. This will include meetings with the insurer and its lawyers, preparation for and attendance at examinations for discovery and preparation for and attendance at the trial of the action. In cases which involve complex situations or a large number of parties, the time requirements can be extensive. As well, if there is a lengthy trial, various members of the surveying firm involved in the project may be tied up as witnesses for an extended period of time. The loss of time and the inconvenience involved in defending an action is not something for which compensation is provided by liability policies; in lengthier and complex disputes the cumulative effect of time lost from the business as a result of legal proceedings can be crippling.

§9.66 While an insurance policy can protect an individual from some of the financial consequences of negligence, it cannot do away with the emotional strain of being accused of professional incompetence and being forced to defend one's actions in a court of law. Legal proceedings can be long and drawn out, and testifying at examination for discovery or at trial can be a draining and frustrating experience. An insurance policy also does not defeat the negative publicity which may be generated within the profession and in the minds of the public at large by a widely reported judgment in which a finding of negligence is made.

§9.67 The insured practitioner must also be careful that when a claim is made he is aware of and follows the provisions of the policy. All professional liability policies have stipulations which require prompt notice of any potential claim. There have been situations in which coverage has been denied because the claim was not promptly reported. There can also be strains when the insurer feels certain steps should be taken in defending the action and the insured practitioner disagrees. For example, the insurer may feel that an action should be settled when the insured practitioner would prefer to see it taken to trial. There are also exclusions in professional liability policies which will be examined by the insurer carefully to see if the claim made falls within the provisions of the policy. If it does not, or if there is some other reason that coverage is denied, the practitioner may be faced with bearing the costs of a long and expensive action without the benefit of insurance.

§9.68 No professional can guarantee that he will not make a mistake. There is also no way to prevent a client from making a claim and bringing an action which may have little chance of succeeding. Professional liability insurance is essential protection in these circumstances, but it is no substitute for a careful, competent and professional approach to the practice of surveying.

10

The Law in Québec

Grégoire Girard, B.A., Q.L.S.
J. André Laferrière, Q.L.S.
Gérard Raymond, B.Sc., LL.B., C.L.S., Q.L.S.

INTRODUCTION

§10.01 Under the Canadian constitution,[1] the provinces have jurisdiction over property and civil matters and so private law varies from province to province. The law in all provinces other than Québec is based on one common source, common law. The laws of Québec, on the other hand, are based on written sources that originated in France. These two legal systems represent, in fact, two different modes of thought — French Cartesianism and English pragmatism. The fundamental difference between written and common law lies in the way the laws are established. In the case of the former, they are the work of legislators, while with the latter, they are established by judges. French law is thus the fundamental source of written law, while common law originates in case law.

§10.02 In reality, the Québec system of private law and the system used in the other provinces are not such complete opposites for to an ever increasing extent English private law is being created by legislators, while case law has also played a significant role in the evolution of Québec civil law. That does not mean, however, that the institutions functioning under these two systems have become particularly alike. In spite of the intrusion of certain common-law concepts, Québec private law has remained essentially civilist. It differs sufficiently from private law in the other provinces, particularly in the domain of survey law, to justify a special chapter on the subject here.

---

1 *Constitution Act, 1982*, being schedule B of the *Canada Act 1982* (U.K.), 1982, c. 11.

§**10.03** In this chapter, we will be looking at certain legal provisions that are of interest to the practising land surveyor. The first section deals with the *Civil Code*, the regulations of which constitute, in essence, Québec private law. The three sections which follow deal with the cadastre, land registration and boundary determination. The fifth section is concerned with land surveying as a profession. Finally, the last section consists of a synthesis of Québec law concerning bodies of water.

THE CIVIL CODE

Introduction

§**10.04** This first section deals with the provisions of the Civil Code which are of interest to the practising land surveyor. To gain a good understanding of the role of the *Civil Code*, it is essential to know something of its history and sources and to be acquainted with the general theory of obligations which serves as a basis for other aspects of civil law. After a brief look at matrimonial regimes, we will turn to property law, particularly the law governing real property and its corollaries, servitudes, privileges and hypothecs, and possession and prescription.

The Historical Roots of Québec Private Law

§**10.05** *The French regime* In contrast to the private law of the other provinces, which originates in British law, civil law in Québec has its roots in French law.

New France became subject to French law in 1534, when Jacques Cartier took possession of the region. The colonists brought with them to their new home their own local or regional compendia of customs and unwritten laws and procedures known as "coutumes". The result was that, for over a century, various coutumes were in use in New France concurrently, the best known of these being the coutumes of Paris, Brittany, Normandy, and the Vexin. In order to avoid a proliferation of different laws, the coutume of Paris was recognized as the official code of New France by the edict of 1664, by which the Compagnie des Indes occidentales was established. Before that, in 1663, French civil law was introduced into the region by means of the newly created Conseil souverain. New France was henceforth organized along the model of the French provinces, and its legal system became one of the best that could be imagined at the time. The laws in effect at the time of the conquest of 1760 originated in the following sources: the coutume of Paris, the laws of France at the time of the establishment of the Conseil souverain, royal ordinances registered with the Conseil souverain, and regulations passed by it. Thus began a critical period in the history of civil law in this country.

§10.06 *The English regime* Capitulation was followed initially by military rule, which remained in effect from 1760 to 1764. During that period of transition, the administration of justice was put in the hands of military authorities who followed English law. The Treaty of Paris, signed in 1763, by which Canada came under English rule, contained no provisions concerning the legal system to be followed. Not until 1764, in an ordinance signed by Governor Murray concerning the establishment of a superior court in 1764, was it specified that justice was to be carried out according to English law. It need hardly be said that this decision, which later was declared illegal, encountered strong opposition from French Canadians who made up the vast majority of the population.

§10.07 *The Québec Act of 1774* The persistent demands of the French Canadian elite and the support of the Governor, Lord Dorchester, forced England to give serious consideration to this state of affairs which, among other things, contravened a principle of international law according to which the mere defeat of a country did not in itself justify a change in its private law. No doubt shaken by the American Revolution, England decided to remedy the situation by adopting the *Québec Act*[2] in 1774, whereby French civil law was re-established in the region and English criminal law was officially introduced. All subsequent constitutional acts, such as the *Constitutional Act*[3] of 1791, the *Act of Union*[4] of 1840, and the *British North America Act*[5] of 1867, ratified these provisions of the Québec Act. And so the civil law of Québec was born and survived the constitutional upheavals of our country.

§10.08 *The Civil Code of Lower Canada* Originally, the laws of New France were totally French. Nevertheless, as the centuries went by, a typically Québécois legal code developed which corresponded more closely with the reality of Québec and the aspirations of its people. Québec private law developed over time from a wealth of different judicial sources. Since this situation led to confusion, it became necessary to consolidate our civil law. In 1857, the government of the Union, which consisted of an equal number of representatives from Upper and Lower Canada, approved a bill calling for the codification of civil law in Lower Canada. The three commissioners entrusted with this task were given the mandate to codify the existing civil law using the *Napoleonic Code* as a model. This explains an apparent similarity between the two codes. However, this similarity is more one of form than of content. With the *Napoleonic Code*, which was

---

2 1774, 14 Geo. 3, c. 83.
3 1791, 31 Geo. 3, c. 31.
4 1840,, 3 & 4 Vict., c. 35.
5 1867, 30 & 31 Vict., c. 3.

imposed after the French Revolution, an entirely new legal code was introduced. This was not the case with the *Civil Code* of 1866, which only gave a modern shape to the old civil law of Lower Canada. The legal systems of France and Québec still have common origins. Both have been influenced primarily by the coutume of Paris, Canon law and Roman law, Canon law constituting a further development of Roman law. In this manner, early French law books have remained as a source of Québec civil law. The *Civil Code* remains the pattern for Québec private law. In spite of numerous modifications of the *Code,* several of its provisions no longer correspond to the realities of today's society. The basic premises under which the codification of 1866 was undertaken are poorly suited to the contemporary values of an urban and industrial world. Under pressure to solve the problems of an ever-changing society, the legislators had no choice but to enact legislation concerning matters regulated by the *Civil Code* which, in some cases, contradicted its provisions. A reform of the *Civil Code* was inevitable.

§10.09 *The revision of the Civil Code* The act respecting the revision of the *Civil Code* was adopted in 1955, but the actual reforms did not get under way until 1962. A proposal concerning the revised *Civil Code* was finally presented to the Minister of Justice in 1978. It was decided to present the proposed revisions to the National Assembly in sections, and the first section, concerning family law, was approved in 1980. This revised part of the *Civil Code* of Lower Canada constitutes Book Two of the new code, called the *Civil Code of Québec.* The remaining parts of the *Civil Code* of Lower Canada will be revised section by section and incorporated into the *Civil Code of Québec* as the revisions are completed.

The present *Civil Code* consists of four distinct books. The first of these concerns persons; the second, property, ownership and its different modifications; the third, the acquisition and exercise of rights of property; and the fourth, commercial law. The *Civil Code* does not, however, contain all the legal provisions of private law. Some are included in the *Code of Civil Procedure*[6] and others in statutary laws, such as the *Consumer Protection Act.*[7]

The Sources of Law

§10.10 *Objective law* Although laws form the foundation of our legal system, justice is not based on them exclusively, but also on custom, case law and law books. Customs acquire the force of law through long and

---

6 R.S.Q. 1977, c. C-25.
7 R.S.Q. 1977, c. P-40.

habitual use. They constitute a supplementary source of law, particularly in the area of commerce, where it is customary, for example, for the vendor to deliver his merchandise to the purchaser. In fact, under article of 1017 of the *Civil Code*, "customary clauses must be supplied in contracts, although they be not expressed." Case law is based on all former decisions of the courts. In civil law, where laws are the work of legislators, this source is not as important as it is in common law. Judges in Québec are, thus, not really bound by the decisions or their colleagues, as is the case under common law. Nevertheless, they do pay considerable heed to them, especially in the case of a judgment handed down by a higher court. And law books, the last of these sources, have been compiled from all the treatises written by experts in the field of law. Thus, positive law is a compendium of all the rules of law taken from all these sources. Here, law is considered, objectively, as the sum of normative rules.

§10.11 *Subjective law* Law can also be understood subjectively, that is, in terms of the person who invokes it for his own benefit. In this context, we are dealing with the rights that form an individual's estate. Furthermore, as set down in articles 1980 and 1981 of the *Civil Code*, a person's estate constitutes the common pledge of his creditors who, if necessary, can have the assets of a debtor seized if payment has not been made. An estate is made up of two main kinds of rights, real rights and personal rights, the latter also called debts or obligations. These rights, by the way, correspond to two distinct domains of law — real law, including in particular property law which applies only to things, and personal law which applies only to people. Property law is an aspect of real law, while the law of obligations deals with personal rights.

Obligations

§10.12 *Introduction* The law of obligations constitutes a very important part of the *Civil Code*, and includes more than a thousand articles. This part of the *Civil Code* is based largely on Roman law and contains numerous fundamental principles which apply beyond the confines of the law of obligations. It is, thus, impossible to have a good understanding of civil law without a general knowledge of the theory of obligations.

§10.13 *Obligations defined* The word "obligation", which can have several meanings, is used here to refer to the obligation of a debtor, in the general sense of the term. An obligation can, thus, be defined as the law which binds the debtor to give something to a creditor or do or refrain from doing something at the request of a creditor. In these terms, a person could contract an obligation either to transfer ownership of something or carry out or refrain from carrying out a certain thing.

§**10.14** *Sources of obligations* Pursuant to article 983 of the *Civil Code*, obligations arise from one of the following sources: a contract, a quasi-contract, an offence, a quasi-offence and the operation of the law. A contract is a legal act resulting from an agreement between two or more persons who give their word to do or not do something. The quasi-contract, as its name indicates, is not a contract properly speaking because, unlike a contract, it does not result from a legal act but from a legal fact, that is, from the fact that one person acts in such a way that another person benefits from the action. This would be the case if, for example, somebody ordered repairs to be carried out on a neighbour's immoveable because a fortuitous event during the neighbour's absence had made such action a necessity. Offences or quasi-offences are also legal facts giving rise to civil liability, a matter which will be discussed later. Obligations also arise, in certain cases, from the sole operation of the law, as indicated in article 1057 of the *Civil Code*. This is the case with certain obligations of owners of adjoining properties, such as the obligation to settle property boundaries.

§**10.15** *Distinction between results and means* As stated in article 1058 of the *Civil Code*, an obligation must have for its object something which a party is obliged to give, do or not do. When determining the weight of a given obligation, a distinction must be made between obligations of result and those of means. In the case of the former, the debtor commits himself to produce a certain result, such as delivering merchandise to the place agreed to in the contract. If the obligation is one of means, however, as is the case with professional endeavours, the debtor is not bound to guarantee a certain result, but he must employ all means at his disposal to achieve the best possible result.

§**10.16** *Kinds of obligations* Obligations are also classified by kind. Several of these kinds are defined in articles 1079 and following of the *Civil Code*, the more common being conditional obligations, obligations with a term, joint and several obligations, and obligations with a penal clause.

§**10.17** *Conditional obligations* According to article 1079 of the *Civil Code*, an obligation is conditional if it is made to depend on a future and uncertain event. There are two kinds of conditions. Suspensive conditions are those which prevent a contract from going into operation until a certain event has taken place. Resolutive conditions are the opposite; while they do not delay the obligation coming into effect at the time that the contract is concluded, they terminate the obligation as soon as they have been accomplished. A contract of fire insurance is an example of a conditional obligation because the payment of benefits depends on a future and uncertain event.

§**10.18** *Obligations with a term* An obligation with a term is defined in article 1089 of the *Civil Code*. Unlike a suspensive condition, a term does not suspend an obligation, but only delays its execution, an example being a credit sale. A life-insurance contract is another example of an obligation with a term because the payment of the benefit depends on a future and certain event.

§**10.19** *Joint and several obligations* It can happen nevertheless that one faces several creditors or several debtors in relation to only one obligation. When this is the case, the obligations, as defined in article 1100 of the *Civil Code*, are said to be joint and several. A joint and several interest can exist among creditors, a joint and several obligation among debtors. A joint and several interest among creditors implies that each has the right to demand of a single debtor the discharge of the whole debt. If there is a joint and several obligation on the part of co-debtors, then a single creditor may pursue one or more of the debtors and claim the full amount of the debt. As indicated in article 1105 of the *Civil Code*, an obligation is joint and several only if it is expressly declared to be so in the contract, except in the area of commercial transactions where the obligation is presumed to be joint and several. An obligation resulting from an offence or a quasi-offence committed by two or more persons is also joint and several, under the terms of article 1106 of the *Civil Code*.

§**10.20** *Obligations with a penal clause* Finally, obligations may have a penal clause attached to them. In such a case, the penalty is enforced in the event of non-fulfillment or late fulfillment of the obligation on the part of the debtor. A penal clause can stipulate, for example, that a contractor who has agreed to complete a construction project by a certain deadline must pay a penalty for each day thereafter. Under the provisions of articles 1131 and following of the *Civil Code*, an obligation with a penal clause allows the parties to determine beforehand the amount of damages to be paid in case the obligation is not performed. However, the creditor cannot claim the agreed amount unless he has given prior notice to the debtor.

§**10.21** *Notice of default* A notice of default is a solemn warning to the debtor reminding him that he must fulfil his obligation and that the creditor intends to claim what is due to him. As a general rule, notice of default must have been given before any court action can be undertaken. However, notice of default need not be given if, for example, the parties have agreed that it is deemed to have been given as soon as the obligation becomes due. According to article 1069 of the *Civil Code*, the same applies to contracts of a commercial nature in which the term is fixed, the debtor being in default by the mere lapse of time in fulfilling his obligation.

Contracts

§10.22 Introduction The two most important sources of obligations are contracts and civil liability. They will be discussed in the following two sections. Since the Industrial Revolution, society and the economy have been based on an exchange of goods and services which, from a legal point of view, is carried out with the help of contracts. The *Civil Code* does not define a contract. In reference to the definition of an obligation, a contract can be defined as a legal act resulting from an agreement of will between two or more persons who commit themselves to give, do or not do a certain thing.

§10.23 *The role of will in contracts* A contract is to be distinguished from other obligations in that it arises from an agreement of will. Thus, the source of the obligation is human will. The theory of contracts is in fact based on the philosophical principle that the human will is autonomous. The principle that all men are free and equal before the law was strongly advocated by the philosophers of the 17th and 18th centuries and became generally accepted at the time of the French Revolution. From a legal standpoint, the doctrine of the freedom of decision translates into the principle that every person is free to conclude contracts, that is, to commit himself by contract to conditions he judges appropriate. This freedom is only limited by article 13 of the *Civil Code*, which prohibits agreements that are contrary to public order and good morals.

§10.24 *Conditions as to the form of contracts* A contract arises, thus, from the decision of the parties involved, whose consent in itself concludes the contract, without any need for particular formalities. Such is the case with sales, which, according to article 1472 of the *Civil Code*, are concluded by the consent of the parties alone. This is a contract of consensus. Under Québec law, the great majority of contracts belong in this category. There are, however, some contracts that are not duly concluded until certain formalities have been carried out. This is the case with solemn contracts and real contracts. A solemn contract must be concluded by a declaration of will of the parties in a notarial deed. Examples include marriage contracts, donations of real property and hypothecs. A real contract is not valid until the thing that is its object, such as loans, deposits and securities, has been remitted. Certain formalities are also required in a few other cases, such as in communication contracts and lease agreements for living accommodations.

§10.25 *Requirements for the validity of contracts* Article 984 lists the following conditions which must be met for a contract to be valid:

Parties legally capable of consenting; Their consent legally given; Something which forms the object of the contract; A lawful cause or consideration.

§10.26 *The legal capacity of the parties* Legal capacity is the capacity of a person to have rights and exercise them. In other words, it involves the capacity to enjoy rights and the ability to exercise them. The capacity to enjoy rights means the aptitude to acquire them, whereas the capacity to exercise them means the competence to do so without the help of a third party. In general, all persons have the capacity to enjoy rights, except for those excluded by law. For example, judicial officers are excluded for they lack the capacity to acquire litigation rights. As to the capacity to exercise rights, article 985 of the *Civil Code* stipulates that all persons are capable of contracting, except those whose incapacity is expressly declared by law. Article 986 lists persons incapable of contracting, the most important of these exclusions being minors and interdicted persons. A minor is any person who is not yet 18 years old. Since, by law, he is considered incapable of performing a legal act, a minor must be represented by a tutor who will exercise his rights on his behalf. A minor over the age of 14 may, however, bring his own actions to recover his wages and a minor engaged in a trade is reputed to be of full age in all matters relating to such trade, as specified in articles 304 and 323 of the *Civil Code*. An interdicted person is a person of full age incapable of distinguishing between right or wrong because of a state of dementia, drunkenness, narcomania or prodigality. To look after the interests of an interdicted person or his possessions, the law provides for the naming of a curator for the person or his possessions, as the case may require. Article 987 adds that the incapacity of minors and interdicted persons is established in their favour which means that only the incapable person or his representative can invoke his incapacity and that a person of full age may not invoke incapacity in his favour in order to have a contract annulled.

§10.27 *Consent* Consent, that is, the expression of the will of the parties, is the most important condition for concluding a contract. It can manifest itself expressly, in the case of a written or verbal contract, or tacitly, if it can be inferred from the conduct of the parties. Consent requires that an offer be made and accepted. The conclusion of a contract is not always instantaneous. An offer may include a term for the acceptance, in which case the person making the offer cannot, in principle, withdraw his offer before the term has lapsed and he is bound to the offer if it is accepted during that time. However, if no term has been fixed for accepting the offer, the person making the offer has generally the power to revoke it, provided he does not abuse that power.

§10.28 *Irregularities of consent* It is important that consent be free and informed. To that effect, article 991 of the *Civil Code* provides for the nullity of a contract if consent was given under a number of irregular circumstances, these being error, fraud, violence, or fear and lesion.

§10.29 *Error* Certain kinds of error cause the absolute nullity of a contract. Such is the case if the error occurs in the nature of the contract itself. This kind of error is so fundamental that it is an obstacle to the actual concluding of the contract. This would be the case, for example, if one of the parties believed that he had agreed to a sale on credit and the other, a sale with deferred payment terms. In that case, no real agreement has been made and the contract has not been concluded. There can also be a mistaken identity concerning the object of the contract. That would be the case if one of the parties believed he was buying a certain property when he was in fact being sold another. Since the consent of the parties has not been given for the same object, there cannot have been any contract. There can also be errors concerning the cause of the contract such as if someone recognizes a debt he has not incurred. There are also some causes of error which, although they do not affect the existence as such of consent, make the contract annullable on the demand of the person who made the error. These are causes of error either in terms of the substance of the thing which is the object of the contract or as to the principal consideration for the contract.

§10.30 *Fraud* Article 993 deals with fraud, that is, any deceit or fraudulent manoeuvre practised by one of the parties to induce the other to agree to a contract. An example would be if a seller of a business showed the buyer false financial statements that indicated gains where there were none. To be considered fraud, however, the artifices practised by one party must be such that, without them, the other party would not have consented to the contract.

§10.31 *Violence and fear* In the case of error or fraud, a consenting party was not informed of all circumstances pertaining to the contract, whereas in the case of violence or fear, as discussed in articles 994 to 1000 of the *Civil Code*, consent is not given freely. It is considered violence if a person has agreed to a contract out of fear of a serious physical or moral danger either to himself or those dear to him. As indicated in article 995, in declaring fear to be a cause of nullity the circumstances, sex and condition of the party are to be taken into consideration.

§10.32 *Lesion* Lesion is another cause of nullity. It consists in prejudice being suffered by one of the parties through a disproportion of the benefits provided. Article 1001 of the *Civil Code* stipulates that lesion can be invoked only by minors or interdicted persons. For example, a minor can cancel an automobile sales contract if the agreed price exceeds by far the value of the automobile as well as his ability to pay for it out of his own financial resources. Article 1012 explains, however, that "persons of the age of majority are not entitled to relief from their contracts for cause of lesion only". Since the *Civil Code* came into force, legislators have decreed exceptions in principle to article 1012. Article 1040c, for instance, permits a person of the age of majority to ask a court to reduce or annul excessive and harsh monetary obligations resulting from a loan of money. The concept of lesion between persons of the age of majority is also admitted by the *Consumer Protection Act.*

§10.33 *Object and cause of contracts* The first two requisites to the validity of a contract concern the capacity of the parties to consent to a contract with each other, while the object and cause of the contract concern the legal mechanisms of same. The object of a contract can be defined as that which the debtor promises to perform. It is, thus, not the object of the contract as such, but rather the object of the obligation arising from the contract. The object of the obligation must be possible, determinate, licit and commercially available. The cause of the contract must likewise be licit, article 989 of the *Civil Code* stipulating that "a contract without a consideration, or with an unlawful consideration has no effect."

§10.34 *The interpretation of contracts* Once a contract has been concluded, it becomes law between the parties. But it can happen that a contract is not clearly worded and that it is necessary to interpret it. Articles 1013 to 1022 provide some rules of interpretation, the purpose of which being to discover the true intention of the parties to the contract and, as set out in articles 1013 and 1020, to do so by interpretation rather than by adherence to the literal sense of the words. The contract must also be interpreted in such a way that it produces the effects set out in articles 1014 and 1015. According to articles 1016 and 1017, it is to be presumed that the parties intended to follow common usage. Furthermore, a contract must be seen as a whole and article 1018 specifies that each clause must be interpreted in relation to all the others, providing each with the meaning that results from the act as a whole. Article 1019 gives the debtor the benefit of the doubt and article 1021 stipulates that a general obligation is not restricted by a single and particular stipulation.

Civil Liability

§10.35 *Introduction* One area of particular interest to the general public is that of civil liability. People are becoming increasingly aware of their rights and responsibilities. This is good news for the insurance business, as land surveyors can testify, for surveyors are required to carry professional liability insurance, the cost of which keeps climbing from year to year.

Since civil liability is the basis for professional liability, a few remarks on elementary principles would seem to be in order here.

§10.36 *General principles* In general terms, responsibility can be considered on three levels — moral, penal and civil. In the case of moral responsibility, a sanction can only be personal and subjective. Penal responsibility, on the other hand, entails a penalty inflicted according to the wording of a law and only prohibited acts constitute infringements. Moreover, it must be proven beyond any reasonable doubt that a criminal infringement has taken place. Not so in the case of civil responsiblity, which is first and foremost a private matter. Action is not brought by a public ministry as in a penal or criminal matter, but by the victim of the prejudice.

§10.37 *Contractual and delictual responsibility* Civil responsibility is either contractual or delictual. It is contractual if it follows from the non-performance of a contractual obligation, and delictual if it arises from a fault of another kind. These two modes of civil liability are based on the same principles and differ only in certain details, such as the terms of prescription. A final distinction concerns the offences and quasi-offences mentionend in the *Civil Code*. An offence is an illicit and voluntary act committed with the intention of causing another person damage, while a quasi-offence is involuntary and done without any intention to cause damage. For example, a defamatory document could constitute an offence, a hunting accident, a quasi-offence. One of the practical consequences of this distinction is that one cannot insure oneself against intentional faults, that is to say against offences. Insurance, for example car insurance, only covers quasi-offences.

§10.38 *Essential elements of civil liability* The general theory of civil liability is outlined in article 1053 of the *Civil Code*. This article imposes on all an obligation not to harm others. It goes without saying that this article has been abundantly interpreted in court, thereby assuring its adaptation to new situations. Civil liability presupposes three essential elements: a fault, a damage and a causal link between the fault and the damage.

§10.39 *Fault* Fault is the first essential element. It is not sufficient that damage has been caused for it is necessary that the author of the damage has committed a fault. Moreover, the person who is liable for the fault must be capable of distinguishing between right and wrong. Incidentally, there is no connection between this and the capacity to conclude contracts for, as indicated in article 1007, minors are as equally responsible for their offences and quasi-offences as other people. While it is true that children who have not yet attained the age of reason are not considered responsible for the damage they cause, after that it becomes a question of fact which is left to the discretion of the judge. Persons who do not have the free use of their reason and will are not considered to be responsible either. Obviously this exception cannot be applied if loss of reason is caused by the consumption of inebriating beverages or the use of narcotics.

Fault is essentially a moral concept. It arises from the violation of the duty of everyone to do no wrong to one's neighbor. Article 1053 lists several ways of committing a fault, including imprudence, negligence and want of skill. Thus, it is taken for granted in the *Code* that everyone must act reasonably. Clearly, fault is a question of fact that must be determined by the court. To that end, case law has established as a model of conduct the prudent head of a family. Fault, thus, stems from the absence of that prudence which characterizes a good father.

§10.40 *Damage* Damage constitutes the second essential element of civil liability. Whereas in penal law responsibility stems from the simple infringement of the law, in civil law the fault must have caused damage. The principle of civil liability is that the author of a fault is obliged to redress all the prejudicial consequences resulting directly from that fault. This redress must cover every prejudice suffered and proven by the victim. Redress is limited, however, to immediate, personal and direct damage.

§10.41 *The causal link* The causal link constitutes the third essential element of reponsibility. It is not enough that a fault results in damage, for there must also be a direct and immediate connection between the fault committed and the damage suffered. In other words, there must be a cause-and-effect link between the prejudicial act and the prejudice suffered by the victim. It was due to the lack of a causal link between the error committed and the damages claimed that the Supreme Court of Canada quashed the decision of the Québec Court of Appeal, which held the land surveyor professionally responsible in the case of *Parrot v. Thompson*.[8]

Matrimonial Regimes

§10.42 *Introduction* Since the capacity of a person to dispose of his goods may depend, among other things, on the kind of marriage contract he has concluded, it is important to have a general understanding of the different matrimonial regimes that are in effect in Québec.

§10.43 *Definition* A matrimonial regime can be defined as the settling of the legal relations between spouses. There are two kinds of matrimonial regimes. The legal regime, as set out in article 464 of the *Civil Code*,

---

8 [1984] 1 S.C.R. 57.

applies to spouses who have not established their matrimonial regime by contract. The conventional regime can be adopted by the spouses in a marriage contract, as authorized by article 463. In accordance with article 472, "marriage contracts must be established by a notarial deed *en minute.*" The articles of the *Civil Code* mentioned here are not part of the *Civil Code of Lower Canada,* but of the *Civil Code of Québec.* In fact, since the adoption of Bill 89 in December 1980, family law is now part of Book Two of the new *Civil Code.*

§10.44 *The legal regime* As of July 1, 1970, the partnership of acquests system has become the legal regime of Québec, replacing that of community of property. However, the community of property regime still applies for all persons who were married before that date unless they change their regime, as is now permitted under article 470 of the *Civil Code of Québec.*

§10.45 *Partnership of acquests* Under the partnership-of-acquests regime, all the property of the spouses is divided into two categories, private property and acquests. As stipulated in article 480,

> The property that each spouse possesses when the regime comes into effect or that he subsequently acquires constitutes acquests or private property according to the rules that follow.

To summarize, the property which either spouse possesses at the time of marriage or receives by succession or donation is private property, whereas all other kinds of property are in principle acquests, particularly salaries and the revenue from all property. In this regard, article 491 specifies that

> [A]ll property is presumed to constitute an acquest, both between the spouses and with respect to third persons, unless it is established that it is private property.

During marriage, each spouse may administer on his own his private property and acquests, but may not, without the consent of the other spouse, give away substantial acquests, as stipulated in articles 493 and 494 of the *Civil Code.* In addition, article 499 stipulates that

> [E]ach spouse retains his private property after the regime is dissolved. He may accept or renounce the partition of his spouse's acquests, notwithstanding any agreements to the contrary.

§10.46 *Separation as to property* Most couples choose a separation as to property regime when concluding a marriage contract. This regime was particularly popular before 1970, since it allowed couples to avoid the legal regime of community of property, the principles of which reflected less and less the contemporary values of Québec society.

By virtue of article 518, "the regime of conventional separation as to property is established by a simple declaration to this effect in the marriage contract." This regime recognizes only two kinds of patrimony, these being the respective properties of the husband and wife. Article 519 of the *Civil Code* affirms that each spouse is entitled to administer, enjoy and freely dispose of all his moveable and immoveable property. To summarize, under this regime each spouse keeps his property and is responsible for his debts. This remains true even after the dissolution of the regime. However, under the terms of article 520, the property over which neither spouse is able to establish his exclusive right of ownership is presumed to be held by both, one half by each.

§10.47 *Community of property* Although this regime has not been retained in the new *Civil Code of Québec*, a couple may still choose it for their marriage contract. But for those married under that regime, the *Civil Code of Lower Canada* still applies and, indeed, has remained in effect for that purpose. The regime of community of property, called henceforth community of moveables and acquests, specifies four categories of property: the private property of the husband, the private property of the wife, the assets of the community and the reserved property of the wife. Under this regime, the larger part of the property belongs to the community, as indicated in article 1272. In summary, the common assets are composed of the moveable property which the spouses possess at the time that the marriage was solemnized and of all the property which they have acquired afterwards, except for certain legacies and gifts as specified in article 1276. As to the administration of the property of the community, article 1292 states that

> [T]he husband alone administers the property of the community subject to the provisions of article 1293 and articles 1425a and following. He cannot sell, alienate or hypothecate without the concurrence of his wife any immoveable property of the community but he can, without such concurrence, sell, alienate or pledge any moveable property other than a business or than household furniture in use by the family. The husband cannot, without the concurrence of his wife, dispose by gratuitous title inter vivos of the property of the community, except small sums of money and customary presents.

On the other hand, the major portion of private property consists of immoveables which the husband owned before the marriage and those which he acquired as replacements, immoveables acquired by succession or as gifts with a clause of exclusion from the community and indemnities, and rights to damages pertaining to either of the spouses. Each of the spouses has the free disposal of his private property, both moveables and immoveables.

As to the reserved property of the wife, article 1425a contains the following provisions,

> The proceeds of the personal work of the wife common as to property, the savings therefrom, and the moveable or immoveable property she acquires by investing them, are, on pain of the nullity of any covenant to the contrary, reserved to the administration of the wife, and she has the enjoyment and free disposal of them. The wife cannot, however, alienate them by gratuitous title, nor alienate or hypothecate the immoveables, nor alienate or pledge the stocks in trade and the household furniture in use by the family without the concurrence of her husband. She may appear before the courts without authorization in any action or contestation relating to her reserved property. Such reserved property shall not include the earnings from the joint work of the consorts.

When this regime is dissolved, the private property remains the exclusive property or either spouse, but the reserved property of the wife is included in the property to be divided, except if she renounces her share of the community. In this respect, only the wife has the choice of renouncing her share in the community. The husband cannot do this, since he was responsible for its administration.

Introduction to Property Law

§10.48 *Distinction between things and property* In addition to personal rights, the estate of a person includes his real rights, and these belong in the domain of property rights. "Property" must not be confused with "things". Things are necessarily material, whereas property may be corporeal or incorporeal, corporeal property being things which may be perceived by the bodily senses and incorporeal property being that which exists only in contemplation. Thus, the category of corporeal property applies to physical things, such as a building, while incorporeal property includes real rights such as usufruct, use and superficiary rights.

§10.49 *Distinction between movables and immovables* Article 374 of the *Civil Code* distinguishes further between kinds of property by stipulating that "all property, incorporeal as well as corporeal, is movable or immovable." Movable property consists of property which can be moved or transported or displaced, while immovable property consists of kinds of property that cannot be transported and cannot move on their own. It is, thus, evident that an automobile is moveable, a house, immovable property. But this distinction is not always so straightforward, as in the case of a trailer which, depending on the circumstances, may be considered either movable or immovable.

§10.50 *Kinds of immovables* The *Civil Code* distinguishes four classes of immovables: immovables by their nature, by their destination, by reason of the object to which they are attached and by determination of law. Immovables by nature comprise land and the things attached thereto. Articles 376 to 378 identify four kinds: land, buildings, windmills or water mills and rooted plants and trees. The word "building" must be interpreted with considerable latitude. Court decisions have determined that installations such as aquaducts and power or natural gas networks are immovables by nature. Crops and fruits, however, become movable as soon as they are detached from the ground or plucked from trees. The same is true for trees which have been felled and buildings once they have been demolished.

Immovables by destination are movables that have been declared as such under the provisions of articles 379 and 380 of the *Civil Code*. A movable thing becomes immovable if it is incorporated into a thing which is immovable by its nature or if its proprietor has placed it permanently on his real property. This is the case with machines and even vehicles that are used in the operation of a factory. The latter remain immovable by destination as long as they continue to serve that purpose. Immovables by nature and immovables by destination both belong to the category of corporeal property, while immovables by reason of the object to which they are attached belong to the category of incorporeal property. Article 381 classifies real rights and real actions as immovables. The final category, immovables by determination of law, defined in article 382, is encountered only rarely.

§**10.51** *Kinds of movables* Article 383 of the *Civil Code* assigns movable property to two categories, that which is movable by nature and that which is movable by determination of law. All movable bodies described in articles 379 and 380 are movable by nature, while incorporeal property, as defined in articles 387 and 388, is movable by determination of law.

§**10.52** *Implications of the distinction between movables and immovables* Although the distinction between movable and immovable property may seem somewhat theoretical, it has some practical consequences. Movable property can be alienated more easily than immovable property. In the case of the latter, it is normally required that a deed be executed by a notary and registered. As indicated in article 776 of the *Civil Code*, this is required in the case of gifts of immovable property, while gifts of movable property become valid upon delivery. The lapse of time required for prescription is longer for immovable property. Moreover, only immovable property can be hypothecated. Finally, the two categories of property are subject to different rules of taxation.

Property Law

§10.53 *Introduction* Property law is a reflection of the social and economic life of a people. In Québec, as in most Western societies, citizens have the right to acquire and own movable and immovable property. Property law has evolved throughout history, originating when mankind gave up its nomadic existence and began to cultivate the soil. Roman law already recognized the right to private property and gave it the attributes it still has today. In Québec, the abolition of seigneurial tenure in 1854 brought with it the right to own property, as understood under Roman law. The same development was made possible in France by the liberation of landed property as a result of the French Revolution. Thus, at the time our civil law was being codified, the right to own property was seen as a guarantee of freedom. The historical context helps to explain the unequivocal terms in which this right is defined in article 406 of the *Civil Code*, "[o]wnership is the right of enjoying and of disposing of things in the most absolute manner, provided that no use be made of them which is prohibited by law or by regulations."

§10.54 *Attributes of ownership* Ownership is the most absolute of the real rights. It confers complete power over a thing, that is, the right to use it, take all it produces and dispose of it as one pleases. Ownership also includes the right of accession, as defined in article 408, according to which ownership in a thing gives the right to all it produces and to all that is joined to it. Moreover, ownership of land includes ownership not only of the soil, but also of what is above and below it, as stated in article 414 of the *Civil Code*.

§10.55 *Traditional limitations of ownership rights* Although the right of ownership is thought of as individual, sovereign and exclusive, it is tempered by an ever-increasing number of legal and regulatory restrictions. These include such traditional restrictions as expropriation and legal servitudes. Article 407, in fact, authorizes expropriation for reasons of public utility and after payment of a just indemnity. Thus, the various levels of government have the right to expropriate property and force its proprietor to relinquish it. Legal servitudes, which will be studied further on, include the obligation to determine boundaries and co-ownership and concern the requirement of maintaining a certain distance between the windows and openings of a property and the boundary line of an adjacent property.

§10.56 *Modern limitations of ownership rights* There are even more restrictions of ownership in modern law. Zoning regulations govern the types of buildings allowed in residential, commercial and industrial zones. There are also new laws that oblige the proprietor to obtain authorization or permits of all sorts to be able to exercise his rights of ownership. An example is the *Act to Preserve Agricultural Lands*,[9] which requires authorization from the Commission de protection du territoire agricole du Québec (CPTQ) if land located in an agricultural zone is to be used for a non-agricultural purpose. Likewise, a proprietor may not make any changes to a building which is protected by the *Cultural Property Act*,[10] without authorization from the minister. Similar controls are exercised under the provisions of several other acts, particularly the *Loi sur la Régie du logement*[11] and the recent *Loi sur l'aménagement et l'urbanisme*.[12]

§10.57 *Abuse of rights and difficulties between neighbours* A person's right to ownership of a property carries with it the obligation not to harm his neighbour. Ownership rights end where those of a neighbour begin. A proprietor who causes someone else any prejudice over and above normal inconvenience is guilty of an abuse of rights. Where this occurs, the court will adjudge that ownership rights have been abused and find the proprietor guilty of annoying his neighbour, even if the proprietor was exercising his lawful rights. An example would be a factory that pollutes the neighbourhood and thus inconveniences the neighbours. In addition, the abuse of a right is often the result of malicious behaviour as, for instance, if a proprietor builds an excessively large and ugly fence for the sole purpose of annoying his neighbour. The abuse of rights is also subject to general rules of civil reponsibility.

Co-ownership

§10.58 *Kinds of co-ownership* Ownership of a property is generally considered to be an exclusive individual right. Co-ownership constitutes an exception to that traditional principle. The *Civil Code* recognizes three kinds of co-ownership: ordinary co-ownership, forced co-ownership, and the co-ownership of immoveable property established by a declaration.

§10.59 *Ordinary co-ownership* Ordinary co-ownership is also called undivided to stress that it applies to the whole material property. The share of each co-proprietor is, however, non-material and is expressed as a fractional part of the whole property. Thus, each of the co-proprietors is the exclusive proprietor of his share of the property and a collective proprietor of all the property. As stipulated in article 689 of the *Civil Code*, no one can be compelled to remain in undivided ownership and either of the co-proprietors may initiate a partition.

---

9 R.S.Q., c. P-41.1.
10 R.S.Q. 1977, c. B-4.
11 R.S.Q., c. R-8.1.
12 R.S.Q., c. A-19.1.

§10.60 *Forced co-ownership* Although essentially on a temporary basis, co-ownership can in certain cases be forced, in which case partition cannot be initiated. Co-ownership will be forced if a piece of real property is an indispensable accessory to two neighbouring family properties, as in the case of certain common structures. The same is true for the common parts of a building where co-ownership has been established by declaration.

§10.61 *Co-ownership of immovables by declaration* A law regulating the co-ownership of immovables was passed in 1969, and its provisions have been integrated into the *Civil Code* by articles 441b to 442p. The *Civil Code* permits the creation of a particular form of co-ownership by means of a declaration, the content and form of which are clearly set out. This declaration must specify the destination of the immoveable, give a detailed description of its exclusive and common portions in conformity with cadastral norms and establish the relative value of each fraction. It must then designate the powers of the co-proprietors' assembly and the procedures to be followed in their meetings and establish the rules for nominating and replacing administrators.

In drawing up a declaration of co-ownership, the correct form must be strictly followed. The declaration must be notarized, signed by all proprietors of the immoveable, accompanied by the written consent of all holders of privileges or hypothecs registered against the immovable and registered. These formalities must be observed before co-ownership becomes valid. The formality of registration, which ordinarily is required only for third parties, serves to establish the rights between the parties.

By virtue of a declaration of co-ownership, ownership of an immovable is divided among the owners into fractions, each of which comprises an exclusive portion and a share of the common portion. Each co-proprietor enjoys at one and the same time two inseparable rights, the right to his individual portion and the right to use the common portions. In terms of the *Civil Code*, this right of co-ownership is subject to all legal acts that apply to ownership of the property. Thus, the co-proprietor can dispose of his right as he pleases and even hypothecate it or transfer its usufruct. Finally, the declaration specifies that the management of the immovable property is to be assured by meetings of the co-proprietors, who are responsible for its direction, and by the administrators, who are responsible for day-to-day business.

Superficiary Rights

§10.62 Superficiary rights constitute another mode of ownership by which the real right of ownership to a structure may belong to somebody other than the proprietor of the soil. This right is not specifically set out in the

Civil Code, but its existence is implied by article 415 which recognizes that the presumption of ownership rights accessory to the land is rebuttable. Superficiary right thus constitutes a particular form of ownership which differs from co-ownership in that title is divided precisely between the surface and all that lies below it. A superficiary right is not to be mistaken for a kind of servitude, as it is subject to all property laws.

Superficiary ownership is seldom established by written contract. A study of relevant court cases reveals that this form of ownership generally implies that permission to build has been granted to a future proprietor of the surface by the proprietor of the property. Permission of this sort may be implied if the proprietor of the land has contributed to the construction of the building or even if he knew of the construction. In the absence of any agreement to the contrary, the courts have ruled that superficiary rights are extinguished if the structure concerned has been demolished.

Corollary Property Rights

§10.63 *Introduction* Co-ownership and superficiary rights are modes of exercising the right of ownership to a property. There are also corollary ownership rights which are specific real rights with certain of the attributes of ownership, that is, the right to make use of a thing, the right to its products and, to a certain degree, the right to dispose of it. These real rights, which are limited in number, are defined by the *Civil Code* which, without stating so expressly, considers them to be servitudes.

§10.64 *The concept of servitudes* A servitude is a real right over someone else's property. There are two kinds of servitudes — personal and real. A personal servitude is a temporary right, such as usufruct, use or habitation, which a person may have over someone else's property. A real servitude is a real right over an estate by another estate belonging to a different proprietor. Unlike personal servitudes, real servitudes are perpetual and persist in spite of a change of ownership.

Usufruct, Use, and Habitation

§10.65 *Definition of usufruct* Usufruct is a corollary of ownership. Article 443 defines it as "the right of enjoying things of which another has the ownership, as the proprietor himself, but subject to the obligation of preserving the substance thereof." The object of usufruct is thus divided between the usufructuary, who has the use and the fruits, and the mere proprietor, who has only the right to dispose of the property with due respect to the unsufruct. According to article 446, usufruct "may be established upon property of all kinds, moveable or immoveable."

§10.66 *Characteristics of usufruct* Usufruct is a temporary right, normally held for the duration of the usufructuary's life and terminating with his death. It may also be granted by contract for a limited period of time. A usufruct that is granted without term to a corporation only lasts 30 years, as stipulated in article 481 of the *Civil Code*. In practice, usufruct is encounterd mainly in wills as, for example, when a husband gives the usufruct of his property to his wife and mere ownership to his children.

§10.67 *Rights and obligations of the parties* The usufructuary has the right to make use of a thing and to enjoy its fruit. He is, however, under the obligation to maintain the thing in good condition. This involves keeping an inventory of it, administering it with care and carrying out routine repairs. The mere proprietor keeps the right to dispose of the thing, but may not do anything that could prevent the usufructuary from the full exercise of his rights. He also remains responsible for major repairs.

§10.68 *Distinction between the rights of a usufructuary and those of a tenant* The rights of a usufructuary are fundamentally distinct from those of a tenant, although both have the enjoyment of a thing of which another has the ownership. The usufructuary has a real right in the thing, whereas the tenant only has a claim against the landlord that forces the latter to carry out his obligation to procure him the enjoyment of the rented thing. However, the tenant may demand that the thing be delivered to him in good condition, whereas the usufructuary takes the thing as it is.

§10.69 *Use and habitation* Use, like usufruct, is the right to enjoy a thing belonging to someone else and to take its fruit, but this right is limited to the needs of the user and his family. When applied to a house, the right of use is called the right of habitation. Use and habitation are different from usufruct in that they may not be granted to a corporation or acquired by the operation of a law or by prescription.

Emphyteusis

§10.70 *Essential elements of emphyteusis* Emphyteusis is the real right most similar to ownership, except that it is not perpetual and does not comprise all aspects of the right to disposal. Article 567 of the *Civil Code* defines emphyteusis as

> [A] contract by which the proprietor of an immoveable conveys it for a time to another, the lessee subjecting himself to make improvements, to pay the lessor for annual rent, and to such other charges as may be agreed upon.

An emphyteutic lease must contain four elements: it must refer to an immoveable, stipulate a term which is longer than nine years but less than 99 years, and it must specify the improvements which the lessee is obliged to carry out and the annual rent he has to pay. Emphyteusis is different from ordinary leases in that it is not subject to tacit renewal, as stipulated in article 579.

§10.71 *Rights and obligations of the parties* By virtue of article 569 of the *Civil Code*, emphyteusis carries with it alienation and the lessee has the rights of a proprietor for the duration of the emphyteutic lease. Article 570 permits him to alienate or hypothecate the immoveable which, according to article 571, may be seized by the lessee's creditors and sold by sheriff's sale. These transactions clearly do not invalidate the emphyteutic lease and whoever takes the place of the lessee must, at the expiry of the lease, return the immoveable to the proprietor. Another characteristic of emphyteutic leases is the option of the lessee to abandon and thereby terminate the lease as stipulated in articles 579 and 580. The lessee can, however, waive that option in the emphyteutic contract. The lessor, in turn, is entitled to the annual rent and can force the lessee to carry out the improvements stipulated in the contract. He is, however, obliged to allow the lessee enjoyment of the thing and to take it back if the latter abandons it. In practice, emphyteusis is made use of mostly by the state and public or religious corporations, in order to retain ownership without having to worry about the management of their immoveables.

Introduction to the Study of Real Servitudes

§10.72 Usufruct, use and habitation and, in a sense, emphyteusis constitute personal servitudes since they are connected to an immoveable for the profit of an individual. A servitude may also be attached to an immoveable in relation to another immoveable that belongs to a different proprietor. This kind of servitude, called a real servitude, can be defined as a legal bond between two immoveables, the servient estate that owes the service and the dominant estate that is the beneficiary. Article 499 of the *Civil Code* defines a real servitude as a charge imposed on one real estate for the benefit of another belonging to a different proprietor.

Kinds of Servitudes

§10.73 Article 500 distinguishes between three classes of servitudes, natural, legal and conventional. Only the last of these constitute servitudes in the full sense of the term, natural and legal servitudes being more correctly described as inherent restrictions on ownership. In fact, neighbours can create a true servitude by laying aside one of these restrictions.

An example of this is the servitude of view, by which a proprietor may be permitted to have a view of the neighbouring property at a distance less than that prescribed by the *Civil Code*. Even if natural and legal servitudes do not constitute true servitudes, they are considered as such by the *Civil Code*.

Natural Servitudes

§10.74 Natural servitudes which arise from the geographic situation of a given property are dealt with in articles 501 to 505 of the *Civil Code* and apply to waters and their sources and the obligation of proprietors to settle their boundaries and make fences or other separations between their respective properties. Article 501 concerns lands situated on different levels and stipulates that the land on the lower level is subject to receiving all the waters that flow naturally from that on the higher level. The proprietor of the land on the higher level, however, may not do anything to aggravate the servitude, for instance, by polluting the water or preventing its flow. Article 502 stipulates that he who has a spring on his land may use it and dispose of it as he pleases. Similarly, article 503 allows the proprietor of land bordering on a private watercourse to make use of it for the utility of his land, on condition that this use does not prevent the exercise of the same right by the other proprietors of land by the river, saving any statutory provisions to the contrary. The settling of boundaries mentioned in article 504 will be discussed in a separate part of this chapter. Articles 505 and 520 concern the obligation of proprietors to make fences or other separations between contiguous lands. Under the provisions of article 505,

> [E]very proprietor may oblige his neighbour to make in equal portions or at common expense, between their respective lands, a fence, or other sufficient kind of separation according to custom, the regulations and the situation of the locality.

However, courts have generally ruled that whoever makes a fence at his own cost may not, after the fact, ask to be reimbursed by his neighbour. Article 520 is hardly ever applied because its provisions have not been adapted to our ways of separating immoveables.

Legal Servitudes

§10.75 According to article 506 of the *Civil Code*, "servitudes established by law have for their object public utility or that of individuals." Servitudes of public interest are governed by particular laws or regulations, as indicated in article 507. Servitudes of private interest concern party walls, required distances which must be maintained when planting or erecting certain structures, the view of one property from another, the eaves of roofs and land enclosed on all sides.

Common Areas or Structures

§10.76 *Introduction* Common property is a particular form of forced co-ownership that applies to walls, hedges and ditches separating two adjoining properties. Firstly, a wall is considered to be a party wall depending on its location in terms of the separating line between two properties. Thus, a wall can be private or common. It is private if it belongs exclusively to one proprietor and common if it belongs at the same time to two proprietors of adjoining properties. According to article 415 of the *Civil Code*, a wall that is situated back from the separating line is necessarily presumed to belong to the proprietor of the land on which it has been constructed. The same applies to a wall that touches the separating line, unless common ownership of the wall has been acquired under the terms of article 518. On the other hand, article 510 stipulates that a wall which straddles the separating line shall be presumed to be a party wall.

§10.77 *Acquisition of common ownership* Common ownership can be acquired in several ways. Firstly, it can be established at the time that the wall is constructed if the two neighbours so agree. Common ownership of a private wall along the separation line may also be acquired subsequently through an agreement between two neighbours. In the latter case, it may even result from the unilateral action of one of the neighbours, under the provisions of article 518 of the *Civil Code*. This, then, becomes an expropriation for the sake of private utility. The neighbour who exercises this option must in this case reimburse the proprietor for half the value of the wall and half the value of the ground on which the wall was built. This article obviously does not apply to a wall situated at a distance from the line of division between the two properties. However, since article 520 stipulates that a neighbour may be forced to furnish nine inches of ground for the building of a party wall, one can infer that, for article 518 not to apply, the wall must be situated more than nine inches away from the line of separation of the two properties.

§10.78 *Presumption of common ownership* Common ownership may also be established by the presumption of same under the provisions of article 510, which reads as follows,

> Both in town and country, walls serving for separation between buildings up to the required heights, or between yards and gardens, and also between enclosed fields, are presumed to be common, if there be no title, mark or other legal proof to the contrary.

This presumption of common ownership can be rebutted by titles or prescription. It can also be overturned by a presumption to the contrary resulting from certain marks mentioned in article 511 of the *Civil Code*, which reads,

> It is a mark that a wall is not common when its summit is straight and plumb with the facing on one side, and on the other side exhibits an inclined plane; and also when one side only has coping, or mouldings, or corbels of stone, placed there in building the wall. In such cases the wall is deemed to belong exclusively to the proprietor on whose side are the eaves or the corbels and mouldings.

Articles 523 to 527 deal with the common ownership of ditches and hedges. The remaining articles concern the rights and charges of common ownership in the case of a common wall.

Spacing Requirements for Plantations and Structures

§10.79 Articles 528 to 531 of the *Civil Code* specify the distances which must be maintained from the separating line when planting. Article 528 reads as follows:

> No neighbour can plant trees or shrubs or allow any to grow nearer to the line of separation than the distance prescribed by special regulations, or by established and recognized usage; and in default of such regulations and usage, such distance must be determined according to the nature of the trees and their situation, so as not to injure the neighbour.

This is supplemented by article 529

> Either neighbour may require that any trees and hedges which contravene the preceding article be uprooted. He over whose property the branches of his neighbour's trees extend, although the trees are growing at the prescribed distance, may compel his neighbour to cut such branches. If the roots extend upon his property, he has the right to cut them himself.

It follows from this that a proprietor may not cut the branches of his neighbour's trees without becoming liable for damages. The provisions of article 532 concerning the distances required between certain structures and the separating line have been replaced by municipal bylaws and certain specific laws.

The View Onto the Property of a Neighbour

§10.80 Articles 533 to 538 of the *Civil Code* regulate the right of a proprietor to make openings which provide a view of the neighbouring property, by establishing the minimum distance that must be kept between such openings and the separating line in order to preserve privacy. However, the restrictions imposed by these articles do not constitute the servitude of view. On the contrary, the servitude is an agreement to waive these restrictions.

§10.81 The rules concerning the right to a view vary according to the position of the wall in relation to the separating line between the two properties. Article 533 stipulates that if the wall is a party wall, neither of the neighbours may, without the consent of the other, make any opening or window in it. Article 534 states that lights may be placed in a wall that is not a party wall which adjoins the land of another, but no view of the neighbouring property is permitted. The distinction between lights and views is important. Lights are apertures with gratings and fixed glass. They cannot be opened and only serve to let light through, not to allow a view or let in fresh air. It is this kind of aperture which is regulated by articles 534 and 535. The views discussed in the following articles are apertures that can allow the passage of air and also provide a view of the neighbour's land.

§10.82 In the case of a wall situated at a distance from the line of separation, articles 536 and 537 stipulate that any windows, balconies or the like which permit a direct view onto the neighbour's land must be at least six feet from the line of separation and any oblique view must be at least two feet from said line. According to the law books, these distances are to be measured using the French system. A view is direct if it is possible to look into the neighbour's house from the axis of the window without turning one's head. If it is necessary to move, the view is deemed to be oblique. Article 538 stipulates the correct procedure for measuring these distances. Such practical considerations for the land surveyor are studied further on.

Right of Way

§10.83 The right of way in reference to land enclosed on all sides is also considered to be a servitude established by law, as outlined in articles 540 to 544 of the Civil Code. The basic premise, as stated in article 540, is that

> [A] proprietor whose land is enclosed on all sides by that of others, and who has no communication with the public road, may claim a way upon that of his neighbours for the use of his property, subject to an indemnity proportionate to the damage he may cause.

Although this article considers a property to be enclosed on all sides if there is no access to a public road, the courts have adopted a wider interpretation to include cases where access to a public road is inadequate or too cumbersome due to the type of terrain. Thus, once the state of enclosure is established, this servitude becomes a full right in favour of the property thus enclosed and a charge on the property through which the right of way passes. However, the proprietor of the neighbouring land may not be responsible for the enclosure, as specified in article 543,

> If the land become so enclosed in consequence of a sale, of a partition, or of a will, it is the vendor, the copartitioner, or the heir, and not the proprietor of the land which offers the shortest crossing, who is bound to furnish the way, which is in such case due, without indemnity.

§10.84 According to article 541, the property encumbered by the servitude is generally the one that offers the shortest crossing from the enclosed property to a public road. It may happen, however, that the right of way must lead over a different property, if the shortest way would seriously inconvenience the proprietor of the enclosed land or the proprietor of the land through which the access would lead. Once the property to be crossed has been selected, the right of way is established along a path that causes the least damage to that property, as specified in article 542. And, in accordance with article 544,

> If the way thus granted ceases to be necessary, it may be suppressed, and in such case the indemnity paid is restored, or the annuity agreed upon ceases for the future.

Conventional Servitudes

§10.85 *Introduction* The servitudes established by the act of man, also called conventional servitudes, are governed by articles 545 to 566 of the *Civil Code*. A conventional servitude is constituted by an agreement between the proprietors of two properties to the effect that one of them owes a service, which might even be a negative one, such as the obligation to abstain from building above a certain agreed height. The first paragraph of article 545 establishes that,

> Every proprietor having the use of his rights, and being competent to dispose of his immoveables, may establish over or in favor of such immoveables, such servitudes as he may think proper, provided they are in no way contrary to public order.

§10.86 *Kinds of conventional servitudes* The *Civil Code* distinguishes three categories of conventional servitudes: urban or rural, continuous or discontinuous and apparent or unapparent. The distinction between urban and rural servitudes, as defined in article 546, is of no great practical interest.

§10.87 *Continuous and discontinuous servitudes* Continuous or discontinuous servitudes are defined in article 547:

Continuous servitudes are those the exercise of which may be continued without the actual intervention of man; such are water conduits, drains, rights of view and others similar. Discontinuous servitudes are those which require the actual intervention of man for their exercise; such are the rights of way, of drawing water, of pasture and others similar.

By virtue of this distinction, a right of view is considered to be a continuous servitude because it is exercised without the actual intervention of man, while a right of way is a discontinuous servitude because it is exercised only when the proprietor of property to which it is due actually crosses over the land that is subject to the servitude. This distinction is important in the case of the extinction of a servitude, as can be seen in article 563:

> The thirty years commence to run for discontinuous servitudes from the day on which they cease to be used, and for continuous servitudes from the day on which any act is done preventing their exercise.

§10.88 *Apparent and unapparent servitudes* According to article 548,

> Servitudes are apparent or unapparent. Apparent servitudes are those which are manifest by external signs, such as a door, a window, an aqueduct, a sewer or drain, and the like. Unapparent servitudes are those which have no external sign, as for instance, the prohibition to build on a land or to build above a certain fixed height.

Clearly, rights of view and rights of way are apparent servitudes whereas an easement for an aqueduct with buried tiles might be considered an unapparent servitude. This distinction occurs again in the context of registration in article 2116a of the *Civil Code*, which stipulates that,

> [I]n default of registration, no real, discontinuous and unapparent servitude, constituted by title has any effect as regards third parties who become subsequent proprietors or creditors, whose rights have been registered.

Article 2116b adds that

> [N]o real servitude, constituted by title, and established on or after the first of January, 1917, can have any effect as regards third parties or subsequent creditors whose rights have been registered, unless such servitude has been registered.

In other words, prior to 1917, only contractual, discontinuous and unapparent servitudes had to be registered, whereas since then all contractual servitudes must be registered in order to be claimed against a third party.

§10.89 *Modes of establishing servitudes* A servitude is established by title, that is, by contract, will or the destination made by the proprietor. As stipulated in article 549, "no servitude can be established without a title; possession even immemorial is insufficient for that purpose." Thus, a servitude cannot be acquired by prescription, but it extinguishes as a result of non-use in the course of 30 years as stipulated in article 562. Should the original title of the servitude be lost, it may be replaced, in accordance with article 550, by an act of recognition proceeding from the proprietor of the land subject thereto. A destination made by the proprietor may also constitute a servitude. Article 551 stipulates,

> As regards servitudes the destination made by the prorietor is equivalent to a title, but only when it is in writing, and the nature, the extent and the situation of the servitude are specified.

A destination made by the proprietor exists when the proprietor of two adjoining properties makes a change that would create a servitude if the properties belonged to two different people. Such would be the case if a proprietor erects on one of his properties a building which does not conform to the provisions of the *Civil Code* regarding views. A servitude cannot exist in this case because the two properties belong to the same person, but the sale of one of the properties can result in a right of view being established if the destination made by the proprietor is in writing. A servitude by destination on the part of the proprietor, since it is not a contractual servitude, is not subject to the provisions of articles 2116a and 2116b of the *Civil Code* and thus, it would seem, does not require registration in order to be valid with regard to a third party.

§10.90 *Use and extent of servitudes* The second paragraph of article 545 stipulates that

> [T]he use and the extent of these servitudes are determined according to the title which constitutes them, or according to the following rules if the title be silent.

In that respect, article 552 states that

> [H]e who establishes a servitude is presumed to grant all that is necessary for its exercise. Thus the right of drawing water from the well of another carries with it the right of way.

But article 558 stipulates that

> [O]n his part, he who has a right of servitude can only make use of it according to his title, without being able to make, either in the land which owes the servitude, or in that to which it is due, any change which aggravates the condition of the former.

Article 553 adds that "he to whom a servitude is due has the right of making all the works necessary for its exercise and its preservation." The person to whom the servitude is due remains nevertheless responsible for any damage that his works may cause to the property owing the servitude. Moreover, according to article 554, "these works are made at his cost and not at that of the proprietor of the servient land, unless the title constituting the servitude establishes the contrary." Finally, article 555 permits the proprietor of the servient land,

> [C]harged by the title with making the necessary works, for the exercise and the preservation of the servitude, [to] free himself from the charge by abandoning the servient immoveable, to the proprietor of the land to which the servitude is due.

§10.91 *Rights and obligations of the proprietor of the servient land* Article 557 sets out the rights and obligations of the proprietor of the servient land. Under the provisions of this article,

> [T]he proprietor of the servient land can do nothing which tends to diminish the use of the servitude or to render its exercise more inconvenient. Thus he cannot change the condition of the premises, nor transfer the exercise of the right to a place different from that on which it was originally assigned. However if by keeping to the place originally assigned, the servitude should become more onerous to the proprietor of the servient land, or if such proprietor be prevented therby from making advantageous improvements, he may offer to the proprietor of the land to which it is due another place as convenient for the exercise of his rights, and the latter cannot refuse it.

Articles 559 to 566 concern the extinction of servitudes established by the act of man.

Privileges and Hypothecs

§10.92 *Introduction* In addition to the principal real rights, such as the right of ownership and its corollaries, there are accessory real rights, such as privileges and hypothecs. In contrast to principal real rights, accessory real rights do not confer on the bearer any of the attributes of ownership. Rather, immoveables affected by these rights serve as guarantees for loans.

§10.93 *Preliminary provisions* The provisions of the *Civil Code* concerning privileges and hypothecs are contained in articles 1980 to 2081a. The first three articles set out the preliminary principles. The first paragraph of article 1980 affirms that

> [W]hoever incurs a personal obligation, renders liable for its fulfilment all his property, moveable and immoveable, present and future, except such property as is specially declared to be exempt from seizure.

Article 1981 rules that

> [T]he property of a debtor is the common pledge of his creditors, and where they claim together they share its price rateably, unless there are amongst them legal causes of preference.

And article 1982 states that "the legal causes of preference are privileges and hypothecs."

§**10.94** *Privilege defined* Article 1983 of the *Civil Code* defines a privilege as

> [A] right which a creditor has of being preferred to other creditors according to the origin of his claim. It results from the law and is indivisible of its nature.

Thus, a privilege is a provision in law which favors a creditor by permitting him to be paid before the other creditors of the same debtor. Thus, in the event of a sheriff's sale, the privileged creditors are paid before any ordinary creditors. The order of preference among privileged creditors depends on whether the property sold is moveable or immoveable.

§**10.95** *Privileges upon movables* Article 1994 establishes the order of preference on movable property, as follows:

> 1. Law costs and all expenses incurred in the interest of the mass of the creditors;
> 2. Tithes;
> 3. The claims of the vendor;
> 4. The claims of creditors who have a right of pledge or of retention;
> 5. Funeral expenses;
> 6. The expenses of last illness;
> 7. Municipal taxes;
> 8. The claim of the lessor and the claim of the owner of a thing lent, leased, pledged or stolen;
> 9. Servants' wages and those of employees of railway companies engaged in manual labor, and sums due for supplies of provisions;
> 10. The claims of the Crown for its moneys.

The last paragraph of article 1994 specifies that the privileges listed under numbers 5, 6, 7, 9 and 10 apply to all the movables of the debtor and that the others are special and affect only certain objects. For example, the privilege of the unpaid vendor applies only to the property sold, and that of the landlord, only to the furniture in the tenant's apartment.

§**10.96** *Privileges upon immovables* Article 2009 lists the privileges upon immovables in their order of preference, as follows:

> 1. Law costs and the expenses incurred for the common interest of the creditors;
> 2. Funeral expenses, when the proceeds of the moveable property have proved insufficient to pay them;
> 3. The expenses of the last illness, when the proceeds of the moveable property have proved insufficient to pay them;
> 4. The expenses of tilling and sowing;
> 5. Assessments and rates, such as the assessments pursuant to the *Loi des fabriques* (Factory Act);
> 6. Seigniorial dues;

7. The claim of the workman, supplier of materials, builder and architect, provided they fulfil the formalities required by law;
8. The claims of the vendor upon the immoveable that he has sold;
9. Servants wages, when the proceeds of the moveable property have proved insufficient to pay them.

Under the provisions of article 2084, the privileges under numbers 1, 4, 5 and 9 are exempt from the requirement of registration. The other privileges must be registered upon the immovable to which they apply in order to be effective.

§10.97 *Definition and characteristics of hypothecs* Article 2016 of the *Civil Code* defines a hypothec as

> [A] real right upon immoveables made liable for the fulfilment of an obligation, in virtue of which the creditor may cause them to be sold in the hands of whomsoever they may be, and have a preference upon the proceeds of the sale in order of date as fixed by this code.

A hypothec, thus, carries with it a right of pursuit and a right of preference. The right of pursuit permits the creditor holding the hypothec to have the immoveable sold by sherrif's sale, even if it no longer is owned by the same person, and the right of preference permits him to be paid out of the proceeds of the sale in preference over all other creditors. Article 2017 of the Civil Code outlines the main characteristics of a hypothec, as follows,

> Hypothec is indivisible and subsists in entirety upon all the immoveables made liable, upon each of them and upon every portion thereof. Hypothec extends over all subsequent improvements or increase by alluvion of the property hypothecated. It secures besides the principal, whatever interest accrues therefrom, under the restrictions stated in the title Of Registration of Real Rights and all costs incurred. It is merely an accessory and subsists no longer than the claim or obligation which it secures.

§10.98 *Kinds of hypothecs* Article 2018 stipulates that "hypothec can take place only in the cases and according to the formalities authorized by law." Article 2019 distinguishes between three kinds of hypothecs, these being legal, judicial, and conventional.

§10.99 *Legal hypothecs* A legal hypothec is one which results from the sole operation of the law. For example, under the provisions of article 2030, a minor has a hypothec upon the immovables of his tutor for the property which the latter has to hand over to him at the end of his tutorship.

§10.100 *Judicial hypothecs* A judicial hypothec is one which the law attaches to any judgment ordering a debtor to pay a fixed sum of money. For example, a judgment granting alimony may be registered upon the immovables of the debtor in order to guarantee payment. Nevertheless, the debtor can ask the court to determine the immovable to which the judicial hypothec can apply.

§10.101 *Conventional hypothecs* A conventional hypothec is one which results from an agreement between the parties, generally a lender and a borrower. As stipulated in article 2037, a conventional hypothec can only be granted by a person capable of alienating his immoveables. Moreover, the hypothec must be granted by a notarized act, as stipulated in article 2040 of the *Civil Code*. Articles 2042 and 2044 add that a hypothec is valid only if the act designates specifically the hypothecated immoveable and mentions the sum for which it is granted. In addition, the last paragraph of article 2130 stipulates that "no hypthec has any effect without registration."

Possession

§10.102 *Introduction* The various modes by which ownership of property may be acquired, as listed in article 583 of the Civil Code, are [O]wnership in property is acquired by prehension or occupation, by accession, by descent, by will, by contract, by prescription, and otherwise by the effect of law and obligations. Possession and prescription will be discussed in the following two sections. Prescription is the subject of the nineteenth title of the *Civil Code* and is dealt with in articles 2183 to 2270. Possession, which is a fundamental element of prescription, is governed by articles 2192 to 2200.

§10.103 *Definition and characteristics of possession* Article 2192 defines possession as "the detention or enjoyment of a thing or of a right, which a person holds or exercises himself, or which is held or exercised in his name by another." Possession must not be confused with the right of ownership, the former being a situation of fact and the latter, a situation of law. In other words, possession is the fact that one exercises personally or through the intermediary of a third party the prerogatives of a right, whether one has that right or not. Thus, a property can belong to a person other than the possessor, although most often the person in possession is also the true proprietor.

§10.104 *Protection of possession* Possession is protected by legislation through such judicial effects as actions that establish possession and prescription. This assures the maintenance of public peace, firstly, by preventing people from taking justice into their own hands and reacting to encroachments in a violent manner and, secondly, by assuring the stability of land transactions by allowing the acquisition of property through prescription. This protection also allows the proprietor to safeguard his right of ownership for it is much easier to establish and prove possession than ownership which involves searching the title.

§10.105 *Area of application of possession* Possession is not limited to the right of ownership. It also applies to other real rights, as set out in article 2192, which defines possession as "the detention or enjoyment of a thing or of a right." Similarly, the fourth paragraph of article 2203 states that "emphyteusis, usufruct and other like proprietary rights are susceptible of a distinct ownership and of a possession available for prescription."

§10.106 *Elements of possession* Possession consists of a material and an intentional element. The fact that characteristic acts of possession are being exercised upon a thing constitutes the material element. Article 2192 indicates that these acts may be exercised by the possessor or by a third party exercising them in his name. The intentional element is characterized by the intention of the possessor to act as the holder of the rights which he exercises. Both the material and the intentional elements are essential to possession. If either one of these elements is absent, possession does not exist. To prove possession, however, only the material element has to be established for article 2194 stipulates that "a person is always presumed to possess for himself and as proprietor, if it be not proved that his possession was begun for another."

§10.107 *Precariousness* If the person contesting possession is able to prove the absence of intention, thereby rebutting the presumption, possession becomes precarious and is presumed to have always been so, for article 2195 states, "[W]hen possession is begun for another, it is always presumed to continue so, if there be no proof to the contrary." Article 2203 adds that

> [T]hose who possess for another, or under acknowledgment of a superior domain, never prescribe the ownership, even by the continuance of their possession after the term fixed. Thus emphyteutic lessees, lessees, depositaries, usufructuaries and those who hold precariously the property of another cannot acquire it by prescription.

However, the same article states further on that "emphyteusis, usufruct and other like proprietary rights are susceptible of a distinct ownership and of a possession available for prescription." To summarize, precariousness implies the recognition of the rights of another person. For example, the state of tenancy produces no legal results, nor can it lead to prescription, since the tenant exercises possession in the name of the proprietor.

§10.108 *Interversion of title* Precariousness can, however, be transformed into possession if the title is interverted as set out in article 2205,

Nevertheless the persons mentioned in articles 2203 and 2204 and also persons charged with a substitution, may, if their title have been interverted, begin a possession available for prescription, dating from the information given to the proprietor by notification or other contradictory acts. Such notification of title and other contradictory acts only avail when made to or in respect of a person against whom prescription can run.

Thus, interversion of title is a change of the title of possession which then ceases to be precarious and becomes available for prescription. Such would be the case if a tenant, after giving notice to the proprietor, ceases to pay his rent but continues to occupy the premises. The cause of interversion can also be brought by a third party as, for instance, if a tenant purchases the immoveable from someone who pretends to be the proprietor.

§ 10.109 *Qualities of possession* Article 2193 states, "[f]or the purposes of prescription, the possession of a person must be continuous and uninterrupted, peaceable, public, unequivocal, and as proprietor." And according to article 2194, these qualities are to be presumed, it being up to the person contesting possession to prove any defect.

§ 10.110 *Possession must be continuous* Possession must be continuous which means that the possessor must perform those acts of possession which are normally performed by the proprietor. This is a question of fact which depends on the thing which is the object in possession. For example, the possession of a summer cabin is continuous, even if the possessor is not there during winter. Article 2199 makes proof of continuity easier by stipulating that "an actual possessor who proves that he was in possession at a former period is presumed to have possessed during the itermediate time, unless the contrary is proved."

§ 10.111 *Possession must be uninterrupted* Neither may possession be interrupted by the actual proprietor for such would constitute an interruptive act of prescription against the possessor. This continuity, however, is not a quality of possession, but a condition for prescription as set out in articles 2222 to 2231 of the *Civil Code*.

§ 10.112 *Possession must be peaceable* Possession must be peaceable, that is, non-violent, as described in article 2197: "[n]or can acts of violence be the foundation of such a possession as avails for prescription." Possession is violent if the possessor has occupied the premises by means of assault or after serious threats against the possessor. But, as article 2198 indicates, "in cases of violence or clandestinity, the possession which avails for prescription begins when the defect has ceased."

§ 10.113 *Possession must be public* Possession must be public, insofar as it must be known to the person against whom prescription is claimed. Clandestinity is also a temporary defect according to article 2198.

§10.114 *Possession must be unequivocal* In summary, possession must be unequivocal and have the appearance of a claim to an exclusive right. For example, article 2196 states that "acts which are merely facultative or of sufferance cannot be the foundation either of possession or of prescription." Merely facultative acts are those which a proprietor has the right to perform because his neighbour is not exercising his right. For example, under the provisions of article 514 of the *Civil Code*, a co-proprietor may place a beam or a joist in a party wall, leaving only about four inches of the whole thickness of the wall, but his neighbour can force him to withdraw the beam to the middle of the wall if he himself wants to put a beam in the same place or build a chimney against it. By placing a beam in the party wall, the co-proprietor has performed a merely facultative act, but the exercise of this faculty cannot lead to prescription. Acts of mere sufferance are those which a proprietor tolerates because they constitute only occasional encroachments, as for instance if a proprietor authorizes certain persons to cross over his land on a particular occasion. The beneficiaries of mere facultative acts and of those of mere sufferance can, thus, not claim protection of possessor's rights or prescription.

§10.115 *Possession must be proprietorial* Finally, possession must be exercised as a proprietor would exercise it which implies the existence of the element of his intention, for want of which possession is precarious.

§10.116 *Judicial effects of possession* The principal judicial effects of possession are possessory action and prescription, the former being action which allows the possessor to defend his possession against those wanting to evict him and the latter being that which allows him to acquire the property which has been in his possession for a certain length of time. The purpose of possessory action is to protect possession against any claim not founded on a proprietory title. This is based on article 2194 of the *Civil Code*, which establishes the principle that the possessor is presumed to be the proprietor. This principle also prevents people from taking justice into their own hands and responding to encroachments by force. It is also to the advantage of a proprietor who cannot easily prove his ownership. Possessory action does not protect the right of ownership in itself, but it does permit the possessor to obtain in court the full enjoyment and peacefulness of his possession. It permits him to defend himself even against the true proprietor who, in order to reclaim the thing, must prove his right of ownership.

§10.117 *Possessory actions* Possessory actions are brought by virtue of article 770 of the *Code of Civil Procedure*, which stipulates that

[A]nyone in possession, for more than a year, by other than precarious title, of an immoveable real right, may bring an action on disturbance against the person who disturbs his possession in order to put an end to the disturbance, or an action for re-possession against any person who has forcibly dispossessed him, in order to be put back into possession.

Obviously, the possessor who brings such an action does not have to prove that he is the proprietor, but his possession must have the qualities discussed above in order to be judicially effective. This article provides for two kinds of possessory actions, an action on disturbance or an action for re-possession. An action on disturbance, which is the more common kind, allows the possessor to put an end to a disturbance such as the non-authorized crossing of his property and to assure his continued possession. Whereas an action for re-possession allows him to recover a lost possession such as a part of the premises in his possession that has been separated from the rest by the construction of a fence. Court procedure also allows for an action of denunciation of new construction, by which the possessor is permitted to have any work halted that bears a risk of prejudice against his possession.

§10.118 *Petitory action* Possessory actions, however, do not prevent the true proprietor from defending his rights of ownership. Article 771 of the *Code of Civil Procedure* states, however, that "the owner of an immoveable or immoveable real right may, by petitory action, have his right of ownership recognized." Unlike a possessory action, which concerns a situation of fact, a petitory action bears only on the right of property. Moreover, article 772 rules that "the possessory and petitory action cannot be joined, nor can a petitory action be continued until the possessory action has been decided and the condemnation satisfied." This constitutes a defense of public order that permits the court to settle the conflict concerning possession first and then, if there is any cause, to hear the case for petitory action.

Prescription

§10.119 *Introduction* Another effect of possession is that it allows for the acquisition of property by prescription. In fact, after a certain length of time, possession, which is a situation of fact, becomes a situation of law through the effect of prescription, by which property is transferred to the possessor. As stipulated in article 2183 of the *Civil Code*,

[P]rescription is a means of acquiring, or of being discharged, by lapse of time and subject to conditions established by law. In positive prescription, title is presumed or confirmed and ownership is transferred to a possessor by the continuance of his possession. Extinctive or negative prescription is a bar to, and in some cases precludes, any action for the fulfilment of an obligation or the acknowledgement of a right when the creditor has not preferred his claim within the time fixed by law.

Since extinctive prescription applies more to obligations, we will only discuss positive prescription which is very useful because it allows a search of title to go back no further than 30 years.

§10.120 *Areas of application* In general, prescription applies to all the major real rights, with the exception of servitudes which, according to article 549 of the *Civil Code*, cannot be established without title. As stated in the fourth paragraph of article 2203,

> [E]mphyteusis, usufruct and other like proprietary rights are susceptible of a distinct ownership and of a possession available for prescription. The proprietor is not hindered by the title which he has granted from prescribing against these rights.

By virtue of this article, rights of use and habitation and superficiary rights are also prescriptible.

§10.121 *Imprescriptible things* In principle, prescription can establish rights over all kinds of moveable or immoveable things. There are, however, certain things that are not prescriptible, such as things which are not objects of commerce, sacred things, state property and public property belonging to municipalities. According to article 2201, "things which are not objects of commerce cannot be prescribed." In other words, whatever cannot be acquired, cannot be prescribed. Sacred things are also considered as not being of commerce for, in accordance with article 2217,

> [S]o long as their destination has not been changed otherwise than by encroachment, [they] cannot be acquired by prescription. Burial-grounds, considered as sacred things, cannot have their destination changed, so as to be liable to prescription, until the dead bodies, sacred by their nature, have been removed.

As to state property, article 2213 stipulates that

> [S]ea-beaches and lands reclaimed from the sea, ports, navigable or floatable rivers, their banks and the wharfs, works and roads connected with them, public lands, and generally all immoveable property and real rights, forming part of the domain of the crown are imprescriptible.

Article 2216 adds that "property escheated to the crown, by failure of heirs, bastardy or forfeiture" is only part of its domain if the Crown accepts said devolution. Until this occurs, the property continues to be subject to prescription like all other property. Articles 2220 and 2221 of the *Civil Code* concern property belonging to municipalities. By virtue of article 2220,

> [R]oads, streets, wharfs, landing-places, squares, markets and other places of a like nature, possessed for the general use of the public, cannot be acquired by prescription, so long as their destination has not been changed otherwise than by tolerating the encroachment.

Article 2221 specifies that

> any other property belonging to municipalities or corporations, the prescription of which is not otherwise determined by this code, is subject even when held in mortmain, to the same prescriptions as the property of private persons.

To summarize, public property belonging to a municipality is not subject to prescription, but all property belonging to its private domain is prescriptible.

§**10.122** *Acquisition of immovables by prescription* Prescription as pertains to immovables may be based on a 30 year or a ten year term.

§**10.123** *Thirty-year prescription* Thirty-year prescription, which constitutes the general rule in law, is governed by article 2242, which reads as follows,

> All things, rights and actions the prescription of which is not otherwise regulated by law, are prescribed by thirty years, without the party prescribing being bound to produce any title, and notwithstanding any exception pleading bad faith.

In other words, for 30-year prescription neither a title nor good faith is required. The only requirement for prescription under these provisions is effective possession, that is, possession according to the criteria of article 2193 of the *Civil Code*.

§**10.124** *Ten-year prescription* In addition to this basic condition, two supplementary qualities are required for the benefit of ten-year prescription, good faith and a translatory title of ownership. Article 2251 stipulates that

> [H]e who acquires a corporeal immoveable in good faith under a translatory title, prescribes the ownership thereof and liberates himself from the servitudes, charges and hypothecs upon it by an effective possession in virtue of such title [during ten years].

As stated in this article, only corporeal immoveables may be acquired by ten-year prescription thereby excluding such real rights as usufruct and superficiary rights while still allowing it to apply to ownership. But, unlike 30-year prescription, ten-year prescription is not allowed to extend beyond its title, as has recently been confirmed by the Supreme Court of Canada in the case of *Lavoie v. Michaud*.[13]

§**10.125** *The condition of good faith* In order to be of good faith, the

---

13 [1981] R.S.C. 445.

possessor must believe that he has acquired the immoveable from the rightful owner. Article 2253 of the *Civil Code* stresses that it is sufficient if the possessor was of good faith at the moment of acquisition. Article 2202 stipulates that "[good faith is always presumed.] He who alleges bad faith must prove it." This presumption can thus be rebutted by proving that the possessor, at the moment of acquisition, knew he was not dealing with the rightful proprietor.

§10.126 *Translatory title of ownership* The possessor must also have begun his possession by virtue of a legal act that would have transferred the property to him, had the consenting person been the rightful owner. This implies a translatory act, such as a purchase, an exchange, or a donation of property. However, in the matter of successions and donations, only wills and donations granting a particular title constitute translatory titles of ownership since, if they are universal or of universal title, the beneficiaries continue the possession of the author and possess by virtue of the same title as the latter.

Whereas the condition of good faith is sufficient with regard to the title of the author, the mere belief of the possessor that he himself has the title is not sufficient. The title must be valid. For example, a revoked will does not constitute a title, even if the possessor believes it to be still valid. Similarly, the gift of an immoveable without a notarized act required in accordance with article 776 of the *Civil Code* is a formal defect that cannot be covered by a ten-year prescription.

§10.127 *The joining of possession* For prescriptive acquisition it is not always necessary that the acts of possession be performed by the same author. The joining of possession allows the possessor to add to the years of his own possession those of his author. Article 2200 states the following rule in this regard,

> A successor by particular title may join to his possession that of his author in order to complete prescription. Heirs and other successors by universal title continue the possession of their author, saving the case of interversion of title.

This article makes a distinction between two kinds of successors, a successor by particular title who has a possession other than that of the author and a successor by universal title who, on the contrary, continues the possession of its author. Thus, only a successor by particular title may join his possession to that of its author.

§10.128 *Arrests in the course of prescription* Prescription is acquired over a statutory period. Article 2240 states that it "is reckoned by days and not by hours. [Prescription is acquired when the last day of the term has expired; the day on which it commenced is not counted.]" Certain events, however, can arrest the course of prescription resulting in suspension or even interruption. Suspension is simply a break in continuity, the term continuing as soon as the obstacle is removed, while interruption puts an end to it by erasing the period already elapsed. Moreover, as stipulated in article 2255 of the *Civil Code*, "after prescription by ten years has been renounced or interrupted, prescription by thirty years alone can be commenced."

§10.129 *Interruption of prescription* According to article 2222, "prescription may be interrupted either naturally or civilly." By virtue of article 2223, "natural interruption takes place when the possessor is deprived, during more than a year, of the enjoyment of the thing, either by the former proprietor or by any one else." During the year that follows depossession, the possessor may nevertheless bring a possessory action, under the terms of article 770 of the *Code of Civil Procedure*. Civil interruption can result either from a judicial claim or by the possessor's acknowledgement of the right of the proprietor against whom he was prescribing. According to article 2224, any judicial claim brings about the interruption of prescription. The same article goes on to say that an extra-judicial notice, even by a notary or a bailiff, does not interrupt prescription. Also, as set out in article 225, "a demand brought before a court of incompetent jurisdiction does not interrupt prescription." Moreover, according to article 2226,

> [P]rescription is not interrupted: if the service or the procedure be null from informality; if the plaintiff abandon his suit, except to avoid the exclusion provided for in article 1008 of the Code of Civil Procedure; if he allow peremption of the suit to be obtained; if the suit be dismissed."

And under the provisions of article 2227, "prescription is interrupted civilly by renouncing the benefit of a period elapsed, and by any acknowlegment which the possessor or the debtor makes of the right of the person against whom the prescription runs."

§10.130 *Suspension of prescription* In contrast to interruption, suspension arises from law and does not require any proprietory act. It only causes a temporary halt of prescription, which subsequently continues to run. Article 2232 establishes the principle that "prescription runs against all persons . . . unless they are included in some exception established by this code, or unless it is absolutely impossible for them in law or in fact to act by themselves or to be represented by others." In other words, prescription does not run against incapable persons, such as those as yet unborn, minors, idiots or persons afflicted with madness or insanity, but does run against those who have been given a judicial counsel and those interdicted for prodigality. Prescription runs against persons who are absent as well as those who are present but, in accordance with article 2233, spouses cannot prescribe against each other.

§10.131 *Judicial recognition of ownership acquired by prescription* Once prescription is acquired, ownership must be confirmed for the purpose of registration. According to article 2183a, "the judicial recognition of the absolute right of ownership acquired by prescription by ten years or that by thirty years may take place by following the formalities provided in this respect by the Code of Civil Procedure" (see articles 805 to 807 of that code). Thus, a person who has become proprietor by prescription may obtain judicial recognition of his right of ownership and the judgment granting his request has the value of a definitive title upon registration. Prescription thus acquired becomes valid retroactively and the possessor is considered proprietor from the day when his possession began.

REFERENCES

Archambault, J.-P. and Roy, M.-A. *Le Droit des affairs.* Montréal: Les éditions HRW, 1981.

Baudoin, J.-L. *La responsabilité civile délictuelle.* Montréal: Les Presses de l'Université de Montréal, 1973.

Baudoin, L.-J. *Les obligations.* Cowansville: Les éditions Yvon Blais, 1983.

Baudoin, J.-L. *Le droit civil de la province de Québec.* Montréal: Wilson et Lafleur, 1953.

Beaulieu, M.-L. *Le bornage, l'instance et l'expertise; La possession et les actions possessoires.* Québec: le Soleil, 1961.

Bouffard, J. *Traité du domaine.* Québec: Les Presses de l'université Laval, 1921.

Caparros, E. *Les régimes matrimoniaux au Québec.* 3d ed. Montréal: Wilson et Lafleur, 1985.

Clermont, B. and Yaccarini, B. *Initiation au droit des affaires du Québec.* 3d ed. Québec: Les Presses de l'Université Laval, 1981.

Demers, C. *Traité de droit civil du Québec,* Vol. 14. Montréal: Wilson et Lafleur, 1950.

Gérin-Lajoie, A. "Introduction de la Coutume de Paris au Canada." 1 R. du B. (1941).

Goulet, J., Robinson, A. and Shelton D. *Théorie générale du domaine privé.* 2d ed. Montréal: Wilson et Lafleur, 1984.

Huot, L.P. *Sûretés.* Cowansville: Les éditions Yvon Blais, 1984.

Marler, W. de M. *The Law of Real Property.* Toronto: Burroughs, 1932.

Martins, S. *Droits et lois du Québec.* Montréal: Guérin, 1983.

Martineau, P. *Les biens.* 5th ed. Montréal: Thémis, 1979.

Mignault, P.-B. *Le droit civil canadien,* Vol. 1. Montréal: Whiteford & Théoret, 1895.

Mignault, P.-B. *Le droit civil canadien,* Vol. 9. Montréal: Wilson et Lafleur, 1916.

Pineau, L. and Ouellette-Lauzon, M. *Théorie de la responsabilité civile.* Montréal: Les Editions Thémis, 1978.

Poupart, A. *Les enjeux de la révision du Code civil.* Montréal: Université de Montréal, 1979.

Rodys, W. *Traité de droit civil du Québec,* Vol. 15. Montréal: Wilson et Lafleur, 1958.

Taschereau, R. "Mignault et son oeuvre." *Les cahiers de droit.* Québec, I, No. 2 (1955).

THE CADASTRE

Introduction

§10.132 The public registration of real estate in Québec is made up of two elements: the cadastre and the system of registering real rights upon immoveables. The cadastre, which is prepared under the direction of the ministre de l'Énergie et des Ressources, is a mechanism for registering real property. Developed originally to identify immoveables for the purpose of registration, it has since become a municipal taxation and planning tool.

The purpose of this registration system which comes under the authority of the ministère de la Justice, is to make public and to protect rights, mainly property rights and real rights upon immoveables, but also real rights upon moveables and certain personal rights. Under this system, the cadastre identifies a site and describes its geometrical dimensions, while the index of immoveables contains the proprietor and the holders of real rights affecting it. As the cadastral number is used to identify property in both public registers, the rules concerning cadastral operations are very

The Law in Québec 397

important. Conflicts concerning property lines are solved by boundary determination, an operation by which monuments are placed along the dividing line between two adjacent properties.

Each of these systems has a different source. Boundary determination has its roots in the Roman origins of Québec law, the cadastre has been taken over from the European system, and the registration of real estate is based on an equivalent system practised in England.

Division and Concession of Land under the French Regime

§10.133 *The seigniories* During the French period, the region was colonized under the feudal system and land was conceded in the form of seigniories of varying sizes along watercourses such as the Saint Lawrence, Richelieu, Ottawa and Chaudière rivers. In order to develop the land within their domains, the seigniors agreed to concede parcels of it to interested colonists. The outside lines of the seigniories along the Saint Lawrence River had a north-west/south-west orientation and those along the Ottawa River a north-east/south-east orientation. The seigniories were divided into concessions of various depths, the separating lines running more or less parallel to the bends in the watercourse on which they were situated. These concessions were further divided into property lots following a plan chosen by the seignior. Since the only communication link at that time was the river, each lot had to have some river frontage, with the result that the lots were usually between ten and 20 times longer than they were wide. Although all the lots in each concession were of the same size, their dimensions generally exceeded what was indicated in their titles, for the seigniors, careful to avoid complaints, measured the lots generously. This extra allowance — called in French "allouance", "excédent", or "ristourne" — varied from one seigniory to the other and was not recorded in the registers; the amount can be determined through sample measurements. The distances and areas within each seigniory are given in French units of measurement.

§10.134 *Plans and land registers* Each seignior generally had a plan drawn up of his seigniory, showing the division into property lots, all of which were numbered consecutively throughout the seigniory. A register, the land-roll, was established on the basis of this plan, showing the size of each lot, the name of the colonist, or *censitaire*, his dues and any transfers that were permitted.

Division and Concession of Land under the English Regime

§10.135 *Townships* After the conquest, and especially after the Treaty of

Paris, the new administration, while respecting the institutions inherited from the French regime, made fundamental changes in land surveying, conceding and tenure in the regions being opened up for colonization. Land surveying came under the authority of a land commissioner and the land was divided into townships with property lots of uniform length and width and each range had a separate numbering system. Thus, a complete lot identification consists of both the number of the range in which it is situated and the lot number itself. The size and orientation of the first townships were dictated to some extent by the location of existing seigniories.

§10.136 *Methods of subdividing townships* There were two prevailing systems of subdividing townships in Québec. According to the first system, all lots were 28.75 chains wide by 73.05 chains deep and had an area of 200 acres which included a portion of five per cent of the total size reserved for roads. According to the second system, which was introduced later in the Abitibi region, the townships were square in shape with sides of ten miles each, were oriented in a north-south and east-west direction and were divided into ten ranges, each of which was 80.80 chains deep with lots 13.00 chains wide. With five per cent of the total space reserved for roads, the lots were 100 acres each. The units of measurement, of course, were English. The same type of township can be found later in the Saguenay/Lac St-Jean region, although the orientation was different.

§10.137 *Concession of lots* Under the English regime, lots were conceded directly by the Crown to the settler and land was held in free tenure, that is to say, no dues had to be paid to a superior domain once the conditions for land tenure had been met. A number of seigniories were still conceded after the conquest and the seigniorial court accorded them the same status and privileges as those established under the French regime.

Origins of the Cadastre

§10.138 *Introduction* The cadastre is governed by three main laws - the 1841 *Ordinance to Prescribe and Regulate the Registering of Titles*, the 1854 *Seigniorial Act* by which seigniorial schedules were established, and the 1860 *Act respecting registry offices* by which the official cadastre was instituted.[14] By making the registration of land titles mandatory, the legislators hoped to protect the public from fraud and eliminate the

---

14 *Ordinance to Prescribe and Regulate the Registry of Titles*, 1841, 4 Vict., c. 30; *Seigniorial Act*, 1854, 18 Vict., c. 3; and *Act Respecting Registry Offices and Privileges and Hypothecs in Lower Canada*, 1860, 23 Vict., c. 59.

uncertainty of land titles for, as stated in the Preamble to the *Ordinance* of 1841,

> [G]reat losses and evils have been experienced from secret and fraudulent conveyances of real estates, and incumbrances on the same, and from the uncertainty and insecurity of titles to lands in this Province, to the manifest injury and occasional ruin of purchasers, creditors, and others.

The registration of titles which, to a certain extent, did help to protect the public from losses incurred through fraud, did not succeed in eliminating all uncertainties as to titles since some ambiguity persisted in the identification of immoveables. While registration made it possible to trace a chain of title, it was not until the cadastre was established that it became possible to identify immoveables correctly.

§10.139 *The ordinance of 1841 to prescribe and regulate the registering of titles* The *Ordinance* of 1841 established a registry office in each judicial district and empowered the governor of the province to name a person "of fit integrity and ability" to act as registrar for each of the districts, "by whom the said office shall be kept, and the duties imposed by this ordinance . . . be performed." The registrar had to keep, as prescribed by the *Ordinance,*

> [A]n alphabetical list or calendar of all parishes, townships, seigniories, cities, towns, villages, and extra-parochial places within the district for which such registrar shall have been appointed, with references, under the respective heads of such local divisions, to all and every the entries of registered memorials relating to real or immoveable estates comprised within the said local divisions, respectively, and the numbers of such entries, and with a designation of the names of the parties mentioned in such entries, and of the real and immoveable estates to which the same may relate, so as to afford, by means of an index to estates, as far as may be practicable, a like easy and ready reference to every memorial to be registered as aforesaid.

Clearly, this index of names served as an early model for the index of immoveables.

§10.140 *The Seigniorial Act of 1854* The *Seigniorial Act* of 1854 named the commissioners whose duty it was to set up seigniorial cadastres and established the procedure they were to follow. These cadastres were intended solely for the purpose of documentation and no land survey data were included in them. It was not even used for the registration of titles, its primary purpose being to determine "the price to be paid by seignior and censitaire for the commutation of the tenure of their property." The commissioners who were appointed had to

> [D]raw up in tabular form in duplicate, a schedule of such Seigniory, shewing: . . .

(a) the total value of the Seigniory ...
(b) the value of the rights of the Crown in the seigniory ...
(c) the value of the lucrative rights of the Seignior ...
(d) the yearly value of Seigniorial rights upon each land ...
(e) the extent of such land according to the title of the owner.

The seigniorial schedule was based on the plans, maps, documents, and land register of the seignior. The book of reference was in the form of a catalogue and included the name of the proprietor, the number and area of the lot, sometimes its dimensions (if they were listed in the title or other documents) and the dues to be paid. It seems that only a few plans were drawn up specifically for the seigniorial cadastres. The law required, moreover, that "each land ... be described in the Schedule by the number, and concession, under which it stands in the land register of the Seignior." Unfortunately, the commissioners did not maintain that numbering system or add the numbers of the new system to those of the old one. This omission has contributed to the considerable difficulties experienced today by anyone consulting the seigniorial schedules. Nevertheless, while the seigniorial cadastres fulfilled their purpose at the time which was to record the dues payable to the seigniors in compensation for the abolition of feudal rights and duties, they were never used for registration purposes nor could they be used as unequivocal evidence in establishing a title. Thus, six years after introducing the seigniorial cadastres, the legislators had to devise a more effective system.

§10.141 *The Act of 1860 and the establishment of official cadastres* The cadastre as we know it today was introduced by the *Act* of 1860. In order to improve the system of land titles registration, an official cadastre was introduced which included a plan and a book of reference for each city, town, village, parish, township, etc. This Act stipulated among other things:

(a) that each Electoral County shall be a County for Registration purposes.
(b) If there be not ... a proper place for the County Registry Office, with a sufficient metal safe or fire-proof vault for the safe keeping of the Books and papers thereof, — the Governor shall direct....
(c) And whereas for the more effectual working of this Act, it is desirable that there should be in each Registry Office correct plans of the Cities, Towns, Villages, Parishes and Townships, or portions thereof, in the County or Registration Division to which such office belongs, which plans should show the sub-division of such localities into lots, and serve as a basis for the description of the property to which the deeds and instruments registered in such office relate, so that the Index to Estates required by this Act may be easily and correctly made and kept....
(d) The Commissioner of Crown Lands shall cause to be prepared, under his superintendence, a correct plan of each City, Town, Incorporated Village, Parish, Township, or part thereof ... with a Book of Reference to each such plan, in which book shall be set forth —

(1) a general description of each lot . . .
(2) the name of the owner . . .
(3) everything necessary to the right understanding of such plan
(4) and each separate lot or parcel of land shewn on the plan, shall be referred to in the said book, by a number which shall be marked on it upon the plan, and entered in the said book . . .
(e) Each of the said Plans . . . shall be made up to some precise date, up to which it shall be corrected as far as possible, and this date shall be marked upon it . . . .
(f) In the Seigniorial portions of Lower Canada, the Schedules made by the Seigniorial Commissioners, and the plans made under their superintendence, shall serve as the basis for the plans and books of reference . . . .
(g) In the Townships, the Commissioner of Crown Lands shall use such maps or surveys or cause such surveys to be made as he shall deem best adapted to ensure the correctness of the plans and books of reference to be made as aforesaid; but the original numbering of the lots and concessions shall always be preserved.

§10.142 *Instructions for establishing cadastres* The Act by which the cadastre was instituted was supplemented by instructions on how to proceed in the case of parishes, towns, villages, townships, etc. According to these instructions, land was to be surveyed only to the extent that was absolutely necessary for the preparation of the documents, but all property lots lying within the designated area were to be described as accurately as possible. To do that, the perimeter of the area had to be determined and the major axis, such as roads and rivers, the outside lines of the seigniories and townships, the range lines and the concession lines, located. In the case of townships, existing property lots were located within that framework by using original survey documents and, in the case of seigniories, by consulting the land registers and diagrams and the "seigniorial schedule." When the original cadastre was established, the subdivisions of the property lots were measured more or less precisely, depending on whether the area to be measured was a rural parish, a village or a town. The numbering of the lots in the seigniorial schedules was supposed to consist of a continuous series of numbers corresponding to those already assigned by the seignior. This correspondence was generally ignored, and there is often no concordance. In the case of new townships, the new numbering system had to include the original numbers and the numbers of existing subdivisions had to include letters or numbers subordinate to the original numbers. Since the original titles upon property lots conceded in the townships by the Crown, such as location tickets and letters patent, were designated by the original survey numbers, it was crucial that those numbers be preserved in order to reconstruct the chain of title. Roadways and paths were not numbered in rural cadastres, but streets, lanes and squares were supposed to be numbered in urban cadastres, although this was by no means always done. The work of setting up the original cadastre, which was begun in 1866, was completed by about 1900 for all inhabited parts of the province. The cadastre is still being expanded as new land is settled and new property identification is required.

§10.143 *The Civil Code of Lower Canada* In 1866, the *Civil Code of Lower Canada* added new provisions concerning the cadastre. These provisions are contained in Title Eighteen, Chapter VI, Section II "Of the Official Plans and Books of Reference and the Matters Connected Therewith". The *Cadastre Act*[15] and these provisions of the *Civil Code* complement each other and occasionally overlap. For example, while the *Cadastre Act* defines the way in which the original cadastre is compiled, the *Civil Code* spells out the manner and the effects of registration, and while the *Civil Code* authorizes the subdivision and redivision of property lots, the *Cadastre Act* deals with registration and its effects.

§10.144 *Cadastral units* In order to put the cadastre into effect, it was necessary to divide the country into cadastral units. It seemed logical to keep the administrative divisions which had been established in the early days of the colony. The first of these groupings was the church parish. As early as 1663, the Conseil souverain empowered the bishop to set up parishes and define their boundaries for ecclesiastical purposes. By the end of the French regime, about 80 such parishes were in existence. The English regime confirmed the powers of the bishop and the establishment of ecclesiastical parishes continued after 1760 in the same manner as before. In 1839, an ordinance of the Conseil spécial pour le Bas Canada, while reconfirming the power of the bishop to create ecclesiastical parishes, also empowered the governor to constitute these parishes as secular districts and to recognize as such their limits and boundaries. As a result, in 1840, when municipal structures were put in place, the new municipalities assumed the same boundaries as the parishes. In 1872, in order to facilitate cadastral operations, it became law that all applications for civil building permits be accompanied by a site drawing which had been prepared by a land surveyor. The boundaries of these parishes do not necessarily correspond to the divisions established by the original survey of the seigniories and townships; a parish might include parts of two seigniories or two townships or parts of both a seigniory and a township.

These parishes, towns and villages constituted the administrative units which later became the first cadastral land units. There were, however, differences in the systems of measurement and the scales used in these plans, depending on whether they referred to seigniories, townships or rural or urban districts. Each cadastre is given an official name when it is proclaimed and it keeps that name in perpetuity. In 1860, there were only

---

15 R.S.Q., c C-1.

a few towns and not many villages, but there were quite a few parishes. Almost all of these parishes were named for saints and the cadastres for those parishes still bear that name. Next to be established were the cadastres for the townships and the number of these has continued to grow to this day as new townships are being settled or developed by mining concerns or other industries. The list of the Québec cadastres, published by the ministère de l'Énergie et des Ressources, now includes 28 town cadastres, not counting the cadastres for individual neighbourhoods in the case of some of these, 410 parish cadastres, 114 village cadastres, 648 township cadastres, 9 municipal cadastres, 7 seigniorial cadastres, 13 cadastres without any particular designation and one general cadastre for the territory of New Québec. Thus, the inhabited part of the province, which comprises only ten per cent of the total area, is covered by 1,230 official cadastres.

The Cadastre Today

§10.145 *Distinction between an original cadastre and a subdivision* A distinction must be made between the original cadastre registered under the provisions of articles 2166 and 2176a of the *Civil Code* and a subdivision carried out by a private citizen as authorized by article 2175. The original cadastre includes not only all the cadastral documents which have been prepared since 1866, when cadastres were first constituted for the whole province, but also any plan or book of reference made in conformity with the provisions of article 1 of the *Cadastre Act* that cover a division into lots entered into a cadastre for the first time. The division of an original lot is a first cadastral subdivision which, under the provisions of article 2175 of the *Civil Code*, is to be carried out by a land surveyor on the request of the proprietor. These first subdivisions can be subsequently re-subdivided under the provisions of the same article, redivided in conformity with articles 17 and 18 of the *Cadastre Act* or replaced under article 2174b of the Civil Code. The original lot can also be stricken from the cadastre following the regulations of article 1274a and re-registered with a new number in conformity with article 1 of the *Cadastre Act*. Or, under the provisions of article 2174b, an original lot can be replaced by another original lot.

§10.146 *Parts of the cadastre* A cadastre is made up of two documents, the plan and the book of reference. The plan shows the shape of each parcel of land and identifies it by its cadastral number; it is represented by a scale drawing without dimensions. All lots are numbered, including railway property. Most of the roads, streets, lanes and squares are shown without numbers. The book of reference gives a general description of each lot represented on the plan and identifies the name of the proprietor, as required by the *Cadastre Act*.

§10.147 *Deposit of an original cadastre and its effects* The original of the cadastre for any locality is housed in the archives of the *ministère de l'Énergie et des Ressources* and the registration offices are simply provided with a copy of the plans and books of reference concerning the lots in their division. Article 2169 of the *Civil Code* requires that a ministerial order (called a proclamation prior to 1980) is required before an original cadastre is put into effect and published in the *Gazette officielle du Québec*. From the date of the ministerial order, all lots included in a legal document must be identified by their cadastral number and, in the case of parts of lots, a declaration must be added stating that they are parts of the said lot and identifying the adjacent parts. Unless this condition is met, the registration of the document can have no effect on the lot in question. For example, under the provisions of article 2042 of the *Civil Code*, a conventional hypothec would be deemed to have never existed. However, under the provisions of article 2168, the filing of a notice completing or correcting the legal description of a lot will validate the original document as of the date of the filing. Likewise, as of the date fixed by the ministerial order published in the *Gazette officielle*, the registrar must ensure that the index of the immoveables in the cadastre is up-to-date, as stipulated in article 2171 of the *Civil Code*. It is also possible to give a legal description of parts of lots by referring to them as part of a cadastral lot without having to designate them by an individual number.

Types of Cadastres

§10.148 *Original cadastres* An original cadastre is established at the initiative of the ministère de l'Énergie et des Ressources in conformity with the provisions of the *Act of 1860*, as reiterated in article 2166 of the *Civil Code*, concerning any parcel of land that receives a cadastral identification for the first time.

§10.149 *Cadastral revision* A cadastre may be revised on the basis of the Act respecting land titles in certain electoral districts. The application of this law, which received assent in 1948, was at first limited to the electoral districts of Gaspé-Nord, Gaspé-Sud, Bonaventure and the Iles-du-Cap-de-la-Madeleine, but was later extended to include the lands along the north shore of the Saint Lawrence River and east of the Saguenay. In these electoral districts the property owners had to do without the services of a local notary; any deeds, such as those of sale, gifts and wills, had to be concluded before the local justice of the peace or a priest or pastor. Moreover, even where there were registry offices, most such deeds were not registered. By virtue of this law, not only can the minister modify the cadastre, he can also issue certificates of ownership to any occupant of an immovable affected by the modification. The first step in revising the cadastre is to compile as complete an inventory of the properties as possible. Once this has been completed, the minister sends the registrar a list of all the uncontested lots and their occupants. As this list is registered, which is done by deposit, each of the property lots is entered into the index of immovables. Under the provisions of article 7 of the Act, this registration entails the awarding of each lot to its occupant, who thereby is recognized as the proprietor. The title to the property is issued as a certificate of property ownership by the registrar once all formalities with regard to possible objections have been completed; the title is also registered by deposit. Forty-four cadastres have been revised in this manner, either completely or in part.

§10.150 *Cadastral renewal* In 1980, article 2176 of the *Civil Code* was amended to authorize the minister to have new plans and books of reference established for a region if required, and to follow, if necessary, the replacement procedure outlined in article 2174b for the protection of the interests of hypothecary and privileged creditors. So as not to confuse this operation with cadastral revision, which may only be carried out in certain regions, it was called cadastral renewal. Unlike cadastral revision, this procedure for updating cadastres applies to the whole province.

§10.151 *Cadastral reform* A new Act, named the *Act to promote the reform of the cadastre in Québec*,[16] was adopted and assented to by the government of Québec on June 20, 1985. Its purpose was to permit a cadastre to be re-worked over a period of about ten years and to require that all cadastres re-worked in this manner be updated continuously to show the latest divisions. The vehicles used for this re-working are renovation or, if applicable, revision, both of which became subject to this law after October 1, 1985. In order to facilitate this continuous updating, the Act requires the deposit of both a paper copy and an electronically stored copy of the new or the updated version. This Act eliminates the book of reference for every cadastral document deposited after October 1, 1985. All the elements of the book of reference from then on must be contained in the plan. The adoption of this Act brought with it the amendment of several articles of the *Civil Code* and the repeal of others. Certain provisions of the *Cadastre Act*, the Act respecting land titles in certain electoral districts and the *Registry Office Act*[17] were also amended.

This new Act was long overdue. In fact, a study carried out in 1974 showed that only half of the properties were recorded specifically in the official cadastre. This incompleteness, despite numerous mechanisms

---

16 R.S.Q., c. C-1.
17 R.S.Q., c. B-9.

designed to keep cadastres up to date, resulted from the fact that legislators accepted as identification of a lot a reference to its metes and bounds. With the enactment of new provisions to the Act permitting cadastral renewal, the continuous updating of cadastres has now been ensured. In fact, deeds involving alienation *inter vivos* in a cadastre renewed or revised after October 1, 1985, will no longer be entered in the index of immoveables unless the immoveables in question are identified by a separate number. A cadastral number identifying the alienated part must be deposited and the residual portion must be given a separate number. Exceptions are allowed if the deed of alienation serves to guarantee an obligation or if the part of the lot is located in an agricultural zone or more than 345 km away from the nearest registry office. The last two exceptions can only be applied by government decree, in which case the Service du cadastre must give all parts of an existing lot a separate number that will be shown on the cadastral plan. Moreover, plans prepared for reasons of public utility in an area that has been subject to cadastral renewal or revision must show a separate number for each part of the lot included in the confines of the area and another number for the residual portion of the lot. The Act stipulates, further, that governmental agencies are also subject to its provisions. The registration of the boundary determination surveys will have full force only if special mention is made in the boundary determination report that the boundaries coincide with those in the cadastre.

Cadastral Operations

§10.152 *Introduction* The original cadastre can be modified in two ways — either by the minister upon the request of a land surveyor who has been appointed to prepare the change in question, such as a correction, addition or annulment, or by the land surveyor himself upon the request of a proprietor as in the case of subdivision, redivision or replacement. These operations, which, of course, can be carried out only in the case of lots already entered in a cadastre, give rise to a second generation of cadastral numbers derived from the first one. All these operations must be prepared and submitted by a land surveyor. The exclusive character of this professional activity is prescribed in article 34 of the *Land Surveyors Act*.[18]

§10.153 *Division* Division is the operation by which a proprietor prepares and deposits at the Service du cadastre and the registration office the official plan and book of reference for any part of an original lot in accordance with article 1 of the *Cadastre Act*.

---

18 R.S.Q., c. A-23.

§**10.154** *Subdivision* Subdivision is the division of a lot into parcels by its proprietor, who to this end prepares a plan and book of reference in which each parcel is given a separate number that corresponds to a new page in the index of immoveables. Despite the phrasing in article 2175 of the *Civil Code* — "whenever the owner of a lot subdivides into lots" — such an act, in fact, like any other cadastral operation, must be carried out by a land surveyor. The wording of this article is intended to specify that the operation is carried out on behalf of the owner rather than a governmental authority. Upon the request of a property owner, the surveyor prepares a proposal for a subdivision plan which he then submits to the Service du cadastre for approval, by producing a memorial of request together with the required documentation. It may be necessary to obtain the approval, for example, of the municipal council or the ministre des Affaires municipales or a certificate of authorization by the Commission de protection du territoire agricole du Québec. The cadastral numbers created by this operation constitute a second generation of numbers. In fact, only lots entered into a cadastre may be subdivided by appending to the original lot number a continuous series of subordinate numbers. Unlike the original cadastre, the cadastre of a subdivision, even though it may be published in the *Gazette officielle*, does not require a ministerial order to come into force. Nor is it necessary to renew the real rights, as required by article 2172 of the *Civil Code*, for the deposit of original plans and books of reference. After verification by the Service du cadastre, the deposit of the documents at the registry office is sufficient. The registrar, upon receipt of the documents, records in the index of immoveables the fact that a lot has been subdivided and he starts a new page for each new number in recording the transactions carried out with regard to the parcels in question. If all of the original lot is subdivided, the subdivision has the effect of rendering the number of the lot inactive for the future, the number thereby being replaced by the numbers of the parcels for the whole area of the original lot and the original lot cannot be re-entered into the index of immoveables without further authorization. If only part of the original lot is subdivided, the remaining part is designated as such and retains its number in the index of immoveables. Once these operations have been completed, the provisions of article 2168 of the *Civil Code* enter wholly into effect and the new number must be used to designate the lot or part thereof.

§**10.155** *Redivision* Redivision is the partial or complete substitution of a new subdivision for an existing one. This cadastral operation, which is based on the second paragraph of article 2175, involves the completion of two formalities that are set out in articles 17 and 18 of the *Cadastre Act*. These formalities, cancelling certain numbers and replacing them in the same documents with a different subdivision, are completed simultaneously. As with any cancellation, every precaution must be taken to protect the rights of hypothecary and privileged creditors. Therefore, it is necessary to obtain from the registrar an up-to-date certificate of privileges and hypothecs in order to determine whether there are any outstanding hypothecs against the lots to be cancelled and, if so, to obtain the written consent of the creditors that their rights be re-registered against the new property lots. Moreover, the cadastral documents must show clearly the concordance of the new numbers with the old ones by an entry to that effect in the plan and, if applicable, in the book of reference. The cancelled subdivision numbers are never to be used again.

§10.156 *Replacement* In order to minimize delays and risks to creditors during the phase of cancellation and redivision, article 2174b was added in 1980 to the *Civil Code*; this article introduced a new cadastral operation, called replacement. By means of this operation, lots can be replaced by a new series of numbers while the previous numbers, instead of being cancelled outright, continue to exist as far as everything in the past is concerned. By this mechanism, delays in the issuing of hypothecary certificates can be avoided, new rights can be registered between the time that a certificate is applied for and the new cadastral documents are deposited and the refusal of a hypothecary creditor to give his consent can be avoided. The hypothecary creditor, thus, does not need to renew the registration of his rights, since they continue to encumber the property as designated by the old number which remains valid for that purpose. However, if a proprietor intends to replace the numbers of his lots, he must advise each creditor with an interest in the lots or parts thereof and each beneficiary of the declaration of family residence. There must be a close concordance between the old and new numbers of the plan and, if applicable, the book of reference, and this concordance must also appear on the pages of the index of immoveables which bear the old or the new numbers. As specified by the last paragraph of article 2174b

> This article has no effect on real rights existing in or on a lot the number of which was replaced and, particularly, the exercise of these rights may be continued against that part of the lot which was encumbered with a hypothec or a privilege.

Replacement allows for the creation not only of new subdivision lots, but also of new original lots.

§10.157 *Cancellation* By cancellation, it is possible to revert to the situation which existed before subdivision took place. There are many reasons for wanting to cancel a cadastral number series; for instance, if a lot has been double numbered or if a proprietor decides to reduce the number of parcel numbers in his property. Under the provisions of article 2174a of the *Civil Code*, the minister, upon presentation of a request to that effect, will strike out any numbers which have become invalid, provided no hypothecs are registered against them. If a hypothec exists which has not yet been discharged, cancellation can take place only after a release of the hypothec affecting the parcel number in question, whereby the hypothec will be stricken from the index of immoveables. The cancelled numbers may not be used again in the same cadastre.

§**10.158** *Addition of an original lot* This cadastral operation involves substituting original lots for parts of original lots or other original lots. The addition of an original lot number to an already existing cadastre may be required under circumstances such as the following:

(1) under the provisions of article 2174 of the *Civil Code*, if a number was omitted when the plan and book of reference of the original cadastre were created;
(2) under the provisions of article 2174a of the *Civil Code*, if one or several lots or parts of lots is cancelled and replaced by a single number, as in the case of land consolidations, for example, with a view to depositing a cadastre of an immoveable under co-ownership;
(3) under the provisions of article 7 of the *Cadastre Act*, in order to register a road no longer used that appears in the plan of the original cadastre without a cadastral number, when it ceases to be public property and becomes private property;
(4) under the provisions of article 8 of the same Act, when a railway corridor has to be designated across land covered by a cadastre after the original cadastre has been drawn up.

As prescribed in article 2174a, however, this cadastral operation can take place only if no hypothecs or privileges are registered against the lots which are to be cancelled; otherwise, it is necessary to obtain from the creditors the release of the hypothecs or privileges.

§**10.159** *Correction* The cadastre does not confer on the owner any rights that are not already included in his title. Thus, the right of ownership cannot be affected by any errors found in the plan or book of reference, as stated in the last paragraph of article 2174 of the *Civil Code*. The registrar, on the other hand, cannot correct these errors on his own authority; he must submit a report to the ministre de l'Énergie et des Ressources, who may make corrections, if so warranted. An error in the name of the proprietor, the dimensions or the description of a lot, stemming from an error of the copyist, a mistaken interpretation of titles or any other cause, may be corrected by the minister under the provisions of article 2174.

§**10.160** ***Vertical cadastres and co-ownership*** The cadastre was at first exclusively horizontal; it became vertical in 1976, following the introduction in the *Civil Code* of the concept of the co-ownership of an immoveable which extends vertically. A third dimension was added to the cadastre of surfaces to make a cadastre of volumes possible. The main cadastral operations apply to this particular case. However, the site on which a jointly owned immoveable is built must bear an original cadastral number. Because these immoveables can only be built on sites of a certain size and generally are built in urban areas where land is divided into small parcels, it is very often necessary to carry out cancellation, addition or replacement procedures in order to provide the site with a single, original cadastral number. Subdivision is then carried out by designating a first subdivision number for all the common parts located outside the building, such as the yard, underground area and air volume, a second subdivision number to the common parts of the building, such as the exterior walls, roofs, stairs and hallways, and a separate subdivision number to each exclusive part, identified by the number of the floor on which it is located. It is also possible to establish a vertical cadastre for purposes other than co-ownership — for example, in order to identify the space occupied by a subway. It is, furthermore, possible to deposit a plan of a jointly owned property that extends horizontally and has common parts and exclusive parts and was established by the subdivision of a new original lot.

Effects of the Cadastre

§**10.161** Once the plans and books of reference have been approved by the Service du cadastre, a copy is deposited at the appopriate registry office. Article 2169 of the *Civil Code* requires that each original cadastre be declared by ministerial order and published in the *Gazette officielle du Québec*. Beginning with the date fixed in the order, the registrar, following the provisions of article 2170, must prepare the index of immoveables by starting a new page for each lot appearing in the cadastre. Thus, beginning with the effective date of the cadastre, the number given to a lot on the plan and in the book of reference becomes its true designation and must be quoted whenever the lot is referred to as stipulated in article 2168. Notarized deeds concerning an immoveable must, therefore, make reference to that designation, without which the act registered cannot have any effect on the lot in question unless a notice is registered indicating that this registration applies to a specific registered lot. Moreover, as stipulated in article 2172 of the *Civil Code*, the registration of any real right upon a lot included in this deposit must be renewed by a notice designating the affected moveable within the two years following the date fixed in the order and published in the *Gazette Officielle*. But, as stated in article 2173, failure to renew registration does not result in the loss of the real rights registered, but rather renders them uninvokable against other creditors or subsequent acquirers whose rights are duly registered. As already mentioned, a cadastre of subdivision does not need to be proclaimed in order to come into in force.

§10.162 To summarize, once a cadastre has come into effect, the cadastral designation must appear on any document to be registered that concerns an immoveable contained in that cadastre; if the designation is not used, the registration can have no effect on the given immoveable, unless a notice is filed in which the number of the property lot affected by the registration is indicated, a step prescribed by article 2168. The legislator has given such importance to the cadastral number that in certain cases, for example, hypothecs, failure to refer to it causes the nullity of the deed as stipulated in article 2042 of the *Civil Code*.

§10.163 However, even if the plans and books of reference are there to identify properties, to assign them a cadastral number and to delimit them with regard to neighbouring lots, the cadastre cannot by itself determine the size or boundaries of a property: "The number given to a lot on the plan and in the book of reference is indeed the true description of the lot, but the boundaries between properties are not established by the cadastre."[19] In the case of *Carbonneau v. Godbout*[20], Sir Mathias Tellier made the following remark:

> The purpose of the cadastre is not to provide information on the location and size of immoveables, but to give each immoveable a number. The location and size appear in the book of reference only to describe the lot and to identify it.

Case law has decided unanimously that it was not the intention of the legislators to give the cadastre the role of fixing the dimensions and boundaries of lots, but simply to facilitate real estate transactions and registration. In fact, paragraph 3 of article 2174 states that errors in the cadastre cannot be interpreted to give a party more rights to a property than those expressed in the title. But even if its role is to formally identify immoveables, the cadastre does not confer on the proprietor any rights that are not included in his titles. The cadastre is, thus, a secondary document with regards to titles, and as Fortunat Lord says in *Termes et Bornes*,[21]

---

19 See *McLennon v. Nova Scotia Steel & Coal Co.* (1908), 18 Que. K.B. 317.
20 (1920), 31 Que. K.B. 69.
21 F. Lord, *Termes et Bornes* (Montréal: Wilson et Lafleur, 1939) at 129.

Secondary documents are useful because of the supplementary role they play. They diminish uncertainties about titles, break the silence. They speak for the titles in their absence. They are invaluable if possession has been established without the concurrence of the neighbours or only in a faulty manner. After all, it is only prudent that the surveyor does not take possession as the only basis on which to operate; he must consult secondary documents for corroborating evidence.

One must not conclude from this, however, that the information contained in the cadastre cannot be used to establish the boundaries between two properties, for the cadastre is presumed to be exact. If the cadastre does not contain errors, the purchaser of a lot designated by its cadastral number acquires the lot as shown on the plan, except for the rights that any third party may have acquired by prescription or in other ways.

REFERENCES

Fréchette, A.-B. *Etude sur les levés cadastraux dans la province de Québec avec considération spéciale des méthodes photogrammétriques.* Quebec: Université Laval, 1966.

Genest, F.R. *Faits chronologiques du cadastre.* Publication de la societé de géodésie de Québec, No. 11 (1942).

Raymond, G. "La réforme du cadastre québecois." *Arpenteur-géomètre*, 12, No. 4. Québec: Imprimerie d'Arthabaska, 1985.

Sasseville, G. *Le cadastre.* Québec: Les Presses de l'Université Laval, 1981.

An Act for the abolition of feudal rights and duties in Lower Canada, 1854, 18 Vict., c. 3.

An Act respecting registry offices, and privileges and hypothecs in Lower Canada, 1860, 23 Vict., c. 59.

An Act to consolidate and amend the Railway Act, 1919, 9-10 Geo. 4, c. 68.

An Act respecting land titles in certain electoral districts, 1948, 12 Geo. 6, c. 37.

An Act to promote the reform of the cadastre in Québec, Bill 40, 1985 (assented to 1985, now R.S.Q. c. 22).

Cadastre Act, R.S.Q., c. C-1.

An Ordinance to prescribe and regulate the registering of titles to lands, tenements and hereditaments, 1841, 4 Vict., c. 30.

REGISTRATION

Introduction

§10.164 *Definition* Registration is an operation of a formal nature by which documents are made public in order to validate the rights established in them.

§10.165 *Origins and evolution* Under the French regime, the registration of real estate was carried out by insinuation, that is, by depositing the acts to be authenticated with the Conseil souverain or the Conseil supérieur. At the beginning of the English regime, there was no registration to speak of other than what was contained in notarial records or the land registers of the seigniors. At the very beginning of the English regime, Governor Murray tried to establish registry offices but, due to the habitants' state of mind after the conquest, he met with no success. The first registry office was established on March 26, 1830, for the township of Drummond. Between 1831 and 1839, about ten more registry offices were set up. Then, by the Ordinance of February 9, 1841, concerning registration, a general system of registration for the whole province was established as of December 31, 1841. By this new system, the 24 judicial districts, not the townships, were selected as registration divisions. The Act was rewritten in 1860 in order to make the cadastre the basis for land registration. A number of statutes were passed authorizing the subdivision of existing registration divisions and the addition of new ones, so that today there are 82 registration divisions.

§10.166 *Laws governing registration* The provisions of the *Civil Code* concerning registration are contained in articles 2082 to 2182. These, plus the provisions of certain statutes, such as the *Cadastre Act* and the *Registry Office Act*, constitute the legal basis for registration. Although together these provisions set out in considerable detail the procedures of registration, they provide only a few general principles. The essence of these principles, therefore, must be deduced from the provisions as a whole. Some authors have rightly remarked that the Cartesian spirit is less evident in this part of the *Civil Code* than elsewhere, being derived as it is from English law.

Documents Subject to Registration

§10.167 Of the documents requiring the formality of registration, written documents must be distinguished from plans. Under the provisions of article 2129a of the *Civil Code*,

> [T]he deposit of a plan in the registry office in virtue of an act requiring

it is considered as a registration of such plan and treated as such. The plan must be accompanied with a notice showing the description of the immoveable contemplated therein in accordance with the prescriptions of article 2168. This provision does not include deposits of plans contemplated in articles 2166 to 2176c and in the Cadastre Act.

Plans may not be registered except where required by law and, as a general rule, unless they have been signed by a land surveyor. Plans of railways are an exception to this rule; pursuant to the *Railway Act*,[22] they must be signed either by the secretary of the railway company who deposits the plan or, in the prescribed manner, by the national railway commission. Plans that require registration include expropriation and public reserve plans, plans that are prepared under the provisions of the *Cities and Towns Act*, the *Navigable Waters Protection Act*, the *Canada Land Surveys Act*, the *Watercourses Act* and the *Mining Act*,[23] as well as plans attached to minutes of operations.

Rights that Require Registration

§10.168 In principle, all acts affecting real property and the rights pertaining thereto, as well as debts guaranteed by immoveables, must be registered in order to be invoked against a third party. In fact, Article 2098 of the *Civil Code* requires that "all acts *inter vivos* conveying the ownership of an immoveable must be registered." And article 2082 decrees that "registration gives effect to real rights and establishes their order of priority according to the provisions contained in this title." All documents transferring property rights *inter vivos*, such as sales, exchanges, gifts and acts relating to the transfer of property rights by intestate or testamentary succession must be registered. The same applies to acts establishing the modalities of property rights upon immoveables, such as co-ownership or superficiary rights, as well as acts concerning corollary rights connected with land ownership, such as usufruct, use, habitation, emphyteusis and servitudes. Accessory real rights, such as hypothecs and privileges upon immoveables, are also subject to the formalities of registration, as well as any transfer of these debts as required under the first paragraph of article 2127. Legislation, however, provides a few exemptions from the registration requirement as set out in article 2084 of the *Civil Code*. These exemptions include certain privileged claims upon immoveables and the original titles by which lands were granted. Although the registration of

---

22 R.S.C. 1985, c. R-3 and R.S.Q., c. C-14.
23 Cities and Towns Act, R.S.Q. c. C-19; *Navigable Waters Protection Act*, R.S.C. 1985, c. N-22; *Canada Lands Survey Act*, R.S.C. 1985, c. L-6; *Watercourses Act*, R.S.Q., c. R-13; and *Mining Act*, R.S.Q., c. M-13.

real rights mainly concerns immoveables, certain real rights pertaining to moveables, such as debts guaranteed by commercial pledge, as set out in articles 1966 and following, must be registered in order to have the same effect.

The Form of Acts Subject to Registration

§10.169 In order to be registered, a document must have been notarized or certified, unless otherwise stated by law. The authentic acts identified in articles 1207 and 1208 of the *Civil Code* include copies of notarial instruments, records and registers of courts of law, copies of minutes of boundary determinations and official documents of municipal corporations. Documents in private writing must be certified by at least two witnesses, one of whom must sign an affidavit as prescribed by article 2131. Other documents specifically mentioned by the law include, in particular, declarations of residence, actions upon privilege and notices of expropriation.

Custody of Documents

§10.170 *Principal registers* The various registers kept in each registry office are listed in article 2161 of the *Civil Code*; these include an entry-book, an index of immoveables, an index of names, a register of adresses and a register of commercial pledges.

§10.171 *Entry-book* When a document is presented for registration, it is first entered in the entry-book; by this means, all acts deposited for registration are entered in this book by the registrar as they come in, the registrar noting the year, day and time when each document was presented for registration, as well as all other particulars stipulated in paragraph 3 of article 2161. The main function of the entry-book is to document which of any two registrations was entered first. However, careful research requires knowledge of either the date when a document was registered or its registration number.

§10.172 *Index of names* The index of names is an alphabetical register of the names of all persons mentioned in the acts registered but not entered in the index of immoveables, as indicated in paragraph 1 of article 2161. By this register it is possible to trace the registration number of an act through the names of the parties. It also makes it possible to trace any acts not mentioned either in the index of immoveables or in the register of pledges, such as marriage contracts, deeds of gifts and renunciations of succession.

§10.173 *Index of Immovables* The index of immovables is the most important of the registers because it serves as a link between the acts registered and the immovables affected by them. In this register, which is based on the cadastre, a separate page is devoted to each immovable that is identified as a distinct lot in the official cadastre. As indicated in article 2170, the registrar prepares an index of immovables as soon as the plans of the official cadastre have been deposited. Likewise, if a new subdivision comes into effect, the registrar devotes one page of the volume to each lot which bears a new number. The index of immovables, which is the register most often consulted, constitutes a true research tool, since it contains all the information concerning a given immovable. The reason for this systemization of information concerning immovables is that immovables must be described on the basis of their cadastral identity and any document affecting an immovable must be registered in the index of immovables.

§10.174 *Other registers* There are other registers in the registry office that serve purposes accessory to the custody of acts. The complementary register stipulated in article 2164a adds to and updates entries made in the margins of registered documents. There is also a register of commercial pledges, a register of farm and forest pledges and an agricultural zoning register. Although the register of addresses was abolished on July 18, 1980, the mechanism of notice of address remains in effect, in conformity with article 2161c of the *Civil Code*, except that the notices are now only indexed or registered in the index of immovables. The notice of address is a mechanism intended to protect hypothecary or privileged creditors as well as the beneficiaries of a declaration of family residence by allowing the registrar to inform them of any notices of judicial proceedings initiated against the immoveable serving to guarantee the loan.

Modes of Registration

§10.175 *Introduction* Under the provisions of article 2131, there are two modes of registration: registration by deposit which is the most usual and, in a great many cases, the obligatory method and registration by memorial which is allowed only in those cases where registration by deposit is not prescribed by law.

§10.176 *Registration by deposit* Registration by deposit, under the provisions of articles 2132 to 2135 of the *Civil Code*, is effected by the deposit in the registry office of two authentic copies of the document to be registered.

Registration is certified upon one of the copies, the minute, hour, date and registry number being recorded on it before it is returned to the party who requested the registration. The other authentic copy, certified in the same manner, remains in the archives of the office after it has been recorded in the entry-book, the index of names and the index of immoveables. A document in private writing must be attested by two witnesses, both of whom must sign it, and it must be sworn by one of them, as stipulated by paragraph 5 of article 2131.

§10.177 *Registration by memorial* Registration by memorial, as governed by articles 2136 to 2145, is carried out in the same manner as registration by deposit except that, instead of the document itself, a summary, setting forth the the real rights which one of the parties wishes to register, is deposited. The Act respecting transfers of property in stock also allows registration by memorial.

Place of Registration

§10.178 Article 2092 stipulates that "the registration of real rights must be made at the registry office for the division in which the immoveable affected is either wholly or partly situated." There are also various other rules, such as that contained in article 2126: "Renunciation of dower, of successions, of legacies, or of community of property cannot be invoked against third parties unless they have been registered in the registry office of the division in which the right accrued." However, if the act concerns an immoveable, it must also be registered in the registry division where the immoveable is situated.

Duties of a Registrar

§10.179 The keeping of a registry office is entrusted to a registrar appointed by order of the ministère de la Justice, as stipulated in article 2159 of the *Civil Code*. The registrar is required to execute the prescriptions contained in the *Civil Code* and in statutes relating specifically to registration. In addition to his responsibility for the day-to-day operation of the office, he must fulfill the following duties in particular:

(1) registration of acts, providing of registers and other documents upon request and preparation of certified copies;
(2) issuing of certificates of search, under the provisions of article 2177 of the *Civil Code*; and
(3) cancellation of real rights, under the provisions of article 2148 of the *Civil Code*, upon request of the parties or in compliance with a judgment, or upon the filing of an application, in conformity with article 2157b of the *Civil Code*.

§10.180 *Cancellation* Cancellation is a procedure by which the registrar remarks on the margin of an entry that the instrument is no longer valid. Cancellation may thus be defined as the judicial revokation of a registration. Most cancellations concern accessory real rights such as privileges and hypothecs. In view of the importance of the operation, the registrar must be satisfied, before he proceeds, that the documents presented to him are sufficient to bring about cancellation. Cancellation is one of the most important of the duties performed by the registrar, whose activities are otherwise limited mainly to ensuring that the correct form has been used on the documents submitted for registration. There are three modes of cancellation of real rights: voluntary cancellation, that is, with the consent of the creditor; judicial cancellation, after a judgment of last resort; and legal cancellation, in the cases prescribed by law.

§10.181 *Certificate of search* This document, which is issued by the registrar, is a summary of the titles registered upon a given immoveable during a certain period of time. As prescribed in article 2177,

> [T]he registrar is bound to deliver to any person demanding the same a statement certified by himself of all the real rights affecting any particular immoveable, or which may affect the whole of any person's property, or of all hypothecs created and registered during a stated period or only against certain proprietors of the immoveable designated in a written requisition to that effect, containing a sufficient description of the owners.

The certificate of search reproduces, in order of the date of registration, the entries in the index of immoveables, giving in each case the names of the parties and, in the case of an authentic act, of the notary, the date of the act, the kind of right, the consideration, if applicable, and the number and date of the entry. A certificate of privilege or of hypothec is a document analagous to the certificate of search, but relates exclusively to uncancelled hypothecs and privileges. The certificate of search is an authentic document issued by the registrar in his capacity as a personally accountable public officer.

Effects of Registration

§10.182 *Introduction* Under the provisions of article 2082, "registration gives effect to real rights and establishes their order of priority according to the provisions contained in this title." The wording of this article makes some explanation necessary for it is not the registration as such which gives effect to real rights, but the acts from which they arise. The true effect of registration of real rights is that they can be invoked against third parties who neglect to register their rights before those of another party whose title comes from the same author. In short, though not essential to the validity of an act, registration is an indispensable complement if the right is to be invoked against third parties.

§10.183 *Order of priority of principal real rights* Article 2083 begins by stating the following general rule: "[a]ll real rights subject to be registered take effect from the moment of their registration against creditors whose rights have been registered subsequently or not at all." This rule is then softened in the statement which follows: "[i]f however a delay be allowed for the registration of a title and it be registered within such delay, such title takes effect even against subsequent creditors who have obtained priority of registration." There are, in fact, several situations, such as the one covered in article 2099 concerning the transfer of mining rights by an authentic act, where registration before the prescribed deadline has a higher priority, retroactive to the date of the act through which the right was acquired. Another exception to the rule of priority of registration, as deduced by an interpretation *a contrario* of article 2098, paragraph 2, would be a case where a donor is obliged, notwithstanding the priority of his registration, to renounce his prior claim in favour of a person who has purchased the property for a valuable consideration. This traditional textbook interpretation, however, is at variance with the economy of the system of registration which tends to affirm the security of land titles. On the other hand, if at the moment of the act conferring his title, an acquirer by gratuitous title knows that an alienation in return for payment has already taken place, he loses his prior claim in favour of the purchaser, as stipulated in article 2085 of the *Civil Code*.

§10.184 *Order of priority of accessory real rights* Privileged debts and hypothecs take precedence according to their respective ranks as set out in article 2130:

(1) in accordance with the first paragraph, privileged rights which, on the basis of article 2084, are not subject to registration are ranked according to the order established by article 2009 and take precedence over hypothecs;
(2) the second paragraph adds that the privileges that must be registered within a certain period of time also take precedence in accordance with the order established by article 2009, if registration has taken place within that period and they take preference over hypothecs;
(3) then come privileges registered without a deadline, followed by hypothecs which are ranked among themselves according to the date of registration as specified in the third paragraph of article 2130;
(4) the fourth paragraph states: "[i]f, however, two titles creating hypothec be entered at the same time, they rank together concurrently";
(5) under the provisions of the fifth paragraph, "if a deed of purchase,

and a deed creating a hypothec, both affecting the same immoveable, be entered at the same time, the more ancient deed takes precedence";
(6) finally, the last paragraph decrees that "no hypothec has any effect without registration."

REFERENCES

Huot, L.P. *Sûretés*, Cowansville: Les Éditions Yvon Blais, 1984.

"La publication des droits." *Répertoire du droit.* Titres immobiliers, Document 2, Soquij, 1984.

Loi sur les bureaux d'enregistrement. R.S.Q., c. B-9.

Bills of Lading Act and Act respecting transfers of property in stock. R.S.Q., c. C-53.

BOUNDARY DETERMINATION

Introduction

§10.185 The marking of boundaries is an act as old as property law itself. In antiquity and throughout much of history it assumed a sacred character and today it is still considered a solemn act which can only be carried out by a land surveyor.

§10.186 Pothier incorporated the provisions of the *Coutume de Paris* concerning boundary operations into his work on civil law. These provisions were later included in article 646 of the *Napoleonic Code*, which states, "[e]very proprietor may oblige his neighbour to settle the boundaries between their contiguous lands. The costs for the settling of boundaries are to be shared." When our civil law was codified in 1866, article 646 of the *Napoleonic Code* became article 504 of our *Civil Code*. After several amendments, article 504 now reads, "[e]very proprietor may oblige his neighbour to settle the boundaries between their contiguous lands. Boundaries may be determined either by mutual consent between neighbours, and by their mere act, or with the intervention of judicial authority."

Nature of Boundary Determination

§10.187 In our *Civil Code*, as in the *Napoleonic Code*, boundary determination is considered to be a real servitude of the type which arises from the location of a given property. The decision of the authors of the

Napoleonic Code to classify boundary determination as a servitude is still being contested today by some law experts, but, as Fortunat Lord writes in *Termes et Bornes*,[24] "[s]upposing that this charge is not completely in order, any objection to it would crumble before the will of the codifier. To change its given nature would mean destroying the law under the pretext of interpreting it." And Mignault, in his *Traité de droit civil*,[25] writes, "[b]ut the law, by classifying as a servitude the obligation to establish one's boundaries, has placed it in the realm of reality, where it has a completely objective status." Thus, each of two contiguous properties must have a boundary for the sake of the other, so that each property is both dominant and servient. The requirement that for each of two neighbouring properties the end of one and the beginning of the other must be established and monuments be placed on the line constitutes a mutual obligation which is a real charge affecting the property.

Definition and Nature of Boundary Determination

§10.188 Boundary determination is an operation by which the separating line between two contiguous plots of land is marked by means of clearly visible and durable signs after the line has first been determined by delimitation. Boundary determination is thus carried out in two distinct steps: first, the dividing line between the two properties is determined, and then it is marked by means of visible, solid and durable signs called monuments. Boundary determination is a right inherent to property, an attribute linked to its very essence. Prescription does not hinder it; the option to have a boundary determined does not become extinguished even if two properties have existed for 30 years without marked boundaries. Boundary determination is a solemn act and can only be performed by a land surveyor. Moreover, when carried out under the supervision of a court, it assumes the quality of a legal act.

Conditions for Boundary Determination

§10.189 Three conditions must prevail before boundary operations can be undertaken: the immoveables affected must be distinct and belong to two different proprietors, they must be contiguous and their boundaries must be unmarked. Obviously, there would be no point in determining the boundary between two immoveables belonging to the same proprietor.

---

24 Lord, *supra*, note 21 at 12.
25 P.B. Mignault, *Droit civil candien*. (Montreal: C. Théoret, éditeur, 1897) Vol. 3 at 39.

The word proprietor is to be understood here in its broadest sense to include "mere proprietor," "usufructuary," "user," and "emphyteutic lessee." In addition, the courts have added the purchasers of property subject to redemption and owners of properties subject to a resolutive condition. Holders of superficiary rights, logging rights, fishing or hunting rights and occupancy permits also have the capacity to determine boundaries. The public and private domains of the State are also subject to the laws governing boundary determination because article 9 of the *Civil Code* does not exempt the Crown from the authority of the law. To be entitled to delimitation, the two properties must be contiguous, which means adjacent and not separated by the property of a third party, a public road, street or lane, a railway track or a watercourse. Furthermore, no boundary markers may be present. Boundary operations may be carried out if the boundaries between the two properties have never been delimited, if monuments which had been placed previously have disappeared or if fences or line structures have been placed incorrectly. The courts have added to this list, as a fourth possibility, the case where the property beyond an old boundary line has been affected by prescription.

Categories of Boundary Determination

§10.190 According to article 504, boundaries can be determined by agreement between neighbours, and thus in an amicable manner, or through the intervention of a judicial authority. Added in 1966, article 762 of the new *Code of Civil Procedure* sets out the preliminary steps which must be carried out for all boundary operations, whether judicial or out of court. The operation becomes judicial from the very outset if the parties, after having been given notice, cannot agree either on the right to have the boundaries determined or on the choice of the surveyor and if one of the parties takes its request to court. An extra-judicial determination of boundaries becomes judicial if one of the parties does not accept the conclusions of the surveyor's report and wants to challenge it in court. In the first case, the boundary is determined entirely under the authority of the court, while in the second case the court intervenes to conclude the matter.

Capacity of the Parties to Determine a Common Boundary

§10.191 In principle, anyone who is entitled to the free administration of his or someone else's property has the capacity to settle his boundaries. Thus, a tutor has the capacity to participate in a purely administrative boundary determination, but if the result is to enlarge or diminish the property, the act becomes similar in nature to an alienation where authorization must be obtained from the family council and the same legal procedures as for an alienation are to be followed.

Formal Request

§10.192 The rules concerning boundary determination are contained in articles 762 and following of the *Code of Civil Procedure*. Civil procedure requires that any demarcation of boundaries be preceded by formal notice. According to article 762,

> [A]n owner who wishes to compel his neighbour to have the boundaries determined, to verify ancient boundaries or to rectify the division line between their contiguous immoveables, must first put him in default to consent to do so and to agree upon a surveyor to carry out the necessary operations. Such putting in default is effected by the service of a notice containing:
>
> (1) a statement of the demand and of the grounds therefor without mentioning disturbances, damages, or other claims;
> (2) the description of the immoveables concerned;
> (3) the name and residence of the surveyor suggested for the operations;
> (4) a notification that the demand will be brought before the competent court unless, within fifteen days, there is agreement on the right to have the boundaries determined and on the choice of a surveyor.

§10.193 It may be that the parties wish to proceed immediately. Legislation has provided for this possibility by adding the following paragraph to article 762,

> If the owners of contiguous immoveables agree to have the boundaries determined and on the choice of a surveyor, the putting in default provided for in the first pragraph may be replaced by an agreement to have the boundaries determined signed by such owners and containing the elements which should be contained in the putting in default that it replaces.

§10.194 The formal notice must be served by a sheriff. It may also be produced in court as an introductory document to initiate an action if the parties cannot agree on the choice of the surveyor or the right to have the boundary settled. The new procedure allows the party that had the notice served to submit it to the clerk of the court together with the report of the sheriff who has served the notice; this step initiates action and is equivalent to a writ of summons addressed to the other party. The court will then decide on the right to determine the boundary and, if the right exists, it will commission a surveyor to commence the necessary operations. In practice, the right to determine a boundary is hardly ever contested and in the majority of cases the parties agree on the choice of a surveyor, so that the boundary is determined out of court. The surveyor, however, must proceed as if he had been appointed by the court. If one of the parties is not satisfied with the conclusions of the report, the surveyor's work is not lost for his report together with all supporting documents is added to the file of the case, just as if the surveyor had been appointed by the court.

Grounds for Recusation

§**10.195** Before accepting an appointment to determine a boundary, the surveyor must disqualify himself if he is liable to a ground of recusation. Grounds of recusation are the same as those for judges and are set out in article 234 of the *Code of Civil Procedure*. If boundary determination is judicial in nature, the surveyor who is liable to a ground of recusation must address a request to the court, which will decide on the matter.

Oath of Office

§**10.196** In accordance with article 763 of the *Code of Civil Procedure*, surveyors must take an oath of office. They are thereby exempt from the obligations to which experts are subject under the provisions of article 418. The *Code*, thus, recognizes them as public officers.

Procedure for Determining Boundaries

§**10.197** The procedure that surveyors must follow is outlined in article 763 of the *Code of Civil Procedure*, as follows,

> If the parties agree to have the boundaries determined and on the choice of a surveyor, the latter proceeds, under his oath of office and in the same manner as an expert, to visit the site, to study the titles and to hear the parties and their witnesses, and does everything which he considers necessary; he prepares a plan of the site indicating the respective pretensions of the parties, and submits to them a report of his operations showing the division line which appears to him to be just.

The authors of this chapter are of the opinion that the surveyor named to the task must fulfil the terms of his appointment by carrying out in the established order the tasks enumerated in this article.

§**10.198** After he has been appointed, the surveyor arranges with the parties the date of the inquiry and visit to the site. He must have the parties notified by a sheriff within the time prescribed by law. He then proceeds with the surveying operations as soon as possible in the presence or absence of the parties; there are no legal provisions in the matter. Each visit and adjournment must be recorded in his report.

§**10.199** The surveyor is not free to hear or refuse to hear the parties or their witnesses; he has no discretionary powers in this regard, unless exempted by the court. He summons witnesses and if any refuse to appear or to make available useful evidence, he may force them to do so by means of subpoenas issued by the prothonotary.

§10.200 The surveyor personally directs the inquiry and hears the testimony, beginning with that of the applicant. He consults all the documents presented and has them filed according to the usual numbering system of courts of law. Through his questions he gains information on possession and prescription and he examines the application of these principles with regard to the law and the facts. He records the objections to be decided upon by the court. In accordance with article 420 of the *Code of Civil Procedure*, testimony must be received in writing, signed by the witnesses and countersigned by the expert, unless it has been recorded by a duly sworn stenographer.

Surveying Operations

§10.201 After having sworn in his assistants, the surveyor proceeds with the operations with a view to locating the claims of the parties. To that end, he carries out an exact survey of the sites in order to make a plan that shows all signs of occupancy, such as fences, ditches, hedges, trees and buildings. If necessary, he extends his operations to the neighbouring properties, even if they are some distance away. Articles 47 and 48 of the *Land Surveyors Act* protect the surveyor against any undue interference in the performance of his duties. Moreover, the law authorizes the surveyor and his assistants to pass over any property or perform any operations thereon which they consider necessary. A surveyor examines the configurations of the soil, its special features and natural or artificial divisions. He searches for any marks which would determine if a wall or ditch is common; he examines fences, hedges, ditches and watercourses in order to determine whether they form part of the property or serve to mark the boundary. In short, he already begins to sort the information that will become useful in making his decision.

Evidence

§10.202 *Introduction* The surveyor first studies the documents which the parties have deposited in the course of the inquiry. He does not limit himself to that evidence only, however; it is his task to collect carefully all information contained in previous titles, surveying reports and land inventories, as well as the cadastre, plans, maps and other pertinent documents. Law books and court cases have established unequivocally that boundaries must be determined in conformity with the rights and titles of the parties. This is the only criterion to guide the surveyor in the evaluation of evidence, a task which he must carry out objectively, duly recognizing the value and importance of all the evidence which is made up of the following:

(1) titles
(2) witnesses
(3) admissions
(4) possession
(5) prescription
(6) secondary documents, and
(7) presumptions.

§10.203 *Titles* The aim of boundary determination is to mark the limits of the property rights which are normally stated in a title. It may happen that one of the parties is proprietor without a title, in which case proof of property rights is required. The main translatory acts affecting property are sales, gifts, wills, exchanges, transfers and division. If these acts are clear and precise they will determine the separation line between two properties unless there has been prescription. Titles are not all of equal value for the determination of boundaries. The nature of each title must be examined and its chain established, it must be determined if titles emanate from a common author or from different proprietors and the wording of the description must be examined for every author, from first to last. When titles originate from the same author, the surveyor must allocate to the party with the oldest title all the land indicated in that title, unless the other party has acquired all or part of the property by prescription.[26] If there are no common titles, the surveyor must study the title of each of the parties and extract any information contained in them. In principle, older titles have priority over more recent ones. It some cases, however, recent titles may indicate modifications in the boundaries and area of the properties, in which case the surveyor must examine carefully all titles where the description matches the actual site. Fortunat Lord has extracted several rules from law books and court records to guide the surveyor in the evaluation of titles. The main rules are as follows:[27]

> [Second rule] Titles are valid only for what they express with certainty; if an act states that a property lot contains between fifteen and twenty *arpents*, then this title is clearly for fifteen *arpents*; beyond that it announces an uncertainty which only possession can dispel (1 Pardessus, p. 305).

---

26 *Biron v. Caron* (1895), S.C.R. 451.
27 Lord, *supra*, note 21 at 111 and following.

[Third rule] If all the titles together result in an area equal to the total area of the lots being surveyed for the purpose of boundary determination, then the area indicated in each title shall simply be accepted as such (Bugniet, p. 54).

[Fifth rule] If it is obvious that someone has everything that is contained in his title he is not necessarily justified in demanding that his neighbour who possesses more than is given in his title should share that excess with him (*Boulet v. Bourdon* (1882), 12 S.C.R. 121).

[Sixth rule] When the title of one of the neighbours expressly gives him a specific amount of land, and if he possesses it in fact, the presumption of usurpation is difficult to admit, particularly if the two properties are used for different crops (1 Pardessus, p. 312).

[Eighth rule] If one of the parties has a title that fixes with certainty, and in an unambiguous manner, the area of his land, and if the titles of the other indicate the area only in an approximate, vague, or ambiguous manner (i.e. "approximately"), the surveyor must first verify the first (Bugniet, p. 58).

[Ninth rule] If the titles show both the size of the lot and a precise boundary line which make anticipation unlikely, one must decide according to the apparent signs rather than by area, which is nearly always indicated in an uncertain or approximate manner in the acts (*Boivin v. La Cie des Eaux et de l'Electricité de Chicoutimi* (1915), K.B. 394).

[Tenth rule] When the titles submitted indicate precise limits between that property and its adjacent parts, the surveyor must consider these limits rather than the area, because the titles have transmitted a well-defined body, thus making the area redundant (*Vallée v. Gagnon* (1909), K.B. 165.)

[Twelfth rule] A proprietor whose property is enclosed within its boundaries and who does not enjoy any land beyond that stated in his title, must be maintained in his possession, although his neighbour may not possess as much land as he has according to his title (*Tétreault v. Paquette* (1891), S.C.R.).

[Thirteenth rule] When after measuring it is recognized that one of the neighbours has more than the amount recorded in his titles and the other has less, one has to supplement the property of the other by taking some from the one who has more (*Boulet v. Bourdon* (1882), S.C.R.).

[Fifteenth rule] When a translatory title specifies that land is to be detached from a contiguous lot or taken out of a larger lot, and if a precise dimension is given, the deficit in this dimension must be made up out of the original lot or land whenever practicable (*Saint-Aubin v. Brunet* (1911), C.R.).

[Sixteenth rule] When there is a discrepancy between the titles of two neighbours, the rule *melior est causa possidentis* gives the advantage to the titles of the possessor (Bugniet, p. 59).

[Nineteenth rule] A title described in terms of its boundaries by metes and bounds shall be preferred to a title only designated by cadastral number (*St Lawrence Light Co. v. Clément* (1915), K.B.).

[Twentieth rule] When a lot registered in the cadastre is given in a title as the boundary of a land, this lot must be understood strictly within the limits given in the plan and book of reference of the cadastre, and not with boundaries modified by prescription (*Vallée v. Gagnon* (1909), K.B. Vol. 19, p. 165).

[Twenty-first rule] When a township line that is marked and legally recognized by the Crown has been given as the boundary between two lots, this line, if there has not been any prescription, must be taken as the dividing line between the lots regardless of their contractual dimensions (*Duguay, v. Vincent* (1893), 2 K.B. 407).

[Twenty-second rule] When the mensuration of the surveyed lands, except for evidence to the contrary, shows a surplus or a deficit in relation to the sizes indicated in the titles, the surplus or deficit must be shared among the parties involved in the determination of boundaries proportionally to the size of each property, as indicated in the respective title, but with due regard to possession (*Fuzier-Herman, see Bornage*, no. 296).

If the area of each of the properties is larger than appears in the titles, but without the surplus being distributed proportionally, the boundaries should be determined according to present possession, because the party with the smallest surplus, but still more land than given in the title, would have no means to force the other to give up any of its own, because of the principle *melior est causa possidentis* (*Marcoux v. Bélanger* (1894), 5 Que. S.C. 538).

If the titles can be interpreted in two different ways or if certain parts of the titles contradict others, the surveyor must refer to the rules of interpretation and mention these in his report.

§**10.204** *Witnesses* The surveyor must evaluate the testimony of the witnesses. Testimoninal evidence often serves to dispel doubts and confirm a presumption or an interpretation of a title; it fills the gaps in the documentary evidence and takes the place of written evidence if there is a lack of documents or titles. According to article 319 of the *Code of Civil Procedure*, "the testimony given by a party, on his own behalf or at the instance of another party, may serve as a commencement of proof in writing against him." Apart from the evidence brought before him by the parties engaged in the boundary determination, the surveyor can, by virtue of article 50 of the *Land Surveyors Act*, "interrogate, under oath, any person whom he thinks capable of giving him information or is in possession of any writings, plans or documents touching on boundaries or limits of the land which he is employed to survey."

§**10.205** *Admission* By admission is meant the acknowledgement of a fact in the neighbour's favour. An admission is made if one of the parties recognizes as true a fact inconsistent with his ownership rights or his claim thereto. Under the provisions of article 1245 of the *Civil Code*, "a judicial admission is complete evidence against the party making it." But to be acceptable, an admission must be clear, precise and complete and there may not be any doubt as to its object. If so made, an admission exempts the adverse party from proving the admitted fact.

§10.206 *Possession* Possession and prescription have been studied in the first part of this chapter as part of the discussion of the *Civil Code*. Possession plays an important role in boundary determination. In doubtful cases, or where there is insufficient documentation, the surveyor must refer to the possession of the parties. Possession is always a good criterion for applying or interpreting titles; it can even be used as a criterion for modifying or overriding them if the characteristics required by article 2193 of the *Civil Code* are present. He who possesses is presumed to possess in good faith. When possession is certain and perfect it renders proof unnecessary; it is up to the party who rejects a delimitation based on possession to justify his claim. However, in the absence of precise facts, it is wise to gather all evidence which could shed light on possession, such as old plans, signs of cultivation and the cadastre.[28] Obviously, it is always best if possession conforms with the title. If possession extends over a larger area than that given in the title it can facilitate the acquisition of property rights upon the surplus by means of acquisitive prescription.

§10.207 *Secondary documents* In the determination of boundaries, titles are the basic documents used by the surveyor to begin his task, but if they are incomplete or refer to other documents that describe the immoveable, the surveyor must consult these as well, in order to understand fully all aspects of the situation. The first of these secondary documents is the cadastre, which consists of a plan and a book of reference: the plan designates immoveables by serial number, and the book of reference gives its dimensions, area and adjacent parts. The cadastre came into use in Québec around 1860 in order to dispel uncertainty about titles and to facilitate real estate transactions and the registration of real property. Paragraph 3 of article 2174 of the *Civil Code* attributes a secondary importance to the cadastre,

> No right of ownership can be affected by any error in the plan or, where such is the case, the book of reference, nor can any error of description, dimensions or name be interpreted to give any person better right to the land than his title gives him.

For titles which predate the cadastre, the seigniorial land-rolls and registers can be useful documentary evidence for the delimitation of immoveables. The surveyor must consult the survey plans, maps, photographs and any

---

[28] See *Lambert v. St-Lauveur*, 20 R.L. 46 and *Rochon and Boudreau v. Gaudreau*, [1968] Que. Q.B. 889.

other document that can help solve the problem. Even if only of secondary value, these documents are very useful and often serve to dispel ambiguity and clarify titles. Moreover, they often corroborate or invalidate the declarations of the testimonial evidence.

§10.208 *Presumptions* By virtue of article 1238 of the *Civil Code*, "presumptions are either established by law or arise from the facts which are left to the discretion of the courts." In the course of the inquiry and the surveying operations, the surveyor will come across a great number of facts, of which some will be proven and others admitted by the parties. Through his analysis of these facts, the surveyor can determine if a presumption applies. It goes without saying that the greatest prudence is required in the exercise of these discretionary powers. Article 1239 establishes two kinds of presumptions, legal presumptions, or *juris et de jure*, and presumptions of fact, or *juris tantum*. Legal presumptions "exempt from making other evidence those in whose favor they exist," while presumptions of fact "may be contradicted by other proof."

The Report

§10.209 *Introduction* Article 763 of the *Code of Civil Procedure* requires that the surveyor submit a report on his operations to the parties concerned in which he indicates the line of division that he deems to be just. However, if the surveyor has been appointed by the court he must submit his report to the clerk of the superior court and notify the parties that he has done so. A key document in the first stage of the boundary operation, the surveyor's report to the parties outlines the task he has been entrusted with. In it, he states the problem and, from his examination of titles and possession, arrives at the basic premises on which to establish the respective rights of the parties. He then reviews the crucial facts and outlines his reasons for deciding on a particular line of demarcation. In general terms, the report must constitute a complete expert appraisal; it must contain the facts and present the grounds for the decision which was reached in a manner that allows the parties or the court to evaluate it.

§10.210 *Content of the report* The surveyor's report is to consist of three parts: the introduction, main body and conclusion. The introduction contains the following information: the names of the parties, their residences and qualities; the name of the surveyor, his quality and his residence; his mandate — whether he is acting on behalf of the parties or of the court — and the date and object of same; finally, the names in full of his assistants and the records of their swearing-in. The body of the report, which contains a step-by-step description of the operation from the moment when the surveyor took charge of the file, must contain the following information:

(1) notice to the parties and witnesses to appear;
(2) the writ of subpoena to the witnesses, their names in full, qualities and domiciles;
(3) the minutes of the visit to the site and the names of the persons present;
(4) statements of the claims of each party;
(5) records of the swearing-in of the witnesses, the names in full of the witnesses heard for the petitioner and for the respondent;
(6) a summary of the evidence;
(7) a description of the survey operations he has carried out, including the evidence collected, the marks of possession and their locations; the measured lots and the immoveables on which these operations were carried out;
(8) the date of preparation of the plan and an explanation of its component parts; and
(9) an analysis of the titles and the testimony given, and the relation of both to the evidence collected on the site.

Finally, the surveyor states his opinion and explains the reasoning behind his decision based on the titles, possession or prescription. Before the adoption of the new *Code of Civil Procedure* in 1967, a surveyor faced with a point of law arising from the titles or possession had to present more than one possible solution to the court which would then make a decision. Today, under the provisions of article 763, the surveyor must submit a report "showing the division line which appears to him to be just." The surveyor will decide on points of law, where necessary, and the court will evaluate this aspect of his decision just as it evaluates the facts presented.

§10.211 *Documents attached to the report* The surveyor must attach to his report all the documents submitted to him as well as all correspondence with the parties, in particular:

(1) a copy of the consent of the parties involved in the boundary determination or of the court decision by which he was appointed;
(2) notice to the parties of the date of the inquiry;
(3) testimony, titles and plans submitted by the parties and their witnesses;
(4) extracts from the cadastral plan and book of reference; and
(5) archival documents collected by the surveyor.

Acceptance of the Conclusions of the Report by the Parties

§10.212 If the parties accept the conclusions of the report, they must inform the surveyor of their decision in writing. The consent of the parties must be formal and unambiguous. The surveyor cannot presume the tacit consent of one of the parties, for if the minutes are not signed, the operation is not complete. As stated in article 51 of the *Land Surveyors Act*, boundary operations must be "agreed to by the parties or ordered by the Court." If one of the parties does not accept the conclusions of the report, articles 765 and 767 of the *Code of Civil Procedure* prescribe the procedure to follow for the completion of the boundary operation.

Demarcation

§10.213 *Notice to the parties* Whether the demarcation of the boundary is accepted in conformity with the conclusions of the report or on the basis of a judgment, the surveyor must notify the parties that he will be proceeding with the monumentation. This notice must be communicated within the period of time prescribed by law, which is five days, and must give the day and time when the monuments are to be placed.

§10.214 *Presence of the parties* After the parties have been summoned, the operation normally is carried out in their presence or in the presence of their authorized representatives. In the absence of either or both of the parties, the suveyor places the monuments in the presence of witnesses and he will record this fact in the minutes. His authority to do so stems from two legal sources. Firstly, according to article 767 of the *Code of Civil Procedure*, "the court determines the boundary line, and appoints a surveyor who places boundary marks in the presence of witnesses." Secondly, article 52(1)g of the *Land Surveyor's Act*, states that

> [A] land surveyor determining the boundaries shall, when he has finished his operations, draw up a minute declaring therein on pain of nullity the presence of the parties at the operations, or their authorized representatives, or in their absence, the names and qualities of the witnesses who were present at the placing of the boundary markers.

When a surveyor proceeds to place monuments by virtue of a judgment he must record all the steps in the form of minutes, which he then submits to the clerk of the court. Article 767 of the *Code of Civil Procedure* stipulates that "the homologation of the minutes by the court makes proof of the complete execution of the judgment." If the surveyor has proceeded to place the monuments with the agreement of the parties, he records the operations in minutes which are signed by the parties.

§10.215 *Placing of monuments* To mark the boundary, the surveyor must place monuments at the two extremities of the line, wherever they may be. If the line is not straight or if the whole perimeter of a site is to be marked, the surveyor must place monuments at all the angles which will then be joined by straight lines. Whatever the length of the line, a monument must be visible from the adjacent monuments. It may happen occasionally that the nature and location of the site does not permit the placing of monuments in comformity with the law. Article 51(4) of the *Land Surveyors Act* stipulates that in such a case the

> [L]and surveyor shall enter such fact in his minutes; he shall fix the boundaries and describe his operations by designating the streets, neighbouring properties and other fixed objects in such manner that any other land surveyor may by means of such minutes repeat the operations and determine the boundaries, points, lines and other details.

Article 52(j) requires, moreover, that the surveyor declare in his minutes "information helpful in locating and ascertaining the identity of the boundary markers which he has placed and the lines which he has established."

Minutes of a Boundary Determination

§10.216 *Contents of the minutes* The minutes of a boundary operation constitute an authentic act stating what the surveyor has seen and done in the course of his official assignment. Article 52 of the *Land Surveyors Act* obliges the surveyor to record minutes of his operations which, to avoid nullity, must contain the following information:

(a) the judicial district where the bounded real estate is situated;
(b) the date on which the boundary operations are carried out;
(c) the names of the parties to the boundary determination, their qualities, and residences;
(d) his name, his right to practice the profession of land surveyor and the address of his office;
(e) the titles and the documents which he has examined;
(f) under what authority he has placed the boundary markers;
(g) the presence of the parties at the operations, or their authorized representative, or, in their absence, the names and qualities of the witnesses who were present at the placing of the boundary markers;
(h) the operations which he has carried out, including the referencing;
(i) the names of the assistants, mentioning their swearing-in, if necessary, their ages and domiciles;
(j) information helpful in locating and ascertaining the identity of the boundary markers which he has placed and the lines which he has established;
(k) the date on which he draws up such minutes, the date and place of signature of the parties, if necessary, and the number he gives his minutes.

Moreover, under the provisions of this Act, the registration of the minutes of a boundary operation, whether judicial or out of court, has been mandatory since 1974.

§10.217 *Legal force of minutes of a boundary determination* The minutes of a boundary determination, recorded in the legally prescribed form, constitute a definite and irrevocable title. After they have been signed by the parties, they assume the value of a contract and can only be annulled for causes of nullity in contracts, such as error, fraud, violence or fear and lesion, under the provisions of article 991 of the *Civil Code*. The minutes are evidence of the area and limits of a property. As such, it is the key document concerning the boundary, for "what would be the purpose of marking boundaries, if the legal existence and force of proof of such boundaries were not shown in any written document."[29] The parties can refer to them in order to know the size of their immoveables and to find out if the monuments have been moved or destroyed. It is, thus, essential that a plan of the site showing the location of the monuments, their referencing to the limits of the original lots, the existing buildings and any other pertinent information be attached to the minutes. To summarize, the plan completes the data mentioned in the minutes and makes them easier to understand.

Costs for Boundary Determination

§10.218 Under the provisions of the *Civil Code*, the costs for determining a boundary are to be shared. These costs include the expenses incurred for an action *ex parte* when the application has been presented in court. If the action was contested, the losing party must cover the expenses unless the court decides otherwise; the relevant provisions are contained in article 768 of the *Civil Code*. In other words, the costs for a judicial boundary determination are left to the discretion of the court. If a boundary determination is carried out extra-judicially, surveyors generally apply the principle of equity in the proportional distribution of the costs.

REFERENCES

Beaulieu, M.-L. *Le bornage, l'instance et l'expertise. La possession et les actions possessoires.* Quebec: le Soleil, 1961.

Lachance, P. *Le bornage.* Québec: Les Presses de l'Université Laval, 1981.

Lord, F. *Termes et bornes.* Montréal: Wilson et Lafleur, 1939.

---

29 See *Ouimet v. Desmarais*, 21 R. de Jur. 96.

THE PROFESSION OF LAND SURVEYING

Introduction

§10.219 *Role of the land surveyor* The land surveyor takes part in the administration of property law by determining and marking the boundaries of plots of land, identifying these boundaries in the cadastre, dividing them into portions and examining their state with regard to titles, the cadastre, charges and servitudes and laws and regulations that might affect them, as well as by the cartographic representation of public or private property.

§10.220 *Scope of the land surveyor's activities* The important tasks that are exclusive to the land surveyor are described in section 34 of the *Land Surveyors Act* as follows,

> A land surveyor is a public officer. The following constitute the practice of the profession of land surveyor:
>
> (a) all surveys of land, measurements for boundary purposes, making boundaries, plotting of plans, making of plans, minutes, reports, technical descriptions of territories, certificates of localization and all documents and operations made by direct, photogrammetric, electronic or other methods connected in any way with bounding, laying out of lots, or establishing the site of servitudes, staking of lots, and scaling of lakes, rivers and other bodies of water in Québec, with the calculation of the area of public and private property, all cadastral operations or compiling of lots or parts of lots, and cartographic representation of territory for the above-mentioned purposes;
>
> (b) establishing and keeping up-to-date the skeleton map of geodetic points of any order of precision and establishing of photogrammetric controls for the purposes of the work enumerated in subparagraph *a*."

Section 35 adds that "no operation defined by section 34 is valid unless performed by a land surveyor and carried out in accordance with the law and the regulations of the Order."

§10.221 *Outline of topics* The following matters will be discussed in this section:

(1) surveying operations on crown property;
(2) delimitation of private property;
(3) determining boundaries between private and crown property;
(4) analysis of the state of immoveables: the certificate of location;
(5) the keeping of the land surveyor's records; and
(6) professional status of the land surveyor.

Surveying Operations on Crown Property

§10.222 *Division of land in Québec* The division of land in Québec was carried out in two distinct phases: during the French period, from 1608 to 1763, concessions were granted in the form of seigniories, and during the English period, from 1763 until now, concessions have mainly taken the form of townships.[30]

§10.223 *Grant of concessions in the form of seigniories* The King of France, in whose name the explorers took possession of North America, granted vast stretches of land in the form of seigniories to persons whom he considered deserving and capable of turning them to good use by allowing colonists to settle there. These first concessions extended along either side of the Saint Lawrence River and inland by about 30 kilometres, as well as along the large rivers of the interior, such as the Richelieu. During the French period, approximately 200 concessions were established.[31]

§10.224 *Survey of seigniorial boundaries* The official surveys carried out by surveyors from France established only the outer limits of the seigniories. The internal divisions were established by persons especially hired by the seigniors; the largest of these divisions were the ranges, which ran more or less parallel to the Saint Lawrence and its tributaries, each range being subdivided into lots that were to be conceded to settlers. In most of the seigniories, especially those along the Saint Lawrence, the outer boundary lines ran in a 45° north-west direction, based on true north. The lateral boundaries of the lots ran parallel to them.

§10.225 *Survey of lateral boundaries of seigniorial lots* In principle, the lateral lot lines ran parallel to the outside lines of the seigniories, as can be seen on existing plans.[32] It is impossible, however, to retrieve the original surveying notes for the seigniories and the plans show neither directions nor dimensions. The law says nothing about the way in which the boundaries of the seigniorial properties were established or how they should be re-established now. Consequently, if a surveyor is asked to re-establish a boundary in seigniorial territory, he must proceed in the same manner as that used to determine the boundaries of private property: he must examine the occupancy of the sites in reference to titles, plans and the book of reference of the official cadastre that was prepared on the

---

30 P. Terlinck, "L'arpenteur-géomètre au Canada," *Publication de l'Ordre des géomètres de Belgique*, No. 43 (Sept./Oct. 1953).

31 J. Bouffard, *Traité du domaine* (Québec: Les Presses de l'Université Laval, 1921) at 12.

32 L. Giroux, "De l'Establissement des lignes latérales de lots dans les seigneuries," *Société de geodésie*, No. 15 (1942).

basis of seigniorial plans and the testimony of persons knowledgeable about the old marks of possession.

§10.226 *Survey of township boundaries* Townships were mentioned in the first imperial instructions to Governor Murray on December 7, 1763. The area of each township was to comprise about 20,000 acres. A ministerial order of October 30, 1794, signed by Lord Dorchester in the name of the King, established a new form of township which was 10 miles and 5 chains long by 10 miles and 3.55 chains wide, amounting to about 64,680 acres altogether. Most of the townships were oriented in a north/south and east/west direction, except for 50 or so situated mainly in the Lac St-Jean region and in the Gaspé Peninsula, where orientation varies from 10° to 20° in the east or west. The townships located on either side of the Saint Lawrence between Montréal and the Gaspé generally form an angle of 45° with the true north.[33] This manner of dividing land into townships of 100 square miles, oriented north/south and east/west, was in use until 1966. Since that year, land has been divided up for special purposes, such as hydrographic basins or blocks for logging and mining concerns.

§10.227 *Survey of lateral boundaries of township lots* The boundary lines of townships surveyed before April 25, 1908 were drawn in accordance with the laws in effect at the time. In several cases, boundaries were established with reference to the range post and the corresponding post of the next range up or down. The land surveyor must employ the same method to re-establish these lines or establish the other lateral lines of the same range. The lateral lines of lots surveyed after April 25, 1908 are drawn in reference to the base line indicated in the surveying instructions for those townships. This base line may be an outer township line, often the centre line and sometimes another line that is shown on the official plan deposited in the surveying archives.[34]

In the other cases, the general rule is that the lateral line begins at the front range post and runs parallel to the base line of that range. Range lines and lateral lot lines established by order of the government to replace those lost through forest fires or any other cause become the true boundaries of the ranges and lots.[35]

---

33 R. Greffard, *Publication de la société de géodésie de Québec*, No 16 (Québec, 1942).
34 *Survey Act*, R.S.Q., c. A-22, s. 1 ff.
35 The volume outlining general instructions for surveyors is issued by the ministère de l'Energie et des Ressources. The most recent revised edition appeared in April 1985.

Delimitation of Private Properties

§10.228 *Procedures for determining boundaries* Since legal regulations for delimiting private property are less stringent, several different procedures may be followed. Firstly, boundaries can be determined on the basis of documents that clearly identify where they are to be located, for example a recent cadastral subdivision with a grid or reference points clearly marked on the property. Secondly, boundaries can be established by a land surveyor on the basis of the information he has collected by examining the site, the titles, the cadastre and any other pertinent document. Thirdly, the property owners can decide upon or give their consent to a boundary; in that case there is no need for the land surveyor to evaluate the evidence.

§10.229 *Monumentation* Once the boundary has been established, it can be made legally valid — even for successors or their assignees — by the placing of monuments and the preparation of boundary minutes signed by the parties and the surveyor and duly registered.[36] Thus, this document is a contract authenticated by the signature of the land surveyor as a public officer. Without signed minutes, the boundary is valid only with the acquiescence of the neighbours concerned. Thus, a neighbour who does not want to recognize a boundary or the monuments after the land surveyor has completed his assignment may demand a boundary determination survey of the contested boundary in the manner prescribed by law.[37]

Determining Boundaries Between Private and Crown Property

§10.230 When establishing boundaries between private and Crown property, the surveyor is bound by the same rules that apply when only private properties are involved. The land surveyor must examine the titles, the plans of the original survey, the cadastre, the marks of occupancy on the land and, if necessary, hear testimony before he can give his opinion on where the limits are to be established. On the other hand, if the boundary under consideration coincides with a boundary or central line of a township, the line can only be re-established with the authorization of the minister in charge of surveys and following the instructions given to that effect.[38] Minutes of a boundary determination survey, if applicable, will then establish definitively the line or lines drawn and monumented and the minister, together with the private owner and the land surveyor, will be required to sign these minutes. When locating such a boundary line between private and crown property, it is generally necessary that the provisions

---

36 *Land Surveyor's Act*, R.S.Q., c. A-23, ss. 52, 53 and 54.
37 *Civil Code*, article 504 and *Code of Civil Procedure*, articles 762 ff.
38 *Supra*, note 34, s. 5.

of the *Code of Civil Procedure* be followed in order to protect the rights of either party, for if a disagreement arises between the parties or if one of them is not satisfied with the manner in which the task is carried out, the work of the land surveyor will not be lost and his report can be submitted to a court of justice for a decision.

Analysis of the State of Immovables: the Certificate of Location

§10.231 The exercise of ownership rights over an immovable carries with it certain duties in regard to neighbouring immovables, in addition to other duties arising from laws and regulations governing community life. Thus, if it is important that the boundaries of an immoveable be clearly defined so that the extent of the occupied space can become a matter of public record, it is also essential that the space within its confines be used in a manner suitable to the well-being and harmony of its surroundings. In appraising the value of an immovable, not only its location and buildings are to be taken into consideration, but also its state in regard to titles, the cadastre, charges and servitudes and laws and regulations that affect it and the neighbourhood. In general, such a study or examination of the value of a property is carried out at the request of an examiner of titles, in most cases a notary, when the immovable becomes the object of a real-estate transaction, such as a sale, transfer, hypothec, servitude or another contract affecting the immovable or, sometimes, at the request of the owner himself, if he wants to determine the quality of his real-estate. This analytical document is called a certificate of location and can only be issued by a land surveyor.[39]

Nature of the Certificate of Location

§10.232 The *Ordre des arpenteurs-géomètres du Québec*,[40] drawing on the accumulated experience of practitioners over the last 30 years, has arrived at the following definition which it has included in its Regulation respecting standards of practice relative to the certificate of location: "The certificate of location is a document consisting of a report and a plan, in which the land surveyor gives his opinion on the present situation and state of a property in relation to titles, cadastre, and the laws and by-laws affecting it. It may only be used for purposes for which it is intended." Thus, right at the start, it is made clear that the document constitutes the opinion of a land surveyor as a land expert. It must be stressed that this document

---

39 *Supra*, note 36, s. 34.
40 R.S.Q., c. A-23.

represents the expert opinion of a person and is not an authentic act.[41] Only the minutes of a boundary determination signed by the parties and the land surveyor or the minutes of a boundary monumentation in compliance with a court decision are authentic acts and must be registered.[42] When a notary issues a report on an examination of titles, he expresses an opinion that is based largely on information obtained from documents listed in the index of immoveables and the certificate of location. In formulating his opinion, he also draws on his experience and skill in the analysis of titles. Likewise, when a land surveyor analyzes the current situation and state of a property in the light of the documents gathered, the laws and regulations that might affect it and his findings at the site, he puts his technical and legal knowledge, skill, experience and judgment at the disposal of his client. He also proceeds under oath of office, thereby ensuring the impartiality of his opinion.[43]

Importance of the Certificate of Location

§10.233 *Importance of the certificate of location for the examiner of titles* A certificate of location provides the examiner of titles with the basic information that allows him to answer the following questions and conduct a better examination:

(1) Is the legal description of the immoveable in the registered deeds obscure or does it give rise to different interpretations? The report and the plan accompanying it are expected to clear up any ambiguity and give detailed information.
(2) Is the identification of the original title of ownership difficult to trace because of the many deeds of purchase registered on the same lot? The report could provide the necessary link by indicating the prior title of the acquired immoveable.
(3) Is there in the chain of title a servitude, the site of which is difficult to locate with regard to the land parcel under consideration? The report

---

41 G. Girard, professional development course, Chambre des notaires (April 1964). See also *Placement Miracle Inc. v. Gérard Larose*, Judge Lawrence A. Poitras presiding, (16 February 1978), District of Montréal 500-05-015927-73 (Que. S.C.). And in *Tourigny v. Park*, [1959] R.P. 385, it was stated at 388:
> Neither the Land Surveyors Act in force at the time (1964, R.S.Q., chapter 263), nor the new act, assented to on July 6, 1973 (Q.S., chapter 61) contains any provision that would give the certificate of location prepared by a land surveyor the status of an official document or authentic deed. In general, a certificate of location constitutes only the expert advice of a professional in this matter.
42 *Supra*, note 36, s. 53.4.
43 Règlement sur l'admission à l'étude et à l'exercice de la profession d'arpenteur-géomètre, *supra*, note 36, r. 1, s. 42.

will indicate whether the immoveable under consideration might be affected by the servitude.
(4) Does the determination of a possible correspondence between new cadastral numbers and earlier cancelled ones require a search beyond the competence of the examiner of titles? A history of the cadastre might show which lot or part of a lot has been thus renumbered.
(5) Is the immoveable located in an agricultural zone set aside under the *Act to Preserve Agricultural Lands*?[44] The report is expected to indicate this fact.

This list could be expanded to include other important details. "The role of the land surveyor is to set down and certify what he has observed. The role of the notary is to judge the findings of the surveyor and draw conclusions."[45] Outside its context, the word "to judge" might seem exaggerated, but the notary is expected to suggest solutions to any problems he comes across while reading the land surveyor's report.

§10.234 *Importance of the certificate of location for investors* Whether an investor is a purchaser or a creditor, the certificate of location helps to prevent the following unpleasant surprises.

(1) If the size of the immoveable is given only approximately, there is a risk that the size may diminish and possibly even that rights with regard to occupancy are lost.
(2) If the immoveable is identified in vague terms, there is a risk that one might invest in a property other than the one intended as lien or purchase (wrong cadastral number). There is also a risk that a hypothecary act may be invalid.[46]
(3) An investment in an immoveable affected by irregularities such as unlawful views, encroachments, apparent charges that have not been regularized by servitudes or non-conformity with laws and regulations which might vitiate the titles (agricultural zoning, land use and planning restrictions, immoveables classified as cultural property, housing complexes, or a reserve of three chains).
(4) An investment in an immoveable is subject to servitudes or utility easements, the locations of which are poorly or insufficiently known or even unapparent and which limit the use of the immoveable, such as power lines, sewer or water pipes, gas or oil pipelines, parcels reserved for public use, and servitudes of non-construction.

§10.235 *Importance of the certificate of location for owners* The owner

---
44 R.S.Q., c. P-41.1.
45 G. Girard, *supra*, note 40.
46 Article 2042 of the *Civil Code*.

of an immoveable, whose only information is that contained in his title of purchase, even if he is not obliged to provide the lender with a certificate of location, might consider it useful to obtain one in order to have more complete information concerning his property.

§10.236 *Importance of the certificates of location for the courts* Certificates of location may be useful, even indispensible, in court, for example if a request is made in court to obtain judicial recognition of ownership rights over an immoveable, to cancel the registration of servitudes due to non-use, or to obtain servitudes of right-of-way in favour of property enclosed on all sides. In each of these cases, the analysis of titles is more extensive and the plan more elaborate in order to show and illustrate the former occupancies in regard to the more recent ones.

§10.237 *Importance of the certificate of location for municipalities* Municipalities which submit a request for the demolition of a building erected in violation of zoning or building regulations sometimes make use of certificates of location in order to have definite proof that a violation has occurred.

§10.238 *Other uses for the certificate of location* Several other requests must be accompanied by a plan of location as a supporting document. These include requests to the Commission de protection du territoire agricole du Québec (CPTAQ) for permission to use land for non-agricultural purposes, to the municipalites to obtain a building permit, to the municipalités régionales de comté (MRC) to modifiy rules of interim control, to the ministère de l'Environnement to obtain a title or lease upon the bed of a river or lake, or to the Service du cadastre with a request for correction, etc. In several of these cases, the certificate of location may be replaced by an expert report or an ad-hoc report, and the land surveyor makes the required adaptations in accordance with the instructions received.

Contents of the Certificate of Location

§10.239 Some of the contents of a certificate of location are mandatory while others are optional. It is mandatory to include all the elements described in the Regulation respecting standards of practice relative to the certificate of location.[47] The optional content consists of answers to the specific questions asked by the client and any details which the land surveyor deems useful for the proper completion of the task.

---

47 *Supra*, note 36, r. 7, Regulation respecting standards of practice relative to the certificate of location, in force since July 25, 1979, amended on August 1, 1983.

Surveying and Locating Operations

§10.240 Before a certificate of location can be issued, the surveying operations prescribed in sections 2.1.01 and 2.1.02 of the above Regulation must be carried out. Section 2.1.01 states,

> In any survey operation carried out in order to prepare a certificate of location, a land surveyor takes the measurements required to check the occupation and the boundaries of the property ensuring that sufficient territory is covered to support his professional opinion. The survey operation must be recorded in clearly drawn up field notes that precisely show and locate the state of the site, particularly any mark of occupation.

And according to section 2.1.02,

> The result of these operations must be verified by an additional source of information, also recorded in the notes, either by measurement by repetition, cross-checking, different ties, research, plan or some other method.

These operations constitute the basis for the certificate of location. The quality of the document depends largely on the care which is taken in establishing or researching the boundaries of the property, since the location of buildings and equipment within or near the property under observation is based on them.

The Report

§10.241 *Mandatory contents* Section 3.01 of the Regulation respecting standards of practice relative to the certificate of location lists the information that must appear in the report which forms part of the certificate of location. Under the provisions of that section,

> The land surveyor shall enter the following data in his report, where applicable:
>
> a) the date of the survey;
> b) a description of the property;
> c) conformity or lack of conformity between occupancy, cadastre and the titles;
> d) active or passive real servitudes registered as such in the index of immoveables concerning the property;
> e) apparent real servitudes or charges which should normally be the subject of a servitude and which might affect the property;
> f) notices of expropriation and homologation as well as reserves for public purposes registered against the property;
> g) classified cultural property, registered as such in the index of immoveables or the fact that the property is classified in a protected area;
> h) encroachments allowed or exercised;
> i) structures, buildings and dependencies located on the property;
> j) the stage of progress of work in the case of buildings, structures and dependencies under construction;

k) the number of floors and nature of exterior facing in the case of completed buildings;
l) the conformity or non-conformity of the position of the structures, buildings and dependencies with respect to municipal by-laws respecting laying out and zoning and the Québec regulations respecting hygiene;
m) the place and date of the closing of the minute.

Some of the information contained in the report was determined at the site. Other parts of the report represent the opinion of the land surveyor which he has reached after analysing his findings at the site on such matters as titles, the cadastre and any laws and regulations that might affect the immoveable. The report will draw particular attention to a number of these.

§10.242 *Non-apparent encumbrances* The chain of title sometimes includes certain servitudes, encumbrances, obligations or notices which, even if they are not apparent, might enhance or diminish ownership rights. Examples include servitudes of non-construction, view or utility easements, notices of expropriation, reservation of land for public purposes, classification as cultural property or agricultural zoning.

§10.243 *Non-apparent encumbrances not mentioned in titles* If, during his visit to the site, the land surveyor discovers apparent encumbrances that are not mentioned in the titles, such as views, utility easements, rights-of-way, encroachments or walls that appear to be party walls, he must research the lot under consideration at the registry office from its first cadastral entry in order to determine whether any of the encumbrances were regularized by a servitude in an earlier contract.

§10.244 *The reserve of three chains*[48] If an immoveable is situated along a watercourse or lake where a reserve of three chains might apply, it is important to mention the fact so that the notary can examine the original deed of concession and form an opinion on the matter.

§10.245 *Views onto neighbouring properties* One of the encumbrances or servitudes that often require the attention of experts, both land surveyors and notaries, is that of a view onto neighbouring property as set out in articles 533 to 538 of the *Civil Code*. It is not the responsibility of the land surveyor to give an opinion on the legality of views. In principle, opinions in this matter come under the jurisdiction of the legal profession. However, a surveyor reports his findings faithfully and raises any relevant questions. It is standard practice to measure direct views at right angles from the wall that has the opening. Side or oblique views, even those from balconies, galleries or other overhanging structures of the kind, are measured at right angles from the property line. It is the generally

---

48 *An Act respecting the Lands in the Public Domain*, R.S.Q., c. T-8.1, s. 45.

recognized opinion[49] that the French system of measurement is to be used in determining the allowance of six feet for direct views and two feet for lateral or oblique views. Does the installation of glass blocks or bricks in a wall adjacent to the property line constitute an unlawful view onto the neighbouring property? The notary public Jean Guy Cardinal reported the following in the *Revue du Notariat* in 1956:

> The honourable Judge Cannon, in the case of *Kert v. Winsberg*, affirmed that glass bricks encased in a wall constitute a light: "We must say that the glass bricks placed in this wall by the appellant constitute a light because they are translucent and allow the light of day to penetrate the respondent's building." The Supreme Court confirmed that opinion.[50]

This decision seems somewhat surprising in view of the fact that French courts have ruled otherwise in interpreting analogous provisions of their *Civil Code*. Given the current state of our law, however, we must follow the direction of our courts, even if, in the eyes of the practitioner, glass brick would appear to meet the requirements of article 534 of the *Civil Code*. In a number of court rulings, the notion of permanence of views has been considered in order to decide on the legality of a view. Thus, a fire escape would not constitute an illegal view, since it is considered more as an accidental view.[51] Would a wood panel that obstructs the view onto the neighbouring property satisfy the law? In our opinion, this would be a temporary remedy, for the view is obstructed only so long as the panel remains in place.

§10.246 *Encroachments suffered or exercised* In observing a site, not only must possible encroachments by buildings be examined, but also those by wall facings, roofs, balconies or other accessories that might overhang the neighbouring property or vice versa. In reporting these encroachments, the land surveyor should suggest the establishment of servitudes or modification or relocation in order to regularize the situation. Obviously, non-apparent foundation footings cannot be located, even if they exceed the boundary.

§10.247 *Encroachments on or above public roads* Encroachments of this kind are treated differently from one municipality to another. Depending on their bylaws, some municipalities choose to tolerate the encroachment by granting a lease for a term from one to 20 years, according to needs and circumstances. Some grant a servitude of tolerance while retaining the option of demanding the removal of the encroachment, if necessary. Others consent to a diminished road allowance by first adopting

---

49 Me R. Comtois, *Revue du Notariat*, 83, Nos. 3 and 4 (Nov./Dec. 1980).
50 Me Jean-Guy Cardinal, *Revue de Notariat*, 59, No. 4 (Nov. 1956).
51 See *Revue du Notariat*, 79, No. 5 (Dec. 1976).

a bylaw to close off the strip affected by the encroachment and then consenting to a deed of sale, thus resolving the problem definitively. Still others demand that the encroachment be removed, or prefer to leave the problem unresolved. This last course, however, is unacceptable to an examiner of titles.

§10.248 *Party walls* Walls that appear to encroach on a neighbouring property or the property to be examined have to be studied separately, for it is possible that the wall is constructed in such a way as to become a party wall. Article 520 of the *Civil Code* stipulates that in cities and incorporated towns, a property owner may force his neighbour to supply up to nine inches of his land for the construction of a separating wall between the houses, yards and gardens, under the condition, obviously, that he supply the same width on his side of the boundary line. However, this is a complex question that must be considered within the context of a certificate of location and must eventually be dealt with as a separate mandate.[52]

§10.249 *Research in the registry office* This stage is of primary importance. The land surveyor, once he is familiar with the site, is in a position to link what he has observed with any deed establishing a servitude or a registered notice. If applicable, he can indicate on his plan the part of the immoveable affected in this manner. It is generally necessary to search back to the first entry in the index of immoveables for the subdivided lot under consideration.

§10.250 *Regulations concerning the laying out of lots* These regulations establish the dimensions of a property starting from the date when it came into effect. It generally includes a provision declaring that the lots identified at the Service du cadastre or divided into smaller parcels by a deed of alienation before the coming into force of the regulation in question are protected because of the rights acquired. If such a provision does not exist, as is sometimes the case, and if lots do not conform with the regulation as they were created before its adoption, the land surveyor should mention this fact in his report.

§10.251 *Zoning regulations* These regulations decree, among other things, the land use and types of buildings that may be constructed in each zone. Some land surveyors restrict their observations to the conformity of buildings with the regulation. Others give their opinion, whenever possible, as to the degree of conformity of buildings with other buildings falling within the range of permitted types. A land surveyor finds it easier

---

[52] See also in this regard the study by P. Béique and W.E. Lauriault, "Le mur mitoyen."

than an examiner of titles to determine the limits of the zones appearing in the zoning plans.

§10.252 *Enforcement of municipal regulations* The enforcement of municipal regulations is the responsibility of the municipal inspector appointed for this purpose. It may happen, however, that the inspector does not have the information necessary to determine that a new building satisfies municipal regulations, or he may fail to ensure its registration in the cadastre or even to ascertain the dimensions of the site before issuing a permit. Consequently, non-conformity with the regulations is not acknowledged. If these irregularities are noted in the certificate of location, the owner is able to correct them, if necessary by changing the municipal regulation. The examiner of titles together with the interested parties will be able to assess the importance of any irregularities which cannot be corrected.

§10.253 *MRC regulations* Interim control regulations of the Municipalités régionales de comté must now be met as well. Several of these serve to regulate lot dimensions and building types. Some establish local zoning requirements for municipalities that do not have their own zoning regulations.

§10.254 *New laws* After the adoption of the Regulation respecting standards of practice relative to the certificate of location, a number of laws came into effect which have a considerable impact on property rights and can even cause the nullity of transactions. Therefore, the notary must be informed by the surveyor whether any of these laws apply to the property which the latter has visited and examined. Section 2.01 of the Regulation respecting standards of practice relative to the certificate of location describes the certificate as the expression of the land surveyor's opinion about the present situation and state of a property, particularly "in relation to titles, cadastre, and the laws and by-laws affecting it." In view of what has been said and of the non-restrictive enumeration of section 3.01 of the Regulation, a land surveyor who fails to determine whether the property in its current state is affected by any recently established laws or regulations might be asked to complete his document.

§10.255 *The* **Act to Preserve Agricultural Lands** The notary will first want to know whether the immoveable is located in an agricultural zone. These zones are not always easy to identify at the registry office and the report should give the notary clear information is this respect. If the lot does lie within an agricultural zone, he will also want to know if any sections of the Act apply to it. For example, the examiner of titles will want to know where the residence is located in order to claim any acquired rights. On the basis of his findings at the site and his knowledge of the law, the land surveyor will be able to formulate certain questions and, if applicable, give an opinion that will certainly be of interest to the client.

§10.256 *The Act respecting the* **Régie du logement**[53] The information contained in section 45 of this Act will be of particular interest to the notary in determining if he is dealing with a housing complex and if an act can be concluded.

> Housing complex means several immoveables situated near one another and comprising together more than twelve dwellings, if such immoveables are administered jointly by the same person or by related persons within the meaning of the Taxation Act . . . , and if some of them have an accessory, a dependency or part of the structure, except a common wall, in common.

If any of these criteria apply, the land surveyor should make mention of it.

§10.257 *Place and date of the closing of the minute* Once it has been dated and signed, the report reflects the condition and state of the property as of the date entered for the surveying of the site. No changes to the report are permitted. However, the report may be completed or supplemented by a document reflecting the subsequent condition of the site after a new study of it has been carried out or based on new laws and regulations which might affect it.[54]

§10.258 *Use of the certificate of location* Section 3.02 stipulates that

> [T]he report of the land surveyor . . . specify the purpose for which the certificate of location is intended and that it may not be used for any other purpose without the written authorization of its author; it must also specify that the report is an integral part thereof.

This section is intended to limit the use of the certificate of location to the purposes declared in the mandate. If other needs arise and the land surveyor is of the opinion that the document which he has prepared can meet those needs, he will act accordingly. On the other hand, he will be more reticent if the owner wants to use the certificate for locating the boundaries of his lot himself or if he wants to use it for enlarging or remodeling buildings. The report and the plan also state that they complement each other and jointly constitute the certificate of location.

---

53 R.S.Q., R-8.1, ss. 45-56.
54 *Supra*, note 47, s. 21; Regulation respecting records under *Land Suveryors Act*, R.S.Q., c. A-23, s. 56.

The Plan

§10.259 Under the provisions of section 4.01 of the Regulation:

> [T]he plan shall, in particular, contain:
>
> a) the graphic representation and the designation of the property;
> b) the adjacent parts;
> c) the dimensions and area of the property;
> d) the dimensions of structures, buildings and dependencies;
> e) an illustration, where possible, of the items mentioned in paragraphs d, e, f, g, h, i and k of section 3.01;
> f) an approximate indication of north by means of an arrow;
> g) the scale of the plan;
> h) the place and date of the closing of the minute.

Section 4.02 adds that

> [T]he plan must specify the purpose for which the certificate of location is intended and that it may not be used for any other purpose without the written authorization of its author; it must also specify that the plan is an integral part therof.

Section 4.03 states that "the plan must be in legal format."

Optional Contents of the Certificate of Location

§10.260 It is good practice to supplement the information given in a certificate of location with the following details: the ownership title, the cadastral history and the location measurements (on the plan).

§10.261 *Ownership title* In examining the chain of title, the notary may have some trouble in linking the immoveable under consideration with the acquired title. This may be the case if an immoveable has been given a new cadastral identification by an owner who has acquired it through more than one contract. It may also occur if an immoveable has been subdivided by a deed of alienation and its title could be confused with that of another. The land surveyor's report can make the connection in a more reliable manner because he is able to establish it in reference to the boundaries of the original immoveable.

§10.262 *History of the cadastre* It happens occasionally that a lot with a cadastral number has been substituted for lots that have been cancelled, thus causing a break in the chain of title. The history of the cadastre can retrace that link by indicating which lots have been substituted in this manner. Replacement lots present similar difficulties.

§10.263 *Location measurements (on the plan)* It is useful for the examiner of titles to know the distances measured between the buildings and the boundaries of the immoveable in order to evaluate the importance of any irregularity mentioned in the report. A number of municipalities require that an application for a building permit be accompanied by this information in order to ensure that the building regulations have been followed.

Co-ownership of Immovables

§10.264 The Regulation respecting standards of practice relative to the certificate of location does not include special standards for jointly owned immovables. The following has been compiled from the experience of land surveyors in this matter.

(1) Preparation of a certificate of location, as set out in the Regulation, for the entire jointly owned immovable, or of a copy of an already existing certificate that meets the requirements of the applicant. This document is necessary because of the rights held by the exclusive proprietor upon all the common portions. The Regulation respecting standards of practice relative to the certificate of location applies.

(2) Preparation of a certificate of location for the exclusive portion:

  (a) the report is to specify, in particular:

    1. the date of the survey;
    2. the cadastral designation of the exclusive portion and its description;
    3. the history of the cadastre;
    4. the conformity of the cadastral identification with the apartment number;
    5. the conformity or lack of conformity between occupancy, the cadastre, and the titles;
    6. any apparent charges concerning the exclusive portion in particular;
    7. the conformity or non-conformity with respect to municipal bylaws governing laying out, zoning, and building;
    8. the place and date of the closing of the document.

  (b) the plan is to include, in particular:

    1. the graphic representation;
    2. the adjacent parts;
    3. the dimensions and area;
    4. an illustration of the whole property or of part of the floor in the building showing the location of the exclusive portion and its geodetic elevation;
    5. the orientation of the plan;

The Law in Québec 451

6. the scale of the drawing;
7. the place and date of the closing of the minute.

It can be seen that certain details mentioned in the Regulation respecting standards of practice relative to the certificate of location are not given for the exclusive portion if they have been covered in the certificate of location for the whole building. The form of the document may vary according to particular needs. What is important is that the examiner of titles finds in it the information he needs and which is not provided elsewhere.

Copies of the Certificate of Location

§10.265 The *Land Surveyors Act* makes a number of stipulations as to the use of copies of certificates of location in concluding an act. Section 64 of the Act reads as follows, "[e]very copy or extract of a document in the records of a land surveyor used in an act respecting a real right must be certified true."

Staking an Immovable

§10.266 One might wonder why the preparation of a certificate of location does not include staking the site. Firstly, it must be remembered that preparing a certificate of location and staking are quite different operations requiring separate mandates. Whereas a certificate of location evaluates the situation and condition of a property at a given time and is a private, even confidential, document, staking identifies the boundaries of a lot through markers and makes them public. If there is lack of conformity between the true boundaries and occupancy, the certificate of location will reveal this fact to the proprietor in question. The latter may then decide to make the true boundaries public through a staking procedure. Since staking must not disturb the possession of a neighbour,[55] the applicant cannot demand of the land surveyor that he place any markers outside the apparent boundaries unless the neighbour concerned gives his consent, preferably in the course of a boundary action.

Updating the Certificate of Location

§10.267 The *Land Surveyors Act* requires that land surveyors preserve as minutes the documents prepared by them and protect them from any

---

55 *Louis Lavoie v. Arthur Chartier*, (18 January 1967), District of Hull 2976, (Mag. Ct.) Judge Arthur Labbé presiding. *Ravid v. Jasmin* (1939), 66 Que. K.B. 279.

alteration or change.[56] The Regulation respecting records of the ordre des arpenteurs-géomètres de Québec stipulates that a land surveyor can modify a minuted document only by preparing a new minuted document.[57] These operations have been regulated in order to preserve from any alteration the documents that describe the state of a site as of a given date. Such information can be essential in proving in court, for example, the beginning of a possession.

Thus, a land surveyor is required by law and by certain applicable regulations to revisit a site if asked to prepare a new certificate of location, using the previous surveying operations as a basis in order to be able to mention in his report that he has completed them. He then undertakes the research stipulated by the Regulation respecting standards of practice relative to the certificate of location and prepares a new report and a new plan that show any changes that have been made or extensions that have been added to the immoveable and any new laws and regulations that might affect it. Even if, at first glance, the buildings and dependencies do not seem to have been changed, it would be unwise simply to issue a letter saying that the old certificate of location is still valid; laws and regulations change constantly, servitudes and notices are filed, boundary lines shift and buildings constructed on neighbouring properties may give rise to irregularities. For the protection of the applicant it is necessary to carry out a new survey of the site, a new search at the registry office, a new occupancy analysis and complete an up-to-date report and plan which reflect the condition of the immoveable at the date when the document is signed.

The Certificate of Location: Concluding Remarks

§10.268 The certificate of location, by providing basic information to the user, is increasingly being used to complement the ownership title. The title of acquisition identifies an immoveable by its cadastral number alone or together with the area and metes and bounds descriptions. The certificate of location provides a detailed description of the immoveable and an analysis of its condition with regard to title, the cadastre, servitudes and the laws and regulations that might affect it. Together, all this information provides a clear picture of the site. So it is no wonder that, if asked to

---

56 *Supra*, note 36, s. 56.
57 *Supra*, note 36, r. 21; *Regulation respecting records of the ordre des arpenteurs-géomètres du Québec*, s. 5:
    A land surveyor may modify a minuted document only by preparing a new minuted document. The new minuted document and the modified minuted document must refer to each other following the method described in section 7. Such reference must be signed by the land surveyor.

produce his title of ownership, an owner will often show his contract of acquisition and certificate of location as well.

Custody of the Land Surveyor's Records

§10.269 *The records of a land surveyor* The *Land Surveyors Act* requires that land surveyors make a note of all their surveying operations and the documents they prepare, keep them safe and protect them from any alteration or change, make a repertory and classify the documents in such a way that they can easily be found and referred to. All the documents thus kept, the repertory and the index respecting, it constitute the records of a land surveyor. These records must be in the custody of a land surveyor. When a land surveyor dies, ceases to practise his profession or becomes incapacitated from acting as a professional owing to interdiction or removal from office, his records must be placed in the custody of another land surveyor or deposited in the office of a prothonotary of the superior court for the district in which he last practised or had his residence. A land surveyor who is readmitted to practice may retake possession of his records. In accordance with section 61 of the *Land Surveyors Act*, his records are unseizable.

§10.270 *Confidentiality of the records* The documents that constitute the records are confidential; a land surveyor may not communicate or send copies except in the following cases:

(1) if ordered to do so by the court;
(2) at the request of the surveyor general;
(3) with the authorization of the person who requested the work to be done or his assignees;
(4) if the document is registered; or
(5) if the document is mentioned in a registered document.

§10.271 *Copies of documents* Every copy or extract of a document in the records of a land surveyor used in an act respecting a real right upon an immoveable must be certified true. The persons authorized to certify copies are the land surveyor himself, the prothonotary who is the custodian of his records or another land surveyor who is his assignee or custodian of records.

§10.272 *Regulation respecting the custody of records* Pursuant to sections 13(*e*) and 56(2) of the *Land Surveyors Act*, the *Ordre des arpenteurs-géomètres* has adopted Regulation no. 21, called the Regulation respecting records of the *Ordre des arpenteurs-géomètres du Québec*. In it, the word "document" is defined as

[A]ny writing, minute, report, plan, map or table into which the land surveyor engages his professional competence, whether or not such a document is part of his exclusive practice as described in section 34 of the Land Surveyors Act.

Included under this term are minuted documents, documents in original form and provisional documents.

§10.273 *Minuted documents* A minuted document is an original document constituting the sole source of any certified true copy; it is self-authenticating and the depositary may not give it up, except in the cases provided for in sections 57 and 58 of the *Land Surveyors Act*. A land surveyor may modify a minuted document only by preparing a new minuted document.

§10.274 *Documents in original form* A document in original form is a document the original of which is intended for the surveyor general of Québec or of Canada and which is kept in the records of the department in question. Such a document in original form shall be the only source of any certified copy.

§10.275 *Provisional documents* A provisional document is one which is periodically brought up to date and includes any project plan, plan of compilation of lots or parts of lots, cartographic representation of land or territory or document related to the establishment of a skeleton geodesic map. Such documents may not be used for official purposes, unless special notice is given to that effect.

§10.276 *Custody of records* The Regulation also prescribes the manner of preparing the repertory and the index and requires that they be satisfactorily protected against fire, water and theft. The records must be kept indefinitely.

Professional Status of the Land Surveyor

§10.277 *The land surveyor as public officer* Section 34 of the *Land Surveyors Act* begins with the statement, "[a] land surveyor is a public officer." The article continues by citing the acts, operations and documents that are exclusive to the professional domain of the land surveyor. According to *Larousse*, a public officer is "a person with the capacity to authenticate acts". Does that mean that a land surveyor acts as a public officer whenever he carries out any of the operations listed in section 34? In fact, he acts as a public officer only in the following cases:

(1) when he monuments boundaries at the request of the parties concerned;
(2) when he places monuments by order of the court;

(3) when he prepares the minutes of a boundary operation and signs them together with the parties concerned;
(4) when he prepares the minutes of a boundary operation for homologation by the court;
(5) when he prepares and certifies as true a copy of an original document kept in his records or of another land surveyor's records in his care, be it minutes of a boundary operation, a report, a description, a plan or any other kind of document listed in the Regulation as part of a land surveyor's records.

§10.278 *The land surveyor as survey expert* The other operations carried out and the other documents prepared by a land surveyor under the provisions of section 34, although they are the work of a land surveyor, who is also a public officer, do not have the quality of authentic acts or documents as do those mentioned above. The courts have established that these acts and documents must be considered the opinion of an expert in the matter and not authentic acts that in themselves give judicial validity to their content. These acts are concluded by the land surveyor, not as a public officer, but as the most competent and best qualified person for the job owing to his thorough knowledge of the field of surveying.[58]

To summarize, a land surveyor acts as an expert in carrying out the surveying operations listed in section 34 and as a public officer when he authenticates, by his signature, any acts or copies of documents completed under his care.

THE LAW RESPECTING BODIES OF WATER IN THE PROVINCE OF QUÉBEC

Jurisdiction of the Province Over Bodies of Water

§10.279 The *British North America Act* of 1867 entrusted the care and administration of the public property of the former provinces to the governments of the corresponding newly-established provinces, with the exception of those matters that were expressly assigned to the jurisdiction of the federal government. However, this Act does not constitute the basis for the jurisdiction of the provinces over their territory: the fathers of confederation worked out between the provinces and the federal government a division of the rights and powers that previously had been exercised by the individual provinces.

---

58 *Tourigny v. Park, supra,* note 41 at 388. *Placements Miracle Inc. v. Gérard Larose, supra,* note 41.

§**10.280** The property rights of Québec are defined in sections 109 and 117 of the *B.N.A. Act.* By virtue of section 109,

> All Lands, Mines, Minerals, and Royalties belonging to the several Provinces of Canada . . . shall belong to the several Provinces of Ontario, Quebec, Nova Scotia, and New Brunswick in which the same are situate or arise,

section 117 adding that

> The several Provinces shall retain all their respective Public Property not otherwise disposed of in this Act, subject to the Right of Canada to assume any Lands or Public Property required for Fortifications or for the Defence of the Country.

These two sections recognize provincial jurisdiction over the administration and ownership of all lands that had not been granted to individual owners or were not transferred to the federal government by virtue of the Act.

§**10.281** In the case of areas covered by water, the right of ownership over the beds and banks gives the owner the rights to their use. The concept of ownership does not include the water as such; only the bed is immoveable and the rights of the owner extend only to the use of the water. Thus, the province can exercise, alienate, control or regulate these rights. The exclusive nature of this power is recognized in section 92 of the *B.N.A. Act*, which reads as follows, "[i]n each Province the Legislature may exclusively make Laws in relation to Matters coming within the Classes of Subjects next herein-after enumerated"; these include: "[t]he Management and Sale of the Public Lands belonging to the Province and of the Timber and Wood thereon" (92[5]); "[p]roperty and Civil Rights in the Province" (92[13]); and "[g]enerally all Matters of a merely local or private Nature in the Province" (92[16]).

§**10.282** Provincial laws, however, cannot apply outside the boundaries of the province, a limitation which is established in section 92 of the *B.N.A. Act.* Moreover, the *Treaty of Westminster* reserved for the federal government the power of legislation over extraterritorial matters. It seems that the courts have established, in the matter of interprovincial and international watercourses, that provincial law applies to those parts of a watercourse situated within the boundaries of the province in question.[59]

Jurisdiction of the Federal Government Over Bodies of Water

§**10.283** Section 108 of the *B.N.A. Act* gives the federal government

---

59 *Arrow River v. Pigeon River Co.*, [1932] S.C.R. 495.

property rights over the domain of the sea. In 1867, the federal government acquired ownership over the following:

(1) canals, with lands and water power connected therewith;
(2) public harbours;
(3) lighthouses and piers;
(4) steamboats, dredges, and public vessels;
(5) rivers and lake improvements.

The extent of this transferral of ownership rights is limited to property that existed in 1867. In order to determine the boundaries of the property which was transferred, it is necessary in each case to prove ownership as of 1867, the burden of proof resting on the federal government since the provinces are in principle the owners.

As owner of the beds and banks of certain bodies of water, the federal government has the power to control their use. Section 91 of the *B.N.A. Act* assigns to the federal government a certain number of exclusive legislative powers, of which those enumerated in the following subsections concern the domain of water:

(9) Beacons, Buoys, Lighthouses, and Sable Island.
(10) Navigation and Shipping . . .
(12) Sea Coast and Inland Fisheries.
(13) Ferries between a Province and any British or Foreign country, or between Two Provinces.

The federal government also has the power to enact whatever legislation is necessary for the exercise of the powers listed under section 91, that is, the power to acquire property by mutual agreement or expropriation, both from the provinces and from private citizens.

§**10.284** In the matter of navigation, federal jurisdiction extends over both tidal and non-tidal watercourses. When a watercourse is not navigable in the judicial sense, the federal government has no jurisdiction over the navigation carried out on it.[60] It has been established by precedence that a watercourse is navigable when it can be used for transport purposes in a practical and profitable way. By ruling on the navigability of a body of water the courts establish whether the watercourse belongs to the crown and, at the same time, whether federal jurisdiction over navigation applies. It follows that no province can decree that a watercourse is either navigable or non-navigable because that would have the effect of defining unilaterally the application of federal jurisdiction.

---

60 *Tellier v. R*, [1967] Que. S.C. 209.

The Law in Québec

§10.285 *The French regime* In the early years of the colony, the territory of Québec and the Crown domain were one and the same: the ownership of the lands discovered by Jacques Cartier fell to the King of France along with sovereign powers over it. The king reigned there by his representatives who, among other things, had the duty to grant land to individuals. Because of the poor success experienced by the viceroys in colonizing the territory, the Compagnie des Cent-Associés was founded in 1627 and received "full ownership, judicial authority, and dominion over the territory of New France as a grand fief.[61] As the representative of the government, the Compagnie introduced seigniorial tenure in New France. But the king was not satisfied with the results of colonization and the Compagnie had to give back its rights in 1663. One year later, the Compagnie des Indes occidentales acquired all the powers and obligations of the Compagnie des Cent-Associés. In 1674, the new company was dissolved and the administration and granting of land was handed over to the government of the colony, which remained in power until 1760, the year of the conquest. Under the French regime, 200 seigniories were granted and the seigniors were, in turn, under an obligation to grant land to colonists.

§10.286 *The English regime* Under the terms of the *Treaty of Paris*, signed in 1763, the rights of sovereignty, property and possession held by New France were handed over to England, while all private land remained the property of individuals. The English authorities decided that in the future land was to be held in free and common soccage. Under this regime, settlers could acquire the inalienable property right to their land by complying with the terms of settlement required for the acquisition of letters patent. They could also obtain a title of ownership free of all charges and obligations. The new type of tenure was not readily accepted, and the two regimes alternated. Thus, from 1763 to 1771, under Governor Murray, the regime of free and common soccage applied. But from 1771 to 1791, under Governors Carleton and Dorchester, seigniorial tenure was resumed. In 1791, the *Constitutional Act* established freedom of choice. Finally, in 1854, with the abolition of seigniorial tenure, free and common soccage became the sole mode in which land was granted in Québec.

§10.287 After the abolition of the seigniorial regime, the seigniorial court, which had been given the task of liquidating acquired rights, introduced the distinction between navigable and non-navigable water in defining the rights granted over watercourses. These first court decisions in Québec were given in the form of answers to questions submitted by the legislature;

---

61 J. Bouffard, *Traité du domaine* (Québec: Les Presses de l'Université Laval, 1921).

the essence of these decisions can be seen in the answers to questions 26, 28, and 29. After the abolition of seigniorial tenure, the legislature established a commission to draft a *Civil Code for Lower Canada*, which was adopted in 1866. The original version of article 400, which was copied from article 538 of the *Napoleonic Code*, stated the following principle,

> Roads and public ways maintained by the state, navigable and floatable rivers and streams and their banks, the sea-shore, lands reclaimed from the sea, ports, harbors and roadsteads and generally all those portions of territory which do not constitute private property, are considered as being dependencies of the crown domain.

This article of law is a codification of decisions of the seigniorial court and, by deduction, the beds of non-navigable watercourses continued to form part of the land conceded along rivers until 1884, when the reserve of three chains was imposed.

§10.288 In 1888, the legislative assembly of Québec adopted a law that obliged the administration to establish, for fishing purposes, a reserve of at least three chains (198.0 feet) along non-navigable rivers and lakes in the province to be withheld when Crown land was sold or granted by the Crown by gratuitous title. In 1918, the following paragraph was added to article 400 of the *Civil Code*, "[t]he same rule applies to all lakes and to all non-navigable and non-floatable rivers and streams and their banks, bordering on lands alienated by the Crown after the 9th of February, 1918." This amendment had the effect of reserving for the public domain the beds of all watercourses regardless of their navigability. To summarize, the property rights of the province, as far as bodies of water are concerned, extend over:

(1) the beds of navigable and floatable rivers and lakes;
(2) the beds of non-navigable or non-floatable watercourses contiguous with riparian property alienated after February 9, 1918, expressly alienated;
(3) the beds of non-navigable or non-floatable watercourses reserved expressly in land grants before June, 1884;
(4) the beds of non-navigable or non-floatable watercourses where the riparian properties are affected by a reserve of three chains.

The Concept of Navigability and Floatability

§10.289 *Introduction* Since the beginning of the colonial period, navigability, a concept of French law, was the criterion for determining whether a watercourse was Crown property. This concept continued to be applied in Lower Canada. Its sources are the *Ordonnance des eaux et de forêts*[62] and the *Ordonnance de la marine*,[63] but ultimately the concept goes back to the famous edict of Louis XIV in 1669. The king, in answer to a request by the legal community for new legislation to satisfy the requirements of commerce and to safeguard public interest, affirmed that navigable waters were Crown property,

> We declare ownership of all rivers that can carry boats without artificial structures or changes made by the hand of man, in our kingdom and in the lands subject to our rule, being part of the crown domain, notwithstanding all titles and possession to the contrary, except for the rights to fisheries, mills, and lakes, and any other use that may be the right of individuals through valid titles or possession, which they will retain as before.[64]

This famous ordinance should not have been applied in New France because it had not been registered with the Conseil souverain. For that reason, the privy council decided in 1838 in the case of *Hutchinson v. Gillespie*[65] that the ordinance did not have the force of law. The principles contained in these, however, and supplemented by local ordinances, were applied under French regime and even under the English regime before the adoption of the *Civil Code*. Moreover, the crown's ownership of rivers was never in jeopardy; only a miniscule portion of the territory had been conceded and nobody could claim ownership who had not been granted land expressly. In the beginning of the English regime, English laws applying to bodies of water were not really enforced because the population was ignorant of them and in 1774 the *Québec Act* reinstated the use of French civil law in the province.

§10.290 *The criterion of navigability and floatability explained* The navigability of a watercourse cannot be defined in absolute mathematical terms, it being a question of fact to be decided upon by a court. In applying the criteria of navigability or non-navigability, the facts to be taken into consideration are those existing at the moment when the land was first conceded, as can be seen in the case of *Hurdman v. Thompson.*[66] In order to be navigable, a watercourse must allow commercial shipping; the mere fact that it is possible to go for a boat ride on it does not constitute navigability. A watercourse must be similar to a public road and suitable for shipping and the conveyance of goods for commercial purposes, as established in the case of *Bouillon v. R.*[67] There is no universally applicable

---

62 *Recueil des anciennes lois Francaises* (Paris, 1829), vol. 23 at 291.
63 *Ibid.*, vol. 19 at 348.
64 J. Bouffard, *Traité du domaine* (Quebec: Les Presses de l'Université Laval, 1921) at 73.
65 (1838), 2 R.J.R.Q. 313.
66 (1895), 4 Q.B. 409.
67 (1913-17), 16 Exch. C.R. 443.

rule, and each case must be analysed separately. One can, however, refer to the basic principles established by the privy council on March 12, 1937 in the case of *St-Francis Hydro-Electric Co. v. R.*[68]

> The courts have established the principal conditions required so that a river can be considered navigable or floatable in terms of the provisions of article 400 of the *Civil Code*:
>
> a) It is not sufficient that it be floatable "for scattered logs"; it must be capable of conveying log rafts; (*Tanguay v. Canadian Electric Light Co.*, 40, Superior Court, and *Maclaren v. the Attorney General for the Province of Québec*, 1914, C.A. 258).
> b) It is not necessary that this be a constant condition of fact; but it is not sufficient that it be the result of exceptional circumstances, such as unusual tides or high water.
> c) A river may be navigable for part of its length, upstream from its mouth, and cease to be so with the first natural obstacles, thus making it definitely unsuitable for continuous navigation, even if further upstream there might be some stretches suitable for limited local navigation (*Lemay v. R.*, 54 Supreme Court, 143).
> d) However, the existence of rapids along the course of a river, navigable up to that point, does not change its status if navigation and floatation can continue on it in a useful and practical manner.
> e) However, it is required that navigation or floatation be possible not only empirically, but in such a manner that they can be useful and profitable to the public. (*Bell v. Corp. of Québec* 5, C.A. 843; *Attorney General v. Fraser* 37, Supreme Court 577).

The criterion of navigability applies to all watercourses, including lakes. Although article 400 of the *Civil Code* does not mention lakes, the seigniorial court, in response to question 28, gave riparian proprietors ownership over non-navigable lakes as well as ponds.

§**10.291** *Artificial navigability* In the case of *Hurdman v. Thompson*, Judge Bossé, citing article 400 of the *Civil Code*, which does not distinguish between naturally navigable watercourses and those that might have been made navigable by the act of man, affirmed explicitly that a watercourse can become artificially navigable.

§**10.292** *The criterion of the high water mark* In order to define the property line between a river bed and any abutting land, courts have recourse to the criterion of the high water mark. In fact, the courts have always referred to French custom law to determine property rights over bodies of water and in the case of tidal waters, they have been applying the rule of high water as defined in the *Ordonnance de la marine* of 1681: "[t]he bank or shore of the sea shall be deemed to be all that the sea

---

68 [1937] 2 D.L.R. 353 (Que. K.B.).

covers and uncovers at new moon and full moon and up to the mark on the beach which is reached by the flood tide in March." In the case of non-tidal waters, the courts have been applying the criterion of the ordinary high water mark, without overflows or floods; it is the mark which is reached by mean or ordinary high tides, regardless of seasonal variations or such factors as torrential rains or unusually high water. This delimitation depends, however, on the particular circumstances of each individual case. The line between aquatic and land vegetation, however, constitutes a firm criterion although sometimes superseded by expert testimony based on high-water statistics resulting from daily readings over periods of ten to 20 years. For practical purposes, there is no single criterion; as a last resort, the courts rule after having considered all the evidence as a whole including expert testimony to help clarify and situate such concepts as "vegetation", "foreshore" and "level".

§10.293 *The beach* The concept of the high water mark implies that there is also a line to which the water recedes at its lowest ebb; the space between these two lines constitutes the beach. Beaches in principle belong to the Crown. According to court rulings, the term "bank" is equivalent to "beach". Thus, by virtue of article 400 of the *Civil Code*, navigable rivers and their banks, and after February, 1981, non-navigable rivers, are considered to belong to the public domain. It should be noted that when a permanent invasion of land by water through natural causes results in the encroachment of the watercourse onto riparian property, then that part of the land that has become permanently covered by water is to be added to the public domain.[69]

§10.294 *The criterion of the middle thread* Before June 1, 1884, whenever the Crown conceded land abutting on a non-navigable watercourse, the concession included the alienation of the bed. The grantee became owner of the whole width of the bed if his property included the land on either side of the watercourse and, if the land on the opposite sides of the watercourse was owned by different grantees, each became owner of part of the bed up to the median line, or "middle thread," in accordance with the judicial concept of *usque ad medium filum aquae*. The same principle applies when the Crown has conceded land on only one side of the watercourse; public property then extends to the middle thread. But after February 9, 1981, the bed of a watercourse must have been conceded expressly by the Crown in order to belong to an individual.

---

69 See *Price v. La Cie de pulpe de Chicoutimi* (1906), 30 Que. S.C. 293.

Individual Rights over Bodies of Water

§10.295 *Introduction* Almost all rights of individuals over bodies of water arise from property rights, but most of these rights are not property rights, but user's rights. In legal terms, four different categories of water are to be distinguished:

(1) rain water and run-off,
(2) underground water,
(3) spring water, and
(4) flowing water.

§10.296 *Rain water and run-off* This is water which falls naturally onto a property in the form of rain or snow. This water belongs first to the property which is situated in such a way that it can collect the water and it may be retained there in spite of the advantage a lower-lying property might otherwise obtain from it. It is rare that water falling onto a property is absorbed or evaporated there completely. As this water must run somewhere, article 501 of the *Civil Code* stipulates that lower-lying properties are servient to those on a higher level in that they are obliged to receive the water running down naturally, even if that water prevents the cultivation of the land or causes other prejudice. It is obvious that the owner of a property situated higher up cannot alter the water, under the pretext of using it at his pleasure, in a way that spoils it or makes it unsanitary. On the other hand, the owner of lower-lying land cannot do anything to make the water flow back to the property above, as indicated in the second paragraph of article 501 of the *Civil Code*. Where water is shed by a roof, article 539 of the *Civil Code* requires that the owner of a building let the water drain onto his own property and not onto that of his neighbour. If the land that ought to receive that water is sloped, the owner of the neighbouring land cannot object to its reaching his property, for the downward flow is considered natural and the water does not fall directly from the roof onto the neighbouring property.

§10.297 *Underground water* Article 414 of the *Civil Code* stipulates that the ownership of land includes the ownership of what is above and below. The same principle of law applies to the ownership of underground water. The owner of the property can, thus, use the underground water at his pleasure, provided the rights of others are respected.

§10.298 *Spring water* A spring is owned by the proprietor of the land on which it is situated, not as an isolated object, but as an accessory to the land. An owner can thus use it and dispose of it as he pleases, as set out in article 502 of the *Civil Code*. The owner of the land has an absolute and discretionary right over his spring, which is not the case with flowing water.

§**10.299** *Flowing water* The bed of a watercourse is part of either public or private property, although such is not the case with the water that covers it. Property rights upon the beds, beaches and banks of a watercourse are in principle no different from those upon any other land. The exercise of these rights is, however, limited by the presence of the water, which does not belong to any one person more than to any other. The use made of the water as it flows past is the only right connected to it, and as soon as this right has ceased to apply because the water has past by, another owner acquires the same right. The exercise of the right of the owner to exploit the beds, banks and beaches by the extraction of useful substances, such as aquatic plants, hay, gravel or sand, is thus subject to the rights of others. The owner may also place structures on the property, such as pillars for a bridge, in order to join the two banks or use the banks to facilitate the exploitation of his property or to make use of the water. The grass that grows on the beaches of the Saint Lawrence River constitutes a special case: as indicated in article 591 of the *Civil Code*, this resource is in some cases attributed by a special law or particular title to the riparian owner and in other cases it belongs by right of occupancy to the person who cuts it.

Rights of Accession

§**10.300** *Introduction* According to article 408 of the *Civil Code*, the right of accession extends ownership in a thing to everything that is joined to it naturally or artificially. Articles 420 and 427 regulate the rights of accession that result from the movements and shifts of watercourses and banks.

§**10.301** *Alluvion* Article 420 defines alluvion as "deposits of earth and augmentations which are gradually and imperceptibly formed on land contiguous to a stream or river." This article then establishes the following rule,

> Whether the stream or river is or is not navigable or floatable, the alluvion which is produced becomes the property of the owner of the adjacent land, subject in the former case, to the obligation of leaving a foot-road or tow-path.

Article 422, however, specifies that

> [A]lluvion does not take place on the borders of lakes and ponds which are private property; neither the proprietor of the lake nor the proprietor of the adjacent land gains or loses in consequence of the waters happening to rise or fall above or below their ordinary level.

§**10.302** *Accretions* Accretions are portions of the bed which are left dry when a watercourse withdraws from one of the banks and encroaches upon the other. Accretions are dealt with in article 421, which reads as follows,

> As to ground left dry by running water which insensibly withdraws from one of its banks by bearing in upon the other, the proprietor of the uncovered bank gains such ground, and the proprietor of the opposite bank cannot reclaim the land he has lost. This right does not exist as regards land reclaimed from the sea, which forms part of the public domain.

§10.303 *Avulsion* Article 423 deals with avulsion, which is the carrying off by the violent action of the water of an appreciable portion of a riparian field to a downstream field or the opposite bank. As prescribed in article 423,

> If a river or stream, whether navigable or not, carry away by sudden force a considerable and distinguishable part of an adjacent field and bear it towards a lower or opposite bank, the proprietor of the part carried away may reclaim it; [but he is obliged, on pain of forfeiting his right, to do so within a year, to be reckoned from the possession taken of it by the proprietor of the land to which it has been united.]

§10.304 *Islands, islets, and deposits* In conformity with article 400, article 424 states the following principle, "[i]slands, islets and deposits of earth formed in the beds of navigable or floatable rivers and streams belong to the crown, if there be no title to the contrary." On the other hand, as prescribed in article 425,

> Islands and deposits of earth which are formed in rivers which are not navigable or floatable belong to the proprietors of the banks on the side where the island is formed. If the island be not formed on one side only, it belongs to the proprietors of the banks on both sides, divided by a line supposed to be drawn in the middle of the river.

Article 426 establishes a different rule in the case of an island which is formed by the sudden creation of an additional channel. This article reads as follows,

> If a river or stream, by forming a new branch, cut and surround the field of a proprietor contiguous to it, and thereby form an island, the proprietor retains the property of his field, although the island be formed in a navigable or floatable river or stream.

§10.305 *Change in the course of a river* Under the provisions of article 427,

> If a navigable or floatable river or stream abandon its course to take a new one, the former bed belongs to the crown. If the river be not navigable or floatable, the proprietors of the land newly occupied take as an indemnity the ancient bed, each in proportion to the land which has been taken from him.

Riparian Rights

§10.306 Riparian rights are linked to property rights and are generally held by the owner, but they can also be exercised by a lessee, a usufructuary or a user. Riparian rights belong, in principle, to a parcel of land and are in that respect accessory property rights. Most riparian rights, as the right of access to the water, are not founded on any legislation, but their existence has been confirmed many times by court decisions. Section 5 of the *Watercourses Act* stipulates that

> [E]very owner of land may improve any watercourse bordering upon, running along or passing across his property, and may turn the same to account by the construction of mills, manufactories, works and machinery of all kinds, and for such purpose may erect and construct, in and about such watercourse, all the works necessary for its efficient working, such as flood-gates, flumes, embankments, dams, dykes and the like.

The application of this section is subject to the obtaining of the appropriate permits from the appropriate government departments. The right of access to the water is not restricted to riparian land owners near non-navigable and non-floatable watercourses, since it applies even to navigable and floatable rivers and lakes. As a result, owners of riparian property have a right-of-way over public land in order to exercise their right of access to the water.[70] However, riparian owners are not obliged to let anybody pass through their property in order to get to a watercourse, even if that watercourse is public. The right of domestic use includes the following components:

(1) defense against obstructions;
(2) reserve of drinking water:
(3) recreational use;
(4) right of view; and
(5) right of navigation.

Public Ownership of Bodies of Water

§10.307 Public ownership of bodies of water depends essentially on the mode in which the land along a watercourse was granted. Under the French regime, concessions of land were governed by the laws in force at the time, while English common law applied to concessions granted after the conquest. Under English law, the presence or absence of tides determines whether a watercourse is public or private property. The determination of what is Crown property has traditionally been perceived in terms of

---

70 See *Bonnasserre v. The National Harbour Board*, [1972] Que. S.C. 713.

private law, as was confirmed in the case of *Cambell v. Hall*:[71] "[t]he laws of a conquered country remain in force, until they are modified by the conqueror." The *Québec Act* of 1774 respected the integrity of the properties that existed before the conquest. Thus, the former French laws determine the division of property rights between individuals and the Crown.

The Reserve of Three Chains

§**10.308** *Introduction* All the land granted in townships by the Crown after June 1, 1884, along non-navigable rivers and lakes ends at a distance of three chains (198.0 feet) from the bank. The Crown has, thus, remained owner of that reserve, thereby becoming a riparian land owner. Incidentally, the Crown has always claimed full ownership of this reserve of three chains.

§**10.309** *Origin of the reserve* The judgment of April 28, 1882, in the case of *Robertson v. R.*[72] before the Supreme Court of Canada confirmed the exclusive jurisdiction of the provinces over fishing rights in waters owned by them. According to this decision, the federal government, in 1867, was given legislative powers only over coastal and inland fisheries, but was not given any right over the water. Fishing rights, which are accessory to the ownership of the bed of a watercourse, are a real property right and not a mere servitude. The fact is that the bed of a watercourse can belong to an individual as well as to the Crown. Thus, while the Crown can regulate fishing as to seasons and species, it cannot issue permits if it is not the proprietor of the bed. On the basis of this judgment, the province concluded that it had fishing rights in the following cases:

(1) in all navigable watercourses where the abutting properties had not been conceded;
(2) in all navigable watercourses where the abutting properties had been conceded but fishing rights had been withheld; and
(3) in all non-navigable watercourses where abutting land had not been conceded.

The decisions of the seigniorial court in this respect applied to both townships and seigniories.

In alienating its land, the province, thus, had to remain the owner of riparian land because, in the case of non-navigable watercourses, fishing rights belong to the riparian owner. According to Jean Bouffard, the administrators of public lands at the time wanted to retain its fishing rights in this manner. In fact, on July 11, 1882, three months after the Supreme

---

71 (1774), 1 Cowp. 204.
72 (1882), 6 S.C.R. 52.

Court judgment, the Commissioner of Crown lands issued instructions not to sell riparian lots where the government was the owner of the fishing rights or to retain a reserve in order to maintain these rights.

§10.310 *Legal provisions* On March 30, 1883, the first fisheries law of the province received assent.[73] That law gave the commisioner of Crown lands the right to lease Crown land abutting on rivers and lakes. Thus, the leases included fishing rights. There was a strong demand for lots by prospective settlers, so the Crown could not suspend indefinitely the sale of land abutting on rivers or lakes. On June 21, 1884, the Commissioner of Crown lands issued a new directive to land agents to withhold from riparian lots a reserve of three chains along the banks to be taken out of the area of each lot. The department would then deduct the value of a strip of three chains from the first instalment. This reserve of three chains, which existed through a regulation adopted by the governor in council, was confirmed by a law, assented to on July 12, 1888, and made retroactive to June 1, 1884.[74] This law stipulated that "the sales, concessions and free grants of public lands are subject to a reserve of at least three chains for the purpose of fishing." The Crown thus remained owner of a strip of land along non-navigable and non-floatable watercourses and the grantee acquired only that part of the land situated outside this reserve of three chains. A few years later, two judges of the superior court, first Judge Champagne, in 1910, in the case of the *Attorney General for Québec v. MacLaren*,[75] followed by Judge Chauvin in 1915, in the case of *Patenaude v. Edward and Co.*,[76] expressed the opinion that, by instituting the reserve, the Crown only retained fishing rights, not the property of the riparian strip of three chains. These judgments led the government to amend its law of 1888 to dispel any doubts in this regard. An Act assented to on March 17, 1919 replaced the words "for the purpose of fishing" by "in full ownership in favour of the crown".[77] There should have been no more doubt as to the crown's ownership of the reserve of three chains. Nevertheless, the status of this reserve for the period between June 1, 1884, and March 17, 1919, is contested because the Act of 1919 was made retroactive. The judgment in the case of the *Attorney General for the Province of Québec v. Healy*,[78] of June 30, 1983, was affirmed by the Supreme Court in 1987. Since 1970, the reserve of three chains also applies to navigable waters.

---

73 *An Act respecting the management of Public Lands adjoining non-navigable streams and lakes in the Provinces of Quebec*, 1883, 46 Vict., c. 8.
74 *An Act to amend and consolidate the laws relating to fisheries*, S.Q. 1888, c. 17.
75 (1911), 21 Que. Q.B. 42, affirmed (1912), 46 S.C.R. 656, reversed [1914] A.C. 258 (P.C.).
76 (1915), 21 R.L. (N.S.) 523.
77 *An Act to amend the Quebec Fish and Games Laws*, S.Q. 1919, c. 31.
78 [1983] Que. C.A. 573, affirmed [1987] 1 S.C.R. 158.

Characteristics of Property Rights of the Province

§10.311 *Introduction* Three important facts characterize Crown property: it is imprescriptible, it enjoys fiscal immunity and it is alienable.

§10.312 *Crown property is imprescriptible* The imprescriptible status of crown property is stated in article 2213 of the *Civil Code*, which stipulates that

> [S]ea-beaches and lands reclaimed from the sea, ports, navigable or floatable rivers, their banks and the wharfs, works and roads connected with them, public lands, and generally all immoveable property and real rights forming part of the domain of the crown are imprescriptible.

This principle also applies to navigable waters, by virtue of the amendment to article 400, which reads as follows, "[t]he same rule applies to all lakes and to all non-navigable and non-floatable rivers and streams and their banks, bordering on lands alienated by the crown after the 9th of February, 1918."

§10.313 *Fiscal immunity of crown property* The fiscal immunity of Crown property is founded on section 125 of the *British North America Act* of 1867, which states that "no Lands or Property belonging to Canada or any Province shall be liable to Taxation."

§10.314 *Alienability of crown property* As to the alienability of Crown property, the seigniorial court in 1856 recognized as valid the concession of beaches and other parts of navigable watercourses, on condition that the terms of the grant be clear and precise. By virtue of section 1524 of the *Watercourses Act* of March 16, 1961,

> It has always been lawful, before the 16th of March, 1916, whatever may have been the system of government in force, for the authority which has had the control and administration of public lands in the territory now forming Québec, or any part thereof, to alienate or lease to such extent as was deemed advisable, the beds and banks of navigable rivers and lakes, the bed of the sea, the sea-shore and lands reclaimed from the sea, comprised within the said territory and forming part of the public domain. From and after the 16th of March, 1916 until 4 December 1974, every alienation or lease of one or more of the properties mentioned in the first paragraph may be effected solely with the express authorization of the Gouvernement, and on such conditions and under such restrictions as it may determine.

Based on the principle of alienability of Crown property, beds and banks are granted under the *Watercourses Act*, the administration of which comes under the jurisdiction of the ministre des Richesses naturelles. The minister cannot grant such property without the express authorization of the Lieutenant Governor in Council.

REFERENCES

Bouffard, J. *Traité du domaine.* Québec: Les Presses de l'Université Laval, 1921.

Lord, G. *Le droit québecois de l'eau,* vol. 1. Montréal: Centre de recherches en droit, Université de Montréal, 1977.

11

The Surveying Profession

G.K. Allred, A.L.S., C.L.S.

CONCEPTS OF PROFESSIONALISM

§11.01 Previous chapters have dealt with the legal aspects of boundary retracement and demarcation. In addition to these legal requirements, the land surveyor as a professional person is bound by a myriad of other obligations many of which are not codified but are merely stated as ethical obligations or concepts of professionalism.[1] As Mr. Justice Belzil stated in an Alberta Court of Queen's Bench decision relating to the *Charter of Rights*: "[professionals] who enjoy these privileges should be subjected to a more rigorous discipline than that which applies to ordinary citizens. This discipline is peculiar to them and is not part of penal law."[2]

§11.02 The surveyor's practice is governed, not just by the laws of three levels of government (federal, provincial and municipal), but also by a fourth level, namely that of his professional association — the government of and by his peers. Above all, however, he is governed by his own integrity and ethical standards.[3]

---

[1] Government of Quebec, "The Professions and Society," Report of the Commission of Inquiry on Health and Welfare, (1970), Part 5 at 54.
[2] *Fang v. College of Physicians and Surgeons of Alta.* (1985), 66 A.R. 352 (C.A.).
[3] T. Shanks, "The Status of the Dominion Land Surveyor," Annual Report of the Dominion Land Surveyors' Assoc. (1923) at 42-48.

§11.03 The land surveyor is a public officer.[4] He does not represent a single client in establishing the client's boundaries, but rather he represents society at large.[5] Every boundary monument marks a boundary between at least two unique properties. The surveyor, therefore, must be fair and impartial to all parties; he cannot give undue consideration to his client's interests in disregard to the interests of his client's neighbour and potential adversary.[6] Here, his responsibilities are quite different than those of a doctor, lawyer, or accountant each of whom normally need act only with the interests of a single individual in mind. "He must preserve in all his work the judicial mind and the impartial attitude of an arbiter, rather than the bias of an advocate."[7]

§11.04 Essentially the land surveyor is a gatherer of facts[8] — a land information sleuth. His duty is to determine the physical and topographic characteristics of a parcel of land and to establish the facts as to the positions of boundaries on the ground.[9] Based on these facts, he must form an opinion as to the location of all boundaries and the extent and shape of the parcel.[10] In searching for evidence of those boundaries, he is obligated firstly, to conduct an exhaustive search for the original location of the boundary monumentation;[11] secondly, to document precisely all evidence and measurements defining those boundaries; and thirdly, to re-monument those boundaries for the benefit of future generations.[12] In exercising these functions, it is submitted that the land surveyor is acting in the capacity of an officer of the state — working to preserve and perpetuate the survey fabric which contributes to the identification of individual land parcels.

§11.05 Having completed all of this research and documentation however, the final monuments re-established by the surveyor are not necessarily conclusive; they are, rather, his considered opinion as to the location of those boundaries. The words of the late Jack Holloway, former Director of Surveys of Alberta, are appropriate:

> The surveyor should remember that he may be called upon to explain and

---

4 *Land Surveyors Act*, R.S.Q. 1977, c. A-23, s. 24.
5 C.H. Weir, "Professional Practice," *The Canadian Surveyor*, (1976), Vol. 40-1 at 3-12.
6 J.H. Holloway, "Principles of Evidence," *The Canadian Surveyor*, (1952), Vol. XI-2, at 31-42.
7 *Supra*, note 3.
8 C. Anderson, "Legal Elements of Surveying in the Establishment of Land Boundaries," *Surveying and Mapping*, (1951), Vol. XI-1 at 40-45; also Holloway, *supra*, note 6.
9 *Heubner v. Wiebe*, [1984] 1 W.W.R. 272 (Man. Q.B.).
10 *Ibid*.
11 Energy, Mines and Resources, Canada, *Manual of Instructions for the Survey of Canada Lands*, 2d ed. (Ottawa: 1979), chapter B7.
12 L. Petzold, "The Survey and the Real Estate Transaction," *Terravue*, (1984) at 47-59.

justify his operations before a court of law, and in order to be able to do so with confidence and assurance, he should always satisfy himself thoroughly that the evidence which he creates or uses in the course of his field operations is as good as the best that any other surveyor can point to. Many people are prone to think that if a registered land surveyor re-establishes a boundary line, his decision as to its location is final and conclusive. That is not so. It is true that very often adjoining owners in dispute over a boundary will voluntarily submit their differences to a land surveyor and will agree to abide by his decision rather than go to law about it. It is also true that the law recognizes the qualified land surveyor as an expert in calculation and precise measurement and allows due credit to be given to his judgement and experience in cases where the evidence requires expert appraisal. But if dispute remains, surveyors have no more authority than any other men to determine boundaries. The location of any boundary or corner in dispute is, in the final legal analysis, a question of fact to be determined exclusively by the courts, and the surveyor's role is not that of a judge or jury but, at best, that of an expert witness skilled in the finding, appraisal and recording of evidence by which the facts in question may be determined.[13]

§11.06 At any time the surveyor's opinion may be challenged and a court, having reviewed his research, evidence and procedures, may either affirm his decision or substitute its own opinion, thus fixing the boundary by court order.[14] As stated above "surveyors have no more authority than any other men to determine boundaries. The location of any boundary . . . is a . . . question of fact to be determined exclusively by the courts."[15]

§11.07 Nevertheless for all intents and purposes, the surveyor's work is usually accepted by landowners and in most cases the surveyor is, in all practicality the final boundary arbiter. Neighbours are in most cases, content to accept the work of a land surveyor or may agree to the appointment of a land surveyor as a mediator.[16] In the Alberta *Arbitration Act*,[17] Dominion and Alberta land surveyors are identified under the definition of "professional arbitrator" presumbaly because of their experience in the settlement of boundary line disputes. Perhaps it is a tribute to Canadian land surveyors to search through the reams and reams of Canadian legal decisions and find how few surveyors' decisions have in fact been set aside by the judiciary.[18]

---

13 *Supra*, note 6.
14 M.J. Sychuk, "Legal Problems in Establishing Boundaries for a Cadastre," *The Canadian Surveyor*, (1975), Vol. 29-1 at 29-38.
15 *Supra*, note 6.
16 T.M. Cooley, "The Judicial Functions of Surveyors," *Surveying and Mapping*, (1954), Vol. XIV-2 at 161-168; also J.H. Holloway, "What is a Legal Survey Mark," *ALS News* (1965), Vol. VII-3 at 20-21.
17 R.S.A. 1980, c. A-43, s. 1.
18 N.C. Wittman, "Professional Negligence," *ALS News* (1982), Vol. XI-3 at 17-20; also note 14, *supra*.

§11.08 Boundaries are not approximate. By definition, a boundary is the line which determines the limit between two parcels of land and, as such, is the exact point of beginning of one parcel and ending of another. It is a line of no width. A surveyor engaged to mark the boundaries between two or more properties must do a complete survey, searching for all available evidence of the original boundaries,[19] marking them on the ground, preparing and filing any necessary plans or documents to record his work in a public repository where applicable. He must then be prepared to stand behind his work and defend it in a court of law if it is called into question. To do any less than a complete job is an abrogation of one's professional responsibility.[20]

§11.09 Land surveyors like other persons in responsible positions usually must swear an oath of office and on occasion also an oath of allegiance as evidenced by the Alberta *Land Surveyors Act*.[21] A surveyor is bound by his oath, as well as by his professional ethics, to uphold the law and to act without prejudice — to act neither in favour of his client nor against his client's neighbour. In all of his dealings, he must act with total impartiality, respecting the rights of all parties. Figure 11.1 is the sample oath of office required by the *Examination and Training Regulation*[22] pursuant to the *Land Surveyors Act* of the Province of Alberta. It reflects the surveyor's responsibilities.

FIGURE 11.1   OATH OF OFFICE

CANADA }
PROVINCE OF ALBERTA }

I, ..., do swear that I will diligently, faithfully and to the best of my ability, execute according to law the office of land surveyor; and that I will, as an Alberta land surveyor, conduct all surveys faithfully and to the best of my ability, giving due consideration to the lawful rights of all persons; I will accurately locate and record all evidence of boundary monumentation truly and accurately to the best of my ability. I will measure and record all data truly without prejudice either towards or against any land owner, but in all things conduct myself truly and with integrity; maintaining and upholding the law and the interests of the public.

So help me God.

---

19 *Huebner v. Wiebe, supra*, note 9.
20 *Supra*, note 18.
21 S.A. 1981, c. L-4.1.
22 Alta. Reg. 326/82.

```
SWORN before me at the..........  ⎫
of ......................in the   ⎬   ...........................
Province of Alberta this ..........  
day of ............,A.D. 19......  ⎭
```
....
Judge of the Court of Queen's Bench

§11.10 The delivery of true, accurate and reliable measurements and observations also extends to the duly appointed assistants of the land surveyor. Several statutes relating to surveying require survey assistants to swear an oath of service.[23] Verification of chains and other measuring equipment is also a common requirement to ensure that survey returns are true and correct.

LEGISLATION GOVERNING PROFESSIONS IN CANADA

Jurisdiction

§11.11 Under the *Constitution Act, 1982*,[24] or *British North America Act, 1867*,[25] as it was originally enacted, authority to pass laws governing "Property and Civil Rights in the Province" is assigned to the provincial legislatures. All provinces of Canada have exercised the authority thereby granted to enact legislation governing the practice of the profession of land surveying within the boundaries of the respective provinces. This power is not new but rather was initially granted under the original *British North America Act* and was affirmed when the Canadian constitution was patriated in 1982.

§11.12 The Canadian concept of professional regulation is to create a private organization made up of all members of the profession and delegate to that organization full responsibility for the governing of all affairs relating to that professional activity.[26] This is arranged through a statute of the provincial legislature. These statutes confer powers of licensure, authority to set standards, to adopt ethical codes, to maintain competency within the profession and to discipline practitioners as is necessary. This concept, according to the Quebec study on *The Professions and Society*,[27] was found

---

23 *Supra*, note 4 at s. 45.
24 (U.K.), 1982, c. 11, s. 92.
25 (U.K.), 30 & 31 Vict., c. 3, s. 92.
26 G.K. Allred, "The Professional Association — Protector of the Public Interest," *Terravue* (Autumn, 1983) at 12-24.
27 *Supra*, note 1 at Part 5 at 24.

to be unique among a group of seven jurisdictions whose professional institutions were examined.

§11.13 The Quebec study summarizes the public role of a professional association to include:

(1) a governmental role when it is empowered to control admission to the practice of a profession in lieu of a diploma awarded by the public education system;
(2) a legislative role when it sets certain norms (regulations, codes of ethics and of deontology);
(3) an administrative role when it is charged with the inspection and examination of professional acts;
(4) a jurisdictional role when it is charged with judging and, as required, sanctioning acts or failings against existing norms.[28]

Under this system of professional regulation, the professional association is responsible directly to the legislature for the conduct of professional affairs.[29] The elected legislature holds the power to amend or repeal the statute at any time. However, so long as the members and the professional association handle their affairs responsibly, the legislators will presumably be content to allow the profession to govern its own affairs. The alternative is for full state control of professional regulation.

§11.14 This procedure, which has become virtually standard practice in Canada, began over 100 years ago. The first professional statute incorporating a provincial body of land surveyors, *An Act Respecting the Profession of Land Surveyors in the Province of Manitoba*, was assented to on May 25, 1881.[30] The list of professions recognized by legislation has grown from the historically recognized professions of divinity, law and medicine[31] to include architecture, land surveying, accounting, engineering, nursing and a host of other occupational groups.

§11.15 It is somewhat ironic that the principal political jurisdiction in Canada, the federal government, is the only jurisdiction that does not have fully delegated professional legislation for land surveyors. Surveys of federal lands, that is to say those lands within National Parks, Indian Reserves, the Yukon and Northwest Territories and Canada's coastal waters, fall under the jurisdiction of the *Canada Lands Surveys Act*.[32] This statute provides

---

28 *Ibid.* at 17.
29 Government of Alberta, "Policy Governing Future Legislation for the Professions and Occupations" (1978).
30 (1881), 44 Vict., c. 29.
31 H.C. Black, *Black's Law Dictionary*, ed. (1979).
32 R.S.C. 1985, c. L-6, as amended.

The Surveying Profession 477

for the examination and appointment of Canada Lands Surveyors[33] under the jurisdiction of an appointed Board of Examiners. Other than the authority to license candidates and a very limited power to cancel or suspend the commission of a grossly negligent land surveyor, the Board of Examiners has no jurisdiction over the continuing affairs of Canada Lands Surveyors.[34] The Association of Canada Lands Surveyors is a voluntary body composed of Canada Lands Surveyors and was founded in 1985. It only has the power of a voluntary organization and hence has no authority to regulate the practice.

TABLE 11.1 LEGISLATION GOVERNING LAND SURVEYORS IN CANADA

Provincial Association	Year of Incorporation	Current Statute	Citation
Corporation of Land Surveyors of the Province of British Columbia	1905	*Land Surveyors Act*	R.S.B.C. 1979, c. 217
Alberta Land Surveyors' Association	1910	*Land Surveyors Act*	S.A. 1981, c. L-4.1
Saskatchewan Land Surveyors' Association	1913	*Saskatchewan Land Surveyors Act*	R.S.S., c. S-27
Association of Manitoba Land Surveyors	1881	*Land Surveyors Act*	R.S.M. 1970, c. 135
Association of Ontario Land Surveyors	1892	*Surveyors Act*	R.S.O. 1980, c. 492
Ordre des Arpenteurs-Géomètres du Québec	1881	*Le Loi sur les arpenteurs-géomètres*	R.S.Q. 1977, c. A-23
Association of New Brunswick Land Surveyors	1954	*New Brunswick Land Surveyors Act*	S.N.B. 1954, c. 97
Association of Nova Scotia Land Surveyors	1955	*Nova Scotia Land Surveyors Act*	S.N.S. 1977, c. 13
Association of Prince Edward Island Land Surveyors	1968	*Land Surveyors Act*	R.S. P.E.I. 1974, c. L-5
Association of Newfoundland Land Surveyors	1955	*Land Surveyors Act*	R.S.N. 1970, c. 198

The comparable federal legislation is the *Canada Lands Surveys Act.*

§11.16 There are therefore 11 separate and different statutes in Canada which govern the professional affairs of land surveyors. Each of the ten provincial bodies is self-governing and the right to practice is determined pursuant to the governing legislation by a council of management composed of practising members.

---

33 Prior to proclamation of the 1976-77 amendments to the *Canada Lands Surveys Act* in 1979, Canada Lands Surveyors were either Dominion Land Surveyors or Dominion Topographical Surveyors.
34 *Supra*, note 32, s. 22.

Policies Relating to Professional Organization

§11.17 During the last two decades, self-governing professions have been called to task by the public in several Canadian and American jurisdictions. While the Americans have seen their sunset legislation, Canadians have participated in several major government studies on the professions and occupations, together with reviews of the rights and privileges granted to them by government.[35] By and large, these studies have been beneficial in that they have confirmed public confidence in professional organizations. Perhaps more importantly, through dialogue between members of the public, governments and professional organizations, policies respecting the responsibilities and privileges of professional bodies have been established and documented. The Québec Commission "sought to preserve the bulk of the system whose operation and efficiency had withstood the test of time in other jurisdictions ... it also sought to reconcile the public interest with the incontestable advantages of a certain autonomy of the professions with regard to public authority."[36] These policies and guidelines have become useful models for the various self-governing professions. In Alberta, the *Policy Governing Future Legislation for the Professions and Occupations*[37] was adopted by the Legislative Assembly in 1978 as a result of hearings and the preparation of two reports of a select committee of the Legislative Assembly in 1973. A model statute, Bill 31, the *Architects Act*[38] was introduced in 1979 and passed in 1980, and was followed shortly thereafter by the *Land Surveyors Act*.

§11.18 The Alberta policy[39] requires that a specific field of practice be established for a profession with the possibility that the field may be exclusive in whole or in part to practitioners of that particular profession. Entrance standards and licensure criteria must be established with appropriate rights of appeal for persons denied entrance. Complaint and disciplinary provisions must be enacted with proper procedures for appeals both within and without the organizational structure of the professional body. Codes of ethics rate high on the list of requirements for professional legislation. Perhaps the central theme in most policy guidelines is that of

---

35 "Professional Regulation" — Ministry of the Att.-Gen., Ontario, 1979; "The Professions and Society" — Report of the Commission of Inquiry on Health and Social Welfare — Government of Quebec, 1970, Vol. 7, Part 5; "Royal Commission on the Inquiry into Civil Rights (McRuer Report)" — Ontario, 1968; "Report I & II on Professions and Occupations" — Alberta, 1973.
36 *Supra*, note 1 at 9.
37 *Supra*, note 29.
38 R.S.A. 1980, c. A-44.
39 *Supra*, note 29.

accountability to the legislative body which created the powers of self-government and which could ultimately take them away.[40]

§11.19 Most legislatures now require that there be public representation on the boards of management of self-governing professions; these public representatives are appointed by and report to a Minister responsible for either specific professions or professions in general.[41] The Alberta *Land Surveyors Act*[42] requires one public member on each of the governing council and the practice review board. In place of or in addition to lay members on the governing bodies, the Surveyor General or other government official responsible for surveys may also be appointed as a public representative as is the case in British Columbia.[43]

§11.20 Hence all professionals are subject to a level of government or regulation in which they, both as individual practitioners and as part of a professional body, play a key role. The sole purpose of this self-regulation must be to ensure that the practice of the profession is carried out in the greater public interest.

The Professional Association

§11.21 "The right to self-government is a privilege granted to a professional or occupational group by the Legislature."[44] In general, this privilege includes the authority and responsibility for administering its own affairs and regulating and controlling the practice of the members of the profession. Self-government is only granted to an occupational group when it can be clearly demonstrated that in the performance of duties, judgments are required which may endanger the health, safety or property of individuals.[45]

§11.22 The land surveyor is a unique breed of professional.[46] Every boundary line determined by him for his client is in fact also the boundary line for at least one of the client's neighbours. The surveyor is truly a public officer.[47] He must conform to established principles of boundary law whether they support or contradict the position of his client. The surveyor

---

40 *Ibid.*
41 *Ibid.*
42 *Supra*, note 21, ss. 11-12.
43 *Land Surveyors Act* R.S.B.C. 1979, c. 217, s. 6.
44 "Report on Professions and Occupations," *supra*, note 35 at 2.
45 *Supra*, note 29.
46 G. Girard, "L'Arpenteur-Géomètre et la Droiture," la revue de l'arpenteur-géomètre (1975), Vol. 3-3 at 190-198.
47 R.S.Q. 1977, c. A-23, s. 24.

480   Survey Law in Canada

must use his judgment and render his decision based solely on his findings.[48] He must not be unduly influenced by his client or his client's interest.[49] He must preserve in all his work a judicial mind and the impartial attitude of an arbiter.[50] Society is in fact the real client of the land surveyor even more so than for most other professionals.

§11.23 It is imperative that the public interest be held foremost in mind in the practice of land surveying.[51] The practising professional owes his livelihood to the public that created his rights and privileges; hence his practice must be devoted to providing a service to that public. The professional association must clearly stand up as the protector of the public interest and must actively operate to ensure that all its members fulfill their obligation to the public.[52]

§11.24 The obligation to protect the public interest includes several specific responsibilities which are inherent in the role of every professional association.[53] Self-government includes the responsibility to license entrants to the profession, to set standards of performance and to conduct and maintain professional competency by the discipline of those members who fail to meet those standards.[54] Self-government is granted to a professional group because it is generally held that the practitioners themselves are best qualified to judge whether a satisfactory level of performance is being provided by their members. Neither the general public nor governmental bodies are sufficiently familiar with the intricacies of practice to make this judgment.[55] Some of the main roles of a professional association are: entrance requirements, registration or licensure, professional standards, codes of ethics, codes of professional conduct, professional advertising codes, professional development, practice review and advisory services, complaint and discipline procedures.[56]

---

48   *Huebner v. Wiebe*, [1984] 1 W.W.R. 272 (Man. Q.B.).
49   W.M. Magwood, "Law and the Surveyor," *The Canadian Surveyor* (1960), Vol. XV-5 at 281-301.
50   T. Shanks, "The Status of the Dominion Land Surveyor," Annual Report of the Dominion Land Surveyor's Assoc. (1923) at 42-48.
51   J.H. Holloway, "Principles of Evidence," *The Canadian Surveyor* (1952), Vol. XI-2 at 31-42.
52   N.R. Mattson, "Brief to the Minister of Advanced Education Relating Further Comments on the Policy Governing Future Legislation for the Professions and Occupations," *ALS News* (1980), Vol. IX-1 at 14-15.
53   C.H. Weir, "Professional Practice," *The Canadian Surveyor* (1986), Vol. 40-1 at 3-21.
54   *Supra*, note 29.
55   "Royal Commission on the Inquiry into Civil Rights (McRuer Report)," *supra*, note 35 at Vol. 3 at 1186.
56   G.K. Allred, "The Professional Association — Protector of the Public Interest," *Terravue* (Autumn, 1983) at 12-24.

§11.25 Given these parameters, it is obvious that the primary role of a professional body is to regulate the practice of the professional in a manner which fully protects the public interest. A secondary role is that of benefit and service to members. In fulfilling this secondary role, however, the professional association must be very careful to ensure that the public interest is always placed ahead of the interest of its members. By providing an adequate service to its membership, the profession serves the public. These membership services often take the form of newsletters, continuing education seminars, investigation into and preparation of proposals for new legislation required by society and generally keeping the membership informed and up-to-date on current practices.

FEATURES OF CANADIAN SURVEY LEGISLATION

Exclusive Field of Practice

§11.26 It is generally conceded that the *exclusive* field of practice of the land surveying profession in Canada is restricted to the practice of cadastral surveying. The Québec *Land Surveyors Act* however, includes quite a lengthy definition which gives the *arpenteur-géomètre* jurisdiction in geodetic surveys and photogrammetry in addition to the general field of cadastral surveying.

> The following constitutes the practice of the profession of land surveyor:
> (*a*) all surveys of land, measurements for boundary purposes, making of boundaries, plotting of plans, making of plans, minutes, reports, technical descriptions of territories, certificates of localization and all documents and operations made by direct, photogrammetric, electronic or other methods connected in any way with bounding, laying out of lots, or establishing the site of servitudes, staking of lots, and scaling of lakes, rivers, and other bodies of water in Québec, with the calculation of the area of public and private property, all cadastral operations or compiling of lots or parts of lots, and cartographic representation of territory for the above-mentioned purposes;
> (*b*) establishing and keeping up-to-date the skeleton map of geodetic points of any order of precision and establishing of photogrammetric controls for the purposes of the work enumerated in subparagraph *a*.[57]

§11.27 The Alberta *Land Surveyors Act*[58] provides for an exclusive field of practice in cadastral surveys and a general field of practice in non-cadastral surveys. An Alberta land surveyor may practise in the general field of practice, but does not have an exclusive right to that practice.

---

57 *Supra*, note 4 at s. 34.
58 *Supra*, note 21 at s. 1.

482 Survey Law in Canada

(i) "practice of land surveying" means
- (i) the survey of land to determine or establish boundaries;
- (ii) the survey of land to determine or establish the boundaries of any right or interest in land or in air space;
- (iii) the survey of air space to determine or establish boundaries;
- (iv) the survey of land to determine the location of anything relative to a boundary for the purpose of certifying the location of the thing;
- (v) the survey of lakes, rivers or watercourses to establish or determine their boundaries;
- (vi) the survey by any means, including photogrammetric, electronic or astronomic methods, of land, water or air space for the purpose of preparing maps, plans and documents connected in any way with the establishment or determination of boundaries delineating any right or interest in land, water or air space;
- (vii) cadastral operations and compiling and recording information related to the matters specified in subclauses (i) to (vi);
- (viii) establishing and maintaining a network of geodetic points of any order of precision and establishing photogrammetric controls for the purposes of the work specified in subclauses (i) to (v) including the preparation of maps, plans and documents and giving advice with respect to any of the matters specified in this clause;

(j) "practice of surveying" means
- (i) the determination, establishment or recording by any means of the positions of points or natural or man-made features on, over or under the surface of the earth.
- (ii) the determination of the form of the earth.
- (iii) the practice of land surveying

and includes the preparation of maps, plans, systems and documents and the giving of advice with respect to any of the matters referred to in this clause.

§11.28 Several provincial statutes do not specifically define a field of practice although the restricted practice is generally deemed to be in the cadastral field. Most cadastral surveyors, however, also offer a broader practice extending into construction surveys, control surveys, hydrographic surveys, mine surveys and occasionally to geodetic and photogrammetric surveys.

Qualifications

§11.29 Until very recently, qualifications for land surveyors in the respective provincial associations varied considerably. In a study done by the Land Surveying Committee of the Canadian Institute of Surveying and Mapping in 1970 a great divergence was found to exist in the qualifying examinations for registration as land surveyors in the different provinces across Canada.[59] A similar comparison done by the Board of Examiners for Canada Lands Surveyors, based on the 1979 syllabus of examinations for Canada Lands Surveyors, confirmed this variation in academic requirements.[60] The main reason for this divergence in qualifying requirements was likely the lack of specific educational programs in surveying available in Canada and perhaps the fairly restricted movement of surveyors between provinces in those days. Laval University has offered a degree program in Forestry and Geodesy since 1907. Until 1960 when the University of New Brunswick program in Surveying Engineering was established, there were no English language survey degree programs available in Canada. In 1972, the University of Toronto survey option in Civil Engineering was dropped in favour of a Survey Science degree program at Erindale College. The University of Alberta also developed a Survey Science specialty in the Department of Geography in 1974. For several years, however, the program was merely an amalgam of courses available in various other faculties with nominal expertise in the major surveying disciplines.[61] In 1979, the University of Calgary developed a program in Surveying Engineering. The addition of these latter four programs in the last 28 years has provided opportunities to many young persons interested in a surveying career. It has also broadened the scope of the Canadian survey profession and provided some degree of uniformity in survey training.

§11.30 In 1979, the Board of Examiners for Canada Lands Surveyors adopted a new syllabus of examinations (see Fig. 11.2) which was equated to a degree program in surveying. Shortly thereafter, the Western and Atlantic Provinces' Boards of Examiners adopted syllabi similar to the first three levels or academic levels of the CLS syllabus.

---

59 Canadian Institute of Surveying and Mapping "Synopsis of Organization and Administration of the Provincial Associations of Land Surveyors". (Report of the Land Surveying Committee, 1970) [unpublished].
60 Board of Examiners for Canada Lands Surveyors, "Table of Exemptions to be Granted to Provincial Land Surveyors" (1979) [unpublished].
61 University of Alberta, Department of Geography (Survey Science), "Guideline on Programme of Studies" (1978), [unpublished].

FIG. 11.2 SYLLABUS OF EXAMINATIONS FOR CANADA LANDS SURVEYORS

CLS Syllabus

SCHEDULE I
1. Mathematics I
2. Statistics
3. Physics
4. Computer Programming
5. Earth Sciences
6. Surveying I
7. Report

SCHEDULE II
1. Mathematics II
2. Adjustments and Data Analysis
3. Survey Astronomy
4. Geodesy
5. Photogrammetry
6. Surveying II
7. Cadastral Studies

SCHEDULE III
1. Applied Photogrammetry and Remote Sensing
2. Hydrographic Surveying
3. Positioning at Sea
4. Cartography and Map Projections
5. Engineering Law, Business Administration and Economics

SCHEDULE IV
1. Acts, Regulations and Instructions for Survey of Canada Lands
2. Survey Law
3. The Surveying Profession
4. Project or Thesis

The Canadian Council of Land Surveyors (CCLS), following this lead, adopted the following policy[62] in 1983 which was in turn ratified by all provincial associations:

Reciprocity Policy

1. The normal entry into the profession will be by satisfying an academic requirement equivalent to a bachelor's degree in surveying and by completing a surveyor-in-training requirement of a maximum of two years articleship.
2. This academic requirement will be fulfilled by registered land surveyors who have completed the requirements of the Atlantic, Québec, Ontario, Western or Canada Lands Surveyors Board of Examiners, or who have

---

[62] Canadian Council of Land Surveyors, "National Policy on Reciprocity for Land Surveyors — Accreditation Program" (1984).

graduated from an accredited baccalaureate program in surveying.
3. Land surveyors who are registered in one province, and who have satisfied the above academic and surveyor-in-training requirements, may seek registration in another province by:
   3.1 sitting one or more examinations on the specific legislative, administrative and professional practice requirements of that province; and
   3.2 by satisfying a surveyor-in-training requirement of a maximum of one year.
4. Applicants who do not meet the academic requirements as set out in paragraph 2 above will be assessed on an individual basis by the appropriate Board of Examiners.
5. The National Surveying Education Committee of the Canadian Council of Land Surveyors will be charged with administering this national policy on reciprocity, and with issuing guidelines, etc. from time to time to assist in its effective implementation.

With this policy in place, as well as an accreditation program which CCLS has developed, one can assume that the qualification requirements will continue to remain relatively stable and revolve around the acceptable university degree plus two years of internship.

§11.31 The Board of Examiners for Canada Lands Surveyors is a creature of the *Canada Lands Surveys Act* and functions obviously to qualify persons as Canada Lands Surveyors. The two regional boards (Western and Atlantic), however, set up entrance examinations which are purportedly equivalent to a university degree[63] which is recognized by the respective provinces as the academic requirement for entrance into the respective associations. The Atlantic Board is composed of representatives of the four Examining Boards plus appointees of the four provincial governments (Newfoundland, Prince Edward Island, Nova Scotia and New Brunswick) and the chairperson and one other faculty member of the Department of Surveying Engineering at the University of New Brunswick. The Western Board is made up of two appointees from each of the four provincial associations/corporation (Manitoba, Saskatchewan, Alberta and British Columbia) plus two members representing the University of Calgary. Eighty percent of the provincial associations are thus represented by these two boards although it must be recognized that the provinces of Ontario and Québec represent over half of the cadastral surveyors in Canada. All provinces also have their own boards of examiners which operate under the authority of the respective provincial statutes. The two regional boards recognize the respective syllabi as common academic syllabi; each provincial board administers its own set of examinations in statute law and other professional subject areas.

---

63 Atlantic Provinces Board of Examiners, "Syllabus of Examinations for Certificate from the Atlantic Board of Examiners for Surveyors" (1979), [unpublished].

FIGURE 11.3 BOARDS OF EXAMINERS FOR LAND SURVEYORS IN CANADA

Association of Ontario Land Surveyors Board of Examiners
Atlantic Provinces Board of Examiners for Land Surveyors
Board of Examiners for Canada Lands Surveyors
L'ordre des Arpenteur-Géomètres du Québec Board of Examiners
Western Canadian Board of Examiners for Land Surveyors

§11.32 Surveyors have become acutely aware of the need for academic upgrading and continuing education particularly during the last three decades. The development of the early electronic distance measuring equipment, satellites and the application of computer technology to the technical aspects of surveying has prompted a reassessment of survey training. Commencing in 1959 with the first *Colloquium on Survey Education*,[64] the most recent held in 1985, there have been several colloquia and studies into the need to keep pace with technological and societal advancements. Two particular projects, the Workshop on Offshore Surveys[65] and the Weir Study Committee on the D.L.S. Qualification,[66] were likely among the major initiatives which resulted in the upgrading of the Dominion Land Surveyors syllabus (as it then was) and the multitude of other events which followed.

Codes of Ethics

§11.33 Professional ethics is "that branch of moral science which treats of the duties which a member of a profession owes to the public, to his professional brethren and to his client."[67]

§11.34 Nothing distinguishes the professional from the businessman more than the reliance on ethics in fulfilling his commitments to colleagues, employers, clients, and society as a whole. Where the businessman's creed is profit, products, people and principle[68] in that order, the professional must exactly reverse the order and importance of these four criteria.

§11.35 The highest and most onerous obligation on the professional, whether doctor, lawyer, engineer or land surveyor is the ethical obligation. He must act not merely for the sake of earning a living, but rather to

---

64 "Colloquium on Survey Education," *The Canadian Surveyor* (1960), Vol. XV-1 at 3-86.
65 Department of Energy, Mines and Resources, "Surveying Offshore Canada Lands for Mineral Resource Development" (1970).
66 C.H. Weir, "Final Report of the Weir Study Committee on the D.L.S. Qualification", [unpublished].
67 C.M. Brown, "Land Surveyors Ethics," *Surveying and Mapping* (June, 1960).
68 J. White, *Honesty, Morality and Conscience* (Colorado Springs: NavPress, 1979).

provide his client with sincere and objective counsel in his field of expertise.[69] The client must have the utmost trust and confidence in the professional, divulging his innermost concerns and closely guarded confidences to a relative stranger.[70] "The professional practitioner deals on an individual basis with clients in a close relationship rather than at arm's length."[71] To attain that degree of trust, the image of the entire profession must be totally without blemish and be characterized by a high degree of honesty, integrity and character. If a profession does not maintain its high ethical standards of care it will lose its privileged position in society to the detriment of all its members, but most importantly, in the long run, to the society it serves.[72]

§11.36 In order to ensure that practitioners provide an adequate service to the public and retain the dignity of the profession, professional bodies often set codes of ethics and codes of professional conduct. The judicious enforcement of these codes is a hallmark of a true professional body. The codes must be more than mere motherhood statements; they must contain reasonable rules to ensure that the practice of the profession is provided adequately to the public and not merely to enforce a monopoly situation or restrict competition among practitioners.[73] Codes of ethics often establish rules of honesty, integrity and character which are considerably more restrictive than codes that would normally be applied to the general public.[74] These strict rules are necessary to ensure that the professional person holds himself out as a reasonable, dependable and upstanding member of society and maintains the confidence of the individual client, as well as the public at large, in the integrity and ethical conduct of the profession as a whole.[75]

§11.37 A professional association can establish a code of ethics, publicize and lecture on it, but it remains largely up to each individual practitioner to interpret and apply it as personal circumstances and professional outlook dictate.[76] Lord Moulton once said: "[t]he real greatness of a nation . . .

---

69 W.E. Wickenden, "The Second Mile," *Electrical Engineering* (May, 1942).
70 G.K. Allred,"Professional Ethics — The Missing Link to a Complete Education," *The Canadian Surveyor* (1985), Vol. 39-4 at 385-390.
71 B. Stephenson, "A Self Governing Profession," *The Ontario Land Surveyor* (1984), Vol. 28-1 at 4-6.
72 A.C. McEwen, "A Professional Surveyors Association" (1981). Address to the Land Surveyors Association of Jamaica [unpublished].
73 "Royal Commission of the Inquiry into Civil Rights (McRuer Report)," *supra*, note 35, Vol. 3 at 1162-1166.
74 *Supra*, note 71.
75 *Supra*, note 35.
76 M. Woods, "A Professional Outlook," *The Canadian Surveyor* (1958), Vol. XIV-6 at 225-231.

is measured by the extent of . . . Obedience to the Unenforceable."[77] A code of ethics is largely unenforceable, yet the observance of a broad strong code can set the tone of the professionalism of the entire membership.[78]

§11.38 All provincial associations plus the Association of Canada Lands Surveyors have adopted codes of ethics. Each code varies in its specific canons; however the model Code of Ethics adopted by the Canadian Council of Land Surveyors (CCLS) in 1981 would appear to contain most of the broad and specific provisions necessary and unique to the land surveying profession.[79] The CCLS code contains eight articles each of which states a general principle. Each of these general canons is expanded by several tenets which deal with specific items relative to survey practice.

Standards of Practice

§11.39 Many of the fields of professional practice are conducive to the establishment of minimum standards or standard procedures. Minimum standards adopted by an association are usually based on the practices of specialists in the field and may often involve outside consultation with knowledgeable clientele. These standards however, must not be overly rigid but must allow for changing or unique circumstances.[80] They should be treated as reasonable guidelines but should neither restrict the judgment of a practitioner nor become the basis of inflexible administrative procedures. A professional association must be cautious in adopting standards to ensure firstly that they are well thought out, and secondly that they are qualified with any reasonable restrictions to their use. Once adopted they may be accepted by disciplinary tribunals[81] to determine standards of conduct or by the courts to determine negligence or malpractice.[82]

§11.40 Specific written standards of practice vary from province to province with standards being adopted by by-law, regulation, statute or good practice resolution. In Saskatchewan for instance, there are very few written standards adopted specifically by the professional association; rather, most standards are pursuant to the respective provincial statute, for example, regulations pursuant to the *Land Titles Act*[83] establish

---

77 J.F. Moulton, "Law and Manners," *Atlantic Monthly* (1924), Vol. 134-1 at 1-5.
78 Brown, *supra*, note 67, Vol. XX-2 at 195.
79 See Appendix A.
80 G.K. Allred, "The Professional Association — Protector of the Public Interest," *Terravue* (Autumn, 1983) at 12-24.
81 "Alberta Land Surveyors' Assoc. v. Hogg", *ALS News*, Vol. XIV-2 at 37.
82 *Supra*, note 80.
83 R.S.S. 1978, c. L-5.

standards for plans of survey to be registered under that statute. Ontario, on the other hand, has an extensive set of standards promulgated and kept up to date by the association's Standards Committee.[84] In Québec standards for certificates of location and staking of structures have been adopted by regulation pursuant to the *Land Surveyors Act*.[85]

§11.41 The *Manual of Instructions for the Survey of Canada Lands*[86] contains standards for surveys which have, in many cases, been adopted by other provincial associations. Section 193 of the 9th Edition of the *DLS Manual* was accepted in evidence by the Supreme Court of Canada in *Clarke v. Edmonton (City of)*.[87] The earlier editions of the *Manual of Instructions for the Survey of Dominion Lands*[88] (as it was then called) are particularly relevant and valuable for surveys in western Canada under the Dominion Land Surveys system.

Discipline

§11.42 Madame Justice Wilson of the Supreme Court of Canada commented recently "[m]embership in a profession should be in and of itself a guarantee of competence."[89] Unfortunately that is not always the case. Like every box of apples, there are always a few bad ones and professionals whether they be lawyers, doctors or surveyors are no exception. Some professionals don't heed their obligation to maintain high standards and ethical conduct once they commence practice and submit to other pressures. On occasion, the aspiring professional may just not have the requisite qualities necessary in the first place. In any event, the disciplinary process is a necessary function of every professional association in fulfilling its mandate to provide an adequate level of service to the public.

§11.43 The essential powers of discipline are the same in every province. All Canadian professional statutes for surveyors require a hearing before a tribunal composed of members of the association, which may or may

---

84 Assoc. of Ont. Land Surveyors, *Standards for Surveys*, 3d ed. (1985).
85 R.S.Q. 1977, c. A-23, r. 7.
86 Energy, Mines and Resources, Canada, *Manual of Instructions for the Survey of Canada Lands*, 2d ed. (1979).
87 [1930] S.C.R. 137.
88 There were ten editions of the *Manual of Instructions for the Survey of Dominion Lands* issued between 1871 and 1946. In addition there were amending circulars, bulletins, instructions, maps and specimen plans and supplements issued throughout this time span. Two editions of the *Manual of Instructions for the Survey of Canada Lands* have been printed in 1961 and 1979.
89 B. Wilson, unpublished convocation address presented at the University of Alberta (May, 1985).

not be the governing council of the organization, with the right of appeal to the courts. Procedural steps, however, vary considerably due to the fact that disciplinary powers are quite specific in individual provincial statutes.[90] The fulfillment of the disciplinary mandate appears to be a function of the size of an association. The larger and more active organizations seem to be more assertive in maintaining a consumer attitude of ensuring the competence of their members. These organizations play an active and vigilant role in monitoring the competence of practitioners in an attempt to assure the public that the standard of practice will meet acceptable requirements. These statements are supported by the findings of the Professional Organizations Committee of Ontario in a recent study of the legal, accounting, engineering and architectural bodies in Ontario.[91]

§11.44 Discipline is a quasi-judicial function and is not to be treated lightly. Disciplinary powers include the right to suspend or cancel a practitioner's license, fine or reprimand a member, or impose other penalties. Penalties are not necessarily always punitive. Some of the curative or rehabilitative powers common to disciplinary tribunals are the right to limit a person's practice, to require him to practise under supervision, to require that he improve his education by taking specific courses or to even refer him for counselling.[92] The power of discipline must only be used to protect the public from unprofessional, incompetent or unskilled practice based on an "independent and impartial application of predetermined rules and standards."[93] The practitioner being subjected to the disciplinary process must be given the full opportunity to respond to all allegations made against him. Although disciplinary tribunals which are normally composed of professional peers and perhaps a public representative are not expected to follow the strict procedures of a court of law, they must act in accordance with the principles of natural justice. In *Elliot v. University of Alberta Governors*,[94] the court held that "[t]he essential characteristic that must be present in the proceedings of a quasi-judicial body is that of fairness. However these bodies are not Courts of law and the strict rules of procedure and evidence that one is entitled to expect in a Court cannot be applied to them. It is sufficient to meet the requirements of natural justice, that a person . . . know the nature of the case, is given the opportunity to

---

90 *Re DeSantis and Alberta Land Surveyors' Act.* (1964), 48 W.W.R. 50 (Alta. S.C.).
91 M.J. Treblicock, C.J. Tuoky and A.D. Wolfson, "Professional Regulation" (1979), a staff study of accountancy, architecture, engineering and law in Ontario prepared for the Professional Organizations Committee.
92 *Land Surveyors Act*, S.A. 1981, c. L-4.1, s. 52.
93 Government of Ontario, "Royal Commission on the Inquiry into Civil Rights (McRuer Report)" (1968), Vol. 3 at 1181.
94 [1973] 4 W.W.R. 195 (Alta. S.C.).

state his case, and that the tribunal act in good faith." The concept of natural justice can basically be defined as the concept of fair play — the accused has the right to hear and challenge evidence produced against him and to cross-examine witnesses. He must be afforded the right to be represented by counsel and time to prepare himself for a reasonable defense. The adoption of the *Canadian Charter of Rights and Freedoms*[95] appears to have clouded many fundamental issues with regard to material justice and basic freedoms. To date, the courts have not had sufficient opportunity to clarify all these issues. Cases, however, which uphold the right of a disciplinary statute to impose more stringent standards of procedure on professionals have been upheld in both the Alberta and Quebec courts. Mr. Justice Belzil of the Alberta Court of Queen's Bench quoted from the Québec case of *Belhumeur v. Discipline Ctee. of Québec Bar Assoc.*[96] as follows:

> The practice of a profession is a privilege. The law grants to certain groups a monopoly to carry on certain well-defined activities and imposes upon the members of those groups an obligation to prevent abuse and to ensure that the monopoly will be exercised for the public good. It is normal that those who enjoy these privileges should be subjected to a more rigorous discipline than that which applies to ordinary citizens. This discipline is peculiar to them and is not part of penal law. In consequence of this, the right to silence preserved in Article 11(c) of the Charter does not apply to professional disciplinary law. One cannot claim in the same breath the so-called right to silence and the privileged status as a professional.[97]

§11.45 The McRuer Commission[98] noted the inconsistency in Ontario professional statutes regarding the right of a practitioner to continue practice pending the disposition of an appeal. The Commission noted that the Ontario *Surveyors Act*[99] did not allow a person whose right to practise had been suspended to continue to practise pending an appeal.[100] The Saskatchewan Court of Queen's Bench has dealt with this issue regarding the *Saskatchewan Land Surveyors Act* and the *Charter of Rights*.[101] The court found section 41(6) of the *Land Surveyors Act*[102] to be contrary to the provisions of the *Charter* and declared it to be "unconstitutional and

---

95 Part I of the *Constitution Act, 1982*, being Schedule B of the *Canada Act 1982* (U.K.), 1982, c. 11.
96 (1983), 34 C.R. (3d) 279 (C.S. Qué.).
97 *Fang v. The College of Physicians and Surgeons* (1985), 66 A.R. 352 at 355 (C.A.).
98 *Supra*, note 93.
99 R.S.O. 1960, c. 389, s. 36(2), this section has since been amended to allow a suspended member to continue to practice pending an appeal "except in the case of professional misconduct constituting incompetence." [R.S.O. 1980, c. 492, s. 28(21)].
100 *Supra*, note 93.
101 *Larson v. Sask. Land Surveyors' Assoc.* (1987), 63 Sask. R. 119 (Q.B.).
102 R.S.S. 1978, c. S-27.

of no force and effect." No doubt the future will bring more challenges to various provisions of professional statutes which may appear to conflict with the *Charter of Rights* and which the courts will be required to rule upon and clarify.

§11.46 The growing trend in professional legislation is to conduct periodic practice reviews. These "audits of the auditor"[103] so to speak, are designed to assist the practitioner or detect problems, or both, at an early stage before they develop into problems which might have a detrimental effect on the public, the practitioner or the profession. In the words of the Ontario Professional Organizations Committee, this " 'active competence' orientation implies that the promotion of competence is a major objective of the discipline process and is actively pursued by a variety of positive strategies designed to identify and eradicate cases of incompetence."[104] Both the Quebec legislation[105] and the Alberta statute[106] contain provisions which authorize a form of 'active competence' review.

§11.47 In all provinces except the Atlantic Provinces, government departments associated with Land Titles Offices or the offices of Surveyors General[107] have established extensive plan examination procedures to monitor the quality of the final plans of survey of private as well as of government survey practitioners. In the recent trend toward reducing government operations some provinces have eliminated this plan examination service. The respective professional associations have rallied to this call and have established post-registration audit procedures which systematically review the work of their members, thus fulfilling the mandate of ensuring the continued competence of all practitioners. Under the Ontario proposal, the Association of Ontario Land Surveyors has a mandate to conduct a systematic review of one plan of survey of every practitioner every year and one comprehensive review of the practice of each practitioner every five years. The systematic review is merely an office inspection while the comprehensive review includes field inspections and visits to the surveyor's office to examine such things as field notes and office procedures.[108]

---

103 V. Dzurko, "The Role of the Professional Association in Insuring that the Practitioner Carries out his Professional Responsibility" (1979), address at Alberta Land Surveyors' Association Professional Practice Seminar [unpublished].
104 *Supra*, note 91.
105 R.S.Q. 1977, c. A-23, r. 10.
106 *Supra*, note 92, s. 13.
107 In some provinces these positions may be referred to by the title Director of Surveys, Examiner of Surveys or other nomenclature.
108 Assoc. of Ont. Land Surveyors, "Survey Review Department" (1986), [unpublished].

THE PROFESSIONAL SURVEY PRACTITIONER

§11.48 Surveyors in Canada primarily practise in the private sector although the ratio of private sector to government varies from province to province.

TABLE 11.2 MODES OF PRACTICE OF CANADIAN LAND SURVEYORS

Province	Private Practice	Government	Education	Industry	Total
British Columbia	270	24	3	5	302
Alberta	239	40	7	6	292
Saskatchewan	90	14	2	2	108
Manitoba	36	30	1	5	72
Ontario	530	120	15	25	690
Québec	567	169	19	34	789
New Brunswick	67	30	3	9	109
Nova Scotia	165	100	10	20	295
Prince Edward Island	10	10	—	1	21
Newfoundland	77	9	1	13	100
Totals	2051	546	61	120	2778

Several provinces have additional members who are employed in other fields or outside the province.

§11.49 With the expansion of educational opportunities, the trend has been for surveyors to practise in multi-disciplinary corporations where land surveying has been only one of many specialties offered. These firms may focus their skills and expertise around land development, mapping and aerial photography, land information management, electronic positioning and measuring systems, or mineral resource exploration and extraction. The examination syllabus adopted in 1979 by the Board of Examiners for Canada Lands Surveyors was part of a major overhaul of the *Canada Lands Surveys Act*.[109] This amendment changed the definition of "public lands" to "Canada Lands" and expanded them to include "any lands under water belonging to Her Majesty in right of Canada or in respect of any rights in which the Government of Canada has power to dispose."

§11.50 This broadening of the definition of Canada Lands had the effect of broadening the scope of the Canada Lands Surveyor to include offshore positioning and in doing so, increased the need for surveyors to be qualified in geodesy, hydrography and photogrammetry. This, in turn, required a broader syllabus of examinations. The 1976-77 amendments also triggered a "grandfathering" process to allow practicing geodesists, hydrographers,

---

109 S.C. 1976-77, c. 30.

photogrammetrists, as well as cadastral surveyors who had the necessary technical expertise but were not Canada Lands Surveyors (or Dominion Land Surveyors as they had previously been known), to receive a commission to practice as a Canada Lands Surveyor upon successful completion of a limited set of examinations.

§11.51 Again, this national process has had a domino effect, causing all provincial associations to examine and consider expanding their mandates to include specialties other than cadastral surveying in the broad sphere of positioning and land management. As previously mentioned, several provinces have a broadened mandate within existing legislation at this time. Land surveyors are thus reaching out into new areas such as environmental and gravimetric studies, design of land information systems, arbitration, land and project management and many fields which have evolved in the recent technological revolution.[110] The future of the surveying profession will however, be discussed in greater depth in Chapter 12.

Powers and Responsibilities

§11.52 Irrespective of his mode of practice, the land surveyor in exercising his professional responsibilities of defining the boundaries of either private or public lands must exercise those responsibilities within the framework of existing legislation. In other words, the cadastral surveyor like everyone else, must operate within the confines of the law. To assist the land surveyor in carrying out his duties to establish boundaries between properties, the various legislatures have given him certain powers which are normally guarded jealously by legislative bodies.

§11.53 The right of entry onto private property, for example, is a right that legislatures are very loathe to delegate to any class of citizens without due regard for the need to grant that right.[111] As a general rule, however, land surveyors are given the right to enter on private property in order to evaluate boundary evidence and to establish boundaries between properties. This right usually only extends to the land surveyor in the exercise of his duty to establish boundary lines. He cannot use this authority for other purposes (such as to select a route for a roadway or pipeline) that are not related to his practice pursuant to the statute that grants that right. He must avoid committing unnecessary damage to the property or leaving the property in such a state that the owner's livestock might escape.[112] The *Nova Scotia Land Surveyors Act*,[113] for instance, states:

---

110 Canadian Institute of Surveying, and Mapping "Report of the Task Force on the Surveying and Mapping Industry in Canada" (1985).
111 *Alberta Hansard* (1987), 21st Legislature, No. 86 at 1790 and No. 91 at 1893-1894.
112 *Taylor v. Pac. Petroleums* (1977), A.R. 200 (Dist. Ct.).
113 S.N.S. 1977, c. 13, s. 16.

Every Nova Scotia Land Surveyor and his assistants when engaged in professional land surveying may enter upon and pass over any land, doing as little damage as possible; and, save as hereinafter provided, no action shall lie against such Nova Scotia Land Surveyor or his assistants for any act done under this Section, provided that such Nova Scotia Land Surveyor shall be liable for any unnecessary damage done by him or by his assistants under this Section.

§11.54 Surveyors may also be given certain other rights of entry under other statutes. For instance, the Alberta *Surface Rights Act*[114] authorizes an operator to enter upon land to make surveys for the purpose of determining the location of a proposed facility. Under the definition of operator, the term would include a surveyor in his capacity as an agent for his client. Despite the rights granted by legislation, surveyors "have a professional obligation to make every reasonable effort to contact the landowner prior to entry on the land and to use all reasonable care when passing over that land."[115]

§11.55 The land surveyor in his search for boundary evidence is also granted the power to examine witnesses and take evidence under oath with respect to all matters relating to the survey of land. In this respect, the surveyor is acting "in the capacity of a judge,"[116] and may examine witnesses as to their recollection of the location of survey markers or other boundary evidence. He may also compel witnesses to appear before him and produce any documents relating to the establishment of a boundary monument.[117]

§11.56 It is a criminal offence for anyone other than a land surveyor to wilfully remove or displace any boundary monument. Even in Biblical times, the corners established by surveyors were intended to be permanent and penalties were established for removal. In *Deuteronomy* it is written "[t]hou shalt not remove thy neighbour's landmark which they of old times have set."[118] This principle has been carried forward to modern times and is codified in the *Criminal Code* of Canada[119] as follows:

Wilful and Forbidden Acts in Respect of Certain Property

442. Every one who wilfully pulls down, defaces, alters or removes anything planted or set up as the boundary line or part of the boundary line of land is guilty of an offence punishable on summary conviction.

443. (1) Every one who wilfully pulls down, defaces, alters or removes

---

114 S.A. 1983, c. S-27.1, s.14.
115 *Supra*, note 111, No. 91 at 1894.
116 D.N. Hossie, "Evidence," Annual Report of the Corporation of Land Surveyors of the Province of British Columbia (1928) at 46-53.
117 S.N.B. 1954, c. 97, s. 40.
118 *Deuteronomy* 19:14.
119 R.S.C. 1985, c. C-46.

(a) a boundary mark lawfully placed to mark an international, provincial, county or municipal boundary, or
(b) a boundary mark lawfully placed by a land surveyor to mark any limit, boundary or angle of a concession, range, lot or parcel of land,

is guilty of an indictable offence and is liable to imprisonment for five years.

(2) A land surveyor does not commit an offence under subsection (1) where, in his operations as a land surveyor,
(a) he takes up, when necessary, a boundary mark mentioned in paragraph (1)(b) and carefully replaces it as it was before he took it up, or
(b) he takes up a boundary mark mentioned in paragraph (1)(b) in the course of surveying for a highway or other work that, when completed, will make it impossible or impracticable for such boundary mark to occupy its original position, and he establishes a permanent record of the original position sufficient to permit such position to be ascertained.

It is interesting to note that as recently as 1798 *An Act to Ascertain and Establish on a Permanent Footing, the Lines of the Different Townships of this Province* was brought into force in Ontario. The statute imposes a penalty of "death without the benefit of clergy" for any person who knowingly and wilfully pulls down, defaces, alters or removes a survey monument.[120]

§11.57 Despite the severity of the penalty for the destruction of survey monuments, there appears to be a reluctance on the part of the courts to uphold the intent of the written law. In a recent Nova Scotia case,[121] a person who willfully removed the monuments on a disputed boundary, was found guilty but given an absolute discharge. The individual had no previous criminal record; no evidence was put before the court as to the cost of damages; recovery of costs was possible through a civil action. One review of the case suggests that perhaps the penalties imposed by statute are unreasonably harsh and society is not prepared to impose extreme penalties or even register a criminal conviction.[122] The suggestion is made that perhaps a fine such as is proposed under the British Columbia *Land Titles Act*[123] would be more appropriate and the courts would be prepared to impose such a lesser penalty.

---

120 C.F. Aylesworth, "The Genesis of our Survey Act," Annual Report of the Assoc. of Ont. Land Surveyors (1928) at 170-179.
121 *R. v. Ross* (1985), 72 N.S.R. (2d) 381 (C.A.).
122 J.F. Doig, "Open Season on Monuments," *The Canadian Surveyor* (1986), Vol. 40-3 at 291-296.
123 R.S.B.C. 1979, c. 219, s. 365.

The Surveying Profession 497

§11.58 In some jurisdictions it is illegal for persons other than land surveyors to be in possession of boundary monuments.[124]

Responsibilities

§11.59 Many of the technical duties and responsibilities of a land surveyor go without saying and hence will not be elaborated upon. Obviously, the land surveyor must measure the extent of the property, plant monuments and prepare a plan showing the details of his survey for the information of his client and all who follow. Several key responsibilities, however, are perhaps less obvious and bear emphasizing at this juncture.

§11.60 In conducting any boundary survey, a land surveyor has an initial responsibility to search for all physical, documentary, written and parol evidence available, in order to conclusively identify the original location of any boundary monument. In normal circumstances, this may only require searching the local plan and title (or deed) repository in order to find the relative locations of existing monumentation. When monumentation is obscure or has been destroyed, the real expertise of the land surveyor must rise to the surface. For in these circumstances he must exercise his judgment in, firstly, determining what sources of evidence may be available and, secondly, searching for and analyzing that evidence in light of the case and statute law to replace the monumentation in its original location.[125] The land surveyor must always remember that his function is to determine and analyze the facts and make decisions in accordance with legal precedent for other surveyors and courts of law may follow in his footsteps.

§11.61 The land surveyor must maintain and preserve for posterity accurate notes of his field observations. Field notes are one of the most valuable sources of documentary evidence. To be of optimum value they must however, have been accurately, legibly and dutifully prepared in the field at the actual time of the observations. Notes about specific monuments are invaluable as is information relative to other physical evidence, meteorological conditions, field crews and instrumentation. Because of the inherent value of these records, several provincial survey associations have taken steps to preserve the field notes of deceased surveyors.[126]

Quasi-Judicial Function

§11.62 The land surveyor in exercising his duty to re-establish the

---

124 *Surveys Act*, R.S.A. 1980, c. S-29, s. 89.
125 *Supra*, note 86 at chapter B7.
126 *Land Surveyor's Act*, S.N.B. 1954, c. 97, s. 43.

boundaries of private or public property must be familiar with statute and case law relative to boundary retracement. His decisions are always subject to challenge and appeal to a court of law. He must therefore put himself in the position of a judge making a decision about a person's boundary. Since a boundary always defines the extent of two or more properties, where there is conflict there will be both a winner and a loser. He is not engaged by his client to take that party's position but rather to establish the common boundary in accordance with the evidence and the laws applicable to the jurisdiction. His primary responsibility is to place the new monument in the location where the original monument was, not where it was supposed to have been.

§11.63 Chief Justice Cooley of the Supreme Court of Michigan (1864-1885) prepared an extensive paper entitled "The Judicial Functions of Surveyors"[127] based on several boundary decisions which he had been party to in his judicial capacity. This paper has been quoted at length in several Canadian cases.[128] Justice Cooley's paper is also quoted at length by Marsh Magwood in his paper "The Law and the Surveyor".[129] This article establishes the responsibilities and quasi-judicial function of the surveyor so well that a considerable quotation from it, as adopted by Mr. Justice Kroft of the Manitoba Queen's Bench, is warranted to conclude this section:

> It is by no means uncommon that we find men whose theoretical education is supposed to make them experts, who think, that when monuments are gone, the only thing to be done is to place new monuments where the old ones should have been, and where they would have been, if they had been placed correctly. This is a serious mistake. The problem is now the same that it was before — to ascertain by the best lights of which the case admits, where the original lines were. The original lines must govern, and the laws under which they were made must govern, because the land was granted, was divided, and has descended to successive owners under the original lines and surveys; it is a question of proprietary right.
>
> The general duty of a surveyor in such a case is plain enough. He is not to assume that a line is lost until after he has thoroughly sifted the evidence and found himself unable to trace it. Even then he should hesitate long before doing anything to the disturbance of settled possessions. Occupation, especially if long continued, often affords very satisfactory evidence of the original boundary, when no other is attainable; and the surveyor should enquire when

---

127 *Surveying and Mapping* (1954), Vol. XIV-2 at 161-168.
128 *Home Bank v. Might Directories* (1914), 31 O.L.R. 340 (C.A.); *Weston v. Blackman* (1917), 12 O.W.N. 96 (H.C.); *Kingston v. Highland* (1919), 47 N.B.R. 324 (C.A.); *Houson v. Austin* (1923), 23 O.W.N. 603 (K..B.); *Muller v. Mamchur*, [1955] 4 D.L.R. 184 (Sask. C.A.); *Bateman v. Potruff*, [1955] O.W.N. 329 (C.A.); *Huebner v. Wiebe*, [1984] 1 W.W.R. 272 (Man. Q.B.).
129 W.M. Magwood, "The Law and the Surveyor," *The Canadian Surveyor* (1960), Vol. XV-5 at 281-301.

it originated, how and why the lines were then located as they were, and whether claim of title has always accompanied the possession, and give all the facts due force as evidence. Unfortunately, cases have happened where surveyors have disregarded all evidence of occupation and claim of title, and plunged whole neighbourhoods into quarrels and litigation by assuming to establish lines at points with which the previous occupation does not harmonize.

It is often the case that where lines or parts of lines are found to be extinct, all persons concerned have acquiesced in lines which were traced by the guidance of some land-mark which may or may not have been trustworthy; but to bring these lines into discredit, when the people concerned do not question them, not only breeds trouble in the neighbourhood, but must often subject the surveyor himself to annoyance, since in a legal controversy, the law as well as common sense must declare that a supposed boundary line or a supposed division line, if long acquiesced in, is better evidence of where the real line should be, than any survey made after the original monuments have disappeared.[130]

Professional and Technical Affiliations

§11.64 Not unlike other professional groups, the surveying profession is fragmented by many professional and technical bodies each with a slightly different mandate. Interests of the different organizations will vary from purely professional endeavours, to technical specialties and to strictly business affairs.

Fédération Internationale des Géomètres

§11.65 The International Congress of Surveyors, or the *Fédération Internationale des Géomètres* (FIG), to give its proper name, is a world-wide organization of approximately 60 member countries dedicated to the advancement of the surveying profession professionally, technically and economically. FIG operates through means of a seven-person bureau which is elected for a period of four years. The bureau conducts the day-to-day affairs of FIG and is responsible for organizing an annual meeting composed of delegates from all countries and a quadrennial congress where a ten-day program of technical papers is presented representing the interests of the nine FIG commissions. The nine commissions are divided into three groups as follows:

Group A — Professional Organization and Activities
    Commission 1 — Professional Practice, Organization and Legal Systems
    Commission 2 — Professional Education and Literature

---

130 *Huebner v. Wiebe, supra,* note 128 at 277.

Commission 3 — Land Information Systems

Group B — Surveys and Mapping
Commission 4 — Hydrographic Surveying
Commission 5 — Survey Instruments and Methods
Commission 6 — Engineering Surveys

Group C — Land Management
Commission 7 — Cadastre and Rural Land Management
Commission 8 — Urban Land System: Planning and Development
Commission 9 — Valuation and Management of Real Estate

Canadian Institute of Surveying and Mapping

§11.66 The Canadian member of FIG is the Canadian Institute of Surveying and Mapping (CISM) which is the technical organization or the learned society of surveyors in Canada. Councillors elected from 18 branches located in various regions across the country, eight technical councillors and six executive members constitute the governing council of the Institute. CISM sponsors educational seminars, a technical journal — the *CISM Journal ACSGC* — and an annual three-day technical conference for its membership of approximately 2,500 from all surveying disciplines. Membership in CISM is voluntary with minimal restrictions. CISM is an excellent forum for professional and technical dialogue.

Canadian Council of Land Surveyors

§11.67 The Canadian Council of Land Surveyors (CCLS) is a federation of the ten provincial professional survey associations. It was constituted in 1976 as an organization to represent land surveying at the national level. CCLS has no professional jurisdiction but rather acts as a liaison and study group to accomplish objectives of national interest. It has concentrated to date on adopting reciprocity, accreditation and model standards of practice for the ultimate benefit of the surveying profession in Canada. A national professional liability insurance program has been adopted with CCLS monitoring and sponsoring a program of loss control. The objective of the CCLS accreditation program is to review, monitor and accredit changes in the educational training available to Canadian surveyors, to ensure that the programs continue to provide a sound education and a progressive future for the surveying profession. CCLS publishes and distributes an annual magazine, *Terravue*, which informs municipalities, lawyers, engineers and architects about surveying issues of interest to them. CCLS represents professional land surveyors, on a national basis, through their provincial associations. The Association of Canada Lands Surveyors has been granted observer status within CCLS. In total, CCLS represents approximately 2,800 land surveyors across Canada.

Association of Canada Lands Surveyors

§11.68 The Association of Canada Lands Surveyors (ACLS) is a new organization established in 1985 as a voluntary body of these professionals in anticipation of achieving full self-governing status for the Canada Lands Surveyor outside the jurisdiction of the Board of Examiners and the Surveyor General of Canada. ACLS is composed of Canada Lands Surveyors under the broadened definition of the 1976-77 amendments to the *Canada Lands Surveys Act*[131] which includes geodesists, photogrammetrists and hydrographers in addition to purely cadastral surveyors. ACLS publishes an annual register of members even though membership is not compulsory in order to practice as a Canada Lands Surveyor.

Provincial Professional Associations

§11.69 The ten provincial associations each have complete professional jurisdiction which has been established by virtue of a provincial statute. Each association varies considerably in function and administrative structure dependent on the size and priorities of its membership. The Association of Prince Edward Island Land Surveyors, due to its small size, is the only association without a permanent office or administrative staff. At the other end of the scale is the Association of Ontario Land Surveyors (AOLS) with a staff of ten and an office condominium which is owned by the organization. The AOLS maintains its own self-funded professional liability insurance program, an index of survey field notes, a survey review department and a practice advisory group.

§11.70 Most associations have a half-time to full-time permanent office staffed by one or more land surveyors and one or more support staff. Quarterly newsletters are a norm, although a few associations operate only on periodic circulars and a report, prior to and subsequent to their annual general meetings. Some have libraries of technical and legal literature pertaining to surveying and some undertake extensive publishing of survey literature. Most associations have a continuing education program.

Other Organizations

§11.71 The foregoing organizations have a specific and direct relevance

---

131 S.C. 1976-77, c. 30.

to the cadastral surveyor. Other organizations such as the Canadian Hydrographic Association, the Geomatics Industry Association of Canada and the Urban and Regional Information Systems Association, are of interest to surveyors with specific concerns within these fields of endeavour, but these organizations have little to do with the field of survey law. Societies of survey technicians and technologists also exist to provide a technical home for the technical staffs of surveyors.

CONCLUSION

§11.72 The land surveyor is a specialist in the application of the technology of measurement to the principles of real property law. He must always remember that his measuring instruments are mere tools to assist in determining relative location. There may be occasions when a Gunter's chain and a staff compass are, in fact, the most appropriate tools to determine the true original position of a boundary corner. Regardless, however, of his instruments of measure, he must be guided by his diligence to search out the best evidence of the original corner. In this endeavour, the spade is often the most useful tool of all. Land surveying is not merely a technical endeavour but rather the application of these measurement techniques to one of the fundamental principles of a capitalistic society — the right to own land.

Appendix "A"

Canadian Council of Land Surveyors
Model Code of Ethics

ARTICLE 1.
A surveyor should assist in preventing the unauthorized practice of his profession.

(a) a surveyor shall not sign a certificate, report, or plan relating to work that was not completed and prepared under his personal supervision and for which the surveyor assumes professional responsibility.

(b) a surveyor shall not enter into any arrangement that would enable any unauthorized person or corporate body to practice the profession of land surveying directly or indirectly.

(c) a surveyor shall not knowingly and willingly become an accessory to a misdemeanor by failing to report any unauthorized practice to his Council.

(d) a surveyor shall prohibit his non-professional staff from performing activities that may be interpreted by the public as professional in nature.

(e) a surveyor shall not establish branch offices unless these offices are under the direction and management of a resident land surveyor.

ARTICLE 2.
A surveyor should assist in maintaining the integrity and competence of the survey profession, and should assist in improving the survey system.

(a) the surveyor shall assume the professional responsibility for all works carried out by his non-professional staff.

(b) the surveyor shall not further the application for admission to his profession of another person known by the surveyor to be unqualified by nature of his character, education or other attributes.

(c) the surveyor shall continually advance his knowledge and skills by participation in his association/corporation activities and any relevant continuing education programmes.

(d) the surveyor shall maintain permanent, clear and concise field notes and records, of all land surveying activities.

(e) the surveyor shall at all times serve his client or his employer to the best of his knowledge and ability.

ARTICLE 3.
The surveyor should assist his pupils and employees to achieve their optimum level of contribution to society.
(a) the surveyor shall instruct his pupils in the practical aspects of land surveying to the best of his ability.
(b) the surveyor shall assist his pupil in obtaining instruction in theoretical aspects of land surveying and in developing professional techniques as required.
(c) the surveyor shall assure his employees and pupils of proper working conditions, and equitable remuneration.
(d) the surveyor shall cultivate into his employees, the utmost integrity and a clear understanding of the professional obligation to society.

ARTICLE 4.
The surveyor should perform his professional services, assess and receive fair and just compensation from his client commensurate with the technical complexity, level of responsibility and liability potential of the services performed.
(a) the surveyor shall not enter into any fee splitting arrangement with any person other than another surveyor engaged in the same works.
(b) the surveyor shall give consideration to any schedule of fees published as information by the surveyors' provincial association or corporation during the preparation of accounts for his client.
(c) the surveyor shall not enter into any competitive bidding practice, but may provide a preliminary non-binding estimate for his clients.
(d) the surveyor shall not make any fraudulent or exorbitant charges for his services.
(e) the surveyor shall make available to his client, on request, copies of details relevant to the assessment for compensation.

ARTICLE 5.
A surveyor should avoid even the appearance of professional impropriety.
(a) the surveyor shall disclose to his client any conflict of interest, affiliation or prior involvement that could impair the qualify of his services to the client.
(b) the surveyor shall prevent his name being used in a professional way in association with any persons or enterprises of a dubious or doubtful character.
(c) the surveyor shall refrain from criticism in public of the conduct or practice of his colleagues.
(d) the surveyor shall not receive compensation for the same service from more than one source, except with the knowledge of the involved parties.

(e) the surveyor shall ensure that his client has knowledge of any available appeal procedures in the event of disputes over accounts or his professional conduct.

ARTICLE 6.
The surveyor shall preserve the confidences of his client and regard as privileged the information the surveyor may obtain regarding the affairs of his client.
(a) the surveyor shall maintain confidentiality with respect to the client's affairs during as well as after the completion of the surveyor's assignment or termination of his employment.
(b) the surveyor shall be responsible for the adherence to this Article by his articled students and staff.

ARTICLE 7.
The surveyor should exercise unbiased independent professional judgment on behalf of a client, and should represent a client competently.
(a) the surveyor shall disregard compromising interests and loyalties, and within the limits of the law, shall confine his activities to those benefitting his client.
(b) the surveyor shall not accept assignments beyond his professional competence.
(c) the surveyor shall not endeavour, by contract or otherwise, to limit his individual liability to his client.
(d) the surveyor shall not accept assignments that are beyond his resources to complete in a reasonable time and to carry out in a professional manner.

ARTICLE 8.
The surveyor should maintain the dignity of the profession through his associations with his clients, colleagues and subordinates.
(a) the surveyor shall limit his advertising to the adequate provision of information to the public.
(b) the surveyor shall refrain from any false or misleading statements or self-laudatory language in any advertising.
(c) the surveyor shall not attempt directly or indirectly to injure the professional reputation, and business prospects of any other land surveyor.
(d) the surveyor shall not solicit knowingly, assignments from the client of another surveyor.

12

The Future

John McLaughlin
B.Sc. E., M.Sc.E., Ph.D., P.Eng., N.B.L.S.

INTRODUCTION

§12.01 The evolving concern for more effective resource management and the concomitant demand for more appropriate land information facilities is leading the Canadian surveying community to reappraise its objectives and functions in a modern, post-industrial society. In a 1985 report on the surveying and mapping industry prepared for the Government of Canada, it was concluded that:

> The demand for more and more information about our physical environment presents an unprecedented challenge to the surveying and mapping industry. Our role, our tasks and our objectives will need to be reassessed, and re-oriented to the combined realities of new technologies and new demands for information. We will need to rethink our role — to decide whether we should cling to the traditional role of collecting and storing information, or whether we should also become involved in the management and utilization of information.[1]

In this concluding chapter we will briefly review what is meant by the management of land resources and the value of information to the management process and then examine both the traditional and potential

---

[1] W.D. Usher (Chairman), *Report of the Task Force on the Surveying and Mapping Industry in Canada* (Ottawa: Prepared for the Department of Regional Industrial Expansion, 1985) at 74.

role of the surveying profession in the resource management environment. We will conclude by identifying some of the legal and institutional issues (both from an information management and from the more traditional professional practice perspectives) which need to be addressed.

THE MANAGEMENT OF LAND RESOURCES

§12.02 Land is a term of many meanings. To some it suggests the platform for human activity, to others merely two-dimensional space. But in a much broader sense, land may be defined as the total life-supporting physical and cultural environment. As such, it is the most basic and important of human resources. The land serves individual needs for shelter and privacy, the economic needs for both the resources and the infrastructure of industrialized society and a variety of social needs in the community. As Daniel Green has noted:

> Civilized, urban man too often forgets that the land is still our matrix, and that we survive, whether we inhabit a mud hut on the banks of the Nile or a penthouse in Manhattan, only as products of the earth from which almost all our requirements come and to which, ultimately, all our bodies must return.[2]

§12.03 A central theme in the rise of civilization has been the introduction and subsequent evolution of institutions for the allocation, development, use and conservation of the earth's basic endowment of land. At any particular time, these land management institutions have reflected the possibilities and limitations for human adaption within specific cultural and environmental settings. Important characteristics of these settings have included the prevailing perceptions of space, territory and the organization of society on the one hand and the nature and availability of resources on the other.

§12.04 The fundamental land management institution in Canadian society has been a system of land tenure based in large measure upon the private ownership of productive resources. This system has provided the basic mechanism both for initially allocating land resources and for administering their subsequent development and use. Based upon the legal theories of private property first enunciated in Greco-Roman jurisprudence and as later developed in the English common law and upon the economic doctrines of the French physiocrats and the English liberal economists, it is a system which has focussed the dominant decision-making powers with respect to land management in the hands of the individual landowner.

---

[2] D. Green, *To Colonize Eden: Land and Jeffersonian Democracy* (London: Gordon and Cremonesi, 1977) at 1.

§12.05 In recent years this system of land tenure has been undergoing a fundamental institutional change, a change marked by "greater direct involvement of various levels of government in regulating the land market, and in prescribing the means permissible in the use of the land resource."[3] This has resulted from a series of social and economic concerns related to the supply of and demand for land. These have included a growing concern for the preservation of prime soils for food and fiber production; for controlling corporate, foreign and absentee land ownership; for managing urban growth and containing urban sprawl; for providing access to recreational lands and the preservation of wildlife habitats; and for increased environmental monitoring and protection. The result has been that the traditional decision-making powers of individual landowners and the traditional role of the market-place in the allocation of resources have been increasingly succeeded by professional planners and managers and by formal, systematic planning and management policies. An important by-product of this change has been a demand for new land information products and services.

THE ROLE OF INFORMATION

§12.06 The management of land, or any other resource for that matter, may be defined as the art and science of making decisions in support of certain perceived objectives. The classical formulation of this decision-making function entails the identification and ranking of goals, the cataloguing of alternative methods for achieving these goals, and the investigation of the consequences of each alternative. In a formal managerial environment, the actual steps in decision-making may include: (1) monitoring the environment for conditions requiring a decision (monitoring phase); (2) developing models which permit the analysis of alternative courses of action (modelling phase); (3) selecting a particular course of action (policy-making phase); (4) administering the course of action (operations phase).

§12.07 The basic intellectual resource in this decision-making process is human knowledge. Knowledge may be defined simply as that which is known and acted upon; it includes not only a store of facts but also the contribution of the mind in understanding and acting upon these facts. Knowledge is largely drawn from one's cultural heritage, but is continually being added to in the form of streams of information. This information may be, and indeed most often is, gathered in an informal, intuitive fashion.

---

[3] E.W. Manning, *Issues in Canadian Land Use* (Ottawa: Lands Directorate Working Paper No. 9, 1980) at 14.

But it may also be derived from a formal, systematic process — the product of an information system. Within this context, a land information system may be described as a combination of human and technical resources, together with a set of organizing procedures, which produces information for some managerial requirements. Traditional examples of land information systems have included topographical and geological mapping programs, land valuation and forest inventory surveys and land survey and registration systems.

§12.08 Recent attention has focussed on the design and implementation of new information systems and on the development of new strategies and procedures for gathering, administering, analysing and disseminating land information. These new strategies and procedures have included innovative organizational arrangements for managing the collection and use of information (at both a policy and an operational level), improved administrative procedures for controlling the flow of information and the introduction of new information technologies. Much of this effort has been devoted to upgrading traditional systems through the introduction of computerized spatial analysis and database management tools. Efforts to automate land valuation files, for example, date back to the early 1960's; more recently, automation of land title files has received similar attention.[4]

§12.09 At the same time, considerable attention has also been devoted to the construction of new multifunctional information systems designed to serve a broad range of users. These systems may be viewed from a number of perspectives. For example, they may be distinguished in the following fashion: (1) environmental information systems (wherein the primary focus is on delimiting environmental zones of some unique physical, chemical or biotic nature); (2) institutional information systems (which distinguish zones wherein specific land rights and responsibilities or restraints or both are recognized); (3) infrastructure information systems (which focus primarily on engineering and utility structures); (4) socio-economic information systems (which are primarily concerned with social and economic geography).

§12.10 Computer-based environmental systems, designed to serve a wide variety of resource and environmental processes, date from the early 1960's. Their appearance may be attributed in part to the growing importance of planning in general, and urban planning in particular, during this period. They were also the beneficiaries of a major technological revolution in both computer hardware and software. First generation systems included

---

[4] J.D. McLaughlin and I.P. Williamson, "Trends in Land Registration", *The Canadian Surveyor*, Vol. 39, No. 2, 1985.

the Canada Land Data System of Environment Canada and the Canadian Soil Information System of Agriculture Canada. Sometimes defined as computerized thematic mapping systems, they were designed to acquire, analyse, manipulate and disseminate land information in a format of value to specialized resource planning communities. The earliest installations, not surprisingly, had only limited geographic representation, file structure, data storage and processing capabilities. During the last decade, however, they have been succeeded by increasingly more powerful and flexible systems. These developments in turn have been made possible by the introduction of relational databases and turnkey geographic analysis systems. From an infrastructure perspective, many of the systems' initiatives have come from municipalities attempting to integrate information processing systems and databases from one or more land-related areas of urban government. The city of Burnaby, British Columbia, for example, implemented a computerized mapping system in 1974 that includes a graphical cadastre compiled from survey plans, attribute files containing parcel-referenced data, a unique property reference code and linkages to planning, infrastructure and terrain data.[5]

§12.11 One of the more ambitious institutional efforts has been in the development of multi-purpose cadastral systems. These may be described as large-scale, community-oriented land information systems designed to serve public and private agencies and individual citizens by: (1) employing the proprietary land unit or cadastral parcel as the fundamental unit of spatial organization and (2) employing a local records office as the fundamental unit for information dissemination. It is a concept built around an accurate spatial referencing and base mapping program, a cadastral overlay, and linkages to a series of parcel-referenced land records. Efforts to introduce the multi-purpose cadastre concept date back to the late 1960's and to such pioneering ventures as the Land Registration and Information Service within the Maritime Provinces. The overriding concern with these initiatives has been to provide information at the parcel level for land administration and resource management purposes.[6]

§12.12 Most recently, attention has focussed on the development of land information networks as a means of promoting efficient and effective use of information resources. A network, or confederation of land information

---

5 H. Christie, "Maps and Facilities Records — The Burnaby Experience", in *Proceedings of Seminar on Land-related Information Systems: Municipal/Provincial Integration* (Sponsored by the Inter-Ministerial Committee on Geographic Referencing and the Ontario Ministry of Natural Resources, Dec. 1982).
6 See, for example, N.R.C., *Procedures and Standards for a Multi-purpose Cadastre* (Washington: National Academy Press, 1983).

systems, may be viewed as an attempt to improve the flow of information among producers and users of land-related information. A network requires more co-ordination in the development of policies affecting land information and of standards and procedures regarding its management; it also requires co-ordination of research and of public and private involvement in matters relating to land information. This implies a more detailed knowledge of the user community and an understanding of its functions and requirements. Recent advances in computer and telecommunications technology, such as specialized database machines, communication value-added networks and second generation computer mapping systems with combined cartographic and geographic analysis capabilities, are providing significant benefits in storage, dissemination and manipulation of land information.[7]

THE SURVEYOR AS AN INFORMATION SPECIALIST

§12.13 Within the environment described above, the role of the professional land surveyor in cooperation with allied professionals, may be seen as providing the physical, social and institutional information necessary for the allocation, development and conservation of man's land resources. From this perspective, the functions of the surveyor may include: (1) the delimitation of human settlements; (2) the delineation of the earth's natural and artificial features; (3) the measurement of man's social and economic impact on the land. The surveyor may be viewed as an information specialist and, in a more general sense, as a participant in the land resource management process.

§12.14 This is not a new perspective. During the period of initial Canadian land settlement and development, the surveyor was an extremely important agent of the state, providing not only the cadastral framework but a wide variety of land-related information. This is illustrated in the instructions to deputy surveyors by an early Surveyor-General of New Brunswick:

> You will carefully describe on your plan the face of the country you pass through, noting or marking where hilly, mountainous, marsh or plains, with the nature of the soil and different kinds and sizes of timber, marking where the River is navigable and where obstructed by rapids or falls. All lakes or ponds you meet with are to be laid down and in general every remark to

---

7 For a review of recent endeavours, see: A.C. Hamilton and J.D. McLaughlin, eds., *The Decisions-Maker and Land Information Systems* (Ottawa: Proceedings of a Fédération Internationale des Géomètres Symposium published by the Canadian Institute of Surveying, 1985); and P. Dale and J.D. McLaughlin, *Land Information Management* (Oxford: Claredon, 1988).

be made that can tend to elucidate your survey and convey a knowledge of the country.[8]

It was exemplified as well by the statement of objectives in the constitution of the Association of Dominion Land Surveyors of Manitoba and Northwest Territories, which included "the advancement of a knowledge of the Geography, Geology and Natural History of the North-West."

§12.15 Not only did the surveyor provide a wide range of land information services, but he was deeply involved in a host of land management activities such as land valuation and site planning. The surveyor was seen in part as an information and measurement specialist, but also in a broader sense as an applied geographer. This view of the surveying profession survived well into this century in most parts of Canada and is still recognized in the frontier regions of the Yukon and the Northwest Territories. For the most part, however, the role of the land surveying profession as contracted in this century until now, by and large, has come to focus almost exclusively on property and technical surveying. Very recently, the land surveying profession has once more become interested in the potential of providing a broader range of land information and land management services to the public.

§12.16 Factors which have led to this renewed interest include: (1) the contribution made by public and private surveyors in building the modern computer-based information systems and the increasing role of the university programmes in surveying — together they have provided the knowledge and experience for a thrust into new fields; (2) the development of new tools (in remote sensing, digital mapping, spatial analysis and information management) with a wide range of potential uses; (3) an increasing demand for land information and land management services; (4) the flattening out and foreseeable decline in the demand for certain traditional land surveying services.

§12.17 To date the significant contribution of the Canadian surveying community to land information management, as noted, has been in the design, implementation and management of land information systems and in the development of concomitant technologies. It is a role which has required a profound understanding of modern information science and computer technology, organizational theory and project management, and consumer assessment and marketing. Surveyors are also beginning to appreciate a role of special professional significance: that of broker between

---

8 Elizabeth Dillon and John McLaughlin, "The Role of Land Surveys in the Early Development of New Brunswick", *The Canadian Surveyor*, Vol. 35, No. 2, 1981 at 132.

these public information utilities on the one hand and individual and corporate information consumers on the other.

§12.18 The Canadian Council of Land Surveyors has described this new role in terms of an expanded survey profession:

> In view of the increasing complexity of problems related to land use and development in a modern society, the provision of integrated land information and professional expertise in land economy have become essential services for ensuring the orderly development of land and the rational management of natural resources. Of all professionals, land surveyors are in the best position and are the most competent people to assume a leadership role in providing these services. This ability stems from their training and experience, their understanding of and feeling for land use, the nature of their work and the strategic role they play by being involved from the outset of any land development scheme. These qualifications, combined with the legal and exclusive right to practice land surveying, mean that the land surveyor can converse effectively with other professionals in providing land-related services to satisfy a wide segment of the population.[9]

EMERGING ISSUES

§12.19 The successful implementation of this concept of an expanded surveying profession will entail a considerable and concerted effort on the part of the public, private and academic surveying communities. Many technical, administrative, legal and institutional problems need to be addressed. In terms of land information management, for example, some of the legal and institutional issues will include: (1) the development and administration of standards for the definition of cadastral parcels and parcel identifiers, the classification of land attributes, and the protocols for the exchange of land data; (2) the determination of the ownership of information contained in land databases and the responsibility for maintaining distributed networks; (3) the assignment of responsibility for errors or omissions in the databases; (4) the development of regulations and procedures with respect to access, security and confidentiality of records.

CADASTRAL SYSTEMS ISSUES

§12.20 The individual parcel is the basic building block in any parcel-related information system. A parcel, at least in theory, is an unambiguously defined unit of land to which particular estates or interests are attached. The parcel should envelop a continuous area (or more strictly volume) of land and have a continuous interest attached to it. However, it is possible

---

9 J. Matthews and G. Raymond, "The Expanded Survey Profession," *The Canadian Surveyor*, Vol. 35, No. 3, 1981 at 211.

within some jurisdictions for the parcel to be spatially fragmented, to be linked to discontinuous interests, to cross municipal or other administrative boundaries and so forth. As well, problems can and do arise concerning how to delimit highway and water parcels, large tracts of public lands and subsurface and airspace parcels.

§12.21 Problems may also arise over the choice and administration of a system of parcel identifiers. The parcel identifier may be defined as a code for recognizing, selecting, identifying and arranging information to facilitate organized storage and retrieval of parcel records. It may be used for spatial referencing of information and as shorthand for referring to a particular parcel in lieu of a full legal description. While the initial selection of a parcel-identifier system is of fundamental concern, attention must also be directed to the allocation and subsequent control of these identifiers. Any parcel-identifier system can only work if one agency has the sole authority for assigning identifiers. The control of the subsequent allocation, re-allocation and withdrawal of parcel identifiers is but part of the larger process of keeping track of changes in rights, interests or estates in land. If the configuration of a parcel is changed (for example, by subdivision), that parcel ceases to exist and new identifiers are assigned to the new parcels. However, the original parcel remains as an historic entity and the description of it that was entered in the various registers and files when it did exist remains coded to it. Indexes that identify such retired parcels must be included in the record system unless provided otherwise by statute.

§12.22 Finally, there is a requirement to develop protocols for classifying and coding the land attributes (fee simple, leasehold, value, use, potential, for example) to be stored in the databases and for exchanging digital data. While considerable effort has been expended on developing proposed protocols and standards,[10] there still remains the challenge of getting them adopted and formally recognized.

RECORDS MANAGEMENT ISSUES

§12.23 The creation and management of land-related databases also raises a number of legal and institutional concerns, few of which have as yet been satisfactorily dealt with. These include such questions as: who should have access to the information contained in these data bases? who owns the information? who is responsible for loss of data or errors in the databases? A fundamental difficulty in addressing these questions (as the

---

10 See for example: Robert Scace, *Land Use Classification Systems: An Overview* (Ottawa: Lands Directorate Working Paper No. 14, 1981).

former French Minister of Industry, André Giraud, once noted) is that "there exists no legal infrastructure to sustain the transition to the information economy."[11] Legal rules which are often said to deal with information in fact usually deal only with the physical objects representing information. For example, intellectual property laws have traditionally focussed on the medium rather than the message and it has not been possible to copyright or patent information as such. Similarly, legislation on access to information, such as the federal *Access to Information Act*[12] is invariably framed in terms of access to records including microfilm and electronic records rather than access to information (thus failing to consider, for example, the question of whether computer data is in a disclosable form).

§**12.24** Access to information is a particularly important issue. As Professor Anne Branscomb has noted:

> In a truly open information society witholding too much data under the guise of secrecy leads to a lack of the public confidence that public consensus is built upon. As more and more citizens have their own microcomputers and their own modeling programs, they will expect to have a more open access to the information available to public officials in order to share the responsibility for right decisions. A collegial form of leadership demands a wide data base open to all participants.[13]

The government of Alberta has addressed this issue with respect to their provincial land-related information system network through the adoption of policies and procedures for electronic access to government land data.[14] The procedures focus on establishing a pricing structure for copies of land-related data in electronic form, developing criteria for copyright and liability, determining the means of access to data and setting out the conditions of use of the data.[15]

§**12.25** Also of significance are of the laws of evidence as they relate to the role of computer information and computer-generated information. As Mr. Justice Michael Kirby of the Australian Law Reform Commission has noted, "clearly it would be intolerable to require that every person who had contributed to a much used and thoroughly relied upon computer record should be available to prove orally his individual contribution to

---

11 Quoted in: Mr. Justice M.D. Kirby, "Legal Aspects of Information Technology" in *An Exploration of Legal Issues in Information and Communication Technologies* (Paris: O.E.C.D., 1983) at 11.
12 R.S.C. 1985, c. A-1.
13 A. Branscomb, "Law and Culture in the Information Society," *The Information Society*, Vol. 4, No. 4, 1986, at 289-290.
14 Government of Alberta, *Policy and Procedures for Electronic Public Access to Government Land Data*, 1987.
15 *Ibid.*

the computer record.[16] Here there is a clear need for an extension to and elaboration of the exceptions to the hearsay rule.

§12.26 As well there may arise the issue of record authentication. Traditionally documents such as deeds and plans of survey have been authenticated by having them signed and often witnessed. While signature analysis in some form can be used with digital records,[17] serious problems remain in attempting to authenticate the original records and changes to these records.

§12.27 There is also the issue of archival standards for computerized records. Land-related records must be maintained for varying lengths of time, perhaps 5 to 15 years for sales and valuation data and indefinitely for land titles and cadastral survey data. While the required period for most public records may be defined in federal and provincial archival legislation,[18] a number of questions remain to be addressed: (1) what specific data in a given record should be retained? (2) what storage medium should be used? (3) how will access to the data be maintained? (4) how will changes in hardware or software, or both, be accommodated?[19]

THE PRIVACY ISSUE

§12.28 The development of computerized databases, and especially the trend towards linking these databases together, also raises vital questions about the right to personal privacy and, of course, how to balance this with the right of access. As the Science Council of Canada has noted, "if privacy is no longer respected, then to what extent can Canadians expect the principle of personal self-fulfilment and political freedom to be upheld in Canada? How can Canadians develop a society in which privacy will be respected as an essential element of life?"[20]

§12.29 Privacy may be defined as "the claim of individuals, groups, or institutions to determine for themselves when, how, and to what extent information about them is communicated to others."[21] Within the context of records management, the invasion of privacy may take the form of

---

16 *Supra*, note 11 at 44.
17 F.M. Greguras and D.S. Sykes, "Authentication in EFT: The Legal Standards and the Operational Reality", *Computer Law Journal*, 1980 at 67-86.
18 See, for example: *Public Records Disposal Act*, R.S.N.S. 1976, c. 254.
19 For a review of these questions, see: G. Hunter, *Archival Data and Land Information Systems* (Parkville, Victoria: University of Melbourne, Technical Report, 1986).
20 Science Council of Canada, *A Workshop on Information Technologies and Personal Privacy in Canada* (Ottawa, May 1985) at 9.
21 A. Westin, *Privacy and Freedom* (New York: Antheneum, 1979) at 3.

an unauthorized access or use of records and the linking of records from different databases. For example, the selective extraction of land registration and property assessment information from a cadastral network could prove invaluable (and potentially very damaging) in developing a credit rating profile. The common law has not specifically recognized a legally protected right to privacy, although to a limited extent the torts of trespass, nuisance and appropriation of personality have been used in privacy-related cases.[22] There are also certain administrative law remedies, such as *mandamus*, "that can be used against officials who have failed to exercise their discretion to protect a citizen's privacy."[23]

§12.30 More recent efforts have focussed on enshrining certain basic principles of privacy in legislation. These principles have included: (1) establishing limitations on the nature and extent of personal data that can be collected; (2) requiring that the data be relevant and accurate; (3) establishing limitations on disclosure; (4) developing procedures for protecting against unauthorized use of the data; (5) providing the right for an individual to review and correct data concerning him.[24] Canadian legislative examples include the *Privacy Act* which has as its basic purpose the extension of the present laws of Canada that protect the privacy of individuals with respect to personal information about themselves held by federal government institutions.[25] As well, several provinces have enacted legislation making it a tort for any person to violate the privacy of another, thus extending and enshrining in statute law the beginnings made under the common law.[26]

OTHER ISSUES

§12.31 Of course not all the significant survey law issues relate directly to land information management. There is, for example, the issue of protecting the commercial value of computer software. The Canadian surveying and mapping community has long been involved in developing software packages for plane surveying and adjustment computations, digital mapping processes, image analysis and so forth. While much of the early software was released in the public domain, more recent efforts have focussed on creating marketable products. To protect its investments,

---

22 Peter Burns, "The Law and Privacy: The Canadian Experience"; *The Canadian Bar Review*, Vol. 54, No. 1, 1976.
23 *Supra*, note 20 at 37.
24 *Supra* note 13.
25 *Privacy Act*, R.S.C. 1985, c. P-21, s. 2.
26 These include: British Columbia (R.S.B.C. 1979, c. 336); Manitoba (R.S.M. 1987, c. 125); Newfoundland (S.N. 1981, c. 6); and Saskatchewan (R.S.S. 1978, c. P 24).

The Future 519

industry has turned to the laws pertaining to patents and copyrights, to the laws of contract and to the laws of equity.[27] In May 1987, the federal government tabled a new *Copyright Act* which "would recognize that computer programs are covered by copyright, whether they are in a form to be read by humans or only by machines. The courts have had a wretched time judging cases of unauthorized copying, and have had to fall back on an unhappy analogy with the copyright on books. The proposed law would resolve their dilemma."[28]

§12.32 There are also many outstanding issues in cadastral survey practice, including problems in delimiting tidal boundaries and proportioning accretion, in delimiting air space and condominium rights, in defining and demarcating administrative boundaries and so forth. There are legal issues related to the introduction of new technologies such as the evidential nature of electronic field notes. There are issues concerning the appropriate use of coordinates in both the onshore and offshore regimes and the legal and economic relevance of the guaranteed boundary concept. As the use of the global positioning system in land surveying evolves, there will also evolve a growing concern about the legal and institutional basis for a satellite-based coordinate framework.

§12.33 At the same time, there is still much work to be done towards developing efficient and equitable procedures for resolving boundary disputes (perhaps with more emphasis on arbitration and mediation techniques). And much remains to be accomplished in the area of developing and administering effective standards and procedures for cadastral survey practice. Finally, the role and responsibility of the professional surveyor, especially as he or she moves into the broader land and information management environment, will undoubtedly also have to be re-examined.

---

27 R.G. Hammond, "The Misappropriation of Commercial Information in the Computer Age," *The Canadian Bar Review*, Vol. 64, 1986 at 342-373.
28 *The Globe and Mail*, May 25, 1987.

Index

References are to paragraph numbers in the text.

ABORIGINAL TITLE
  concept 2.21
  change anticipated in law 2.143
  federal responsibility 2.23

ABSTRACT INDEX 3.25, 3.26

ACCRETION
  apportionment 5.88, 8.54
  concept 5.52, 6.87
  land titles system 5.58
  proportional shorelines 8.57
  representative baseline 8.56
  requirements 6.88
  restatement of limits 6.92
  vertical 5.54

ACCURACY 4.44, 4.94, 4.103

AD MEDIUM FILUM RULE 5.19, 6.18, 6.29

ADMINISTRATIVE LAW 1.36

ADVERSE POSSESSION
  concept 4.28, 4.78
  conditions 4.82
  constructive possession 4.83
  setting monuments 4.82
  supporting a claim 4.87

ALLODIAL OWNERSHIP 2.07, 2.10

ASSOCIATION OF CANADA LANDS SURVEYORS 11.68

ASSURANCE FUND
  essence 3.50
  liabilities 3.51

BANK 6.28, 6.36, 6.52, 6.70

BASELINES
  closing 5.99
  normal 5.98
  straight 5.100

BENEFICIAL INTEREST
  concept 2.40
  mortgage 2.41

BEST EVIDENCE
  rule 7.47
  surveyor to use 4.114, 8.17

BORNAGE. *See also* BOUNDARY DETERMINATION, QUEBEC.
  appropriate circumstances 8.93
  definition 8.91
  delimitation 8.95
  demarcation 8.97
  necessary conditions 8.92
  origins 8.89
  report 8.96
  stages 8.94

BOUNDARY
  ambulatory 5.83, 5.88
  artificial 5.09
  Canada's maritime 5.107
  conventional line 8.30

BOUNDARY — *continued*
  conventional boundary 5.70
  creation of 4.05, 4.24, 4.29, 4.37, 4.42, 4.61
  definition of 4.01, 11.08
  equidistance principle 5.103
  fences 8.61
  fixing 4.30, 4.53
  lost 4.113, 4.117
  maritime international 5.102
  marked on ground 4.32, 4.46
  mean high water 5.71
  natural 5.05, 5.84, 6.74
  obliterated 4.144
  ordinary high water 5.67
  principles applying to 4.36
  resolution of disputes 8.20, 11.06
  re-definition, resurvey 4.137
  riparian 6.25, 6.76
  shore allowances 6.78
  terms for water boundaries 6.44
  uncertainties 8.04, 8.06
  water 5.03

BOUNDARY DETERMINATION, QUEBEC. *See also* BORNAGE.
  acceptance by parties 10.212
  capacity of parties 10.191
  categories of 10.190
  conditions of 10.189
  costs 10.218
  definition 10.188
  demarcation 10.213
  evidence 10.202
  formal request 10.192
  minutes of 10.216
  nature of 10.187
  oath of office 10.196
  procedures 10.197
  recusation 10.195
  report 10.209
  surveying operations 10.201

BRITISH NORTH AMERICA ACT
  basis of legal system 1.34
  constitutional statute 1.55, 1.57
  federal powers 1.63, 4.19, 10.283
  provincial powers 1.64, 10.280

CADASTRE, QUEBEC
  division, English regime 10.135
  division, French regime 10.133
  effects 10.162
  introduction 10.132
  modern 10.145
  operations 10.152
  origins 10.138
  types 10.148

CANADIAN COUNCIL OF LAND SURVEYORS
  composition 11.67
  expanded profession 12.18

CANADIAN INSTITUTE OF SURVEYING AND MAPPING 11.66

CAVEAT
  concept 3.72
  transfer 3.79

CERTIFICATE OF TITLE
  land titles system 3.32
  quieting of title action 8.77

CERTIORARI 1.52

CHARTER OF RIGHTS AND FREEDOMS
  application 1.69, 1.72
  concept 1.68
  equality 1.73
  exceptions 1.70

CIVIL CODE
  civil liability 10.35
  common areas and structures 10.76
  contracts 10.22
  conventional servitudes 10.85
  co-ownership 10.58
  corollary property rights 10.63
  emphyteusis 10.71
  historical roots 10.05
  matrimonial regimes 10.42
  obligations 10.12
  possession 10.102
  prescription 10.119
  privileges and hypothecs 10.92
  property law 10.48, 10.53
  real servitudes 10.72
  right of way 10.83
  sources of law 10.10

CIVIL CODE — *continued*
   spacing requirements 10.79
   superficiary rights 10.62
   usufruct, use, habitation 10.65
   view onto property 10.80

COLOUR OF TITLE 2.93

COMMON LAW 1.30

CONCURRENT OWNERSHIP
   joint tenancy 2.106
   tenancy in common 2.106

CONDOMINIUM 2.128

CONSTITUTION
   concept 1.54
   early written 1.56
   source of judicial authority 1.74
   unwritten component 1.60
   effect on water law 5.60

CONSTITUTION ACT 1867
   formerly BNA Act 1.57

CONSTITUTION ACT 1982
   amending formula 1.67
   concept 1.66
   formerly BNA Act 1.57

CONTIGUOUS ZONE 5.96

CONTINENTAL SHELF
   concept 5.96
   delimitation 5.106

CONTRACT LIABILITY
   concept 9.03, 9.55
   extent of 9.10
   standards of performance 9.09
   third-party 9.48

CONVENTIONAL LINE
   application 8.31
   boundary discoverable 8.36
   concept 4.118, 8.30
   Crown cannot be a party 8.34
   dependence on oral evidence 8.32
   essentials 4.120, 8.37
   evidence of 8.32, 8.38
   fence 8.67
   lines never run 8.45
   occupation 4.120
   owners only can agree 8.35
   reduction to writing 8.32
   successors bound 8.40
   surveyed line 8.41, 8.43
   time element 8.39
   variance with description 8.44

CONVENTIONAL BOUNDARY 5.70

CONVEYANCING
   abstract of title 2.81
   marketability of title 2.81
   preliminaries 2.75
   representation by solicitors 2.79
   verification of title 2.81

COPYRIGHT 12.31

COURTS, HIERARCHY 1.76

COVENANT
   benefit 2.70
   burden 2.70
   concept 2.68
   enforceability 2.69

CROWN
   as owner of lands 4.07
   granting lands 4.17
   original surveys 4.23, 4.52
   owner of shore 5.45

DEEDS
   ambiguity in 4.116
   defective or missing 3.07
   essentials 3.18
   quit claim 3.45, 8.32, 8.85
   tax 8.87

DELIMITATION
   bornage 8.95
   see also BOUNDARY
      DETERMINATION, QUEBEC
   maritime boundary 5.90, 5.106
   tidal boundary 5.66, 5.78, 5.81

DESCRIPTIONS
   interpretation 4.64, 4.92
   need for measurements 4.92

524   Survey Law in Canada

DIRECTION
   early surveys   4.104
   orientation   4.106

DISCIPLINE
   plan examination   11.47
   plan examination, Ontario   3.100
   practice reviews   11.46
   quasi-judicial function   11.44

DIVISIONS OF LAW
   administrative   1.41
      survey association an example   1.42
   substantive   1.35
      private   1.38
      public   1.35

DOMINANT TENEMENT   2.43

EASEMENT
   appurtenance   4.71
   by grant express or implied   2.49, 2.50
   by prescription   2.54
   creation   4.73
   concept   2.43, 4.67
   defining extent   4.72
   examples   4.69
   heriditable   2.44
   involves two separate parcels   2.46
   prescription   4.74
   public   2.47
   records   4.73
   termination   2.56
   way of necessity   2.51

ENCROACHMENTS
   concept   8.58
   improvements   8.58

EQUITY   1.32, 2.13

EROSION   5.52

ESCHEAT   2.136

ESTATES
   concept   2.30
   fee simple   2.32
   fee tail   2.33
   life estate   2.34

ESTOPPEL
   acquiescence   8.53
   concept   8.46, 8.48
   definition   4.125
   elements   4.127, 4.132, 4.135, 8.49
   effect of improvements   8.50
   effect on land titles   4.129
   occupation only   8.52
   surveyor's role   4.134

EVIDENCE
   admissibility   7.11
   adverse possession   4.87
   best evidence   7.47
   burden of proof   7.61
   documents   7.63
   examination of witnesses   11.55
   exclusionary rules   7.18
   expert opinion   7.54
   extension of rules   12.25
   hearsay rule   7.20
   hearsay rule exceptions   7.23
      admissions   7.35
      ancient documents   7.33
      declarations against interest   7.23
      declarations as business duty   7.24
      declarations as to reputation   7.30
      privilege   7.38
      public documents   7.34
   need to extend rules   12.25
   opinion   7.51
   parol evidence   7.43
   precedence   4.43, 4.64
   presumptions   7.59
   real evidence   7.69
   standards of proof   7.67
   statutory declarations   8.86

EXCLUSIVE ECONOMIC ZONE   5.96

EXPERT OPINION
   essence   7.54
   report of survey   7.56

EXPROPRIATION
   compensation   2.104
   concept   2.99
   mechanics   2.103
   prerequisites   2.102

FEDERATION INTERNATIONALE DES
   GEOMETRES   11.65

FENCES
conflict with monuments 8.70
conventional line 8.67
evidence 8.63
fence viewers 8.61
limits 8.64
long acceptance 8.66
of convenience 8.65
presumed boundary 4.59, 4.60
protraction 8.68
purpose 8.62

GRANT
significance of date 6.82
lost 2.97, 4.74

HARBOURS. See PUBLIC HARBOURS.

HEARSAY
rule 7.20
exceptions 7.23

HIGH SEAS 5.96

HIGHWAYS
acceptance and dedication 2.57
shore road allowances 6.78

IMPROVEMENTS
concept 8.58
encroachments 8.58

INDEXING
grantor/grantee 3.08
parcel based 3.10

INLAND WATERS
ad medium filum rule 6.29
boundaries 6.25, 6.73
boundary terminology 6.33, 6.35, 6.83
high and low water mark 6.34
interpretation of boundary terms 6.44
natural boundary 6.65, 6.83
navigability 6.10
ownership of bed 6.24
ownership of bed, Ontario 6.10, 6.13, 6.15
ownership of bed, prairie provinces 6.18
water levels 6.38, 6.67

INSURANCE, LIABILITY
analysis of claims 9.61
effectiveness 9.60
loss control 9.61
not complete protection 9.62, 9.64
policy stipulations 9.67

INTEGRATED SURVEYS 4.110

INTERNAL WATERS 5.96

INTERNATIONAL BOUNDARIES
equidistance principles 5.103
geographic anomalies 5.104
resolution of disputes 5.102
special circumstances 5.105

JOINT TENANCY 2.106

JUDICIAL AUTHORITY
concept 1.74
structure 1.76

LACHES 2.88

LAND
a commodity 12.02
government regulation 12.05
private ownership 12.04
parcel 4.03
ownership 4.07

LAND PARCEL 4.03

LAND TITLES SYSTEM
accretion 5.58
appeals from registrar's decision 3.37
caveat 3.72
certificate of title 3.32
correction of the register 3.35
estoppel 4.129
first registration 3.38
fundamentals of the system
  assurance fund 3.50
  completeness of register 3.48
  historical searches 3.47
  indefeasibility 3.39
  invalidation by fraud 3.43
  misdescription 3.46
  two certificates for same land 3.45
Ontario 3.91
registration of instruments
  easements 3.58

LAND TITLES SYSTEM — *continued*
  registration of instruments — *continued*
    mortgages and encumbrances   3.56
    removal of interests   3.60
    restrictive convenants   3.57
    statutory rights of way   3.59
    transfers and leases   3.52
    transmission of title   3.54
  registration of mineral interests   3.81
  registration of plans   3.61
    condominiums   3.70
    expropriations   3.69
    subdivision   3.62
  rejection of instruments   3.36
  search facilities   3.71

LAW, NATURE OF
  natural law   1.07
  positivism   1.03
  realism   1.05

LAW OF THE SEA
  development   5.92
  International Court of Justice   5.102
  UN Conference   5.99

LAW, SOURCES OF
  case law   1.15
  common law   1.15
  Quebec   10.01
  statute law   11.16

LIABILITY
  in contract   9.03, 9.55
  in negligence   9.04, 9.14
  insurance   9.59
  limitation of actions   9.06
  negligent misrepresentation   9.50, 9.51, 9.54
  standard of conduct   9.05, 9.56
  survey plans   9.54
  third party   9.48

LICENCE
  concept   2.59
  heriditament   2.63

LIMITATION OF ACTIONS
  adverse possession   2.91, 2.92
  colour of title   2.93
  concept   2.88
  liability   9.06
  time period   2.90

LIS PENDENS   3.60

LOST BOUNDARIES   4.113, 8.30

LOST GRANT   2.97, 4.74

MANAGEMENT
  land resources   12.02, 12.06
  land information   12.08, 12.16, 12.17, 12.23
  standards   12.19, 12.27

MANDAMUS
  definition   1.52
  breach of privacy   12.29

MARITIME BOUNDARIES
  Canada   5.107
  Gulf of Maine   5.109
  International Court of Justice   5.102
  surveyor's responsibilities   5.111

MARITIME ZONES   5.96

MATRIMONIAL LEGISLATION   2.116

MEAN HIGH WATER   5.32

MEAN SEA LEVEL   5.35

MEASUREMENT
  accuracy and precision   4.94
  astronomic control   4.104
  Dominion Land Survey   4.97
  magnetic control   4.104
  original surveys   4.96
  standards   4.89, 4.103
  weight accorded   4.44, 4.91

MECHANICS LIEN   2.110

MINERAL INTERESTS
  certificate for   3.84
  meaning of   3.87
  petroleum and natural gas   3.88, 3.90
  registration of   3.81
  reservation of   3.83

MISDESCRIPTION   3.46

Index 527

MONUMENTS
  adverse possession 4.82
  coordination 4.111
  removal 11.56
  water boundary 5.84
  weight accorded 4.43

MORTGAGE
  concept 2.85
  charge in land title system 2.86
  effect of 3.19

NATURAL JUSTICE
  concept 1.47
  denial 1.48, 1.50

NATURE OF LAW
  natural law 1.07
  positivism 1.03
  realism 1.05

NAVIGABLE WATERS
  public navigation 5.18, 6.05, 6.98, 10.284
  tidal waters deemed 5.21, 6.15
  vesting of bed and foreshore 6.05

NAVIGABILITY
  affects title 6.96, 10.284
  established by facts 5.18
  ownership of bed 6.10
  survey cannot determine 6.99

NEGLIGENCE
  definition 9.14, 9.26, 9.34, 9.40
  extent of liability 9.10, 9.25, 9.29, 9.31, 9.43
  insufficient investigation 9.17, 9.24, 9.27
  liability 9.04, 9.14, 9.34, 9.40, 9.41
  mistake 9.08
  negligent misrepresentation 9.50, 9.51, 9.54
  reasonable care 9.14, 9.30, 9.46
  reliance on others 9.17
  survey plans 9.54
  third party liability 9.48

NEWFOUNDLAND
  registration system 3.101
  quieting title 8.78

OBITER DICTA 1.13

OBLITERATED BOUNDARY 4.144

ONTARIO
  land information system 3.91, 3.98
  Land Registration Reform Act 1987 3.97
  plan examination 3.100
  registration systems 3.91

OPTION 2.65

OVERRIDING INTERESTS
  concept 8.11
  examples in Nova Scotia 8.12
  registry system 3.06
  land titles system 3.48

PARCEL
  definition 4.03
  referencing 12.10, 12.20, 12.21
  retirement 12.21

PAROL EVIDENCE 7.44

PLANNING CONTROLS
  zoning legislation 2.123
  indirect legislation 2.125

PLAN, EXAMINATION OF
  general 11.47
  Ontario 3.100

POLARIS PROJECT 3.91, 3.98

PRECEDENT
  concept 1.10
  distinguishing 1.12

PRECISION 4.94

PRESCRIPTION
  concept 2.54, 4.74
  elements of 4.75
  statutory 2.55
  relation to adverse possession 2.94
  user of right 4.76

PRESUMPTION 7.59

PRIORITY OF REGISTRATION 3.12

528 Survey Law in Canada

PRIVACY
  concerns for   12.28
  Privacy Act   12.30

PROFESSIONAL PRACTICE
  concept   11.01, 11.12
  discipline   11.42
  ethics   11.33
  examining boards   11.31
  legislation governing   11.15, 11.26
  organizational policies   11.17
  practice a privilege   11.44
  Quebec   10.219
  reciprocity   11.30
  regulation of   11.12
  standards of practice   11.39

PROFESSIONAL PRACTICE, QUEBEC
  certificate of location
    contents   10.239
    co-ownership   10.264
    copies   10.265
    importance   10.233
    nature   10.232
    operations   10.240
    optional items   10.260
    plan   10.259
    report   10.241
    staking   10.266
    updating   10.267
  Crown property
    boundaries   10.230
    operations   10.222
  private property delimitation   10.228
  surveyor
    as expert   10.228
    as public officer   10.227
    records of   10.269
    role   10.219
    scope of   10.220

PROFIT A PRENDRE
  concept   2.62
  heriditament   2.63

PROOF
  burden of   7.62
  concept   7.61
  standard of   7.67
    civil action   7.68
    criminal action   7.68

PUBLIC HARBOUR
  ownership   4.21, 5.60
  improvements   5.61

QUEBEC
  boundary determination   10.185
  cadastre   10.132
  civil code   10.04
  land surveying   10.219
  registration   10.164
  sources of law   10.01
  water law   10.279

QUIETING TITLE
  action for   8.30
  statutes   8.77

QUIT CLAIM
  concept   8.85
  conventional line   8.32

REAL EVIDENCE   7.69

RECEPTION OF LAW
  concept   2.16
  Ontario   6.07
  prairie provinces   6.20

RE-DEFINITION. *See* RESURVEY, RETRACEMENT.

REGISTER
  completeness of   3.48
  correction of   3.35

REGISTRATION
  land titles system   3.34
    correction of register   3.35
    easements   3.58
    mortgages and encumbrances   3.56
    plans   3.61
    rejection of instruments   3.36
    removal of interests   3.60
    restrictive covenants   3.57
    statutory rights of way   3.59
    transfers and leases   3.52
    transmission of title   3.54
  Quebec
    definition   10.164
    documents subject to   10.167
    duties of registrar   10.179

REGISTRATION — *continued*
  effects 10.182
  form of acts 10.169
  keeping documents 10.170
  laws governing 10.166
  modes 10.175
  origins and evolution 10.165
  place of 10.178
  rights requiring 10.168
  registry system
    instruments which may be registered 3.18
    legal effect of instrument 3.03
    priority of registration 3.12

REGISTRAR
  land titles system 3.33
  Quebec system 10.179
  registry system 3.17

REGISTRY SYSTEM
  advantages and disadvantages 3.21
  basic principles 3.03
  defective or missing deed 3.07
  functions of registrar 3.17
  indexing methods 3.08
  instruments which may be registered 3.18
  legal effect of instrument 3.03
  Newfoundland 3.101
  notice 3.12
  overriding interests 3.06
  priority 3.12

RESERVATION, SHORE 6.78

RESURVEY, RETRACEMENT
  best evidence 4.114, 8.17
  by coordinates 4.151
  concept 4.138, 4.144, 4.145, 4.151
  evidence 4.141, 4.142, 7.01, 11.05
  water boundary 5.86

RIGHT OF FIRST REFUSAL 2.65

RIPARIAN RIGHTS
  accretion 5.52
  boundaries 4.65, 6.25
  classification 5.51
  constitutional provisions 6.05
  development 6.01

  existence of 5.50, 6.30, 6.60
  ownership of bed 5.20
  title records 6.89

SEABED 5.63

SEISIN 2.38

SERVIENT TENEMENT 2.43

SERVITUDES
  conventional 10.85
  real 10.72

SHORE 6.36, 6.52, 6.70

SOURCES OF LAW
  case law 1.15
  common law 1.15
  statute law 1.16

STATUTE LAW
  codification of law 1.16
  concept 2.14
  implementation of social policies 1.16
  reception date 2.16

STATUTE OF FRAUDS
  agreement in writing 4.25, 8.04
  conventional lines 8.36

STATUTES
  boundary uncertainties 8.73
  N.S. title clarification 8.79
  quieting title 8.77
  replotting 8.83
  resurveys 8.80, 8.84
  survey systems 4.47

STATUTORY DECLARATION 8.86

STATUTORY INTERPRETATION
  absurdity 1.24
  grammatical 1.24
  literal 1.21
  mischief rule 1.27
  presumption 1.22

STARE DECISIS
  concept 1.10
  distinguishing 1.12

SUBSTANTIVE LAW   1.35

SURVEYING
  employment within   11.48
  ethics   11.33
  examination syllabus   11.30
  examining boards   11.31
  expanded profession   12.18
  fields of practice   11.27
  liability insurance   9.59
  profession within Quebec   10.219
  professional association   11.21, 11.64, 11.69
  reciprocity   11.30
  regulation of   11.12
  standards of practice   11.39

SURVEYOR
  areas of employment   11.48
  authority   8.21, 11.05, 11.52, 11.59, 11.62
  applied geographer   12.15
  creator of fresh evidence   4.41
  information specialist   12.14
  liability
    in contract   9.03, 9.55
    in negligence   9.04, 9.15
  oath   11.09
  profession within Quebec   10.219
  qualifications   11.29
  responsibilities   4.88, 4.157, 5.110, 8.15, 11.52

SURVEYS
  accuracy   4.96
  coordination   4.110
  directions   4.104
  early   3.28, 4.90, 4.96, 4.153, 8.02
  first demarcation   4.54, 4.57
  future   4.158
  modern   4.147, 4.155
  original by Crown   4.23, 4.52
  records   4.38
  related to registration systems   3.29
  upland riparian parcels   6.95
  water boundary   5.73

SURVEY SYSTEMS
  coordinated   4.110
  Dominion Lands   4.23, 4.47, 4.97, 6.17, 6.25, 6.83
  Ontario   4.24, 4.50, 4.106, 4.144
  integrated   4.110
  statutes   4.47, 4.56

SYSTEMS
  access   12.24
  archival standard   12.27
  authentication   12.26
  cadastral   12.11, 12.20
  cadastral problems   12.32
  contribution by surveyors   12.16, 12.17
  environmental computer-based   12.10
  extension to evidentiary rules   12.25
  land information   4.161, 12.07, 12.12
  parcel referencing   12.10, 12.20, 12.21

TENANCY IN COMMON   2.106

TENURE
  concept   2.26
  free tenure   2.28
  leasehold tenure   2.29, 2.36

TERRITORIAL SEA   5.96

TIDES
  boundary delimitation   5.66, 5.78, 5.81
  chart datum   5.29
  daily   5.24
  foreshore   5.42, 5.47, 5.48
  measurements   5.78, 5.81
  neap   5.25, 5.38
  ordinary   5.38, 5.40
  range   5.35
  reference surfaces
    chart datum   5.29, 5.37
    elevations   5.31, 5.33, 5.36
    higher high water large tides   5.29, 5.30
    higher high water mean tides   5.29, 5.31
    lower low water large tides   5.29
    mean high water   5.32
  shore   5.42, 5.45, 5.47, 5.48
  spring   5.25, 5.38
  variations   5.35, 5.80

TIDE TABLES   5.31, 5.33, 5.79, 5.80

TIME IMMEMORIAL
  concept   2.54
  applicability in Canada   2.96

TITLE
   abstract of   2.81
   concept   8.08
   defects in   8.10
   documentary   8.10
   marketability of   2.81
   overriding interests   8.11
   possessory   4.82, 8.10
   registration systems   3.30, 4.148
   severances creating boundaries   4.37
   tax deed   8.87
   verification of   2.81

TORT   1.39

TRESPASS
   action in   8.26
   dummy   8.27

TRUST   2.113

ULTRA VIRES   1.59

WATER BOUNDARY
   ambulatory   5.83, 5.88
   artificial   5.09
   classification   5.04
   foreshore   5.42, 5.47, 5.48
   interprovincial   5.63
   jurisdictional   5.10, 5.42
   limit of vegetation   5.74
   mean high water   5.71
   natural   5.05, 5.84
   ordinary high water   5.67
   property rights   5.10, 5.42
   shore   5.42, 5.45, 5.47, 5.48
   shore reserve   5.56
   surveys   5.73
   tidal   5.08
   tidal data for   5.78

WATERCOURSE   5.03

WATER LAW, QUEBEC
   accretion   10.302
   alluvion   10.301
   avulsion   10.303
   beach   10.293
   Crown property
      alienability   10.314
      fiscal immunity   10.313
      imprescriptible   10.312
   English regime   10.286
   French regime   10.285
   high water mark   10.285
   jurisdiction
      federal   10.283
      provincial   10.279
   middle thread   10.294
   navigability   10.290
   public ownership   10.307
   reserve   10.308
   rights
      accession   10.300
      individual   10.295
      riparian   10.306

WATERS
   access to shore   5.56
   classification   5.14
   inland. See INLAND WATERS.
   internal   5.96
   navigable   5.18, 5.21, 6.05, 6.15, 6.98
   nontidal   5.17
   private rights   5.58, 5.59
   public rights   5.55, 5.58
   tidal   5.15

WILL   2.133

ZONING   2.123